EVOLUTION AND HUMAN BEHAVIOUR

Evolution and Human Behaviour

Darwinian perspectives on
the human condition

THIRD EDITION

John Cartwright

Senior Lecturer in Biological Sciences
at the University of Chester

 macmillan education   palgrave

First edition 2000
Second edition 2008
Third edition 2016

Published by
PALGRAVE

Palgrave in the UK is an imprint of Macmillan Publishers Limited, registered in England, company number 785998, of 4 Crinan Street, London, N1 9XW.

Palgrave Macmillan in the US is a division of St Martin's Press LLC, 175 Fifth Avenue, New York, NY 10010.

Palgrave is a global imprint of the above companies and is represented throughout the world.

Palgrave® and Macmillan® are registered trademarks in the United States, the United Kingdom, Europe and other countries.

ISBN 978–1–137–34799–2 hardback
ISBN 978–1–137–34800–5 paperback

This book is printed on paper suitable for recycling and made from fully managed and sustained forest sources. Logging, pulping and manufacturing processes are expected to conform to the environmental regulations of the country of origin.

A catalogue record for this book is available from the British Library.

A catalog record for this book is available from the Library of Congress.

Printed in China

Brief Contents

Full Contents

List of Figures

List of Tables

List of Boxes

Preface to the Third Edition

The writing of this new edition has provided a welcome opportunity to refresh and reorganise some previous chapters, remove altogether chapters whose concerns now seem less compelling, and to introduce completely new material which tracks developments in this endlessly fascinating field. A subtle indication of the change of emphasis and scope resulting from this updating is signified by the revised subtitle which now reads: 'Darwinian Perspectives on the Human Condition'. This time-honoured phrase, 'human condition', is better, I think, than the dangerously essentialist concept of 'human nature' (used in the subtitle of the second edition), in that it more adequately conveys the central idea of this book of an evolved genetic programme shaped by the unwavering forces of natural and sexual selection, unfolding in, and moulded by, constantly varying environmental circumstances and influences. Furthermore, the term 'condition' captures the fact that our evolutionary heritage underpins many modern predicaments that humans struggle to deal with and understand. In this sense, the new volume has a slightly more applied focus – an indication, perhaps, of the fact that the field is theoretically mature and empirically grounded enough to begin to understand the roots of pressing social problems (such as violence and criminality) as well as greatly illuminating the ubiquitous experiences of disease, ill health and senescence.

Part I (Historical and Methodological Issues) retains its previous focus on the history of Darwinian approaches to animal and human behaviour, followed by an analysis of the types of modern (and often rival) approaches to understanding how contemporary human behaviour can be located in an evolutionary framework. In keeping with research developments in this area, I have devoted slightly more space here to considering gene–culture coevolutionary models and the principles and methodologies of human behavioural ecology.

Part II (The Two Pillars of the Darwinian Paradigm), as in the previous edition, lays out how natural and sexual selection shape the appearance and behaviour of humans and other animals. I have added new material on single nucleotide polymorphisms (since genetic research increasingly points to their importance); the concept of heritability is made more explicit; and I have touched on the group selection controversy which has experienced a brief (but probably unproductive) revival. In Chapter 4 in this section I have added material on the sex ratio and show how this is a real issue in some cultures (see Box 4.1: The missing women of Asia). The concepts of inter- and intrasexual selection have made major inroads into explaining the mating behaviour of non-human animals and yet do not transfer so easily to our own taxon. To assist this important application of some central ideas, I have tried to clarify the factors and circumstances that incline both human males and females to be choosy about their partners. This is a topic picked up and developed in Chapter 13.

Part III (Human Evolution and its Consequences) deals with material often neglected within evolutionary psychology texts: the forces that shaped our physical evolution, how we got here, and what species-typical characteristics have made us such a globally dominant and unusual species. Recent developments have led to some revisions here about the dating of the 'out of Africa' diaspora of some humans, and I have also alluded to the discovery of a new species: *Homo naledi*. The main change to this section, however, is a slightly reduced emphasis on brain enlargement and a consideration of other typical human attributes such as bipedalism, relative hairlessness and a slow life history. Recent ingenious work on the dating of clothing (put at around 72,000 years BCE, but with wide error margins) is considered in Box 6.1. I have again looked at possible causes of encephalisation but now have given more space to the climate change hypothesis, for which considerable supportive evidence now exists.

Part IV (Adaptations and Developmental Plasticity) represents a completely new and important section. A research paradigm that places a massive emphasis on the adaptive nature of past and/or current human behaviour needs a clear and common understanding of what adaptations are and how they arise. I have tried to do this in Chapter 7 where I look at the various uses of the word adaptation, consider the problem of avoiding teleological language and examine the importance of developmental plasticity. Remarkable new findings in the field of epigenetics have pointed to a whole new way in which organisms can inherit characteristics from their parents and potentially respond in adaptive ways to short-term changes in the environment, and so I have tried to assess the impact of epigenetic inheritance on this field. It is also timely, I think, to pay more attention to life history theory (LHT) and the problem of individual differences. In its early days the evolutionary approach to human behaviour tended to focus on human universals but it is time to correct this imbalance by looking at individual life histories (albeit in a general predictive framework) as well as exploiting the findings of gene sequencing technologies that give a good idea of genetic variation between humans (once considered to be trivial) and the possibility of significant adaptive changes since the Neolithic revolution.

In Part V (Cognition and Emotion) I have reduced the emphasis on brain structure and cognitive adaptations for cheater detection and instead discuss error management theory and the fascinating light it throws on the way humans make biased decisions. I have also added more material on sex differences in cognition as well as the type of learning mechanisms favoured by different environments (in keeping with the growing realisation of the importance of the timescale of environmental changes during our evolutionary history).

In Part VI (Cooperation and Conflict) I have added new material on costly signalling as a potential source of altruism, the importance of onlookers and reputation, and Darwinian grandparenting. In the chapter on conflict and crime, I have looked at the age–crime curve and sexual selection as a potential cause of crime and conflict between the sexes.

Part VII (Mating and Mate Choice) deals with the area where Darwinian approaches to human behaviour have met with most predictive success. This is familiar territory, but I have re-evaluated the role of testosterone in mate choice and tried to clarify how sexual selection theory predicts both displays and choosiness in human mating behaviour. Slightly more space has also been given to neoteny. But perhaps the more significant addition here is a whole new chapter on (male) homosexuality. It is clear to me that in a world where gay rights are being given increasing recognition, for Darwinian psychology to focus exclusively on heterosexual relations seems both scientifically and socially remiss; more so given that work on the possible adaptive basis of (at least) male homosexuality has featured in the literature for several years now. This is a sensitive area and I can only hope that I have done justice to the tangled issues involved.

Part VIII (Health and Disease) is witness to some major departures from, and additions to, the second edition. I have added a long chapter on Darwinian medicine and offer a framework for understanding how evolutionary approaches can inform the understanding of the causes of disease and illness. This is applied in Chapter 19 through three case studies: diet, cancer and mental disorders. I have tried to face head-on one of the major paradoxes of many mental disorders: namely, that the heritability of such disorders is high yet they substantially reduce fitness, raising the question of why such alleles persist in the gene pool. Looking back, I think the search for an adaptive basis for mental disorders (explored in the second edition) is not likely to make much progress. It looks increasing likely that mental disorders really are (and always were) fitness debilitating (in both an individual and inclusive sense) and that their origin may lie in harmful mutations.

In the final section (Part IX: Wider Contexts) I have expanded the material on gene–culture coevolution and take a look at the puzzling phenomenon of religious belief. In the final chapter I once again look at ethics but have given extra space to 'trolleyology' and explore the Moral Foundations Theory of Haidt and Joseph.

In the preface of the last edition I spoke approvingly of the Enlightenment goal of the unity of knowledge. Since then the world in many respects has become a darker place, and the dictum of Alexander Pope, coined in a very different age, that 'the proper study of Mankind is Man' has never seemed more urgent.

Supplementary material

The third edition also comes with a companion website containing resources for students and lecturers. Most of the chapters and sections listed above have additional online material on the website that can be used alongside this volume. These resources can be found at he.palgrave.com/companion/cartwright.

Acknowledgements

I owe thanks to Paul Stevens of Palgrave for persuading me to write this third edition, and for offering wise counsel during its production. The fact that readers will not need a wheelbarrow to carry around this book is also due to Paul's friendly insistence, rightly made I now concede, to considerably reduce the length of the original manuscript. This partly explains why there is abundant additional material on the companion website and why the book you are now reading is such a nice slim volume. Thanks are also due to Isabel Berwick and Cathy Scott, also at Palgrave, for helping me navigate the tortuous currents of seeking rights and permissions; and to Sreejith Govindan, lead project manager at Integra Software Services Pvt. Ltd, for his friendly and efficient communications during the editing stages of the book's completion. I am also grateful to two anonymous reviewers who provided insightful and positive comments on the first manuscript. Heartfelt thanks are also extended to the Graphic Services Department at the University of Chester, and especially to Angela Bell and Gary Martin, who assisted with several diagrams in this book.

Finally, if I wore a hat I would take it off to Cathy Tingle of DocEditor whose superlative professionalism and meticulous copy-editing have spared the unwary reader an embarrassingly large number of typographical errors, inconsistencies and inelegant turns of phrase. I take responsibility, of course, for any errors that remain.

The author and publisher would like to thank the following organisations and individuals for their help and permission in obtaining the following images and diagrams:

Wikimedia Commons for Figure 1.1, Figure 1.2, Figure 8.6 and Figure 17.1; Riccardo Draghi-Lorenz for Figure 1.2; Beth Maynor Young and Edward Wilson for Figure 1.3; SAGE Publishing for Figure 1.6, adapted from Cornwell, R. E., C. Palmer, et al. (2005), 'Introductory psychology texts as a view of sociobiology/evolutionary psychology's role in psychology', *Evolutionary Psychology*, 3: 355–74; the American Psychological Association for Figure 2.3, modified with permission from Geary (2010), *Male, Female,* copyright © 2010 by the American Psychological Association, Table 8.2, adapted with permission from Ellis, B. J., M. Del Giudice, T. J. Dishion, A. J. Figueredo, P. Gray, V. Griskevicius, P. H. Hawley, W. J. Jacobs, J. James, A. A. Volk and D. S. Wilson (2012), 'The evolutionary basis of risky adolescent behavior: implications for science, policy, and practice', *Developmental Psychology*, 48(3): 598–623, copyright © 2012 by the American Psychological Association, Figure 12.9, modified with permission from Hilton, N. Z., G. T. Harris and M. E. Rice (2015), 'The step-father effect in child abuse: comparing discriminative parental solicitude and antisociality', *Psychology of Violence*, 5(1), copyright © 2015 by the American Psychological Association, Figure 14.6(a), adapted with permission from Singh, D. (1993) 'Adaptive significance of female attractiveness', *Journal of Personality and Social Psychology*, 65: 293–307, copyright © 1993 by the American Psychological Association, Figure 14.6(b), adapted with permission from Singh, D. (1995a), 'Female judgement of male attractiveness and desirability for relationships: role of waist-to-hip ratios and financial status', *Journal of Personality and Social Psychology*, 69(6): 1089–101, copyright © 1995 by the American Psychological Association, and Figure 21.3, adapted with permission from Bleske-Rechek, A., L. A. Nelson, J. P. Baker, M. W. Remiker and S. J. Brandt (2010), 'Evolution and the trolley problem: people save five over one unless the one is young, genetically related, or a romantic partner', *Journal of Social, Evolutionary, and Cultural Psychology*, 4(3): 115, copyright © 2010 by the American Psychological Association. The use of APA information does not imply endorsement by APA; Cambridge University Press for Figure 7.2, adapted from Frisancho, A. R. (2010), 'The study of human adaptation' in M. P. Muehlenbein (ed.), *Human Evolutionary Biology*; Gary Martin for Figure 7.8, Figure 8.12, Figure 10.3, Figure 11.3, Figure 14.10, Figure 14.12, Figure 15.1, Figure 15.7, Figure 15.8, Figure 15.9, Figure 18.6, Figure 18.7, Figure 19.3, Figure 20.2 and Figure 21.2; The Royal Society for Figure 7.12, adapted from Faurie and Raymond (2005), 'Handedness, homicide and negative frequency-dependent selection', *Proceedings B* 272, and Figure 11.12, Bateson et al. (2006), 'Cues of being watched enhance cooperation in a real-world setting', *Biology Letters*; Edward Adelson for the figure in Box 9.1; Springer for the third figure in Box 9.3, from Silverman, I., J. Choi and M. Peters (2007), 'The hunter-gatherer theory of sex differences in spatial abilities: data from 40 countries', *Archives of Sexual Behavior*, 36(2): 261–8; Patricia Wynne for Figure 11.7, adapted from Wilkinson, G. S. (1990), 'Food sharing in vampire bats', *Scientific American*, 262: 76–82; Wiley for Figure 11.13, from K. L. Powell, G. Roberts and D. Nettle (2012), 'Eye images increase charitable donations: evidence from an opportunistic field experiment in a supermarket', *Ethology*; Cavendish Press (Manchester) Ltd. for Figure 11.14; the American Sociological Association and Satoshi Kanazawa for Figure 12.11, modified with permission from Kanazawa, S. and M. C. Still (2000), 'Why men commit crimes (and why they desist)', *Sociological Theory*, 18(3): 434–47; Dr Lisa DeBruine for Figure 15.3; Professor Janez Lobmaier for Figure 15.11; www.fotosearch.com for Figure 16.6, © Fotosearch.com; the London School of Economics and Political Science for Figure 17.2, from LSE Library's collections, IMAGELIBRARY/265; John van Wyhe for Figure 17.3, from John van Wyhe (ed.) 2002–. *The Complete Work of Charles Darwin Online* (http://darwin-online.org.uk/).

PART I

Historical and Methodological Issues

Historical Introduction: Evolution and Theories of Mind and Behaviour, Darwin and After

In the distant future I see open fields for far more important researches. Psychology will be based on a new foundation, that of the necessary acquirement of each mental power and capacity by gradation. Light will be thrown on the origin of man and his history.

(Darwin, 1859b, p. 458)

Charles Darwin published his two greatest books on evolutionary theory, *On the Origin of Species by Means of Natural Selection*, and *The Descent of Man and Selection in Relation to Sex*, in 1859 and 1871 respectively, and he was convinced that a revolution in psychology would shortly follow. But for the first three-quarters of the 20th century, while biology became more securely based on deepening evolutionary foundations, psychology failed lamentably to exploit the potential of Darwinian thought. There were some exceptions – William James being the most notable – but many psychologists either ignored Darwinism or, more damagingly, misunderstood the message that it held. Psychology was poorer as a result.

Borrowing the terminology of Thomas Kuhn (1962), we might say that for most of its history psychology has lacked a unifying **paradigm** – a set of procedures, assumptions, methodologies and background theories that all its practitioners can agree upon. There are those who would argue that evolutionary psychology has the potential to supply this missing paradigm, and this book is partly an attempt to explore the strength of this claim. This belief is neatly summarised by the late Margo Daly:

> The reason why psychologists have wandered down so many garden paths is not that their subject is resistant to the scientific method, but that it has been

inadequately informed by selectionist thought. Had Freud better understood Darwin, for example, the world would have been spared such fantastic dead-end notions as Oedipal desires and death instincts. (Daly, 1997, p. 2)

The project of Darwinising humanity, however, is not just an activity taking place inside the discipline of psychology: it is to be found in, and draws upon, a whole range of academic fields such as animal behaviour, behavioural ecology, physical and cultural anthropology, genetics and medicine. This book takes a broad-based and catholic approach to this enterprise and draws upon these diverse fields as required.

Darwinism began in 1859 when Darwin (Figure 1.1), in his fiftieth year, finally published his masterwork *On the Origin of Species by Means of Natural Selection*. The book, originally intended as an abstract of a much larger volume, contained concepts and insights that had occurred to Darwin at least 15 years earlier, yet he had wavered and delayed before publishing. The larger volume never appeared, and Darwin was forced to rush out his *Origin* following a remarkable series of events that began in June of the previous year. It is 1858, therefore, that serves as a convenient starting point.

1.1 The origin of species

On 18 June 1858, Darwin received a letter from a young naturalist called Alfred Russel Wallace, then working on the island of Ternate in the Malay Archipelago. When Darwin read its contents, he felt his world fall apart. In the letter was a scientific paper in the form of a long essay entitled 'On the tendency of varieties to depart indefinitely from the original type'. Wallace, innocent of the irony, wondered whether Darwin thought the paper important and 'hoped the idea would be as new to him as it was to me, and that it would supply the missing factor to explain the origin of species' (Wallace, 1905, p. 361). The ideas were far from new to Darwin: they had been an obsession of his for half a lifetime. Wallace had independently arrived at the same conclusions that Darwin had reached at least 14 years earlier, and the demonstration of which Darwin saw as his life's work. Darwin knew that the essay must be published and, in a miserable state, exacerbated by his own illness and fever in the family, wrote for advice to his geologist friend and scientific colleague Sir Charles Lyell, commenting that he 'never saw a more striking coincidence' and lamented that 'all my originality, whatever it may amount to, will be smashed' (Darwin, 1858).

Fortunately for Darwin, powerful friends arranged a compromise that would recognise the importance of Wallace's ideas and simultaneously acknowledge Darwin's previous work on the same subject. A joint paper, by Wallace and Darwin, was to be read out before the next gathering of the Linnean Society on 1 July 1858. The reading was greeted by a muted response. The president walked out, later complaining that the whole year had not 'been marked by any of those striking discoveries which at once revolutionise, so to speak [our] department of science' (Desmond and Moore, 1991, p. 470). At his home, Down House, Darwin remained in an abject state, coping with a mysterious physical illness that plagued him for the rest of his life and nursing a nagging fear that it might seem as if he had stolen the credit from Wallace. He was also grieving: his young son Charles Waring had died a few days earlier. As the Linnean meeting proceeded, Darwin stayed away and attended the funeral with his wife Emma. By the end of the day, the theory of evolution by natural **selection** had received its first public announcement, and Darwin had buried his child.

After the Linnean meeting, Darwin set to work on what he thought would be an abstract of the great volume he was working on. The abstract grew to a full-length book, and his publisher, Murray, eventually persuaded Darwin to drop the term 'abstract' from the title. After various corrections, the title was pruned to 'On the Origin of Species by Means of Natural Selection', and Murray planned a print run of 1,250 copies.

Darwin, amid fits of vomiting, finished correcting the proofs on 1 October 1859. He then retired for treatment to the Ilkley Hydropathic Hotel in Yorkshire. In November, Darwin sent advance copies to his friends and colleagues, confessing to Wallace his fears that 'God knows what the public will think' (Darwin, 1859a). Many of Darwin's anxieties were unfounded. When the book went on sale to the trade on 22 November, it was already sold out. It was an instant sensation, and a second edition was planned for January 1860. Thereafter, man's place in nature was changed, and changed utterly.

FIGURE 1.1 Charles Darwin (1809–82) from a photograph taken by Maull and Fox around 1854.
SOURCE: Public domain Wikimedia Commons.

1.1.1 New foundations

In *On the Origin*, Darwin was decidedly coy about the application of his ideas to humans, but the implications were clear enough and, in the years following, both Darwin and Thomas Huxley began the process of dissecting and exposing the evolutionary ancestry and descent of man. It was towards the end of *On the Origin*, however, that Darwin made a bold forecast that psychology would be placed on a new foundation and that light would be thrown on the origin of man and his history (Darwin, 1859b, p. 458). On the origin of man, Darwin was right, and light continues to be thrown with each new fossil discovery. On psychology, the new foundation that Darwin foresaw has been slow in coming. Over the past 20 years, however, there have been signs that a robust evolutionary foundation is being laid that promises to sustain a thoroughgoing Darwinian approach to understanding human nature. The bulk of this book is about those foundations. It is fair to say, however, that initially evolutionary principles were most successfully applied to the study of the behaviour of non-human animals.

1.2 The study of animal behaviour: ethology and comparative psychology in the 20th century

A number of disciplines have laid claim to providing an understanding of animal behaviour, including ethology, comparative psychology, behavioural ecology and, emerging in the 1970s, sociobiology. The problem for the historian is that these terms were not always precisely defined and the disciplines frequently overlapped. It is appropriate, therefore, to consider the origins of comparative psychology and ethology together.

1.2.2 Ethology 1900–70

One of the giants of 20th-century ethology was the Austrian Konrad Lorenz (Figure 1.2). Lorenz originally trained as a doctor but was influenced by the work of Oscar Heinroth at the Berlin Zoological Gardens on the behaviour of birds. Heinroth exploited the analogy between animals and humans in both directions. Animals could be understood using concepts drawn from the mental life of humans, and this understanding could then be reapplied to understand the human condition. Lorenz frequently expressed his debt to Heinroth's approach.

By conducting experiments at his home on the outskirts of Vienna, Lorenz observed numerous features of animal behaviour that have become associated with his name. In one classic study, he noted how a newly hatched goose chick will 'imprint' itself on the first moving object it sees. In some cases, this was Lorenz himself, and chicks would follow him about, presumably mistaking him for their mother. Lorenz stressed the importance of comparing the behaviour of one **species** with another related one and argued for the importance of understanding evolutionary relationships between species. In this respect, Lorenz unashamedly drew parallels between the behaviour of humans and other animals. In his most popular work, *King Solomon's Ring*, for example, he suggested that the 'war dance of the male fighting fish… has exactly the same meaning as the duel of words of the Homeric heroes, or of our Alpine farmers, which, even today, often precedes the traditional Sunday brawl in the village inn' (Lorenz, 1953, p. 46).

FIGURE 1.2 Nikolaas Tinbergen (1907–88) (left) and Konrad Lorenz (1903–89) (right), photograph taken about 1978.
Konrad Lorenz and his Dutch colleague Nikolaas Tinbergen helped to found the science of ethology.
SOURCE: Wikimedia Commons (http://commons.wikimedia.org/wiki/File:Lorenz_and_Tinbergen1.jpg, accessed 16 December 2015). Reproduced with kind permission of Riccardo Draghi-Lorenz.

One of Lorenz's early concepts was that of the **fixed action pattern**, which referred to a pattern of behaviour that could be triggered by some external stimulus. Using a term that was later to prove so troublesome for ethology, Lorenz regarded these action patterns as 'instincts' forged by natural selection and common to each member of the species. Fixed action patterns have the following characteristics:

● Their form is constant: that is, the same sequence of actions and the same muscles are used
● They require no learning
● They are characteristic of a species
● They cannot be unlearned
● They are released by a stimulus.

Evidence of a fixed action pattern that is often cited is the observation that a female greylag goose (*Anser anser*) will retrieve an egg that has rolled outside her nest by rolling it back using the underside of her bill. Lorenz noticed that this action continued even when the egg was experimentally removed once the behaviour had begun. Once it started, the behaviour had to finish whether it was effective or not. The stimuli that trigger fixed action patterns became known as sign stimuli or, if they were emitted by members of the same species, releasers. An interesting example is to be found in the behaviour of the European robin (*Erithacus rubecula*), documented by the British ornithologist David Lack in the 1940s. Lack showed that the releaser for male aggression in this species is the patch of red found on the breast of the bird. A male robin will attack another male that it finds in its territory, but it will also attack a stuffed dead robin and even a tuft of red feathers (Lack, 1943).

Once the essence of the stimulus has been identified, it becomes possible in some cases artificially to exaggerate its characteristics and create supernormal stimuli. If a female oyster catcher (*Haemotopus ostralegus*), for example, is presented with a choice of egg during incubation, she will choose the larger. Even if an artificial egg twice the size of her own is introduced, the oyster catcher still prefers the larger one, even though common sense (to an outsider) would indicate that it is unlikely to be an egg actually laid by the bird.

Lorenz had little interest in the individual variation of instincts displayed by different members of the same species, a subject now of great interest to behavioural ecologists. Burkhardt suggests that this neglect of intraspecific variation in behaviour was partly a reflection of the fact that Lorenz wished to distance himself from animal psychologists and their work on captive animals. Lorenz distrusted inferences from laboratory and domesticated animals on aesthetic grounds and also as a consequence of his concern that captive animals showed too much variability in the behaviours they had learned. This variability was, to Lorenz, a hindrance (Burkhardt, 1983).

It was one of Lorenz's students, Nikolaas Tinbergen (1907–88), who finally completed the establishment of ethology as a serious scientific discipline (Figure 1.2). Tinbergen joined Lorenz in 1939 and helped to develop methods for studying behaviour in the wild. In 1949, he moved to Oxford and led a research group dedicated to the study of animal behaviour. Tinbergen studied how fixed action patterns interact to give a chain of behavioural reactions. In his classic study of the stickleback, Tinbergen (1952) showed how, during the courtship ritual, males and females progress through a series of actions in which each component of female behaviour is triggered by the preceding behaviour of the male, and vice versa, in a cascade of events. The culmination of the sequence is the synchronisation of gamete release and fertilisation.

Both Tinbergen and Lorenz developed models to conceptualise the patterns of behaviour they observed. Lorenz interpreted his observations as consistent with a psycho-hydraulic model, sometimes called, somewhat disparagingly, the flush toilet model. If behaviour is interpreted as the outflow of water from a cistern, the force on the release valve can be interpreted as the trigger. The model was more sophisticated than suggested by its comparison with a domestic flush toilet, but its essential feature was the accumulation of 'action-specific energy' in a manner analogous to the accumulation of a fluid in a cistern. Sigmund Freud employed similar hydraulic metaphors in his thinking about drives and repression. Despite their obvious shortcomings as accurate analogues of mental mechanisms, they are still commonplace in everyday speech. To 'explode with rage' or 'let off steam' are both echoes of the type of models utilised by Lorenz and Freud.

Tinbergen developed an alternative model that, while retaining the concept of accumulating energy that drives behaviour, suggested a hierarchical organisation of instincts that are activated in turn. The models of Lorenz and Tinbergen met with much subsequent criticism. It proved difficult, for example, to correlate the features of the model with the growing body of information from neurobiology about real structures in the brain.

One of Tinbergen's most lasting contributions was a clarification of the types of question that animal behaviourists should ask. In 1963, in a paper called 'On the aims and methods of ethology', Tinbergen suggested that there were four 'whys' of animal behaviour:

1. What are the mechanisms that cause the behaviour? (causation)
2. How does the behaviour come to develop in the individual? (development or **ontogeny**)
3. How has the behaviour evolved? (evolution)
4. What is the function or survival value of the behaviour? (**function**)

It has been noted that a useful mnemonic for these is 'ABCDEF: Animal Behaviour, Cause, Development, Evolution, Function' (Tinbergen, 1963).

To appreciate the application of these questions, it may be useful to consider an example. In many areas of the northern hemisphere, birds fly south as winter approaches. One such species is the wheatear (*Oenanthe oenanthe*). Wheatears migrate to Africa in the winter even though some groups have moved from their European breeding grounds and have established new populations in Asia and Canada. We could ask of this behaviour: What triggers flocks to take to the air? How do they 'know' when the time arrives to depart and which way to fly? These questions address the **proximate causes** of the behaviour and relate to Tinbergen's first question of causation. The answer lies in specifying the physiological mechanisms that are activated by environmental cues, possibly day length, temperature, angle of the sun and so forth.

Probing further, we could ask how the ability to fly over such vast distances in a species-typical manner is acquired by an individual. Do animals know instinctively which way to fly and how far to travel,

or do they have to learn some components from a parent or an older bird? These questions belong to Tinbergen's second category dealing with the development and ontogeny of behaviour.

We could also ask about the evolution of the behaviour to its present form. Is the behaviour found in related species? If so, has it been acquired by descent from a common ancestor? A particular question concerns why even Canadian and Asian wheatears travel to Africa. If the aim is simply to move south, those in Canada and Asia could save themselves thousands of miles of travel. Is the move to Africa by the new populations outside Europe a 'hangover' from when the wheatear only lived in Europe? These questions refer to the evolutionary origin of the behaviour as raised in the third 'why' question above.

Finally, in relation to the movement to Africa, we could ask questions about **ultimate causation**. Why do birds make such arduous and perilous journeys? How does flying to Africa increase the survival chances of those making the journey? It must carry some advantage over not moving, otherwise a mutant that appeared and did not fly away would leave more survivors, and non-migration would gradually become the norm. The type of answer given to this last functional 'why' of Tinbergen would presumably show that the benefits of travelling in terms of food supply and then securing a mate outweigh the drawbacks in terms of risks and energy expenditure. Ultimately, we would have to show that migrating is a better option than staying in one place as a means of leaving offspring. We will then have demonstrated the function or adaptive significance of the behaviour.

Taking an overview of Tinbergen's four 'whys' as a broad generalisation, psychologists (dealing with humans or animals) have tended to be interested in the 'whys' of proximate causation and ontogeny, and less interested in questions about evolution and adaptive significance. The growth of sociobiology and evolutionary psychology is a direct attempt to reverse this trend and supply a unifying paradigm for the behavioural sciences based on the understanding of ultimate causation and evolutionary function (Barkow et al., 1992).

The work of Lorenz and Tinbergen is usually classified as being central to the tradition of classical ethology. Classical ethology was forced to adjust its ground as a result of telling criticisms that appeared in the 1950s and 60s from comparative psychologists and, to be fair, discoveries by the ethologists themselves. Lehrman (1953), in particular, was a forceful critic of the use of the term 'innate' in ethology. The attempt to classify behaviour as either innate or learned was soon seen to be too simplistic (Archer, 1992). The deprivation experiments suggested by Lorenz, in which an individual is reared in isolation from other individuals and hence sources of learning, simply isolate an individual from its social environment rather than from temperature, light and nutrition. Isolation experiments beg the question: From what has the animal been isolated? Peking ducklings (*Anas platyrhynchos*), for example, are able to recognise the call of their species if it is heard during embryonic development while the chick is still in the shell (Gottlieb, 1971). Moreover, individuals create their own environment as a result of their actions. Aggressive or assertive people create a different environment from shy people, resulting in a different set of feedback. It came to be realised that all behaviour must be the result of both influences.

Important work by Thorpe (1961) on song development in chaffinches demonstrated the mutual interdependence between heredity and the environment. Thorpe demonstrated that whereas the ability of a chaffinch to sing was to some degree 'innate', the precise song pattern depended upon exposure to the song of adults at critical times during the development of the young bird. The form the song took also depended on the ability of the chaffinch to hear its own song. Chaffinches would not, however, learn the songs of other birds, even if exposed from birth. Furthermore, once song development occurred, a chaffinch would not learn other variants. Work such as this showed that the interaction between innate templates of behaviour and the environment is more complex than hitherto thought.

Eventually, the classical ethologists had to accept that behaviour that they had often labelled as innate could be modified by experience. It did not follow from this of course that all behaviour was learnt and unconstrained by genetic factors, as some of the behaviourists seemed to imply. The history of comparative psychology is also one in which fundamental assumptions had to be revised.

1.2.2 Comparative psychology 1900–70

An early exponent of the methods that came to be associated with comparative psychology was Ivan Petrovich Pavlov (1849–1936). Pavlov, the son of a priest, began his career studying medicine, and was awarded the Nobel prize in 1904 for his work on digestion. Pavlov showed that if the sound of a bell accompanied the presentation of food to a dog, the dog would learn to associate the sound with food. Eventually, the dog would salivate at the sound of the bell even in the absence of food. Pavlov thereby produced the first demonstration of what later became known as classical conditioning. By focusing on the observable reactions of animals without presupposing what went on in their minds, Pavlov stressed the objectivity and rigour of these methods in contrast to a psychology that dwelt upon putative inner experiences. Pavlov's work on conditioning became known to Western psychology around 1906 and, although his more ambitious claims for the establishment of a new brain science went unheeded, his methodology proved highly influential.

The focus on the observable reactions of animals and humans under controlled conditions came to be regarded as the hallmark of comparative psychology and what later became known as **behaviourism**. It is easy to overestimate the influence of behaviourism on 20th-century psychology. Smith (1997) notes that it served the polemical interests of cognitive psychologists in the 1960s to represent psychology between 1910 and 1960 as a behavioural monolith from which they were seeking liberation. As a broad generalisation, however, a pattern was emerging. By the middle of the 20th century, the European approach to animal behaviour and extrapolations to humans was dominated by ethology, while in the USA the experimental approach using laboratory animals was pre-eminent.

One figure who more than any other seemed to symbolise the behaviourist approach in its early days was John Broadus Watson (1878–1958). In his later years, Watson came to be reviled by ethologists as the architect and archetype of an alien approach to the study of behaviour. By adopting a positivist approach to knowledge, Watson claimed that psychology would be retarded in its development unless it ditched its concern with unobservable entities such as minds and feelings. Similarly, both animal and human psychology must abandon any reference to consciousness. A psychology that deals with inner mental events is clinging, he claimed, to a form of religion that has no place in a scientific age. For Watson, the brain was a sort of relay station that connected stimuli to responses.

Watson issued his manifesto in a series of lectures delivered in 1913 at Columbia University. Before the First World War, the reaction was lukewarm: some welcomed his objective approach but warned against its excesses. Others feared that a sole concern with behavioural phenomena as opposed to human consciousness would reduce psychology to a subset of biology. It was after the war that behaviourism became more deeply embedded in American scientific culture. The war had demonstrated the value of objective tests applied in the classification of military personnel. By 1930, behaviourism had become the dominant viewpoint in experimental psychology. In stressing the importance of environmental conditioning, Watson's whole approach was profoundly anti-evolutionary and anti-hereditarian. He denied that such qualities as talent, temperament and mental constitution were inherited. Perhaps his most famous remark on the effect of environmental conditioning, and one of the most trenchant and extreme statements of **environmentalism** in the literature, is his claim for the social conditioning of children:

> Give me a dozen healthy infants, well-formed, and in my own specified world to bring them up, and I will guarantee to take any one at random and train him to become any type of specialist I might select – doctor, lawyer, artist, merchant-chief and yes, even beggarman and thief, regardless of his talents, penchants, tendencies, abilities, vocations and race of ancestors. (Watson, 1930, p. 104)

In this respect behaviourism fitted in well with American culture. It appeared to be a practical science that promised to deliver socially valuable answers to such things as how to raise children and turn them into effective citizens. It also arrived on the scene during the First World War when anti-German sentiment made the notion of a distinctly American (as opposed to German) science of psychology more appealing.

Behaviourism sought philosophical credibility by allying itself with a philosophy of science known as logical **positivism** and articulated by a group of philosophers known as the Vienna Circle. The logical positivists argued that statements are only meaningful, and hence part of the purview of science, if they can be operationally defined. A statement about the world then only becomes meaningful if it can be verified. The aims of this approach were to outlaw religious and metaphysical claims to knowledge. It was in this approach to epistemology that American behavioural psychology, with its emphasis on empirical, quantifiable and verifiable observations, found a natural ally.

The decline of behaviourism in the 1960s coincided with the downfall of logical positivism as a reliable philosophy of science. Philosophers such as Popper and historians such as Kuhn showed that the verifiability criterion of meaning espoused by the Vienna Circle was untenable both in theory and as a realistic description of the way in which science actually worked. The irony of the linkage between behaviourism and positivism is neatly summed up by Smith (1997, p. 669): 'It appeared as if the behaviourist enterprise had emptied psychology of its content in order to pursue an image of science that was itself a mirage.'

One movement in animal psychology related to behaviourism was B. F. Skinner's operant psychology. Skinner was professor of psychology at Harvard from 1958 to 1974 and was particularly influenced by the work of Watson and Pavlov. Skinner's programme adopted a number of key principles. One was that he believed that science should be placed on a firm foundation of the linkages between empirical observations rather than speculative theory. For Skinner, and in this respect he resembles Watson, theoretical entities such as pleasure, pain, hunger and love were meaningless and should be expunged from laboratory science. Another essential feature of Skinner's work was that he thought that all behaviour could be resolved and reduced to a basic principle of reinforcement. One typical schedule of reinforcement devised by Skinner was to reward pigeons in a box with grain. By rewarding some forms of behaviour and not others, he was able to make the rewarded behaviour more probable, an approach that became known as **operant conditioning**.

While early behaviourists in America such as Watson were feverishly attempting to jettison excess metaphysical baggage from psychology, in Europe, Freud was weaving a psychology replete with rich and colourful complexes, emotions and subconscious forces. Skinner was both an admirer and critic of Freud. For Skinner, Freud's great discovery was that human behaviour was subject to unconscious forces. This accorded well with the view of behaviourism that conscious reason was not in the driving seat of human behaviour. Freud's mistake was that he encumbered his theory with unnecessary mental machinery, such as the ego, the super ego, the id and so on. For Skinner, such entities were not observable and hence were not justified in scientific enquiry.

By 1960, a large number of psychologists in America had been trained in Skinner's methods. Their output was influential in some areas, such as the inculcation of desirable habits in the development of children, but few behaviourists were willing to go as far as Skinner in suggesting that organisms were empty boxes. Skinner faced his most difficult hurdle when, in his book *Verbal Behavior* (1957), he attempted to interpret language development in terms of operant conditioning. There were already signs that many behaviourists were realising that language threatened to be the 'Waterloo' of behaviourism. Skinner marched on, however, and argued that there was nothing special about language, denying any fundamental difference in verbal behaviour between humans and the lower animals. Skinner had now pushed behaviourism too far, and its weaknesses were fatally exposed.

In 1959, a linguist called Noam Chomsky, then relatively obscure, reviewed Skinner's *Verbal Behavior* and, in showing that behaviourism failed lamentably when tackling language, also undermined some of its basic pretexts. Chomsky argued that behaviourism could hardly begin to account for language acquisition. He showed that Skinner's attempts to apply the language of stimulus and response to verbal behaviour lapsed into vagueness and finally hopeless confusion. For Chomsky, behaviourism could not be improved or modified: it was fundamentally flawed and had to go (Chomsky, 1959). In their efforts to avoid unobservable mentalist concepts, the behaviourists had indeed transformed psychological terminology; so that memory became learning, perception

became discrimination and language became verbal behaviour. As Chomsky remarked, defining psychology in this way was like defining physics as the science of meter-readings. Chomsky's review, and his own positive programme for linguistics stressing the creativity of language and its foundation on inherent, deep-seated mental structures, sparked off a revolt against behaviourism that tumbled it into terminal decline. This roughly coincided with increasing reports from animal researchers that animals trained according to operant conditioning methods would occasionally revert to behaviours that seemed to be instinctive.

1.3 The rise of sociobiology and evolutionary psychology

1.3.1 From sociobiology to evolutionary psychology

In the 1970s and 80s, the contribution of ethology to the natural sciences was recognised by the award of Nobel prizes in 1973 to Konrad Lorenz, Niko Tinbergen and Karl von Frisch for their work on animal behaviour. Then in 1981, Roger Sperry, David Hubel and Torsten Wiesel were awarded the prize for their work in neuroethology. The classical approach to ethology associated with Lorenz was continued in Germany by Irenaus Eibl-Eibesfeldt (1989). In the UK, the Netherlands and Scandinavia, where the influence of Tinbergen was more dominant, a more flexible approach to study of animal behaviour was emerging. At Cambridge in the 1950s, for example, under the leadership of William Thorpe, ethologists became particularly interested in behavioural mechanisms and ontogeny, while at Oxford Tinbergen headed a group more concerned with behavioural functions and evolution. In a sense, these two groups divided up Tinbergen's four 'whys' (Durant, 1986).

Meanwhile, aided by some fresh ideas from theoretical biology, a new discipline called sociobiology was emerging that applied evolution to the social behaviour of animals and humans. Sociobiology, like behavioural ecology, is concerned with the functional aspects of behaviour in the sense raised by Tinbergen. It drew its initial inspiration from successful attempts by biologists to account for the troubling problem of altruistic behaviour. Of all the problems faced by Darwin, he considered the emergence of altruism and, more specifically, the existence of sterile castes among the insects as two of the most serious. Darwin provided his own answer in terms of community selection but also came tantalisingly close to the modern perspective when he suggested that if the community were composed of near relatives, the survival value of altruism would be enhanced. The British biologist J. B. S. Haldane came even closer when, in his book *Causes of Evolution* (1932), he pointed out that altruism could be expected to be selected for by natural selection if it increased the chances of survival of descendants and near relations.

Sociobiologists such as Barash (1982) suggested that the new discipline represented a new paradigm in the approach to animal behaviour (Barash, 1982). Others such as Hinde (1982, p. 152) concluded that 'sociobiology' was an 'unnecessary new term' since behavioural ecology covered the same ground. Behavioural ecology is an established and uncontroversial epithet for those who study the adaptive significance of the behaviour of animals. It would be a mistake, however, to treat 'sociobiology' and 'behavioural ecology' as interchangeable terms. Behavioural ecologists tend to focus more on non-human animals than humans and have a particular concern with resource issues, game theory and theories of optimality. Sociobiology itself deals with human and non-human animals, although it started out with a particular concern with inclusive **fitness**, and can be thought of as a hybrid between behavioural ecology, population biology and social ethology. The term 'human behavioural ecology' comes close to that of 'sociobiology'.

A book that served in some ways as a seed crystal for the new approach – at least in the sense that it galvanised its opponents – was *Animal Dispersion in Relation to Social Behaviour*, written by V. C. Wynne-Edwards and published in 1962. In this work, Wynne-Edwards advanced a position that was already present in the literature but had aroused no real opposition, namely that an individual would sacrifice its own (genetic) self-interest for the good of the group. It was attacks on this idea that catalysed the emergence

of a more individualistic and gene-centred way of viewing behaviour.

One example of the new approach that countered group selectionist thinking came in 1964 when W. D. Hamilton published two ground-breaking and decisive papers on inclusive fitness theory (see Chapter 3). Then in 1966, G. C. Williams published his influential *Adaptation and Natural Selection*, in which he argued that the operation of natural selection must take place at the level of the individual rather than that of the group, and in a similar vein exposed a number of what he thought were common fallacies in the way in which evolutionary theory was being interpreted. In the 1960s and 70s, the British biologist John Maynard Smith pioneered the application of the mathematical theory of games to situations in which the fitness that an animal gains from its behaviour is related to the behaviour of others in competition with it. Then in the early 1970s, the American biologist Robert Trivers was instrumental in introducing several new ideas concerning reciprocal altruism and parental investment (see Chapters 3 and 11). It is a sad reflection on the specialisation of academic life and the disunified condition of the social sciences that while this revolution was occurring in the life sciences, psychology initially remained aloof from these ground-breaking developments.

The book that encapsulated and synthesised these new ideas more than any other was E. O. Wilson's *Sociobiology: The New Synthesis*, published in 1975. The book became a classic for its adherents and a focus of anger for its critics. In this work, Wilson irritated a number of scientists by forecasting that the disciplines of ethology and comparative psychology would ultimately disappear by a cannibalistic movement of neurophysiology from one side and sociobiology and behavioural ecology from the other. Others were alarmed that Wilson (Figure 1.3) extended biological theory into the field of human behaviour. Although only one chapter out of the 27 in Wilson's book was concerned with humans, a fierce debate over the social and political implications of Wilson's approach ensued. It was as if Wilson had stormed the citadel that had been fortified from biology by social scientists over the previous 40 years, with predictable repercussions.

One of these took place in February 1978 when Wilson attended a meeting of the American

FIGURE 1.3 The American biologist Edward Wilson (born 1929).
Wilson did pioneering work on the social insect before publishing his massive and controversial work *Sociobiology* in 1975. His subsequent contributions to evolutionary thinking have been profound.

SOURCE: Photograph taken by Beth Maynor Young in the Red Hills, Alabama, in 2010. By kind permission of Kathleen Horton (assistant to E. O. Wilson).

Association for the Advancement of Science held in Washington DC. As he prepared to deliver his paper about ten people charged onto the stage accusing Wilson of genocide. One of them poured a jug of water over his head.

As sociobiology developed its theoretical tools, mainstream theoretical psychology, as epitomised by the journals *Psychological Review* and *Psychological Bulletin*, kept its distance and seemed reluctant to engage with this fledgling discipline. Meanwhile, some anthropologists were beginning to explore the application of sociobiological concepts to their field. In 1979 there appeared a ground-breaking work edited by Chagnon and Irons: *Evolutionary Biology and Human Social Behavior: An Anthropological Perspective*, a book containing articles by Richard Alexander, William Irons and Napoleon Chagnon.

It is significant that, following Wilson's book, a number of new journals, such as *Behavioral Ecology and Sociobiology* and *Ethology and Sociobiology*, appeared to cater for this growing area. The latter journal eventually changed its name to *Evolution and Human Behavior*, reflecting a new diversity of approach to the study of human behaviour. For many, the term 'sociobiology' is redolent of the painful debates that

surrounded Wilson's early work. Partly to avoid associations with the vitriolic debates of the 1980s and partly to reflect a change in emphasis and ideas, the term sociobiology is now rarely used.

The legacy of sociobiology now lives on, in modified form, in the discipline called evolutionary psychology. Evolutionary psychologists focus on the adaptive mental mechanisms possessed by humans that were laid down in the distant past – the so-called **'environment of evolutionary adaptedness'** (EEA). To be fair, sociobiology too had emphasised the importance of adaptive mechanisms forged in the geological period known as the Pleistocene. There are those who insist that evolutionary psychology and sociobiology are really the same thing. Robert Wright, in *The Moral Animal: Evolutionary Psychology and Everyday Life* (1994), for example, suggests that 'sociobiology' was simply dropped as a name (for political reasons) although its concepts carried on under new labels:

> Whatever happened to sociobiology? The answer is that it went underground, where it has been eating away at the foundations of academic orthodoxy. (Wright, 1994, p. 7)

The rise of evolutionary psychology was facilitated by the cognitive revolution in psychology. One factor in promoting this revolution was perhaps the Second World War and the stimulus it gave to research on information processing by machines or human subjects in conjunction with machines. Cognitive psychology is based on the idea that mental events could be conceived as the processing of information within structures in the brain. As Tooby and Cosmides (2005, pp. 14–16), two of the world's leading evolutionary psychologists, observe:

> The brain's evolved function is computational – to use information to adaptively regulate the body and behaviour ... the brain is not just like a computer. It is a computer – that is, a physical system that was designed to process information.

And just as a good computer has lots of applications and sub-routines separate from each other (for example, word processing, playing music, handling images) so the brain, in this view, consists of a variety of modular-based capabilities. Parts of the brain

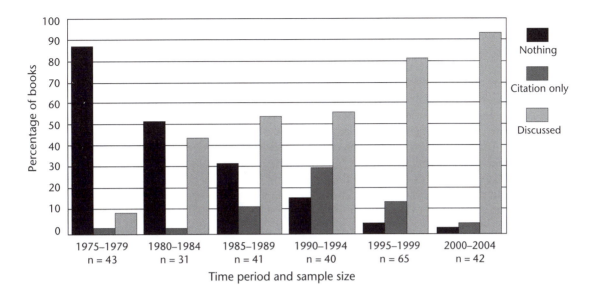

FIGURE 1.4 Coverage of EP in introductory psychology texts 1975–2004.

SOURCE: Data adapted from Cornwell et al. (2005), Figure 1, p. 361.

useful for finding a fertile mate, for example, will not be called upon or be of much use in the task of identifying nutritious foods.

1.3.2 The influence of evolutionary psychology

An interesting question is whether or not evolutionary psychology has had much impact on mainstream psychology. After all, it was the prediction of Darwin and Wilson that the evolutionary approach would eventually transform psychology. Barkow et al. (1992) were equally sanguine when they stated that evolutionary psychology had the 'potential to draw together all of the disparate branches of psychology into a single organised system of knowledge' (p. 3). At the time of writing it is probably fair to say that the evolutionary approach has not yet unified psychology in the way its protagonists hoped it would, and mainstream psychology remains in a state of, to put it charitably, 'conceptual pluralism'.

In an effort to examine what influence evolutionary psychology has exerted, Cornwell et al. (2005) examined 262 introductory psychology textbooks spanning the period 1975–2004. The influence of

evolutionary psychology (EP) was measured using a variety of criteria including coverage in the books, accuracy of portrayal and the type of topics discussed. Figure 1.4, for example, shows how the attention given to evolutionary psychology has increased over the last 40 years.

The authors also note that this increasing exposure was accompanied by a generally more accurate treatment, a more positive evaluation and fewer undefended criticisms. One troubling feature, however, was that many textbooks had narrowed the coverage of evolutionary psychology to mating strategies.

In view of the rather sordid history of some aspects of evolutionary theorising about humans it is necessary, even at the risk of repetition elsewhere, to make some statements about how the Darwinian paradigm now stands on some of the issues raised in this chapter. Nearly all sociobiologists and evolutionary psychologists now assert the psychic unity of humankind; this is done not out of political correctness – a poor foundation for knowledge anyway – but simply because the biological evidence points in that direction. Where Darwinians profoundly disagree with the 'nurturists' is the view of the latter that our evolutionary past has no bearing on our present condition; that human behaviour and the human mind and are moulded and conditioned solely or largely by culture; and, crucially, culture bears no relation to our genetic ancestry and hence can only be explained in terms of more culture.

In contradistinction to this, Darwinism asserts the existence of human universals upon which the essential unity of humankind is predicated. Some cultures may amplify some of these universals and suppress others, and culture itself may in some as yet unclear way reflect the universals lurking in the human **gene pool**, but the crucial point is that these universals represent phylogenetic adaptations: they have adaptive significance. This is not to suggest of course that there are 'genes for' specific social acts; genes describe proteins rather than behaviour. It will be genes in concert with other genes and environmental influences (meaning both the cellular and extracellular environments) that are seen to shape the neural hardware of the brain that forms the ultimate basis of human universals. Moreover, there is increasing evidence that in addition to distinct human universals one characteristic of humans is their behavioural plasticity: the ability to adapt

during development in functional ways that further enhance fitness. This is considered in more detail in Chapters 7 and 8.

In retrospect, perhaps a better way to understand the historical status of sociobiology is in phylogenetic terms. In this light we might suppose that sociobiology is the ancestor of such descendants as human behavioural ecology, evolutionary psychology and gene–culture evolution theory. To continue the metaphor these descendants have drifted apart and now tend to recruit different types of scientists, emphasise different aspects of the evolutionary project and vary in their toolkits of models, theories and methodologies. Yet beneath this diversity they belong to the same genus and share the same overarching aims. These different approaches, all members of Darwinian paradigm, are considered in the next chapter.

Summary

- Darwin and some of his immediate followers sought an explanation of animal and human minds based on the principle that the mental and physical life of animals belong to the same sphere of explanation and that both have been subject to the force of natural selection.

- In the 20th century, two distinct approaches to animal behaviour emerged: ethology and comparative psychology. The ethological tradition in Europe was based around the work of such pioneers as Heinroth, Lorenz and Tinbergen, entailing the study of a variety of animal species in their natural habitat using a broad evolutionary framework. Comparative psychology took root particularly in America. Watson and Skinner were extreme representatives of this tradition. With the passage of time, comparative psychology became increasingly concerned with the behaviour of a few species in laboratory conditions, to the neglect of an evolutionary perspective.

- The theory of human instincts begun by Darwin was continued in America by James and in Europe by Heinroth, Lorenz and other ethologists. The experimental problem of demonstrating the action and existence of human instincts, as well as the reactionary associations of biological theories of human nature, led most social scientists and anthropologists in the middle years of the 20th century to reject biological explanations of human behaviour and assert the primacy of culture.

- Seeking explanations for human behaviour in terms of adaptations lies at the heart of Darwinian psychology and provides 'ultimate' style explanations for current behaviour. Such explanations are responses to two of the four 'whys' of animal behaviour set out by Tinbergen: how the behaviour has evolved ('evolution') and the survival value of the behaviour ('function').

▶ In the 1960s and '70s, a number of fundamental papers and books set in motion the sociobiological approach to animal and human behaviour. Sociobiology is predicated on the view that animals will behave so as to maximise the spread of their genes. Behaviour is therefore examined largely with functional (in the sense of Tinbergen) questions in mind. When applied solely to humans, the approach is also sometimes called evolutionary psychology (EP), some regarding this movement as a new paradigm. This new movement seeks to revive the Darwinian project of demonstrating the evolutionary basis of human behaviour and of many facets of human culture.

▶ The cognitive revolution in psychology precipitated the demise of behaviourism and allowed evolutionary psychology to flourish.

▶ Evolutionary psychology is slowly making headway into mainstream psychology and increasingly features in introductory textbooks. Its promise to transform psychology and provide it with a much-needed unifying base in an ongoing task.

Key Words

- Behaviourism
- Environment of evolutionary adaptedness (EEA)
- Environmentalism
- Fitness
- Fixed action pattern

- Function
- Gene pool
- Ontogeny
- Operant conditioning
- Paradigm
- Positivism

- Proximate cause
- Selection
- Species
- Ultimate causation

Further reading

Degler, C. N. (1991). *In Search of Human Nature: The Decline and Revival of Darwinism in American Social Thought*. Oxford, Oxford University Press.

True to its title, this book examines the period 1900–88. A penetrating sociological analysis of the fate of Darwinian ideas in America.

Gangestad, S. W. and J. A. Simpson (2007). *The Evolution of Mind*. London, UK, The Guildford Press.

Excellent set of articles on the whole field of evolution and human behaviour. Several articles address fundamental theoretical and methodological issues.

Plotkin, H. (2004). *Evolutionary Thought in Psychology. A Brief History*. Oxford, Blackwell.

An excellent, readable and balanced discussion of evolutionary thinking in psychology leading up to EP.

Richards, R. J. (1987). *Darwin and the Emergence of Evolutionary Theories of Mind and Behaviour.* Chicago, University of Chicago Press.

A detailed and thorough work. Covers the ideas of Darwin, Spencer, Romanes, Morgan and James. Lays particular emphasis on the evolutionary origins of morality. Contains the author's contemporary defence of using evolution as a basis for ethics.

Segerstrale, U. (2000). *Defenders of the Truth: The Battle for Science in the Sociobiology Debate and Beyond.* Oxford, Oxford University Press.

A highly detailed account of the characters and controversies involved in the rise of sociobiology.

Thorpe, W. (1979). *The Origins and Rise of Ethology.* New York, Praeger.

An inside account of ethology from someone who helped to shape the discipline. Plenty of anecdotal information.

Workman, L. (2013). *Charles Darwin: The Shaping of Evolutionary Thinking.* Basingstoke, UK, Palgrave Macmillan.

Clear and thoughtful book about the influence of Darwin on thinking about behaviour.

Foundations of Darwinian Psychology

Hereafter we shall be compelled to acknowledge that the only distinction between species and well-marked varieties is, that the latter are known, or believed, to be connected at the present day by intermediate gradations, whereas species were formerly thus connected.

(**Darwin**, *The Origin of Species*, 1859b, **p. 485**)

We noted in Chapter 1 how the eventual resurgence of evolutionary thinking in psychology was linked to the cognitive revolution in psychology coupled with theoretical developments in biology that re-energised the study of animal behaviour to give rise to behavioural ecology and sociobiology. Within the broad church of Darwinian approaches to psychology, however, a number of variants exist, sitting, sometimes uncomfortably, side by side. In this chapter we will examine the plurality of Darwinian perspectives on human nature and the theoretical and methodological problems the discipline has to face.

2.1 Testing for adaptive significance in evolutionary psychology

The whole of evolutionary psychology is based on the premise that behaviour is driven by **adaptations**. To say that a feature or behavioural trait is adaptive is to say that it promotes, or once promoted, reproductive success. To demonstrate this effectively, we need to show how a feature in question confers some reproductive advantage. This is by no means easy. Giraffes with long necks may have an advantage over rivals with shorter necks in terms of grazing from tall trees, but they also probably have a better view of approaching predators. They may also have the edge in aggressive disputes with rivals of the same sex. It is all too easy to jump to conclusions: a given trait may be advantageous in a number of different ways in a given species or even in different ways in different species. Rabbits may have large ears to detect predators, but the large ears of an African elephant probably have more to do with heat regulation than sound detection.

It is also easy to find adaptations that are not there. Consider for a moment balding in human males; what adaptive function could it have? You may suggest that it helps exposure to sunlight and the synthesis of vitamin D. It could show that the male has high levels of testosterone and is thus virile. It could be an adaptive response to the need to lose heat on the African savannah plains (this, after all, is probably why humans lost their body hair). Bald men especially will be good at devising flattering and functional explanations. Most of them are probably false and amount to what Gould, after Kipling, has called 'Just So Stories'. This particular trap is examined below.

2.1.1 Pitfalls of the adaptationist paradigm: 'Just So Stories' and Panglossianism

In his *Just So Stories*, Rudyard Kipling (1967) gave an amusing account of how animals came to be as they are. The basic structure of the stories is that when the world was new, animals looked very different from today's types. Something then happened to these ancestral species that left them in the form we see now. The elephant, for example, once had a short nose, but after a tussle with a crocodile its nose was pulled and stretched into a trunk (Figure 2.1). In evolutionary biology, the 'Just So Story' has become a metaphor for an evolutionary account that is easily constructed to explain the evidence but makes few predictions that are open to testing.

A similar trap is what Gould and Lewontin have referred to as 'Panglossianism' (Gould and Lewontin, 1979). In Voltaire's book *Candide*, Dr Pangloss is the eternal optimist who finds this world to be the best of all possible worlds, with everything existing or happening for a purpose, our noses, for example, being made to carry spectacles. In evolutionary thinking, Panglossianism is the attempt to find an adaptive reason for every facet of an animal's morphology, physiology and behaviour. Panglossian explanations are fascinating exercises in the use of the creative imagination. Consider why blood is red.

FIGURE 2.1 How the elephant acquired a long trunk.
There was a time when elephant trunks were short but an ancestral crocodile stretched the trunk of the first elephant.
SOURCE: Kipling (1967).

It could help to make wounds visible, it could indicate the difference between fresh and stale meat and so on. Yet blood is red simply as a consequence of its constituent molecules, for example haemoglobin, and has probably never been exposed to any selective force. The ease of devising adaptive-style explanations for behavioural phenomena is illustrated by Ramachandran's article 'Why do gentlemen prefer blondes?' (Ramachandran, 1997). He proposed that blonde hair (and associated light skin) enabled males to better judge the age of a woman and her health status. Pale skin, he argued, better revealed the effects of anaemia, skin infections and jaundice; whilst at the same time making the signs of ageing (dark spots) easier to detect. The article was published in *Medical Hypotheses* – a journal designed to encourage adventurous hypotheses in the knowledge that many of them may prove to be false but very interesting if they are supported. Ramachandran then revealed that the paper was a hoax designed to expose the gullibility of some evolutionary psychologists.

The evolutionist must be prepared to accept that just as some genetically based traits may no longer be adaptive, so some adaptive features may not be directly genetically based – although learning mechanisms will themselves have a genetic basis (Box 2.1).

It was Williams who, in 1966, helped to clarify what is meant by an adaptation. An adaptation is a characteristic that has arisen through and been shaped by natural and/or sexual selection. It regularly develops in members of the same species because it helped to solve problems of survival and reproduction in the evolutionary ancestry of the organism. Consequently, it can be expected to have a genetic basis, ensuring that the adaptation is passed through the generations. Williams suggested that three criteria in particular should be employed to ascertain whether the feature in question is truly an adaptation: reliability, economy and efficiency (Williams, 1966). The first criterion is satisfied if the feature regularly develops in all members of the species subject to normal environmental conditions. Economy is satisfied if the mechanism of characteristic solves an adaptive problem without a huge cost to the future success of the organism. Finally, the characteristic must also be a good solution to an adaptive problem; it must perform its function well. If these three criteria are satisfied, it looks increasingly unlikely that the feature could have arisen by chance alone.

Box 2.1 Behaviour that is genetic but not adaptive, or adaptive but not genetic

We must be wary of interpreting the basis of all behaviour as genetic adaptation. There may be non-adaptive or non-genetic explanations for the phenomenon we are investigating. Some such alternative explanations could be:

a) Genetic but not adaptive

(i) Phylogenetic inertia

Organisms may show signs of an ancestry they are unable to completely escape from even though the features in question are no longer adaptive. It is not optimal, for example, for a hedgehog to curl into a ball as a defence against oncoming traffic. The human skeletal frame is not an optimum form for vertical posture, as anyone with a bad back will confirm (see Chapter 18).

(ii) Genetic drift

Some genetic polymorphisms may exist in a population as a result of chance mutations that are neither advantageous nor disadvantageous or have not yet had time to be weeded out by natural selection. One special case of **genetic drift** is the **founder effect**. If a new population is formed from a few individuals then alleles may be fixed in the population that were once only a partial sample of a larger population. The new populations may look different but not for adaptive reasons – simply the effect of the founders being a limited sample from a larger and more diverse gene pool. The fact that the blood group B is virtually absent from North American Indians is probably a result of genetic drift rather than adaptive change.

(iii) Mutations

The genotype organism is just a snapshot in time of genes travelling through the generations. Some of them may contain recently acquired mutations (giving rise to maladaptive behaviours) that are in the process of being removed by natural selection. In general we have to accept that there will be a balance between the rate of accumulation of new mutations and their rate of removal (see Chapter 20).

(iv) Adaptive landscapes

It may be that a feature of an organism appears suboptimal in comparison to what it could be. In such cases a feature may be trapped on a local adaptive peak and unable to rise to a more successful design since any movement involves a reduction in fitness. The classic case is the human eye (see Chapter 18).

b) Adaptive but not genetic

(i) Phenotype plasticity

The phenotype of an organism can often be moulded by external influences during ontogeny to suit the prevailing environmental conditions. Bone, for example, grows in such a way to resist adequately the pressures applied. The growth of corals and trees is well adapted to the direction of water and air currents. We could say, of course, that the mechanism to so adapt is genetic and therefore heritable but the adaptation itself is not.

(ii) Learning

Humans, in particular, have a great capacity to learn from each other, from experience and from their culture. If humans in widely dispersed and different cultures show similar patterns of behaviour which appear well adapted it may because of similar shared genes but it could also be that they have come to the same conclusions as to how to behave by parallel social learning.

(iii) Cultural adaptations

In many ways culture is an expression of social learning. But in the view of gene–culture coevolutionary theory (see Chapter 20) culture itself may be subject to selective forces and evolve in ways parallel to and sometimes in harmony with biological evolution. Hence we may observe cultural practices that have evolved over time, have no direct basis in genetics, and yet serve human adaptive ends. Cooking, clothing, and tool making are examples.

In searching for adaptations, we must be wary of the pitfalls of Panglossianism and try to avoid them by making precise predictions about how a feature or behavioural pattern under investigation confers a competitive advantage. Some of the specific procedures that can be used to test **hypotheses** are considered in the next section.

2.1.2 Adaptations and fitness: then and now

The terms 'adaptive' and 'fitness' are troublesome concepts. An adaptive character should obviously palpably assist the survival and reproductive chances of its bearer, but this criterion is not always easy to apply. 'Fitness' is a term with its own difficulties. As Badcock (1991) points out, human males would probably be fitter in the sense of living longer and enjoying a reduced susceptibility to disease if they were castrated or had evolved lower levels of testosterone, but this may not be the best way to produce offspring. Darwinism is ultimately about the differential survival of genes rather than about the fitness of the gene carriers.

We must also not be tempted to hybridise Darwin with Pangloss and expect every adaptation to be perfect for the task in hand. In some cases, the environment can change more rapidly than natural selection can keep up with, and an adaptation is left high and dry, looking imperfectly designed.

An adaptation often involves a trade-off between different survival and reproductive needs. A big body may be helpful in fighting off predators, but big bodies need lots of fuel and time to grow. It is also important to consider how behaviour leads to fitness gains over the whole lifespan of the animal. An animal must devote resources to growth, repair and reproduction. Over a single year behaviour may not seem to be optimal, whereas over a lifespan a different picture may emerge. This selective allocation of resources is often known as a 'life history strategy' and will be considered in Chapter 8.

Reverse engineering and adaptive thinking

In evolutionary psychology two important forms of thinking are adaptive thinking and **reverse engineering**, both sometimes lumped together under adaptationism (Griffiths, 2001). Put simply, adaptive thinking claims to infer the solution from the problem, whilst reverse engineering claims to be able

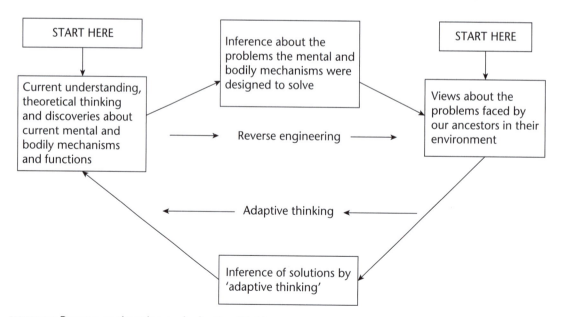

FIGURE 2.2 Reverse engineering and adaptive thinking.
Reverse engineering starts from what we know about the human mind and human behaviour and infers backwards what adaptive problems had to be solved. Adaptive thinking starts from a knowledge of what problems our ancestors faced and predicts likely mechanisms that solved them. The trick is not to go round in circles.

to suggest the problem by examining the solution. As an example of the former, our male ancestors must have faced the problem of paternity certainty and so, by inference, we probably have mental circuitry to incline us to guard our mates and react jealously to situations where our reproductive interests are threatened by a rival. We can then proceed to study the fine detail of this mechanism and how it is activated in a variety of contexts. As an example of reverse engineering, we might note that we have lost our bodily hair and so it is a plausible hypothesis to suggest the environment in which this happened was open grassland savannah where humans hunted by day and needed to keep cool through sweating as we expended energy (see Chapter 6).

But reverse engineering does have its pitfalls. Engineers normally start with a problem (a gorge to be spanned, for example) and design an artefact or system to solve it (in this case a bridge). The problem with biology is that we are left with an abundance of solutions (the living world) but have ambiguous information about the problems to which the physiology and behaviour of contemporary organisms are responses. In the case of cognitive phenomena the system to be explained or reverse engineered might call forth a whole bundle of hypotheses about the nature of the ancestral environment, many of which are difficult to refute because of the paucity of information. One obvious example concerns the growth in the size of the human brain over the last 2 million years (encephalisation; see Chapter 6). Just what was the problem(s) to which the enormous growth of the hominid brain was a response? In answer to this there are a variety of competing hypotheses that have proved infuriatingly difficult to judge between.

A problem inherent in adaptive thinking is the temptation in evolutionary psychology to believe that if a particular trait makes adaptive sense then it must exist. To be sure, adaptive thinking can serve a positive heuristic: it suggests likely areas of research since a hypothesis that makes adaptive sense is worth pursuing. But, as Griffiths (2001) suggests, adaptive hypotheses can have a negative effect if they tend to lead to a neglect of empirical evidence that does not quite fit.

Another problem is that of historicity or contingency. A weakness of naïve adaptationism is the belief that natural selection acts like some all-seeing eye that can scan all possible solutions and select the best one. But the hands of natural selection are tied by various contingencies and unpredictable events. The lineage of any organism moves through an adaptive landscape accumulating baggage along the way. As it encounters a new problem, its range of solutions is constrained and the outcome may not seem that optimal long after the event.

2.2 Darwinian methodologies: evolutionary psychology, human behavioural ecology, Darwinian anthropology and gene–culture coevolution theory

Suppose we identify some physical or behavioural trait of humans, such as hairlessness (compared at least with other primates) or the specific mating preferences of either sex, and attempt to demonstrate that they have some **adaptive significance** and have been the subject of natural or sexual selection. Among other problems, one serious question to address is 'adaptation to what?'; to current conditions, or to conditions in the past? The problem is the same for non-human animals: a trait that we study now may have been shaped for some adaptive purpose long ago. The environment may have changed so that the adaptive significance of the trait under study is now not at all obvious; indeed, it may now even appear maladaptive. When human babies are born, they have a strong clutching instinct (one of several newborn reflexes) and will grab fingers and other objects with remarkable strength. This may be a leftover from when grabbing a mother's fur helped to reduce the number of accidents from falling. It is not, however, clear that it helps the newborn in contemporary culture.

The problem is especially acute for humans since, over the last 10,000 years, we have radically transformed the environment in which we live. We now encounter daily conditions and problems that were simply absent during the period when the human genome was laid down. It could be expected that we have adaptations for running, throwing things, weighing up rivals and making babies but not specifically for reading, writing, playing tennis or coping with jet lag. One crucial question is whether the human mind was only designed to cope with specific problems found in our **environment of evolutionary adaptedness** (the EEA – a phrase coined by John Bowlby) before the invention of culture (roughly the period between two million and 40 thousand years before the present) or whether those selective conditions gave us minds now flexible enough to give rise to behaviours that still maximise reproductive fitness in current environments. Another vexed question concerns the impact of culture on behaviour. Is culture something we erect around us to serve our reproductive interests? Does culture have its own evolutionary momentum? Does culture itself exert selective pressures on our genome?

These questions are profound and have led to several basic schools of thought in the application of evolutionary theory to human behaviour: that of the evolutionary psychologists, who argue that the human mind is replete with domain-specific modules adapted to solve ancestral problems that in current contexts do not necessarily enhance reproductive success; a cluster of related perspectives, including Darwinian anthropology, human sociobiology, biological anthropology and human behavioural ecology, that argue that human behaviour can still be analysed on the assumption that it strives to maximise fitness in current contexts; and finally a growing school of what is sometime called gene–culture coevolutionary theory that asserts and explores the interactions between genes and culture. Modern cultural evolutionary theory has sought to apply evolutionary models to understanding how cultural artefacts spread and change; how socially transmitted information is selected, assimilated, modified and transmitted; how cultural change responds to and drives genetic change; and, more generally, how culture is a potent force in explaining human behavioural diversity. This is considered more fully in Chapter 20. In this book we will take these three approaches as complementary rather than necessarily competing, although obviously for specific problems one perspective might be more fruitful than another. Table 2.1 shows a summary of these approaches and they are also discussed in more detail below.

2.3 Evolutionary psychology

Evolutionary psychologists would argue that human behaviour as we observe it today is a product of contemporary environmental influences acting upon ancestrally designed mental hardware. The behaviour that results may not be adaptive in contemporary contexts. We should focus then on elucidating mental mechanisms rather than measuring reproductive behaviour. We should expect to find mind mechanisms that were shaped by the selection pressures acting on our distant ancestors. An analogy is often drawn with the human stomach. We cannot digest everything we put in our mouths; the human stomach is not an all-purpose digester. Similarly, the mind is not a blank slate designed to solve general mental problems because there were no general mental problems in the Pleistocene age, only specific ones concerning, hunting, mating, travelling and so forth.

A mundane example concerns our food preferences. As humans, we are strongly attracted to salty and fatty foods high in calories and sugars. Our taste buds were probably a fine piece of engineering for the Old Stone Age when such foods were in short supply and when to receive a lot of pleasure from their taste was a useful way to motivate us to search out more. Such tastes are now far from adaptive in an environment in developed countries where fast food high in salt, fat and processed carbohydrates can be bought cheaply, with deleterious health consequences such as arteriosclerosis and tooth decay.

At this point, it is worth stressing the difference between behavioural adaptations and cognitive adaptations. A pattern of behaviour may be adaptive without any cognitive component. The greylag goose noted in Chapter 1 that rolls its egg back to the nest is a case in point. The behaviour is highly adaptive – millions of eggs must have been saved by this device – but the fact that the behaviour continues when the egg is removed from beneath the bird's feet shows that it is rather simply constructed, inflexible and not 'thought about'. In contrast, as Tomasello and Call (1997) argue, a cognitive adaptation:

- involves decision-making among a variety of possible courses of action
- is directed towards goals or outcomes
- probably involves some sort of mental representation that goes beyond the information immediately presented to the senses.

2 FOUNDATIONS OF DARWINIAN PSYCHOLOGY   23

Table 2.1 Contrast between the methods and assumptions of some approaches within the human evolutionary behavioural sciences

Features	DARWINIAN ANTHROPOLOGY (Human sociobiology, Darwinian social science, human behavioural ecology, human ethology)	EVOLUTIONARY PSYCHOLOGY	GENE–CULTURE COEVOLUTIONARY THEORY
Basic approaches and assumptions	Behaviourist approaches Humans are flexible opportunists and so optimality models (for example, for foraging and birth intervals) can be used. Life history theory is central. Concentration on behavioural outcomes and strategies, not beliefs, values, emotions or motives. No assumption about whether behaviour demonstrates adaptiveness due to genes or socially transmitted information.	Cognitivist approaches Studies should look for mental mechanisms that evolved to solve problems of the Pleistocene, the environment of evolutionary adaptedness (EEA), not contemporary environments. Key concept is that of adaptive lag, hence current fitness is irrelevant. Design (by natural selection) is manifested at the psychological not the behavioural level.	Interactionist approaches Culture can evolve in a Darwinian manner alongside and in tandem with genetic evolution. Cultural evolution can sometimes help fitness maximisation (for example, clothes, tools, socially transmitted useful knowledge and practices) but some forms of culture can evolve independently.
What to study and measure	Measure lifetime reproductive success of individuals in relation to their environments. Count current babies. These are methods typical of those of behavioural ecologists. Proxies such as energy can be used as indicators of fitness and success.	Focus on survey work (questionnaires) and laboratory tasks typical of mainstream psychology. Subjects are often undergraduates.	Measure changes in the evolution of cultural artefacts or the spread of memes. Use of mathematical modelling. Look for associations between genetic variants and cultural practices (for example, lactose and starch digestion (see Chapter 20)).
Views on adaptations	Ancestral adaptations have given rise to domain-general mechanisms. Use of the behavioural and phenotypic gambits – eschewing assumptions about internal mechanisms.	Ancestral adaptations have given rise to 'domain-specific' modules designed to solve specific problems. Mind is like a Swiss army knife consisting of discrete tools or problem-solving algorithms. Analogy with organs of the body: the brain has mental organs. Such modules may now function in maladaptive ways.	The mind is equipped with domain-general biases (for example, the conformist bias) that help to shape how culture evolves. Cultural learning and evolution can solve newly arising adaptive problems.
Views on genetic variability	Genetic variability can still exist and is particularly influential in mate choice. More generally great emphasis on behavioural plasticity as humans adaptively adjust their behaviour to different ecological conditions.	The evolved mental mechanisms we now possess show little genetic variability; they point to a universal human nature. Focus on human universals.	Since culture can drive genetic change we can observe how culture has impacted on the human genome (for example, jaw size reduction following the invention of cooking) and is a source of current genetic diversity between populations (for example, amylase gene copy number and lactose tolerance).

In a sense, cognitive adaptations are the result of evolutionary processes that have relinquished hard-wired solutions to directing behaviour optimally in favour of judgements made by an individual organism. In such cases, an organism (although not necessarily consciously) has goals, makes decisions and calculations according to context and its own life experiences, and hence chooses an appropriate strategy.

Daly and Wilson (1999) argue that the use of the term 'evolutionary psychology' only for humans introduces an unnecessary species divide. The divide, they say, is unwarranted by the history of the subject since many ideas from animal behaviourists have been incorporated into human behavioural studies and, moreover, the principles that apply to the human animal must also apply to non-human animals. Consequently, they prefer the term 'human evolutionary psychology' when referring only to humans.

2.3.1 The modular approach of Tooby and Cosmides

The evolutionary approach to psychology has been much influenced by the work of Tooby and Cosmides and the powerful manifesto on evolutionary psychology that they issued in *The Adapted Mind* (Barkow et al., 1992). Many of the issues, in particular the extent to which the brain can be thought of as a set of discrete problem-solving modules, remain controversial, but the approach has met with considerable success as a heuristic model. For Tooby and Cosmides, psychology is to be seen as a branch of biology that studies the structure of brains, how brains process information and how the brain's information-processing mechanisms generate behaviour. The dominance of this American approach with its emphasis on domain-specific mental modules initially pioneered by Tooby and Cosmides led to the appellation of this group as 'the Santa Barbara church of psychology' (Laland and Brown, 2002, p. 154).

Some potential candidates for domain-specific modules

Some modules that might be expected to form part of the human mental toolkit are mechanisms for cooperative engagements with kin and non-kin; means by which to detect cheats; parenting; disease avoidance; object permanence and movement; face recognition; learning a language; anticipating the reactions and emotional states of others (theory of mind); self-concept; and optimal foraging – to name but a few. A central problem with all of this, of course, is knowing when to stop. At what level of discrimination and finesse have we reached an indivisible module? A useful, albeit conjectured, hierarchical organisation of modules is provided by Geary (1998) (Figure 2.3). The two basic sets of inputs – social and ecological information – give rise to folk biology and folk physics. These are systems of thought that provide quick common-sense interpretations to the world. They may turn out later to be incorrect by scientific standards. Folk physics, for example, is Aristotelian in nature: a force is needed to keep an object moving; whereas the more reliable Newtonian physics suggests that net forces cause acceleration.

One overall point stressed strongly by Geary is that such modules do not constrain people to behave in fixed patterns along the lines of stimulus–response thinking; predators would soon home in on organisms that behaved in predictable ways. We should instead regard the modules as conferring a flexible response according to the context.

In this view, one can see the purpose of development as enabling the calibration of these modules (setting the start-up conditions) to local social, biological and physical conditions. The modules constrain and bias the type of experiences to which a growing child should attend, but then use the result of the experiences gained to fine-tune the inherent functional mechanisms. Thus, for example, we are born with a disposition to attend to verbal sounds and organise utterances. The language that we finally speak is a result of this cognitive bias coupled with the actual evidence obtained about syntax and vocabulary. This ontogenetic development of modules in relation to the local environment can be thought of as an 'open genetic program', an open programme being one that takes on board instructions from the environment (Geary, 1998, p. 205).

It is important to realise that Darwinian psychology does not stand or fall on the existence of discrete modules or their degree of independence and domain specificity. Before Tooby and Cosmides announced their view of the mind, David Marr (1982) made the

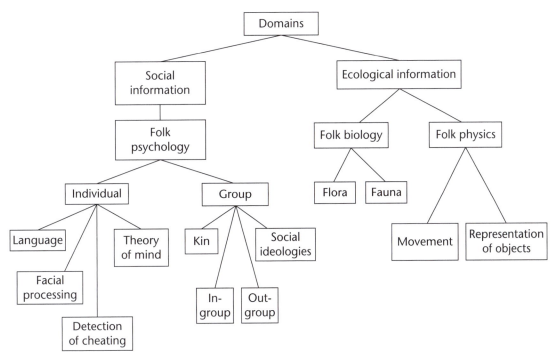

FIGURE 2.3 Some possible evolved domains of the mind.
SOURCE: Modified from Geary (2010). *Male, Female*. Washington DC, APA, Figure 9.2. Copyright © (2010) by the American Psychological Association. Adapted with permission.

useful distinction between levels of understanding of how the brain drives behaviour. As Table 2.2 shows, traditional cognitive psychology with its emphasis on computational systems and problem-solving algorithms stands midway between the proximate/lower level of neurophysiology (the wiring diagram) and the ultimate/high level of adaptive responses to selective pressures.

Table 2.2 Levels of explanation in evolutionary neurophysiology (see Marr, 1982)

Level	Aims of study at this level	Associated discipline
High level	To elucidate the tasks that the system was 'designed' to achieve	Darwinian psychology
Intermediate level	How the system computes information to achieve these tasks	Cognitive psychology
Lower level	How the circuitry is arranged in the brain	Neurophysiology

In Chapter 9 we will examine how this evolutionary modular approach can throw light on patterns in human reasoning and how thinking at the 'high level' of Table 2.2 can help make sense of the 'intermediate level' below.

2.3.2 Challenges to the modular view of evolutionary psychology and the concept of the EEA

The early founders of evolutionary psychology in the 1980s stressed the importance of the premise of a universal human nature. This view was based on several arguments: firstly, the assumption that the human genome was largely a product of adaptations in the Pleistocene; secondly, the well-established fact that, compared to other apes, humans show relatively little genetic diversity; and thirdly, the belief that the last 10,000 years or so since the end of the Pleistocene was too short a time frame for significant genetic change to have occurred. This latter point gave rise to the idea of genome lag: our adaptations to the Pleistocene may not serve us so well in contemporary environments. Given a commitment to a universal nature (or two given male/female differences) evolutionary psychology has always faced the difficulty of explaining human behavioural diversity in the light of a universal cognitive system. It has done so, successfully in many cases, by the postulation of context-dependent strategies: the mind flips between repertoires of pre-programmed behavioural outputs depending on the specific environmental inputs. But these central premises of evolutionary psychology are open to challenge.

The EEA: the land of lost content

Objections to the approach of evolutionary psychology have focused on the mysterious EEA. Much clearly depends on the EEA, so what was it like, and how far do we go back? Betzig (1998) makes the point that for the past 65 million years our ancestors existed as some sort of primate, so do we look to the selective pressure on primates over this period as a clue to the human psyche? We could narrow it down by suggesting that we have spent the last six million years as one species or other of hominin (*Homo habilis*, *Homo erectus* and so on; see Chapter 5), so should we consider those environmental conditions? Archaic *Homo sapiens* and the subspecies *Homo sapiens sapiens* (that is, us) have been around for about 200,000 years, mostly spent in a hunter-gatherer lifestyle – or rather lifestyles, for even among contemporary hunter-gatherers, there are large differences in mating behaviour, paternal investment and diet – so perhaps we should focus on this period.

Moreover, over the past two million years, there have been a series of EEAs rather than just one. The EEA has acquired an almost fabled status, but Tooby and Cosmides themselves point out that there was not one single EEA. They argue that it is 'a statistical composite of the adaptation-relevant properties of the ancestral environments encountered by members of ancestral populations, weighted by their frequency and their fitness consequences' (Tooby and Cosmides, 1990b, p. 386). This is a fine definition in theory, but integrating such factors across time for human evolution is extremely difficult.

If we are only allowed to speculate about adaptations to an EEA, and given that establishing the features of EEA will be difficult enough, we are restricted in how far we can understand human nature with any precision. Moreover, it has been pointed out that, as humans, we know the probable responses of our own species to questionnaires and other measures of thinking, so there arises the temptation to select features from what we imagine to be the EEA to predict correctly outcomes that are already foreseen (see Crawford, 1998a). It would be better if palaeontologists and palaeogeographers supplied the conditions of EEA and an intelligent chimp made the predictions about human behaviour.

Another major problem with the concept of the EEA is the assumption that it was a stable environment exerting steady long-term selection pressures that have given us the human genome we have today. Recent evidence for paleoecology, however, shows that the period 1.8 million to 10,000 years ago (the Pleistocene) was far from stable with significant and rapid climatic changes. The potential impact of this on something as important as brain size is discussed in Chapter 5. The focus on the Pleistocene as the defining EEA also risks neglecting facets of human intelligence and cognition that were already in place then. By analogy, if we wish to understand the human skeleton we would have to consider it in relation to adaptation well before the Pleistocene – about 4 million years ago, for example, to understand bipedalism and well before that to understand more basic structures such as the pentadactyl limb and the backbone. We must be wary also of treating the EEA as some sort of 'land of lost content' when the human genome was in total harmony with its surroundings. In reality, no organism is perfectly adapted to its environment. Adaptations are compromises between the different requirements of an animal's life.

Genome lag and modularity

Recent developments in human genetics have also challenged the notion of genome lag. Since the 1980s molecular geneticists have devised ways not only to map the human genome but to identify regions that have been subject to recent selection pressures. This subject is beyond the scope of this book but a brief review of several studies will give a flavour of what can now be done. Sabeti et al. (2007) looked at 3 million SNPs (see Box 3.1) and found 300 regions where there was evidence of recent strong selection. They also found evidence for recent regional positive selection in genes relating to viral infections, skin pigmentation and hair follicles. Williamson et al. (2007) analysed a data set of 1.2 million SNPs and observed evidence of recent selection in genes involved in pigmentation pathways, olfactory reception, the immune system and heat shock proteins. This group concluded that 'recent adaptation is strikingly pervasive in the human genome' with possibly 10 per cent of human genes affected by genetic change over the last 50,000 years. In the light of this and other evidence, Laland et al. (2010) concluded that many cultural changes in the Holocene (that is, the last 10,000 years), such as the domestication of animals and plants and the rise of high-density human settlements, have exerted selection pressures to which the human genome has responded. The arrival of this relatively recent genetic evidence has coincided with the development of gene–culture evolutionary models and the idea of 'niche construction': the suggestion that humans, by altering their world through agriculture, cities, architecture and social institutions are to a significant degree shaping their own evolution (Bolhuis et al., 2011). Many researchers now feel that it is time for evolutionary psychology to abandon its emphasis on a single universal human nature and become more open to the use of evolutionary theory and the study of developmental process to explore human behavioural diversity.

There have also been challenges to the idea that the mind is full of domain-specific modular systems. Comparative psychology has demonstrated the existence of domain-general mechanisms of thinking and learning found also in many other animals. One of these is surely associative learning: that is, attending to patterns and regularities in experience and forming useful conclusions. One particularly helpful and influential example has been the Rescorla–Wagner model, which has been successfully applied to animal and human behaviour (see Siegel and Allan, 1996).

In the modular view adaptive behaviour is seen as the result of cognitive processes. But must this be the case? Because behaviour has to solve a complex problem and provide an adaptive solution, this does not imply that it is driven entirely by mental processes. A physiological system might suffice. The heart, for example, is a highly adaptive organ that has to solve numerous problems to do with pumping rates and the coordination of pressure changes according to the activity of the organism, yet its behaviour is hardly a product of cognition.

The debate will continue but it is clear that the modularity thesis will have to accommodate communication between modules and some degree of phenotypic plasticity. A single genotype might produce a range of phenotypes differing as a result of the environmental cues received during development. It would in fact be a highly efficient and adaptive mechanism that allowed development to be structured by experience in ways to achieve an adaptive fit to local conditions. There is also a growing need within evolutionary approaches to behaviour to take account of the growing evidence on real genetic diversity between humans. These subjects are further explored in Chapters 7 and 8.

2.4 Darwinian anthropology and human behavioural ecology

2.4.1 The phenotypic gambit

For the sake of convenience, we will include human sociobiology, human behavioural ecology (HBE) and human ethology under the heading of Darwinian anthropology. Darwinian anthropologists argue that ancestral adaptation was not so specific and that we possess 'domain-general' mechanisms that enable individuals to maximise their fitness even in the different environment of today. They suggest that different contemporary environments will give rise to different fitness-maximisation strategies. The way to look for adaptations is not to try to find ancient mental mechanisms but to look at current behaviour in relation to local environmental conditions. In this sense, their approach is similar to that of the behavioural ecologists who study non-human animals. The assumption is that the behaviour of modern humans is still governed by the iron logic of fitness maximisation.

Darwinian anthropologists focus on 'adaptiveness' to different environments, implying that humans will tend to maximise their reproductive output in the circumstances in which they find themselves. In contrast, evolutionary psychologists focus on 'adaptations' – discrete mechanisms and traits that an organism carries as a result of past selective pressures. An important point, however, is that behavioural ecologists have shown that non-human animals do display adaptiveness: they have a range of strategies that are evoked in different environmental conditions, and moreover are capable of behavioural plasticity. It would be a simple or a foolish and very 'un-Darwinian' animal that only had a single behavioural strategy to be expressed in all circumstances. With respect to humans Baker and Bellis (1989) provide evidence that human males adjust the number of sperm in their ejaculate according to the likelihood of sperm competition (see Chapter 13). In the view of evolutionary psychology, the modification of behaviour according to context is acknowledged but the role of the environment is conceived as something that initiates a selection of pre-adapted behavioural outputs: hence behavioural diversity results from different environments selecting for different outputs. An analogy suggested for this was that of a typical 1960s juke box where a variety of records was neatly stacked awaiting selection by a specific set of buttons to be pressed.

A key feature of Darwinian anthropology or human behavioural ecology is the borrowing of methods from behavioural ecologists. Behavioural ecology is a thriving sub-discipline within the field of animal behaviour. Behavioural ecologists base their work on the premise that by and large behaviour will tend to maximise the reproductive fitness of an individual over its lifetime in the environment in which it lives. Hence they have developed various **optimality** models to show how behaviour responds to different environments to yield optimal outcomes. Human behavioural ecology was initially a small field that emerged within anthropology and is still sometimes known as Darwinian anthropology. Early pioneers in this field included Napoleon Chagnon, William Irons and Richard Alexander who all explored the idea that humans in contemporary environments still behave so as to maximise their reproductive success. Typical concerns to which this has been applied include optimum group sizes for hunting, types of marriage systems (for example, in relation to resource availability) and the length of inter-birth intervals.

In some ways, human behavioural ecologists are agnostic about whether the adaptive responses and the match between phenotype and environment that they claim to demonstrate are the products of genetically based strategies or socially transmitted information: the crucial point to them being that, within limits, adaptive responses will result.

They subscribe to a view that is sometimes called the 'phenotypic gambit', which is the idea that humans possess considerable phenotypic plasticity in their decision-making and the assumption that different behavioural phenotypes correspond to readily available and simple genetic systems, a position that is quite controversial (van Oers and Sinn, 2011). The phenotypic gambit assumes that the genetic basis to produce adaptive behaviour in a variety of environments is present in all humans, and that human genetic variation is small and has an insignificant effect on behavioural variation. In a sense behavioural ecologists are environmental rather than genetic determinists. As an example of this, Bobbi Low at the University of Michigan has provided evidence that the training of children is related to the type of society in a way that tends to maximise fitness. Low found that the more polygynous the society, the more that sons were reared to be aggressive and ambitious. The logic here is that, in polygynous societies, successful males stand to secure more mating opportunities. If Low is right, the behaviour of adult males is strongly influenced by their childhood training, but such training is a response to local social and ecological conditions (Low, 1989). It is easy to criticise the phenotypic and behavioural gambit as showing a cavalier disregard for internal mechanisms, or naivety in thinking that behavioural plasticity is unconstrained. One advantage, however, is that that a comparison of real outcomes with those expected from optimality assumptions can point to interesting proximal constraints, genome lags, or unidentified trade-offs that are illuminating in themselves. One of the strengths of HBE is that it is able to make predictions about behavioural variation within and between societies. In addition it offers a degree of consilience in that its foundation discipline, behavioural ecology, employs methods and theories that apply to species other than humans. Central to HBE is life history theory and this is considered in Chapter 8.

2.4.2 The problem of fitness measurement

One of the key problems faced by HBE is the contrast between the assumption that humans still strive to maximise fitness and the obvious observation that many couples in industrialised societies choose to have small families or no children at all.

Consider how a male could maximise his fitness in the late 20th century. A scenario envisaged by Symons (1992, p. 155) is both amusing and instructive:

> In a world in which people actually wanted to maximise inclusive fitness, opportunities to make deposits in sperm banks would be immensely competitive, a subject of endless public scrutiny and debate, with the possibility of reverse embezzlement by male sperm bank officers an ever-present problem.

But when the UK opened its first national sperm bank in the UK in 2014 it attracted only nine donors over its first year of operation – despite the inducement of £35 per donation.

One answer, of course, is that natural selection did not provide us with a vague fitness-maximisation drive. The genes made sure that fitness maximisation was an unconscious urge; like the heartbeat, it is too valuable to be placed under conscious control. Instead, males and females were provided with sexual drives that could be moderated according to local circumstances. Counting the size of the queue outside a sperm bank would be a fruitless way of assessing a male's fitness – counting partners and real sexual opportunities might be better. If sperm banks were set up to allow males to deposit sperm more naturally (*in vivo*), I suspect that the queues would be longer. The dominance of the ancestry of our genes and the ancestral function of gene-driven emotions is also illustrated by jealousy. If we focus on males, in the past the emotion of jealousy arose as a way of ensuring paternity confidence: it would have motivated mate guarding, hostility to rival males, and coercive behaviour over the female. It still has this adaptive function of course but few men today would be reconciled to adultery if their wives told them they were taking the contraceptive pill and so paternity confidence was not an issue.

More generally applying the idea of optimal fitness to modern cultures where contraception is used

on a massive scale is a real problem for these type of approaches; the problem of low and falling fertility in modern societies, the so-called demographic transition, remains acute for HBE and we return to it in Chapter 8.

2.5 Gene–culture coevolutionary theory

Gene–culture coevolutionary theory, sometimes called dual inheritance theory, first came to prominence in the 1980s. A number of landmark publications contributed to its rise; these include: *Genes, Mind and Culture* by C. Lumsden and E. O. Wilson (1981), *Cultural Transmission and Evolution: A Quantitative Approach* by L. Cavalli-Storza and M. Feldman (1981), and, perhaps most influential of all, *Culture and the Evolutionary Process* by R. Boyd and P. Richerson (1985). Despite these early starts the output of work from this perspective has been small compared to evolutionary psychology and HBE.

The gene–culture coevolutionary perspective is based on three assumptions:

1. The ability of humans to produce culture is a specific and evolved biological adaptation, selected because it enhanced reproductive success.
2. Since humans generate, acquire, modify and transmit culture this amounts to a second system of inheritance and evolution running parallel to biological evolution.
3. The two inheritance systems (biological and cultural) interact: evolved psychological biases in learning and cognition shape the transmission of culture; whilst cultural change sets up new selective pressures causing genetic change (see Chapter 20). Culture is like the beaver's dam: it is both an expression of our natures and the environment in which we live.

Whereas evolutionary psychology argues that the mind is full of domain-specific mechanisms, gene–culture theorists tend to take a more flexible approach to the suggestion of innate mechanisms. However, cultural evolutionists often refer to 'biases' in the human mind such as the prestige bias (copying high-status members of a group) or the conformist bias (copying the most common behaviour), but these context-based biases (also known as social learning strategies) apply across a variety of content domains and so are more akin to domain-general rules than the specific rules envisaged by evolutionary psychology.

One crucial difference between gene-culture theories and HBE is that the latter is predicated on the assumption that behaviours will tend towards optimality, delivering the best possible fitness outcomes in the circumstances encountered, whereas the former argue that cultural units (artefacts, ideas and so on) can themselves evolve in ways that may or may not overlap with biological fitness. Gene–culture co-evolutionary theory is considered in more depth in Chapter 20.

2.6 Prospects

On the question of evolutionary psychology, HBE or gene–culture coevolution this book takes the approach that there is room for all three sets of methodologies. The human brain is complex enough and powerful enough to accommodate behaviours that are learned or unlearned, behaviours that are soft wired and hard wired, behaviours that adjust to local conditions and behaviours that are invariant. The importance of adaptive behavioural plasticity is also suggested by recent advances in epigenetics (see Chapter 7). Moreover, the logic of natural selection is such that there are good reasons for supposing it is a universal mechanism that drives change in a much wider range of entities than DNA-based life forms (see Chapter 20). In subsequent chapters, we offer significant findings and successful predictions from all three of these broad perspectives (see also Sherman and Reeve, 1997; Daly and Wilson, 1999), which perhaps should not be seen as competing paradigms but milestones and useful boundary markers within a broader and ongoing evolutionary synthesis.

But what of Darwin's prediction that in the 'distant future ... [p]sychology will be based on a new foundation'? Despite the availability of the evolutionary concepts and perspectives outlined in this chapter mainstream psychology still remains a highly segregated and divided field of knowledge

with numerous sub-disciplines such as cognitive psychology, developmental psychology, behaviourism, comparative psychology and social psychology. But there are signs that evolutionary principles are creeping into and informing virtually every one of these sub-disciplines, holding out the prospect, at least, that evolutionary psychology (in one of its guises) may eventually provide the unifying framework about which psychology can cohere (Fitzgerald and Whitaker, 2010). It is arguable, for example, that cognitive neuroscience and evolutionary psychology have already merged to form evolutionary cognitive neuroscience – a subject using new brain imaging techniques (such as fMRI) to inspect how brains react to experience and employing evolution as a meta-theory to guide hypothesis formulation and testing (about sex differences in cognition, for example). In developmental psychology, evolutionary theory both informs behavioural genetics and, crucially, underpins life history theory which in recent years has illuminated the study of child development, puberty, adolescence and adult health using adaptive behavioural plasticity as a key principle. Social psychology has for some decades been of interest to evolutionary biologists and behavioural ecologists. The initial guiding premise of social psychology was that the environment (especially the interaction between the self and others) was the key to understanding social behaviour. Yet as early as the 1960s, sociobiology and behavioural ecology were demonstrating how such biological concepts as kin selection and sexual selection could inform social behaviour. Sadly, it remains the fact that social psychology textbooks often convey erroneous interpretations of kin selection and altruism (see Park, 2007). In clinical psychology evolutionary approaches have met with a mixed reception. To date, evolutionary psychology has not discovered any new treatment for a mental disorder and so its usefulness is viewed by many as suspect. On the other hand, clinical psychology itself is beset with multiple and often conflicting theories. The fact that many disorders are highly heritable, however, demonstrates the need for the inclusion of evolutionary genetics in any understanding of mental illness. Several promising lines of enquiry are discussed in Chapter 19. Although mental illness has proven to be a difficult problem for the evolutionary paradigm, Darwinian perspectives applied to other disorders such as pregnancy sickness have met with some success, providing persuasive arguments to see this condition not as an illness in need of treatment but a natural and indeed healthy adaptive response to protect both mother and foetus (see Chapter 18).

But taking a broader view of academia, there are now signs that evolutionary approaches are permeating, informing and even transforming a whole range of disciplines, many of which were once bastions of sociocultural explanations of behaviour, such as: economics (Hoffman et al., 1998; Friedman, 2005); criminology (Walsh and Beaver, 2009); literary theory (Carroll, 2004; Gottschall and Wilson, 2005);

medicine (Trevathan et al., 2008; see also Chapter 18); neuroscience (Webster, 2007); ethics (Richards, 1993; Ruse, 1993); political science (Rubin, 2002); the law (Masters and Gruter, 1992); women's studies (Browne, 2002); aesthetics (Voland and Grammer, 2003); and even management theory (Nicholson, 2000), to name but a few.

If all these perspectives are varieties, stemming from the same root stock of Darwinian organic evolution, then there is clearly a need to examine the processes that are common to all. In the next two chapters, therefore, we examine the twin pillars of Darwinism: natural and sexual selection.

Summary

▶ We should expect the behaviour of animals to be adaptive in the sense that it has been selected by natural and sexual selection to help confer reproductive success on individuals. There are various ways in which the adaptive significance of behaviour can be demonstrated and investigated. Some of these involve experimental manipulation of the natural state, and some involve looking for correlations between behaviour and environmental factors. In such studies, we must be constantly wary of finding convenient but spurious explanations designed post hoc to fit the facts.

▶ Applying evolutionary thinking to the human mind makes use of reverse engineering and adaptive thinking. Reverse engineering tries to infer what problems our ancestors faced in the light of what we know about how current minds operate. Adaptive thinking starts with knowledge of the problems faced by our ancestors and predicts what adaptations humans should possess as a result.

▶ A debate exists over the correct way to apply Darwinian reasoning to human behaviour. Darwinian anthropologists, sometimes called human sociobiologists, human behavioural ecologists or human ethologists, suggest that current human behaviour measured in terms of reproductive success shows signs of adaptiveness. Evolutionary psychologists argue that the correct Darwinian approach is to look for adaptations to ancestral environments that can now be identified with discrete problem-solving modules in the brain. A new school of gene–culture coevolutionary theory has also risen which examines the interaction between biological and cultural evolution.

Key Words

- **Adaptation**
- **Adaptive significance**
- **Environment of evolutionary adaptedness (EEA)**
- **Founder effect**
- **Genetic drift**
- **Hypothesis**
- **Optimality**
- **Reverse engineering**

Further reading

Barkow, J. H., L. Cosmides and J. Tooby (1992). *The Adapted Mind*. Oxford, Oxford University Press.

See Chapters 1 and 2 for discussion of the approach of evolutionary psychology. Discusses some complex methodological issues.

Buss, D. M. (2005). *The Handbook of Evolutionary Psychology*. Hoboken, NJ, John Wiley and Sons.

Numerous chapters by leading authorities on the whole field of evolutionary psychology.

Laland, K. N. and G. R. Brown (2002). *Sense and Nonsense: Evolutionary Perspectives on Human Behaviour*. Oxford, Oxford University Press.

A frank look at some of the misconceptions surrounding evolutionary psychology and an assessment of its potential.

Gangestad, S.W. and J. A. Simpson (2007). *The Evolution of Mind*. London, UK, The Guildford Press.

Excellent set of articles on the whole field of evolution and human behaviour. Several articles address fundamental theoretical and methodological issues.

Miller, A. S. and S. Kanazawa (2007). *Why Beautiful People Have More Daughters: From Dating, Shopping, and Praying to Going to War and Becoming a Billionaire: Two Evolutionary Psychologists Explain Why We Do What We Do*. London, UK, Penguin.

A lively book in the genre of popular science. A provocative account of how evolutionary psychology illuminates everyday life.

Whitehouse, H. (ed.) (2001). *The Debated Mind*. Oxford, UK, Oxford University Press.

Useful series of articles that tackle debates about the proper application of evolutionary theory to cognition. Covers arguments for and against modularity.

Workman, L. (2014). *Darwin: The Shaping of Evolutionary Thinking*. Basingstoke, UK, Palgrave.

Short but accessible introduction to Darwin's ideas as they apply to thinking about human behaviour.

PART

II

Two Pillars of the Darwinian Paradigm: Natural and Sexual Selection

Natural Selection, Inclusive Fitness and the Selfish Gene

It may be said that natural selection is daily and hourly scrutinising, throughout the world, every variation, even the slightest; rejecting that which is bad, preserving and adding up all that is good; silently and insensibly working, whenever and wherever opportunity offers, at the improvement of each organic being in relation to its organic and inorganic conditions of life.

(**Darwin**, *Origin of Species*, **1859b, p. 84**)

This chapter outlines the ideas central to Darwinism and examines some of the difficulties that Darwin faced with the concept of natural selection. Many of the problems that confronted Darwinism in the 19th century have been largely resolved by work over the past 75 years. A successful theory, however, has not only to confront empirical evidence, but also to show that it can do so better than alternative accounts. With this in mind, we will contrast Darwinism with Lamarckism as alternative ways of explaining how organisms become adapted to their environments. The chapter also looks at how Darwin's ideas on fitness were extended into the concept of inclusive fitness, a development that helped solve many of the problems that puzzled Darwin. The chapter concludes with an examination of the gene-centred view of evolution and natural selection. Our understanding of the genetic basis of behaviour and the whole operation of natural selection is also greatly assisted by a clarification of what can properly be considered to be the unit of natural selection, and by the crucial distinction that must be made between the replicators (genes) and the vehicles of these replicators (bodies). The concept of the selfish gene has been much maligned, but properly understood it does not suggest that all individuals must behave selfishly in the pejorative sense. This chapter will demonstrate how altruism can possibly arise in a world of selfish replicators.

3.1 The mechanism of Darwinian evolution

The essence of Darwinism can be summarised as a series of statements about the nature of living things and their reproductive tendencies:

- Individuals can be grouped together into species on the basis of characteristics such as shape, anatomy, physiology, behaviour and so on. These groupings are not entirely artificial: members of the same species, if reproducing sexually, can by definition breed with each other to produce fertile offspring.
- Within a species, individuals are not all identical. They will differ in physical and behavioural characteristics.
- Some of these differences are inherited from the previous generation and may be passed to the next.
- Variation is enriched by the occurrence of spontaneous but random novelty. A feature may appear that was not present in previous generations, or may be present to a different degree.
- Resources required by organisms to thrive and reproduce are not infinite. Competition must inevitably arise, and some organisms will leave fewer offspring than others.

- Some variations will confer an advantage on their possessors in terms of access to these resources and hence in terms of leaving offspring.
- Those variants that leave more offspring will tend to be preserved and gradually become the norm. If the departure from the original ancestor is sufficiently radical, new species may form, and natural selection will have brought about evolutionary change.
- As a consequence of natural selection, organisms will become adapted to their environments in the broadest sense of being well suited to the essential processes of life such as obtaining food, avoiding predation, finding mates, competing with rivals for limited resources and so on.

We can conceive of the essential elements of this process as a sort of 'Darwinian wheel of life' (Figure 3.1) where organisms (or rather their genes) are propelled around a relentless cycle of reproduction, variation and differential survival.

One of the crucial points to appreciate about Darwinian thinking is that evolution is not goal directed. Organisms are not getting better in any absolute sense; there is no end towards which organisms aspire. Creatures exist because their ancestors left copies (albeit imperfect ones) of themselves. One of the great triumphs of Darwinism was that it offered an

explanation of the structure and behaviour of living things without recourse to any sense of purpose or teleology. **Teleology** (Greek *telos*, end) is the doctrine that things happen for a purpose or are designed with some express end. It was a cast of mind widespread in the early 19th century that natural phenomena and living things were designed with some motive or purpose in mind. The teleological way of thinking was systematised and popularised by Archdeacon William Paley. In his classic *Natural Theology* (1802) – a work Darwin knew well and once even admired – he advanced the famous watchmaker analogy. Just as a watch could not arise by chance but immediately suggests a designer and maker, so too the intricate organisation of living things suggests a cosmic designer. The whole of creation could then be seen as God's 'Book of Works', a book, moreover, that provided ample evidence for a benign Creator. It followed that creatures appeared to be fitted to their mode of life because they were designed that way by the supreme artificer we call God.

Purpose, teleology and the whole concept of providential design were all to be swept away by what Daniel Dennett called Darwin's 'dangerous idea' (Dennett, 1995). For Darwin, there was no grand plan, no evidence that life forms were placed on Earth by a Creator, no ultimate purpose or inevitable progress towards some goal. The watchmaker is blind.

3.1.1 The ghosts of Lamarckism

One person to suggest the possibility of the transmutation of species before Darwin was the French thinker Jean Baptiste Lamarck (1744–1829). His views are now virtually totally discredited, but they once served as the only serious alternative to Darwinism as a way of explaining the adaptive nature of evolutionary change. Lamarck argued that the characteristics (**phenotype**) acquired by an individual in its lifetime could be passed on to subsequent generations through the germ line (**genotype**). This view is usually parodied by the suggestion that the large muscles of a blacksmith, acquired through use during his lifetime, will result in slightly larger than average muscles in his son (and presumably daughter). Darwin is known to have mostly rejected **Lamarckism**, but what he was really rejecting was the

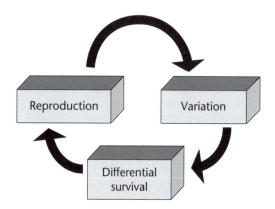

FIGURE 3.1 A Darwinian wheel of life. The birth and death of thousands of individual organisms results in gradual modification of their genes through differential survival. The result is the gradual formation of new species, and, crucially for Darwinian psychology, the emergence of adaptations.

additional notion within Lamarckism that creatures possessed an inherent tendency to strive towards greater complexity. It was this teleological idea of purpose, rather than Lamarck's mechanism of adaptation, that Darwin wisely attacked. Indeed, it comes as a surprise and shock to many to learn that Darwin accepted the possibility of 'the effects of use and disuse' as a mechanism for influencing the characteristics of the next generation.

The ghost of Lamarck was virtually (but not quite) laid to rest by August Weismann (1839–1914). Weismann distinguished between the germ line of a creature and its body, or 'soma'. Characteristics acquired by an individual affected the somatic cells (all the cells of the body other than the sperm or eggs) but not the germ line (information in the sperm or eggs). Weismann's essential insight was that information could flow along the direction of the germ line and from the germ line to the somatic cells, but not from somatic cells to the germ line (Figure 3.2).

In establishing his case, Weismann (largely between 1875 and 1880) cut off the tails of mice over a number of generations and showed that there was no evidence that this mutilation was inherited. This is in fact a rather unfair experiment – to both Lamarck and the mice – since, in strict Lamarckian terms, the mice were not striving in any way to reduce their tail length. It turns out that Weismann performed the experiment as a counterblast to those who argued at the time that it was well known that if a dog's tail was docked, its pups often lacked tails (Maynard Smith, 1982). For an equally forceful source of evidence, consider the fact that many Jewish male babies are still circumcised despite the fact that they are the products of a long line of circumcised male ancestors.

In recent years Lamarckian notions have enjoyed something of a revival with discoveries concerning epigenetic inheritance mechanisms. These complex and potentially transformative developments are considered in Chapter 7, but briefly it now seems that the experiences of an individual in their lifetime can influence the silencing or expression of specific regions of **DNA** which is inherited by descendants. So that although the basic information on the genotype is not affected by the environment, the expression of this encoded information on DNA can be altered in future generations. Inheritance can be understood in terms of the flow of information within individuals and between generations carried by macromolecules; this is shown in Figure 3.3. The diagram also shows the potential for epigenetic modification of the expression of DNA considered more fully in Chapter 7.

3.1.2 Darwin's difficulties

There were four notable areas in which Darwin's theory faced serious problems:

1. the mechanism of inheritance
2. the means by which novelty might be introduced into the germ line
3. the existence of altruistic behaviour
4. the fact that some features of animals (for example, the tail of a peacock) ostensibly seem to place the animal at a disadvantage in the struggle for existence.

Concerning the first two problems, Darwin knew nothing of course of the schema shown in Figure 3.3. His own theory of inheritance (sometimes called the theory of pangenesis involving 'gemmules' circulating around the body) is so wide of the mark that it is not worth considering. With respect to the last two problems, Darwin made more headway, but a thorough examination of these problems will be covered in Chapters 11 and 4 respectively. Modern genetics provides an answer to the problem of how spontaneous novelty arises and allows natural selection to take effect. Much of

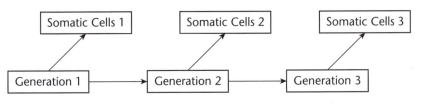

FIGURE 3.2 The germ line and information flow, according to Weismann.

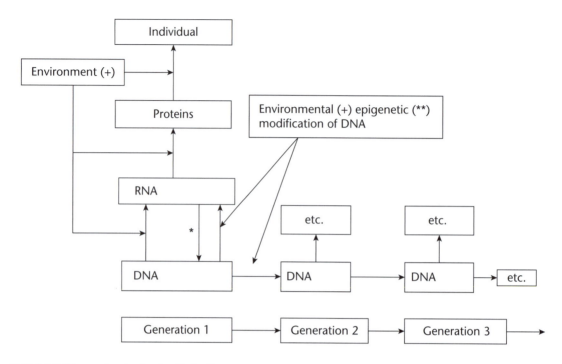

FIGURE 3.3 The molecular basis of inheritance.

* The reverse flow between DNA (Deoxyribonucleic acid) and RNA (Ribonucleic acid) is to account for the fact that some viruses carry RNA that codes for an enzyme (reverse transcriptase) that can copy base sequences of its RNA onto the DNA of the host.

(+) Note that here the 'environment' can be considered as the cellular environment of DNA and the environment in which the organism lives.

(**) Epigenetic modification refers to the discovery that certain environments can imprint (that is, mask or switch off) regions of DNA and, crucially, this modification can be inherited. It is discussed further in Chapter 7.

the apparent novelty that arises in any one generation comes from the shuffling of genes during **meiosis** and sexual reproduction, as discussed above. Fundamental changes must, however, be caused by changes in the **base** sequence of the DNA. We now know that chemical and physical agents such as high-energy radiation can have a mutagenic effect on DNA and cause alterations in its structure. Mutagenesis can also occur spontaneously by errors in replication. It seems likely that sexual reproduction may have begun as a way of reducing the number of these spontaneous errors. One important class of **mutations** is single nucleotide polymorphisms (Box 3.1)

3.1.3 Genes, behaviour and heritability

Our current knowledge of the molecular basis of inheritance, gene expression and the inheritance of genetic disorders is considerable, but it is important to point out that a simple one-to-one correspondence between genes and behaviour is hard to find. Part of the reason for this is that most behavioural characteristics are polygenic; that is, they are the result of the expression of many genes rather than just one.

It should be pointed out, therefore, that when we say 'gene for', we are not implying a simple one-to-one correspondence. The trait in question, be it morphological or behavioural, could be the result of many genes interacting with the environment. It would perhaps be better to say that the behaviour in question has some genetic basis and that genetic similarities exist between individuals displaying this trait.

So there are no 'genes for' behaviour since genes simply prescribe specific polypeptides, but given that behaviour is under the control of a nervous system, and that the nervous system is built following instructions from the information to be found (ultimately) on DNA, it would be surprising if there were to be found no genetic influences on even complex aspects of human behaviour.

Box 3.1 Mutations and single nucleotide polymorphisms (SNPs)

An SNP is where a single base pair change has occurred (compared to a reference sample) in any region of DNA. SNPs are most common in non-coding regions, partly because these regions contain more DNA and partly because they are non-coding and fitness neutral and so natural selection is unable to weed them out. They occur in the human **genome** about once every few hundred base pairs. They arise either through mistakes in DNA replication or as a result of mutations caused by ionising radiation or mutagenic chemicals. By knowing when two species diverged and comparing base sequences on similar non-functional areas of DNA an estimation of the rate of mutation can be made. By comparing human DNA with equivalent sequences in chimpanzees, for example, the rate of base substitution is estimated at between 1.3×10^{-8} and 3.4×10^{-8} per site per generation. Since, as noted earlier, the human genome contains about 6.3 billion base pairs we can expect between 82 and 214 mutations per generation, with one or two of these occurring in protein-coding regions (Nachman and Crowell, 2000). If SNPs occur in a coding region they fall into two groups; synonymous or non-synonymous. Synonymous SNPs do not affect the **amino acid** sequence of the protein (since there are often several different base triads of **codons** for the same amino acid); but non-synonymous SNPs can affect protein structure through changes to amino acids or more dramatically by affecting the regulation of gene expression. Three examples of SNPs are lactase persistence (and hence lactose tolerance) cystic fibrosis and sickle-cell anaemia (see Chapters 20 and 18).

If we pick any two same-sex individuals at random from across the entire human population they will be about 99.9 per cent identical in respect of the pattern of their base pairs. One convention is that when a mutant form of a gene reaches a frequency of greater than 1 per cent in the population it deserves the name **allele** (a shortened form of allelomorph, or 'other form'). The more general term is polymorphism, which is a term for genetic variants at any frequency in the population.

Studies on identical twins are often useful in identifying a potential genetic basis to behaviour. Dizygotic or fraternal (non-identical) twins result from the separate fertilisation of two eggs by two sperm, but identical twins are monozygotic; that is, a single female cell is fertilised by a single sperm, but as the zygote grows, for reasons that are unclear, it separates into two. An inspection of the process of meiosis reveals that, on average, fraternal twins share half their genome, whereas truly zygotic twins have their entire genome in common. It is now possible to use DNA fingerprinting to establish zygosity with some confidence. Research has shown that even though environmental influences have an effect on the behaviour of these people, zygotic twins are more alike both physically and behaviourally than fraternal twins.

Moreover, zygotic twins, even when separated and raised in different environments, reveal bizarre similarities (Plomin, 1990).

A crucial concept in this context is that of **heritability** (see Box 3.2). Heritability estimates refer to the variation between individuals that can be accounted for by differences in the genome. For the evolutionary psychologist looking for species-typical behaviour patterns, it is features that have low heritability that are interesting since these point to similarities in anatomy and behaviour that are common to humans and that may be explained in adaptive terms. This is because, through time, natural selection will tend to favour advantageous genes, and a population will become increasingly homogeneous. In other words, the genetic variation will be used up. The key point here is that heritability is not a property of a gene or a trait

Box 3.2 Heritability

To understand heritability in its technical definition we need to apply the concept of variance. Variance is a measure of the 'spreadoutness' of the data or more formally a measure of the average spread of the difference between a measured value and the mean value of that population. If a phenotypic character (P) has a variance of Vp then this can be thought of as made up of two contributions: the variance due to genetic factors (Vg) and the variance due to environmental factors (Ve). If we take height as an example, the variation in height between you and your peer group will depend upon differences caused by the genes you inherited from your parents, and from the fact that you were all raised in different environments (some fed well some not so well fed). Even identical twins (with zero genetic variance) will reach different heights if raised in different environments. So we can write:

$$Vp = Vg + Ve$$

The heritability of a trait is the proportion of the variance that is genetic. It is written as:

$$h^2 = Vg/(Vg + Ve)$$

If the value of h^2 varies across different environments, we may want to question the very usefulness of the concept itself. Visscher et al. (2008) make a spirited defence of the value of heritability. They argue that the heritability of diseases is important to study in order to gauge the relative influence of genetic and environmental factors. Artificial breeding programmes also rely on reliable heritability estimates. For traits that have a high heritability and are easy to measure, such as height and weight, it is easy to see that the best way to predict outcomes is to look at the phenotypes of the breeding pair. A high heritability implies a strong correlation between phenotype and genotype.

We should also note that understanding how heritability changes with environment is an important component of understanding evolutionary change. If phenotypes, for example, are strongly decoupled from genotypes due to high environmental variance then the force of natural selection acting on genes is correspondingly diminished. Estimates of heritability also throw up interesting scientific questions. The heritability of body size is a case in point since it is surprisingly similar across a whole range of species.

A measure of heritability (even a high measure) does not tell us if a trait in an individual is primarily caused by genes or not. The development of fingers on your hand, for example, is under direct genetic control but the number of fingers that you have has a low heritability. Most people have five but a few have six due to rare genetic effects and many more have fewer than five due to accidents (that is, an effect of the environment). Hence the contribution of environment is very large compared to genetic influences and so h^2 is very low. Applying this to evolutionary psychology, in very simple terms we might say that in dealing with universal and species-typical or sex-typical traits we are dealing with low heritabilities: all humans have a similar genetic determination underlying the trait and the environment is largely responsible for differences. If we are looking, however, at individual differences, as in the case of, for example, genetic diseases, or variation in skin tone around the globe, then we are looking at high heritabilities.

It is also important to realise that although evolutionary psychology posits a universal human nature (with sex-related variants) and a common genetic architecture underlying universal adaptations, such adaptations can exhibit differences that are heritable. Running, for example, is a universal human adaptation yet people do run at different speeds and with varying degrees of skill and efficiency.

Given that the value of heritability varies according to the environment, then generally low-stress and favourable environments tend to increase heritability. This is because good environments allow the flourishing of genotypes to their full potential and so differences that remain in the population are due to genetic factors. Figure 3.4 illustrates this point with references to the heritability of the height of humans and the weight of salmon reared in different environments.

On the vexed subject of IQ we see the same effect. Siv Fischbein (1980), for example, studied a Swedish sample of 87 monozygotic (MZ) and 126 dizygotic (DZ) siblings from a sample of 12-year-old school children. She divided the parents into two groups: high socioeconomic status (SES) and low SES. Heritabilities were higher in the high SES group compared to the low SES group. It would be hasty, however, to jump to the conclusion that the differences between the two SES groups were solely environmental in origin. It is possible that SES is itself influenced by genetic factors (see Turkheimer et al., 2003).

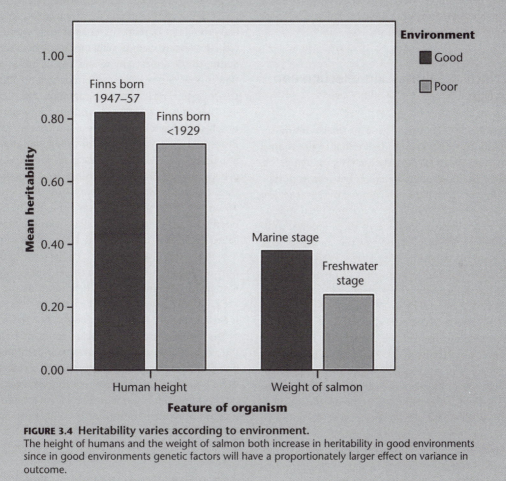

FIGURE 3.4 Heritability varies according to environment.
The height of humans and the weight of salmon both increase in heritability in good environments since in good environments genetic factors will have a proportionately larger effect on variance in outcome.
SOURCE: Salmon data from Visscher et al. (2008); human data from Silventoinen (2003).

but instead reflects the distribution of alleles in a population and the state of the environment at any time. In highly genetically homogeneous populations, such as may be the case for many psychological adaptations common to humans, phenotypic variation will largely be caused by environmental influences, so heritability will be low. Suppose, however, that the environmental influences to which people are exposed become more similar. Differences will then largely be due to genetic differences, so heritability will increase. The interpretation of heritability estimates must be carried out with great caution (see Bailey, 1998). One area where high heritabilities do underlie differences in human behaviour, however, concerns physical disease and mental disorders; this topic is explored in more detailed in Chapters 18 and 19.

3.2 Inclusive fitness, kin selection and altruism

To ensure its survival, a gene normally impresses itself on its vehicle in ways that enhance the chance of the vehicle, and ultimately itself, reproducing. It is because of this strong association between genotype and phenotype that the behaviour of the *Hymenoptera* (the group that includes ants, wasps and bees) appears extremely puzzling. In some of these species, individuals care for the offspring of the queen, defend and clean the colony and, in short, devote their lives to the survival and reproduction of other individuals in the colony while they themselves remain sterile. For this and other reasons, Darwin declared that the insects posed a 'special difficulty, which at first appeared to me insuperable, and actually fatal to my whole theory' (Darwin, 1859b, p. 236). In terms of the language of vehicles and replicators, the problem becomes how to explain the existence of a replicator that instructs an individual to behave altruistically if the individual that carries it does not reproduce to pass it on. In a world where the 'fittest survive', why do some individuals forego their own reproductive interests in favour of others? How could neuter wasps and bees have evolved that leave no offspring but instead slave devotedly to raise the offspring of their queens? Why does the honeybee die when it stings?

It is tempting to respond to this difficulty by reference to what Cronin (1991) has called 'greater

goodism' – the idea that individuals will serve the greater good at a cost to themselves. The **altruism** of insects described above could thereby be explained by suggesting that such behaviour serves the interests of the hive, the group or even the species. This line of argument is often associated with V. C. Wynne-Edwards (1962), who argued that the dispersal of animals in relation to food supply is such that the final population density reached is optimal for the group. In this view, groups possess some mechanism whereby the selfish inclinations of individuals to overgraze are restrained in favour of the longer-term interests of the group. Wynne-Edwards' book, *Animal Dispersion in Relation to Social Behaviour*, had the major effect not of converting biologists but of rousing several prominent Darwinians, chief among whom were George Walden, George Williams and John Maynard Smith, to the attack. The whole idea of **group selection** is now rejected by most biologists even though the concept is revived periodically (Box 3.3).

The answer to the conundrum of apparently unselfish behaviour lies in examining the fate of genes that incline an organism to help another one also possessing these genes. The great statistician and geneticist R. A. Fisher (1890–1962) had touched on the idea in his classic book of 1930: *The Genetical Theory of Natural Selection*. It was also outlined (albeit briefly) by J. B. S. Haldane (1892–1964) in an article called *Population Genetics* in a Penguin book titled *New Biology* published in 1955. He gave the example of two adults watching one of their children drown. One has a gene that inclines it to jump in and save the child, the other has not. If we suppose that there is a risk to life of 1 in 10 for the adult, then in the case of the adults possessing altruistic genes, as Haldane put it, 'five such genes will be saved in children for one lost in an adult' (Haldane, 1955, p. 44) since there is a 50 per cent chance that the same gene will be present in the child). But such ideas were put into a rigorous mathematical form by William Hamilton (1936–2000).

3.2.1 Hamilton's rule and inclusive fitness

In 1964, Hamilton laid down the conditions required for gene coding for social or altruistic actions to spread. This theory is also known as the inclusive

Box 3.3 The group selection controversy

Following Williams' withering critique of group selection (Williams, 1966) most biologists have been sceptical about the explanatory power of group-level thinking. Yet it is an idea that refuses to lie down and in recent years has made something of a comeback.

It is worth noting that Darwin himself struggled with the apparent tension between individual and group selection in his discussion of the evolution of the moral virtues. The problem is illustrated by the brave and courageous man who fights to save the group at the cost of his own life; he has increased the fitness of the group but how can he leave offspring that also will be inclined to sacrifice their own interests for the good of the group?

The debate was re-energised recently by a remarkable paper from Nowak, Tarmita and Wilson (2010) where Wilson at least seemed to recant his previous conversion to inclusive fitness models and propose a group-level type of explanation for the emergence of eusociality in insects. The result was that over the year following publication there appeared a flurry of articles in the journal *Nature* attacking the authors' group selectionist views (see especially Abbot et al., 2011).

Although Wilson's initial revival of group selection was aimed at explaining insect societies, a year later the same ideas appeared in a book, *The Social Conquest of the Earth*, where they were also applied to human behaviour (Wilson, 2012). The ensuing commentary has been heated. Richard Dawkins observed of the book that 'one is obliged to wade through many pages of erroneous and downright perverse misunderstandings of evolutionary theory' (Dawkins, 2012, p. 66). Meanwhile, numerous authors have reasserted the power of **kin selection** theory – something that Wilson implied was of limited application (see Bourke, 2011).

Group selection is positively protean in its ability to periodically reinvent itself. Often, however, what are proposed as new group selection models are simply other ways of conceptualising the more basic and mathematically robust processes of kin selection and inclusive fitness maximisation taking place within the group. Limited dispersal, for example (see Chapter 11) may result in the appearance of individuals assisting the group whereas in fact they may simply be following the implications of the high r values (see Section 3.2.2) between each other (West et al., 2011). One of the very few situations where group selection remains a possibility is where selection within a group (that is, between group members) has been suspended and there is only competition between groups. This could arise, for example, if the reproduction of each member of the group is suppressed to be the same, or if all individuals in the group were identical clones.

---

fitness theory. The mathematics of Hamilton's original papers is extremely complex. It was West-Eberhard (1975) who showed how Hamilton's rule could be simplified, and it is the simplified form that is now commonly encountered.

Consider two individuals X and Y, which are related in some way, and that X helps Y. An altruistic act can be defined as one that increases the reproductive success of the beneficiary (Y) at the expense of the donor (X).

Let b = benefit to recipient

c = cost to donor

r = the **coefficient of relatedness** of the recipient to the donor. This is the same as the probability that the gene for helpful behaviour is found in both the recipient and the donor.

The condition for assistance to be given and thus for the gene to spread is: 'help if $rb - c > 0$', which is the same as $rb > c$. Figure 3.5 shows an example of this. It should be clear that although the reproductive success of the helping gene in X is reduced, this is more than compensated for by the potential increased success of the gene appearing in the offspring of Y.

Hamilton's work forces us to redefine what we mean by fitness. Following Darwin, **fitness** was interpreted as something that could be measured by the number of direct offspring left by an organism. Since Hamilton showed that an organism can also have **indirect fitness** (by assisting another organism to reproduce that also carries the 'assisting tendency' genes), we now have the new notion of **inclusive fitness** – defined as the number of direct plus 'r' times the number of indirect offspring. Figure 3.6 illustrates this idea.

3.2.2 **Coefficient of relatedness (r)**

The situation presented in Figure 3.5 is highly simplified. Figure 3.7 shows a more detailed example of two **diploid** individuals I and J, with alleles aA and mM respectively, that mate sexually to produce four types of offspring: am, aM, Am and AM.

The coefficient of relatedness can be understood in a number of ways. The important thing to remember is that we are concerned with how closely individuals are related to each other rather than how similar they are. There are three ways of conceptualising this:

1. r is a measure of the probability over and above the average probability (which is determined by the gene's average population frequency) that a gene in one individual is shared by another.
2. r is the probability that a random gene selected from I is identical by descent to one present in J.
3. r is the proportion of genes present in one individual that are identical by descent to those present in another.

The terms 'identical by descent' and 'probability above average' are there to take into account the fact that a gene may be present in two individuals because it is common in the population (that is, it has a high average probability) or because they are siblings and the gene came from the father or mother; it is the latter that we are interested in in calculations of r.

Figure 3.7 shows how two unrelated parents produce siblings with an average value of $r = 0.5$. It follows that it may pay siblings to help each other if the gene for helping thereby increases in frequency.

Relatedness tends to be high between family members but this is not the only way for two organisms to have a high r value linking them. If individuals in a group breed amongst themselves, with only limited dispersal, over time the r value between any two group members rises. West et al. (2011) point out that a group of 100 individuals where only 1 per cent disperse before breeding will result in a population where the average r value between any two individuals will be about one-third. Obviously, then, the standard r values between two relatives will be even higher than this.

Table 3.1 shows the r values associated with diploidy.

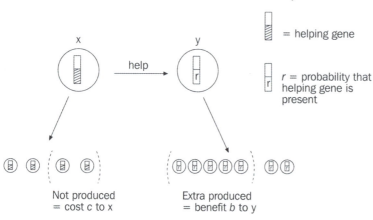

Hamilton's rule is: help if $rb - c > 0$ or $rb > c$. Let $b = 5$, $r = 0.5$, $c = 2$.
If X loses two offspring (produced in this case by asexual reproduction), then since $5 \times 0.5 > 2$, it 'pays' the gene in X to give assistance to Y.

FIGURE 3.5 Conditions for the spread of a helping gene.
Notice that the focus is on the helping gene and not just any gene shared in common since it is precisely the spread of this helping gene that has to be explained.

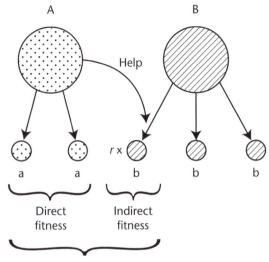

Inclusive fitness = 2a + rb

FIGURE 3.6 Inclusive fitness is the sum of direct and indirect fitness.
Organisms A and B are related by a coefficient of genetic relatedness of magnitude *r*. Indirect fitness is indicated by the number of additional offspring produced by a relative due to help given multiplied by the coefficient of genetic relatedness (*r*). Hence the inclusive fitness of A (as a result of helping B) is *2a + rb*.

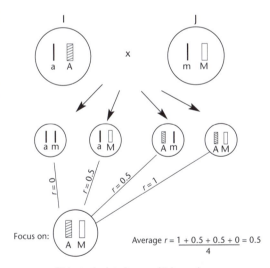

Average $r = \dfrac{1 + 0.5 + 0.5 + 0}{4} = 0.5$

r = coefficient of relatedness, which can be thought of as the probability that a gene sampled at random in one individual is also present in another. In this case, *r* between siblings is 0.5.

FIGURE 3.7 Coefficient of relatedness between siblings.

Table 3.1 Coefficients of relatedness *r* between kin pairs in humans

Relationship	*r* value
Parent and offspring	0.5
Full siblings (i.e. same parents)	0.5
Half siblings	0.25
Identical twins	1.0
Grandparent and grandchild	0.25
Two cousins	0.125

Gene types and gene variants

The phrase 'genes in common', so often used in discussions of the coefficient of genetic relatedness, is a potential source of confusion. Facts about DNA and genes have so permeated public consciousness that many people appreciate that we share about 98 per cent of our genes with chimpanzees. Students who have studied genetics will also confidently state that we share 50 per cent of our genes with a brother or sister (that is, $r = 0.5$). Comparing the two figures suggests something must have gone wrong, or at best we need more clarification. The answer is that we share 98 per cent of the same *kind* of genes with a chimpanzee (genes for various types of enzymes and other proteins), but using this measure we share 100 per cent of genes with our brothers or sisters since we are of the same species (very slightly less if we compare a brother and sister since they are different sexes). The 50 per cent figure that we do share with a brother or sister refers to genes that are identical (that is, the same base pair patterns) by descent. It is this latter figure that is used in Hamilton's equations.

3.2.3 Application of Hamilton's rule and kin selection

We are now, post Hamilton, in a position to answer Darwin's problem about the **eusocial** insects and their extreme form of self-sacrifice. The fact that worker bees have barbed stings that, when used to attack predators, remain embedded in the victim, causing the death of the worker, is understandable when we realise that other members of the nest are close relatives. This extreme form of self-sacrifice is maintained as a behavioural and morphological trait in the face of natural selection because the survival of the genes responsible is ensured by their presence in the rest of the colony.

Although human altruism is examined more thoroughly in Chapter 11, it is worth noting here that humans are more inclined to behave altruistically towards kin than unrelated individuals. Human kin make powerful claims on our emotional life. It is significant that movements that extol the 'brotherhood of man' or group solidarity, such as the world's great religions or organised labour, often resort to kin-laden language. A trade union leader, for example, may typically use the term 'brothers and sisters' when addressing fellow members.

3.2.4 Kin recognition and discrimination

For kin selection to operate, it requires animals to be able to recognise or at least discriminate between kin and non-kin. The evidence for **kin recognition** or 'reading *r* values' is necessarily indirect since it is an internal process, but if we observe animals treating kin differently from non-kin, this kin discrimination could be used as evidence for kin recognition. In a meta-analysis of a whole range of studies on helping among vertebrates, Griffin and West (2003) showed that kin discrimination is an active process in many animal groups driving the level of help given to co-specifics.

There are probably two basic reasons why it is in the interests of an animal to recognise kin. First, kin selection requires that acts of altruism be directed according to Hamilton's formula $rb > c$. This requires some assessment of costs, benefits and *r* values. Second, it is important for sexually reproducing organisms not to mate with close relatives, otherwise deleterious gene combinations may result. It follows that evidence for kin discrimination does not automatically imply the existence of an altruistic gene. We will outline some kin recognition mechanisms and evidence in their favour and then return to this second and important point.

There are probably at least four mechanisms that are available to animal species: location, familiarity, phenotype matching and recognition alleles ('green beards'). In the case of location, for animals living in family-based groups, in, for example, burrows or groups of nests, there is a good chance that neighbours will be kin. In these circumstances, a simple mechanism such as 'treat anyone at home as kin' may suffice. Cuckoos of course exploit this.

It is also important, however, for many animals to avoid inbreeding since close relatives may be **homozygous** for deleterious recessive alleles. It is estimated that each human probably carries between three and five lethal recessive alleles. Mating with close relatives increases the likelihood that the **chromosome** will be homozygous at the loci for these recessive and defective alleles. This could be the genetic basis for the strong taboos in incest found in numerous human societies.

Olfactory cues have some significance in human mating. Work by Claus Wedekind and his colleagues in Switzerland has shown that human females actually prefer the smell of males who are different from them in terms of the major histocompatibility complex (MHC) region of their genome. This region is deeply involved in self-recognition and the immune response. Differences in this region between individuals can be tested by measuring the antigens produced in their body fluids. It is important for a female to choose a mate who will differ in the MHC region since this provides a cue for genetic relatedness: close relatives will be similar in this region. Moreover, differences in the MHC between a woman and her partner may allow females to produce offspring with a more flexible response to parasites. Evidence has been provided that male odours are related to an individual's MHC. In controlled conditions, women were more likely to find the odours produced by males pleasant if they differed in their MHC. Significantly, the effect was reversed if the women subjects were taking oral contraceptives (Wedekind et al., 1995). The avoidance of inbreeding is discussed in depth in Chapter 17.

3.3 Levels of altruism

3.3.1 Selfish genes and compassionate vehicles

So far, we have implicitly adopted a working definition of altruism as 'an act that enhances the reproductive fitness of the recipient at some expense to that of the donor' but we did not specify whether the donor was a gene or a vehicle. If we consider our definition at the level of the gene, we are looking for a gene that helps another non-identical gene at some cost to itself. In kin selection or kin altruism, the gene responsible is merely increasing its own multiplication rate; it is not assisting another gene. The fact that the gene is present in another vehicle is irrelevant. It becomes clear that there can be no truly altruistic genes. A gene that helped others at expense to itself would quickly become extinct when confronted with a variant that simply helped itself. Following this line of thought, it becomes clear that kin altruism is true altruism only at the level of vehicles. The genes responsible may be increasing their propagation in another vehicle, but, by definition, the host vehicle (since we are at this level) is making a sacrifice in favour of another host vehicle.

So despite the reductive approach employed in much of this chapter we can affirm that altruism does exist. Genes, of course, are not 'selfish' in a conscious or pejorative sense: the term is used merely as a metaphor to succinctly capture their apparent behaviour. But genes, driven only by the remorseless logic of survival, competition and multiplication rates, can direct the growth of vehicles such as human bodies with a complex emotional life, displaying empathy and sympathy for others. What started as a system for ensuring gene survival by recognising and helping relatives, has led to human subjects capable of compassion for others whether they are related or not.

Another insight offered by inclusive fitness theory is the role played by replicators in the 'major transitions in evolution' (Maynard Smith and Szathmary, 1995). In very broad and simple terms it is likely that life on Earth (in terms of complexity and sociality) has gone through the following major transitions:

● Freely floating replicators (for example, strands of RNA).
● Replication in cells (for example, DNA in Procaryotes).
● Replication in more complex cells (Eucaryotes).
● Reproduction by cloning then a transition to sexual reproduction.
● Reproduction in individual multicellular organisms (plants, animals, fungi).

● Reproduction in colonies and groups of eusocial organisms (for example, the ants).
● Reproduction of ideas by culture – language and memes.

Viewed in the light of inclusive fitness, we can see multicellular organisms as groups of cells with identical DNA clustered together but united by common genetic self-interest. They behave like a eusocial society with sterile castes: all the cells have a common interest in reproduction and the production of **gametes** even though most cell types (for example, liver, kidney, brain) do not reproduce at all. Instead they serve to enable to gonads to reproduce, secure in the knowledge (metaphorically) that their own genes are also reproducing by this means.

3.3.2 The stupidity of genes

At this point it is also worth pointing out that genes as well as being metaphorically selfish are also metaphorically 'stupid'. Since sibling altruism is strongly favoured by natural selection the relevant 'helping genes' will spread to fixation. This means of course that these altruism-directing genes will become the norm in the gene pool – all organisms of the species will have them. This is where the gene 'stupidity' resides: we still have systems that incline us to treat our kin more favourably than non-kin even though the very genes that benefited from such actions are now possessed by all. If we did want our altruism to be based on hyper-rationality of gene frequencies then it might make more sense, once our own offspring's basic requirements are met, to give more gifts and help to suffering strangers of reproductive age than our own children, since copies of the loving genes inside us will do much better by the act of helping strangers who otherwise might not reproduce. From a gene-centred point of view, the persistence of biasing our loving care to close relatives is now irrational; from a human-centred point of view, that's the way we are.

Darwin realised that natural selection alone could not explain the forms and behaviour of plants and animals. This process was complemented by another potent selective force which Darwin called sexual selection, and this is the subject of the next chapter.

Summary

- The theory of evolution provides a naturalistic account of the variety, forms and behaviour of living organisms. In constructing his theory, Darwin jettisoned the idea of purpose in nature (teleology) and the notion that organisms conform to some abstract and pre-existing blueprint or archetype.

- Evolution occurs through the differential reproductive success of genes. To understand the mechanism of natural selection, we need to distinguish between genotype and phenotype. The genotype consists of the genes that carry the information needed to build organisms. The phenotype is the result of the interaction between genes and the environment within an individual. Whereas environmental factors may strongly influence the way in which genes are expressed within an organism, the outcome of this interaction was traditionally not thought to be communicated to the genotype. The new science of epigenetics suggests a modification to this view (further explored in Chapter 7).

- Darwin was unable satisfactorily to explain the mechanism of inheritance and how novel and spontaneous differences between offspring and parents (which form the raw material for evolution) could arise. Modern genetic theory supplies answers to these problems.

- The existence of altruistic behaviour posed a problem for Darwin that he did not satisfactorily resolve.

- Heritability is a measure of the phenotypic variation between individuals that can be ascribed to differences at the genetic level. Heritability is not a fixed property or value: heritability is higher in good environments where environmental constraints are weak, since this allows genetic differences to become more manifest.

- Although environmental stresses and selection pressures act upon groups and individuals, it is logically more consistent to consider the gene as the fundamental unit of natural selection.

- Kin recognition is important both to ensure that altruism is effectively directed and also to avoid inbreeding.

- Altruism can be understood biologically in terms of kin selection. In this process, genes direct individuals to help other individuals that are related and thus share genes in common. Kin selection offers insights into the biological basis of altruism in humans.

Key Words

- Allele
- Altruism
- Amino acid
- Base
- Chromosome
- Codon
- Coefficient of relatedness
- Diploid
- DNA
- Eusocial
- Fitness
- Gamete
- Genome
- Genotype
- Group selection
- Heritability
- Homozygous
- Inclusive fitness

- **Indirect fitness**
- **Kin recognition**
- **Kin selection**
- **Meiosis**
- **Mutation**
- **Lamarckism**
- **Phenotype**
- **Teleology**

Further reading

Cronin, H. (1991). *The Ant and the Peacock*. Cambridge, Cambridge University Press.

Excellent historical account of the theories of kin selection (the ant) and sexual selection (the peacock). Closely argued and packed with references. A book for the serious historian of ideas.

Ridley, M. (1993). *Evolution*. Oxford, Blackwell Scientific.

A good overview of the whole theory of evolution.

Dawkins, R. (1976). *The Selfish Gene*. Oxford, Oxford University Press.

Now a classic, this is probably the best account of gene-centred thinking ever written. A more recent edition was published in 1989.

Dawkins, R. (1982). *The Extended Phenotype*. Oxford, W. H. Freeman.

Shows how the effects of genes reach outside their vehicles.

Dugatkin, L. A. (1997). *Cooperation Among Animals*. Oxford, Oxford University Press.

A thorough work, with numerous empirical examples. Dugatkin accepts a role for a model of group selection as proposed by D. S. Wilson. An excellent book for the review of altruism among non-human animals, but contains no discussion of humans.

And nothing 'gainst Time's scythe can make defence,
Save breed to brave him, when he takes thee hence.
(Shakespeare, 'Sonnet 12')

For individuals of sexually reproducing species, finding a mate is imperative. It is through mating, essentially the fusion of gametes, that genes secure their passage to the next generation; without it, the 'immortal replicators' are no longer immortal. It is hardly surprising, then, that sex is an enormously powerful driving force in the lives of animals and is attended to with a sometimes irrational and desperate urgency. At a fundamental level, sex is basically simple – a sperm meets an egg – but it is in the varied forms of behaviour leading to this event that complexity is to be found and needs to be understood. It was once thought convenient to classify sexual behaviour in terms of mating systems, and the terminology of such systems is introduced here. It will be argued, however, that a better approach is to focus on the strategies of individuals rather than the putative behaviour of whole groups. This individualistic approach will reveal that sex is as much about conflict as about cooperation, each sex employing strategies that best serve its own interests.

4.1 Describing mating behaviour: systems and strategies

There is as yet no universally accepted precise system to classify patterns of mating behaviour. This largely results from the fact that a variety of criteria can be employed to define a mating system. Commonly used criteria usually fall into two groups: mating exclusivity and pair bond characteristics. In the former, a count is made of the number of individuals of one sex with which an individual of the other sex mates. In the latter, it is the formation and duration of the social 'pair bond' formed between individuals for cooperative breeding that is described.

A more individually focused approach to the study of mating reveals as much conflict as cooperation between the sexes, and the so-called 'pair bond' for many species could be seen with equal validity as a sort of 'grudging truce'. Bearing in mind these reservations, Table 4.1 shows a simplified classificatory scheme combining both sets of criteria.

The fact that naturalists have found it difficult to devise hard and fast definitions for mating systems need not concern us too much, but the attempt to match a species with a particular mating system does face some fundamental problems.

4.1.1 Problems with the concept of mating systems

The 'systems' approach to understanding mating is not entirely satisfactory for a number of reasons:

● **The label is sex specific**. In the case of polygyny, for example, males and females are behaving differently although they are of the same species. In the case of elephant seals, some males are highly polygynous and successful ones may mate with dozens of females in a season, whereas females are monogamous and mate with only one male.

Table 4.1 Characteristics of mating systems

System	Mating exclusivity and/or pair bond character	Examples
Monogamy	Mating with one partner. Some animal behaviourists now make the distinction between social monogamy (living with one member of the other sex) and genetic monogamy (DNA evidence that the monogamous father really is the parent). Many species that were once assumed to be socially and genetically monogamous (for example, swans) have been shown to engage in extra pair copulations and are therefore not genetically monogamous.	About 90 per cent of all bird species are socially monogamous, but a much smaller percentage (possibly as low as 10 per cent) are genetically monogamous. Monogamy is rare in mammals (about 3 per cent of all species). About 19 per cent of pre-industrial human societies are monogamous.
Polygamy	One sex copulates with more than one member of the other sex. There are two types:	
Polygyny	Where a male mates with several females, females with only one male.	Most mammalian species are polygynous. Examples include lions, gorillas and elephant seals. About 80 per cent of pre-industrial human societies allow a man to behave polygynously (i.e., take more than one wife).
Polyandry	One female mates with several males, males with only one female.	Very rare in human societies, only a few cultures documented where this is the recognised system of mating and marriage, but of course women can behave in a polyandrous fashion. Found in birds (for example, the Galapagos hawk, the northern jacana), some primates (for example, marmosets), and insects (for example, field crickets – *Gryllus bimaculatus*).
Polygynandry or 'promiscuity'	No stable pair bonds, males will mate with several females and females with several males. Promiscuity is a rough shorthand for this behaviour but avoided by most biologists since it carries moralistic overtones.	Chimpanzees. Dunnock. Occasionally found amongst human groups (for example, communes); anthropologists use the term 'group marriage'.

- **Species in themselves do not behave as a single entity**; it is the behaviour of individuals that is the raw material for evolution. Natural selection can only act upon individuals, ruthlessly exterminating those that are unsuccessful and allowing those that are better adapted to pass their genes to the next generation. To understand behaviour we need to consider how that behaviour helps or retards the chances of individuals reproducing.
- **Within any species, individuals even of one sex may differ and utilise different strategies.** Where

there is a genetic basis to this, it is sometimes known as a *polymorphism* (many shaped). Hence, in the *gene pool* of a species there may be genes that predispose some individuals more towards monogamy and others towards polygyny.

Some animal behaviourists prefer to describe the social arrangement that facilitates mating in terms of the number and sex of the individuals. Table 4.2 shows how this works. Hence we would say that the mating behaviour of gorillas is uni-male, multi-female since they live in polygynous groups where

Table 4.2 Four basic mating systems. The table shows alternative formulations for the traditional mating systems approach. Hence polyandry becomes uni-female, multi-male since one female shares several males.

	Uni-male	Multi-male
Uni-female	Monogamy	Polyandry
Multi-female	Polygyny	Promiscuity/ Polygynandry

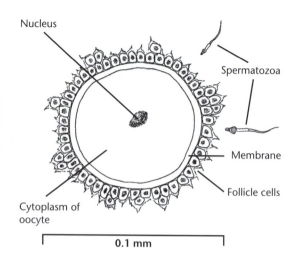

FIGURE 4.1 Relative dimensions of an egg from a human female and a sperm from a male.

one male has access to a 'harem' of females. The mating behaviour of chimps would be described as multi-male, multi-female since they live in mixed groups where both males and females may have several sexual partners.

4.2 The sex ratio: Fisher and after

4.2.1 Why so many males?

The ancestral state of life on Earth must have been that of primitive, single-celled asexual organisms. The first sexually reproducing organisms probably produced gametes that met other gametes by external fertilization (much as fish do today). These early sexual gametes were probably of equal size – a condition known as **isogamy**. Today, however, in virtually all cases of sexual reproduction the size of the gametes from males and females is vastly different. Males produce plentiful, small, highly mobile gametes, whilst female produce fewer, larger and less mobile gametes – a condition known as **anisogamy**. Figure 4.1 shows how great the discrepancy is for humans.

Parker et al. (1972) suggest one probable scenario. Their argument is essentially that an ancestral state of equally sized gametes quickly breaks down into two strategies: providers and seekers. Parker et al. were also able to show that these two strategies are stable in the sense that they can resist invasion from other strategies. Once the archetypal female hit upon the strategy of provisioning her gamete (egg) with nutrients to give it a head start after fertilisation, the archetypal males responded by producing lots of sperm (smaller gametes) in an effort to be the one to successfully fertilise this valuable prize. So, early in

the history of life males and females became locked into their separate strategies.

Some of the facts of human anisogamy, which are in many respects typical of mammals as a whole, are startling. Each ejaculate of the human male contains about 280 x 10[6] sperm; moreover, they are produced at the phenomenal rate of about 3,000 each second (Baker and Bellis, 1995). In contrast, the human female only produces about 400 eggs over her entire reproductive lifetime of 30–40 years. The longer period of fertility experienced by the male, the fact that females are incapable of ovulating when bearing a child or breastfeeding, and the heavy demands of childbearing that fall unevenly on females, all imply that a single male could, in principle and in practice, fertilise many women.

The obvious question that follows from this is why nature has bothered to produce so many men. It would seem that a species would do better in terms of increasing its number by skewing the sex ratio in favour of women, thereby producing fewer men. Men who remained would then be destined to mate polygynously with more women. Yet unfailingly, the ratio of males to females at birth for all mammals is remarkably close to 1:1. The statistics of polygynous mating seem ever more wasteful. In cases where a few males fertilise the majority of females, such as in lekking and polygynous species, given a 1:1 sex ratio at birth it follows that some males are not successful at all. In evolutionary terms, it seems as if their lives have been pointless and, for the parents that produced them, a wasted expenditure of paternal effort. It was Ronald Fisher (1890–1962) who pointed a way out of this conundrum.

4.2.2 Fisher's argument

A superficial answer to the question of why roughly even numbers of human males and females are born is that every gamete (oocyte) produced by the female contains an X chromosome but that gametes produced by the male contains either a Y or X chromosome, these two types being produced in equal numbers. Consequently, there is an equal probability of an XX and an XY fusion, and it follows that boys (XY) are just as numerous as girls (XX). This is, in fact, the mechanism used for all mammals and birds (except that in birds the females are XY and the males XX).

This is of course only part of the answer. The X/Y chromosome system provides a proximate mechanism for sex determination, but we know that this is subject to some variation. In humans, it is estimated that, three months after conception, the ratio of males to females is about 1.2:1 and that because of the higher in utero mortality of male embryos, the ratio falls to 1.06:1 at birth. It draws closer to 1:1 at age 15–20. What we are looking for of course is an ultimate evolutionary argument that explains the adaptive significance of the proximate mechanism. The argument that is now widely accepted was first provided by Fisher in his *The Genetical Theory of Natural Selection* (1930). Fisher's reasoning can be expressed verbally in terms of negative feedback. First we must rid ourselves of the species-level thinking that lies behind the view that species would be better off with fewer males. Species might be better off, but selection cannot operate on species. Selection acts on genes carried by individuals, and what might seem wasteful at a group level might be eminently sensible at an individual level.

Consider the fate of a mutant gene that appeared and caused an imbalance of the sex ratio in favour of females. This could take the form of a gene influencing the probability of fertilisation or survival of the XX zygote in a positive way. Or a gene that

influenced the number of X and Y gametes produced by the male. Let us further suppose that, for some reason, this gene gained a foothold and shifted the ratio of males to females to 1:2. Consider now the position of parents making a 'decision' (in the sense of the selection of possibilities over evolutionary time) of what sex of offspring to produce. In terms of the number of grandchildren, sons are more profitable than daughters since, in relative terms, a son will on average fertilise two females every time a daughter is fertilised once. More grandchildren will be produced down the male line than down the female line. It therefore pays to produce sons rather than daughters. In genetic terms, the arrival of a gene that now shifts the sex ratio of offspring in favour of males will flourish.

The argument also works the other way round. In a population dominated by a larger number of males, it is more productive of grandchildren to produce a female since she will almost certainly bear offspring whereas a male (given that there is already a surplus) may not.

The logic of the argument also works for polygamous mating. Suppose only one male in ten is successful and fertilises ten females. It still pays would-be parents to produce an equal number of males and females even though nine out of ten males may never produce offspring, because the one male in ten that is successful will leave many offspring and the gamble is worth it. This one successful male will have ten times the fertility of each female. In human populations where the sex ratio drifts appreciably from 1:1 there may be other, darker, factors at work (see Box 4.1).

4.3 Sexual selection

4.3.1 Natural selection and sexual selection compared

Darwin's idea of natural selection was that animals should end up with physical and behavioural characteristics that allow them to perform well in the ordinary processes of life such as competing with their rivals, finding food, avoiding predators and finding a mate. The life and death of thousands of ancestors should have ensured that by now the characteristics

Box 4.1 The missing women of Asia

Across whole areas of Asia, but also including the Middle East and North Africa, women and girls are so often neglected, abused and killed that, compared to the rest of the world, sex ratios in these populations are strongly male biased. This box examines the facts, possible causes, consequences and potential for amelioration of this problem from an evolutionary angle.

Facts

In societies without sex-biased abortion the sex ratio at birth varies quite stably between 105 and 107 (with the median at 105.9) males for every 100 females (CIA, 2010). As children grow up males tend to experience a higher mortality at virtually all ages (see Section 12.3.1) and so the ratio (number of males/number of females) across all age groups falls to 1.016. Hesketh and Xing (2006) compared sex ratios in countries where sons are preferred and girls neglected with sex ratios from countries where girls and women are treated more equitably to estimate the number of females that should be alive but are not. They concluded that in India there are somewhere between 27 and 39 million missing females, and in China this figure rises to around 34–41 million. Figure 4.2 shows sex ratios in selected countries. The male-biased sex ratios at birth could represent an underreporting of girls or selective abortion or infanticide. The change in the figure for 0–14 year olds could indicate the neglect of girls after birth or changes in sex ratios with time.

Causes

Variations in sex ratios across place and time has been a topic of extensive research over the last few decades. A variety of factors such as paternal and maternal ages, birth order, ethnicity, climate, chemical pollution and economic stress have been implicated or posited as influencing factors (see James, 1987; Rosenfeld and Roberts, 2004; Mathews and Hamilton, 2005). Diseases are also known to strongly affect the sex ratio.

Mothers suffering from Hepatitis B, for example, can produce sex ratios of about 1.5; and infection with the parasite *Toxoplasma gondii* can send the rate as high as 2.6 (Kaňková et al., 2007). But for the Asian countries described in Figure 4.2 the higher than world median sex ratios are probably a consequence of son preference and the undervaluing of girls. The proximate mechanisms are sex-biased abortion and the killing or neglect of girls following their birth. The advent of amniocentesis in the 1970s and the wider use of ultrasound scanning from the 1980s have both allowed prenatal sex testing, paving the way for sex-selective abortion (Hesketh and Xing, 2006). In their review of the problem, Hesketh and Xing identify a number of reasons why sons are preferred over daughters:

1. Sons have a higher earning capacity.
2. It is usually sons that continue the family line since on marriage women are absorbed into their husband's family.
3. Many areas of China, India and South Korea are rigidly patrilineal and patrilocal, meaning that sons inherit land and wealth and remain in or near the family home.
4. In societies where a dowry system operates girls are considered to be an economic burden since they require a dowry in order to be able to marry.
5. Where girls, after marriage, leave the family home they cease to have responsibility for looking after their parents in old age or during illness.

Evolutionary perspectives

So far proximate causes and reasons for female foeticide, infanticide or neglect have been examined. We have also noted how Fisher provides a compelling selection-based argument for the maintenance of the sex ratio at about 1:1, so are there any adaptive-style explanations for departures from this ratio? There are in fact many

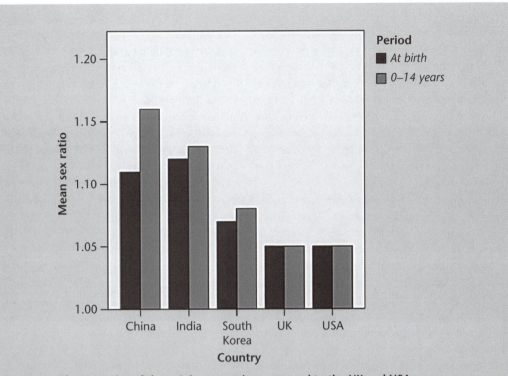

FIGURE 4.2 The sex ratios of three Asian countries compared to the UK and USA.
The natural sex ratio at birth is estimated to lie somewhere between 1.05 and 1.06. The high sex ratios in China and India at birth point to selective abortion or some other means of sex determination of live births. The rise in the sex ratio after birth suggests neglect of girls such that they experience a higher mortality than boys.

SOURCE: CIA World Factbook, https://www.cia.gov/library/publications/the-world-factbook/fields/2018. html, accessed November 2014. By permission of CIA (2013). *The World Factbook* 2013–14, Washington, DC, Central Intelligence Agency, https://www.cia.gov/library/publications/the-world-factbook/, accessed 2 February 2016.

ideas on this topic. In 1989 Maynard Smith, for example, pointed out that departures from a 1:1 ratio could be expected if sons and daughters carry unequal costs of production. If one sex costs more than another to produce the total fitness of a mother could be enhanced by producing more of the cheaper sex: true, she faces counterbalancing selection due to the more expensive sex becoming rarer but initially it is still worthwhile producing more of the cheaper sex since she can produce more offspring overall (Maynard Smith, 1989). These and similar ideas were applied to the sex ratio problem of Asia by Robert Brooks (2011) and he identified a number

of possible approaches. One of these is the 'local resource enhancement hypothesis'. A basic evolutionary premise of parenting is that parents should invest in offspring that give the greatest overall fitness returns. If the philopatric sex (that is, the one that stays in the family area) helps the parents to protect their resources and rear future offspring then this sex will be favoured and the dispersing sex produced in lower numbers or discriminated against. Patterns of female neglect are consistent with this hypothesis (since it is sons that usually remain near the natal home) but work has yet to be done actually measuring the fitness benefits of sons and daughters. We might also note that if this is an ultimate reason then it is mediated through cultural rather than biological evolution. It would seem unlikely that mothers could make such a dramatic facultative adjustment in their biology to give rise to the ratios observed; instead the cultural practice (itself possibly an evolved motivational response to local conditions) of selective foeticide and neglect brings about the same effect of a bias towards sons.

Another potentially useful set of ideas comes from the Trivers–Willard effect. Robert Trivers and Dan Willard predicted that in mammalian species where males compete aggressively to win a harem or territory to facilitate multiple mating opportunities, then mothers in good condition ought to produce sons and mothers in poor condition daughters. The logic of this is that a mother in good condition can invest considerable resources into a son as a means of enhancing his potential to secure many mates. A mother in poor condition with a low reservoir of resources to invest who produces a son faces the risk of him not competing well and not reproducing at all. But if she produces a daughter there is more likelihood that she will reproduce (Trivers and Willard, 1973). Whilst studies on many mammalian species have produced evidence supporting the Trivers–Willard effect, studies on human societies have produce mixed findings and the question remains open (see, for example, Almond and Edlund (2007) and Keller et al. (2001) for contrasting findings). But the idea could, in principle, account for some of the sex ratio bias observed in India. Female infanticide in early colonial India, for example, was largely perpetrated by those in the highest caste; but even in contemporary India there is a correlation between wealth and having a son (Brooks, 2011).

This whole effect may be enhanced by hypergyny – the practice where women are expected to marry into families wealthier and of a higher social standing than their natal families. In this situation, girls in wealthy families face extreme competition for eligible wealthy men from the numerous layers below. Girls at the bottom of the social pile, in contrast, have more prospects in marrying up into one of the many layers above; whereas sons in these poor families will find it hard to find a wife (since girls are looking to marry into wealthier families). Under such circumstances it might be expected that poor families would prefer and invest more into daughters, and rich families invest more into sons (who will have good prospects of marriage from the supply of competing women in castes below them). As noted earlier, there is some evidence that sex ratio bias is wealth related. It has been noted, for example, that landowning and urban families have more skewed male-bias sex ratios than poorer rural families (George and Dahiya, 1998). One problem here, however, is that wealthy families will also have greater access to sex-determining technologies and abortion clinics (George, 2006).

Consequences

Male-biased sex ratios can be expected to lead ultimately to a group of males frustrated by their inability to find a wife and start a family. One possible outcome is that the competition for status and resources that exists among men anyway will be intensified, leading to violence and even homicide. There is some evidence that these outcomes, together with gambling, drug abuse, the trafficking of women and the use of prostitutes are on the rise in India and China. In their study of rising crime rates in China over the period 1988–2004, Lina Edlund and colleagues concluded that a male-biased sex ratio among 16–25 year olds probably accounted for about one-sixth of the rise in property-based and violent crime during this time (Edlund et al., 2013).

Corrective measures

Female foeticide, infanticide and neglect in certain areas of the world is a tragedy stemming from a complicated toxic mixture of factors such as economic circumstance, cultural traditions and evolved parenting and mating strategies and biases. One ray of hope that these issues can be tackled comes from the example of South Korea. The availability of sex testing and abortion in South Korea meant that by 1990 the sex ratio had risen to 1.18. The government had addressed this problem in 1987 with a ban on sex-selective abortion and soon this began to take effect with physicians losing their licences for performing sex-determination tests. This was coupled with a public awareness campaign ('Love your daughters'), greater workforce participation by women, improvements in the legal status of women and the encouragement of parents to save for their old age (and so not be economically reliant on the earning capacity of sons). As a consequence, by 2007 the sex ratio had fallen to 1.07 (Hesketh et al., 2011). The success of the campaign is illustrated by the fact that over a similar period sons were seen as essential by a decreasing fraction of women (Figure 4.3).

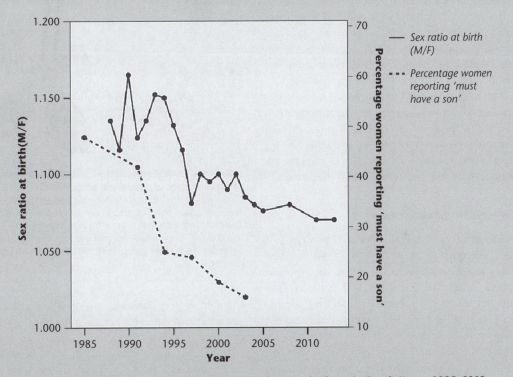

FIGURE 4.3 Declining sex ratios and expressed preference for boys in South Korea 1985–2013. Over the period 1990–2013, probably as a result of a public awareness campaign and measures to improve the status of women, the sex ratio in Korea has declined to a level similar to European countries. The results of a survey of attitudes to children carried out by the Ministry of Health and Welfare also demonstrate that the requirement 'must have a son' has become progressively less important.

SOURCE: Data from Chung and Gupta (2007), Figures 2 and 3. See also www.nationmaster.com/country-info/profiles/South-Korea/People/Sex-ratio (accessed 30 October 2014) for more recent sex ratio data.

of animals are finely tuned to growth, survival and reproduction – nature should allow no extravagance or waste. So what about, for example, the spectacular train of the peacock? It does not help a peacock fly any faster or better. Neither is it used to fight rivals or deter predators – in fact, the main predator of pea-fowl, the tiger, seems particularly adept at pulling down peacocks by their tails. Nor is the peacock's tail an exception: many species of animals are characterised by one sex (usually the male) possessing some colourful adornment that serves no apparent function (or that even seems dysfunctional) whilst the other sex, like the peahen, seems much more sensibly designed. Such features seem, at first glance, to challenge the power of natural selection to explain the behaviour of animals.

It was Darwin himself who provided the answer to this seeming paradox. In his *Descent of Man and Selection in Relation to Sex* (1871), he gave the explanation that is still accepted (with refinements) today. Darwin realised that the force of natural selection must be complemented by the force of sexual selection: individuals possess features which make them attractive to members of the opposite sex or help them compete with members of the same sex for access to mates. In essence, the train of the peacock has been shaped for the delectation of the peahen.

Inter- and intrasexual selection

We should really distinguish between two types of sexual selection: 'intra' and 'inter'. Even amongst species where polygamy is the norm, the sex ratio (males:females) usually remains close to 1:1, in other words, the number of males and females in a population is roughly the same. So where conditions favour polygyny, males must compete with other males for access to females; this follows from the obvious reason that if each male is intent on mating with two or more females to the exclusion of other males then there simply are not enough females to go round. This leads to **intrasexual selection** (intra = within). Intrasexual competition can take place prior to mating or, in the case of **sperm competition** (see below) after copulation has taken place. On the other hand, for many species a female investing heavily in offspring, or only capable of raising a few offspring in a season or lifetime, needs to make sure she has made

FIGURE 4.4 Sexual selection as found in the humming bird species *Sparthura underwoodi*.
Darwin used this illustration in his second edition of *The Descent of Man and Selection in Relation to Sex* (1874). The male is the one with the long and highly ornamented tail. Darwin realised that the tail grew like this to please the female and so help the male gain a sexual partner.

the right choice. There will probably be no shortage of males clamouring for her attention but the implications of a wrong choice for the female are graver than for the male, who may, after all, be seeking other partners anyway. A female under these conditions can afford to be choosy and pick what she thinks is the best male. This leads to **intersexual selection** (inter = between).

Figure 4.6 shows a comparison of inter- and intrasexual selection. We now need to consider how these concepts help us understand human sexual behaviour and how the type of selection that takes place can be understood in relation to the biology of our species.

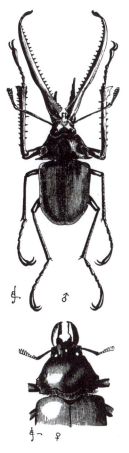

FIGURE 4.5 Sexual dimorphism in the beetle species *Chiasognathus grantii.*
The male has long protruding mandibles, which it uses to fight and intimidate other males. The female, illustrated in the lower half of the diagram, lacks these characters. Darwin concluded that although these devices were selected for fighting they seemed to be excessive in size even for this purpose and noted that 'the suspicion has crossed my mind that they may in addition serve as an ornament'.
SOURCE: Diagram from Darwin, C. (1874). *The Descent of Man and Selection in Relation to Sex.* London: John Murray.

4.3.2 Predicting the direction of inter- and intrasexual selection

An important set of ideas designed to tackle the differences in sexual behaviour between male and female animals came from work on fruit flies carried out by the British geneticist Angus Bateman (1919–96). His

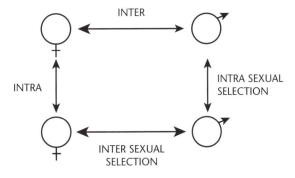

FIGURE 4.6 Inter- and intrasexual selection.

work led to a series of postulates known as Bateman's principles:

(i)   Males experience a higher variance in reproductive success than females.
(ii) Males have a higher variance in number of mates than females.
(iii) Males have a higher correlation slope (that is, regression constant) between number of mates and number of offspring than do females (that is, increasing the number of mates has a greater impact on a male's reproductive success than it does on a female's).

It may seem rash to generalise from fruit flies but these principles were found to be successful in conceptualising the behaviour of mammals as well. Indeed, it was quickly realised that they could form the basis of the theory of sexual selection and fill in some of the explanatory gaps in Darwin's account of inter- and intrasexual selection.

As a very rough guide, the degree of discrimination exercised by an individual in selecting a partner is related to the degree of commitment and investment that is made by both parties. Male black grouse that provide no paternal care will mate with anything that resembles a female black grouse. Female black grouse, however, that bear the brunt of the consequences of fertilisation in terms of egg production and incubation, are rather more careful in their choices, and select what they take to be the most suitable male from those displaying before them. Among humans, both males and females have a highly developed sense of physical beauty and this aesthetic sensibility is consistent with a high degree of both maternal and paternal investment. The more investment an individual makes the more important it becomes to choose its mate carefully – genes that direct for poor choices (for example, infertile or unhealthy individuals) are destined to be eliminated from the gene pool. All this decision-making results in the selective force of intersexual selection. It follows that if a high degree of investment by one sex sets up the force of intersexual selection, then the lack of investment by the other sets up competition within that sex for access to the other sex that invests most. These ideas were formalised by Robert Trivers into the concept of **parental investment**. Trivers (1972) defined parental investment as:

> any investment by the parent in an individual offspring that increases the offspring's chance of surviving (and hence reproductive success) at the cost of the parent's ability to invest in other offspring. (p. 139)

Using this definition, Trivers concluded that the optimum or ideal number of offspring for each parent would be different. In the case of many mammals, a low-investing male will have the potential to sire more offspring than a single female could produce. A male could, therefore, increase his reproductive success by increasing the number of his copulations. On the other hand, females should prefer quality rather than quantity and so a female is predicted to be choosy.

The logic is clear but in practice it has proved difficult to measure such terms as increase in 'offspring's chance of surviving' and 'costs to parents'. Consequently, deciding which sex invests the most is not always easy. One concept that may help to circumvent these difficulties is that of potential reproductive rates.

Potential reproductive rates: humans and other animals

Clutton-Brock and Vincent (1991) have suggested that a fruitful way of understanding the mating behaviour of animals is to focus on the potential offspring production rate of males and females, rather than trying to measure mating effort or parental investment. In this view it becomes important to identify the sex that is acting as a 'reproductive bottleneck' for the other. Applying these ideas to human

mating, it should be clear by now that there are large differences between the potential reproductive rates of men and women. Harems may have been common in ancient civilizations but there are no examples recorded of female rulers guarding a company of male studs or 'toy boys'. Biologically, what would be the point?

It is noteworthy then that the record often claimed for the largest number of children from one parent is 888 for a man and 69 for a woman. The father was Ismail the Bloodthirsty (1646–1727), an emperor of Morocco with privileged access to his harems. The mother was a Russian lady who between 1725 and 1765 experienced 27 pregnancies with a high number of twins and triplets. Most people are more astonished by the female record than the male. Both figures are difficult to substantiate accurately and have been the subject of some discussion (Einon, 1998; Gould, 2000).

We should also note that most men in history have not been emperors, and the harems enjoyed by Ismail and his like would not have been a regular feature of our evolutionary past. It is probably true to say, however, that in *Homo sapiens* the limiting factor in reproduction resides marginally with the female. This by itself would predict some male v. male competition, and both intra- and intersexual selection can be expected to have moulded the human psyche.

The operational sex ratio (OSR)

The potential reproductive rate and the operational sex ratio are closely related concepts. Although for most mammals there are roughly equal numbers of males and females, not all of the males or females may be sexually active and there may be local variations in the ratio of the sexes. This idea is contained in the concept of the **operational sex ratio**:

$$\text{Operational sex ratio} = \frac{\text{Fertilisable females}}{\text{Sexually active males}}$$

When this ratio is high, the reproductive bottleneck rests with males and females could compete with other females for the available males. When the ratio is low, the situation is reversed and males will vie with other males for the sexual favours of fewer females.

At first sight, it would seem that females are always the limiting resource for male fecundity.

Consider the following facts. If you are a young fertile male, while you are reading this you are producing sperm at the phenomenal rate of about 3,000 per second. If you are a young fertile female, you are holding onto a lifetime's supply of only 400 eggs. In addition, a man could impregnate a different woman every day for a year, whereas over this same period a woman can become pregnant only once. We need to consider this with caution, however. Imagine a male who mates with 100 different women over the course of a year and a female who mates with 100 different men over the same period. The woman is likely to become pregnant and bear one offspring in the same year. Now sperm can survive for about six days in the genital tract of the female, so if the male avoids the time of menstruation, he has about a 26 per cent chance (6 days out of 23) of impregnating a woman during her fertile period. Only about 70 per cent of the female ovarian cycles will be fertile, and implantation will only take place about 50 per cent of the time. The number of women a man could expect to make pregnant is about nine (100 x 0.26 x 0.7 x 0.5). Women are in one sense a limiting resource but not to the extremes that might be indicated by the differences in the size or rate of production of gametes.

The operational sex ratio (females/males) for a group of humans with a 1:1 sex ratio (that is, equal numbers of men and women in the population) will be less than one if we measure it in terms of males or females that are fertile. It is likely that there will be more sexually fertile males than females. This arises from the fact that men experience a longer period of fertility compared to women. It is counterbalanced somewhat by the higher mortality rates for men than women, but not entirely.

In virtually all cultures it is significant that men engage in competitive display tactics and are more likely to take risks than women are. It is also men who tend to pay for sex, this being one way of increasing the supply of the limiting resource.

A generalised model

The combination of ideas advanced by Darwin, Bateman, Trivers and Parker has been called by Tang-Martinez (2010) the 'Darwin–Bateman–Trivers–Parker paradigm'. Figure 4.7 shows how these ideas can be used to predict the outcome typically found in mammals that males compete with males and females exercise choice.

Table 4.3 provides a rough guide as to how the forces of sexual selection will arise differently in each sex. But Brown et al. (2009) caution against jumping too readily to conclusions from this line of reasoning and notes other layers of complexity. An individual of a sex in short supply and so inclined to be choosy, for example, may increase the quality of a selected mate by setting the bar for its minimum requirement high; but this comes at the expense of mating rate since there will probably be fewer higher-quality mates around. The solution to this trade-off problem will in turn depend upon searching costs, rates of encountering mates and the variation in mate quality. There is not much point, for example, in waiting for the perfect Mr Right if the variance in the quality of mates is low and the searching time is high. These considerations lead to some general predictions about the choosiness of each sex (Table 4.3). The whole phenomenon of human mate choice is further explored in Chapter 13.

Table 4.3 Factors impinging on the direction of sexual selection

Females will be choosy if:	Males will be choosy if:	Both sexes will be choosy if:	Neither sex will be particularly choosy if:
Male-biased operational sex ratio that is, M/F > 1	Female-biased operational sex ratio that is, M/F <1	High encounter rates High parental investment from both sexes	Low population density and hence low encounter rates
Low paternal investment (since this increases breeding cost to females)	High paternal investment		Balanced OSR (M/F approx. = 1)
High variance in male quality	High variation in female quality	High variation in mate quality	Little variation in mate quality

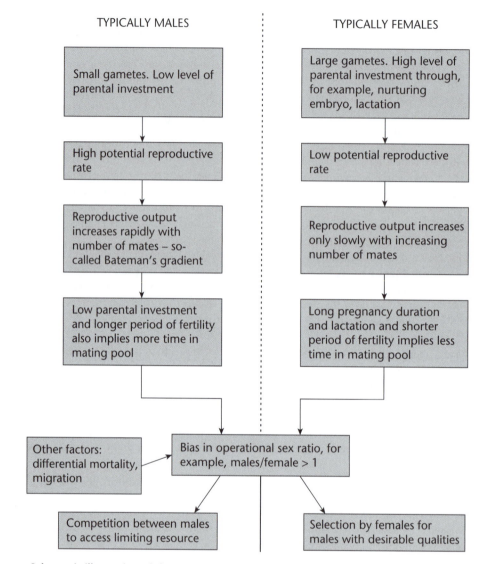

TYPICALLY MALES

Small gametes. Low level of parental investment

High potential reproductive rate

Reproductive output increases rapidly with number of mates – so-called Bateman's gradient

Low parental investment and longer period of fertility also implies more time in mating pool

TYPICALLY FEMALES

Large gametes. High level of parental investment through, for example, nurturing embryo, lactation

Low potential reproductive rate

Reproductive output increases only slowly with increasing number of mates

Long pregnancy duration and lactation and shorter period of fertility implies less time in mating pool

Other factors: differential mortality, migration

Bias in operational sex ratio, for example, males/female > 1

Competition between males to access limiting resource

Selection by females for males with desirable qualities

FIGURE 4.7 Schematic illustration of the Darwin–Bateman–Parker–Trivers paradigm of sexual selection.

4.4 Consequences of sexual selection

Just as natural selection has left humans with bodies and brains suited to the processes of finding food, avoiding predators and resisting disease, so sexual selection has left its mark on our bodies and our sexual inclinations. So much so, that once we understand the basic ideas of sexual selection we can examine humans and make predictions about what type of sexual behaviour patterns may be typical of our species.

4.4.1 Sexual dimorphism in body size

Darwin argued that intrasexual selection was bound to favour the evolution of a variety of special adaptations, such as weapons, defensive organs, sexual differences in size and shape and a whole range of devices to threaten or deter rivals. The importance of size is illustrated by a number of seal species. During the breeding season, male northern elephant seals (*Mirounga angustirostris*) rush towards each other and engage in a contest of head butting. These seals belong to a polygynous mating system. The high-value prize of a mating monopoly over a harem of females has led to a strong selection pressure in favour of size, and consequently male seals are several times larger than females.

Humans show **sexual dimorphism** in a range of traits. Men, for example, have greater upper body strength and more facial and body hair than do women. Men also have deeper voices, later sexual maturity and experience a higher risk of infant mortality than do females. The pattern of fat distribution between men and women is also different. Women tend to deposit fat more on the buttocks and hips than do men. These of course refer to average tendencies. It is likely that many of these are the results of sexual selection. The implications of this for human mating behaviour are further explored in Chapters 14 and 15.

4.4.2 Post-copulatory intrasexual competition: sperm competition

At first glance it may seem that once copulation has taken place then intrasexual competition must cease: one male must surely have won. But the natural world has more surprises in store. Some females mate with many males and retain sperm in their reproductive tracts; sperm from two or more males may then compete inside the female to fertilise her egg. Geoff Parker, Robert Nabours and Ojvind Winge have all been suggested as the first to identify and study the phenomenon of sperm competition (Magurran, 2005). But the most likely candidate is Johannes Schmidt, who studied sperm storage and competition in guppies in the Carlsberg Laboratory in Copenhagen between 1917 and 1918 (Schmidt, 1920).

The concept of sperm competition illuminates many features of male and female anatomy in non-human animals. Male insects have evolved a variety of devices aimed at neutralising or displacing sperm already present in the female. The male damselfly (*Calopteryx maculata*), for example, has evolved a penis designed to both transfer sperm and, by means of backward pointing hairs on the horn of the penis, remove any sperm already in the female from a rival male.

Many animals have also evolved other tactics to outwit rivals in sperm competition. When a male garter snake mates with a female it leaves behind a thick sticky mass, technically known as a copulatory plug, that effectively seals off the reproductive tract of the female from other would-be suitors. When a male wolf mates with a female its penis becomes so enlarged that even after ejaculation it remains stuck in the vagina of the female for up to half an hour after impregnation. The male and female remain fixed like this in some apparent discomfort, but the mechanism ensures that the sperm of the successful male gets a head start over any rivals.

We should not think of females as passive in this process of sperm competition. The female may exercise choice over the sperm once it is inside her (Wirtz, 1997). Many female insects store sperm that they use to fertilise their eggs at oviposition (egg laying) as the eggs pass down the female's reproductive tract. It has been suggested that the function of the female orgasm in humans is to assist the take up of sperm towards the cervix (Baker and Bellis, 1995). Randy Thornhill and his workers carried out a study to show that the bodily symmetry of the male is a strong predictor of whether or not a female will experience a copulatory orgasm. Symmetry is thought to be an indicator of genetic fitness and the possession of a good immune system (Thornhill and Gangestad, 1994). The orgasm therefore ensures that sperm from exciting and desirable males, who presumably are genetically fit and unlikely to be transmitting a disease, stand a good chance of meeting with the female's egg. In this way the human female may be extending her choice beyond courtship (Baker and Bellis, 1995).

The more sperm produced the greater is the chance of at least one finding the egg of the female: 50 million sperm are twice as effective as 25 million and so on. In species where sperm competition is rife, we would expect males to increase the number of sperm produced or ejaculated compared to species where sperm competition is less intense. Between species, this prediction has been supported indirectly by measurements on levels of sperm expenditure as measured by testis size. Species facing intense sperm competition have larger testes than those where sperm competition is less pronounced (see Chapter 13).

Baker and Bellis (1995) at Manchester University provide evidence to support the idea that the number of sperm in the ejaculate of men is adjusted according to the probability of sperm competition taking place. In one study, when couples spent all their time together over a given period the male was found to ejaculate about 389 x10^6 sperm during a subsequent sexual act. When the couples only spent 5 per cent of their time together men typically ejaculated 712 × 10^6 sperm. Baker and Bellis interpret this as consistent with the idea that the male increases the number of sperm in the latter case to compete more effectively against rival sperm which may have entered the female had she been unfaithful. Baker and Bellis have been successful in generating new ideas in an area of research that faces innumerable experimental and ethical difficulties.

In the 'sperm wars' of post-ejaculate intrasexual competition, males can adopt various tactics: they can produce sperm in large numbers, attempt to displace rival sperm, insert copulatory plugs or produce sperm that actively seek out to destroy rivals. Baker and Bellis (1995) have developed this latter idea into a 'kamikaze sperm hypothesis', claiming that a wide variety of animals, including humans, produce sperm whose function is to block or destroy rival sperm. In a careful analysis of the evidence, however, Harcourt (1991) concluded that kamikaze sperm did not in all probability exist.

Other facets of human sperm competition such as sperm size and swimming speed will be examined in Chapter 13 in the context of what this tells us about human mating. Long before sperm competition takes place, however, a male has to be accepted by a female or vice versa. Passing this quality control procedure has also left its mark on the anatomy and behaviour of humans and it is to this process we now turn.

4.4.3 Good genes and honest signals

Darwin had difficulty in explaining in adaptationist terms why females find certain features attractive. A peahen may have forced male peafowl to sport long trains to please her, but why, in functional or ultimate terms, should she be pleased by a long rather than a short tail? If, as Darwinism informs us, beauty

is in the eye of the genes, what genetic self-interest is served by finding long trains or colourful tails beautiful? If we can crack this problem then perhaps the basis of human physical beauty can be understood. Answers to this puzzle tend to fall into two camps: the 'good sense' and the 'good taste' schools of thought (Cronin, 1991).

The good taste school of thought stems from the ideas of Fisher, who investigated the problem in the 1930s. Consider a male characteristic such as tail length that females once found attractive for sound evolutionary reasons, such as that it indicated the species and sex of the male, or it showed the male was healthy enough to grow a good-sized tail, or that a long tail was indeed advantageous to a male in flying or catching food. Fisher argued that under some conditions a runaway effect could result, leading to longer and longer tails. Such conditions could be that sometime in the past an arbitrary drift of preference led a large number of females in a population to prefer long tails, or possibly possession of a slightly longer tail than average did confer fitness benefits and so it was advantageous for females to prefer this feature. Once this preference took hold it could become self-reinforcing. Any female that resisted the trend, and mated with a male with a shorter tail, would leave sons with short tails that were unattractive. Females that succumbed to the fashion would leave 'sexy sons' with long tails and daughters with the same preference for long tails. The overall effect is to saddle males with increasingly longer tails, until the sheer expense of producing them outweighs any benefit in attracting females. In short, sexual selection driven by female choice may extend the magnitude of a feature beyond what is optimal for natural selection.

The 'good sense' view suggests that an animal estimates the quality of the genotype of a prospective mate through the signals he or she sends out prior to mating. These ideas are now some of the most promising lines of enquiry in sexual selection theory with many suggestive applications to human mate choice (Table 4.4).

The '*good genes*' dimension of good sense would explain why in polygynous mating systems females share a mate with many other females, even though there may be plenty of males without partners, and despite the fact that the males of many species

Table 4.4 Mechanisms of intersexual competition

Category	Mechanism
Good taste (Fisherian runaway process)	Initial female preference becomes self-reinforcing. A runaway effect results that leads to elaborate and ultimately dysfunctional (in terms of natural selection) traits, for example a peacock's train.
Good sense (genes and/or resources)	One sex uses signals from the other to estimate the quality of the genome on offer. Such signals may indicate desirable characteristics such as resistance to parasites or general metabolic efficiency. One partner may also inspect resources held by the other and the likelihood that such resources will be made available.

contribute nothing in the way or resources or parental care. Females are in effect looking for good genes. The fact that the male is donating them to any willing female is of no concern to her. The important point here is that the female is able to judge the quality of the male's genotype from the 'honest signals' he is forced to send. In this respect, size, bodily condition, symmetry and social status are all signals providing the female with information about the potential of her mate.

Males and females can send signals about their health and reproductive status in a variety of ways. Consider the time-honoured principle of fashion that 'if you have it flaunt it, if you haven't, hide it'. This applies to cosmetics as much as clothes. This leads to a distinction between honest and dishonest signals. Hiding signs of genetic weakness or false advertising are really *dishonest signals*. Dishonest signals are rare among non-human animals since they are likely to be spotted and eliminated in favour of honest ones. Humans, however, with their clever brains and sophisticated culture are particularly adept at sending both honest and dishonest signals about themselves to others.

A particularly fruitful line of research that has emerged in recent years is costly signalling theory (CST). This theory suggests that honest signals will

emerge and be attended to by the receiver if one or two of the following conditions hold:

● The signal must be honestly linked to the quality of the trait it is trying to advertise. When this is the case imitation by inferior individuals is impossible since the linkage would ensure that the advertisement actually revealed the poor quality of the signaller.

● The signal must be a handicap; that is, it must impose a cost on the signaller. In this way only high-quality individuals can afford the handicap and hence advertise.

Smith and Bliege Bird (2000) are anthropologists in the USA who have applied this theory with some success to turtle hunting among the Meriam islanders of Torres Strait, Australia. The Meriam people live on the barrier reef island of Mer about 100 miles from New Guinea. They have numerous public feasts during which time men engage in competitive dancing, hunting, diving and boat racing. During one type of feast involving funerary rites, large amounts of turtle meat are consumed. To provide this meat turtles have to be hunted, and turtle hunting seems to qualify as an example of a costly and honest signal since:

a) Hunting involves taking a small boat out to sea, locating a turtle and then jumping on its back with a harpoon – a procedure that requires strength, agility and is physically dangerous.

b) When the turtles are captured, they are shared communally during a public feast. In fact the hunters take virtually no share of the meat and have to bear the full cost of the hunt themselves.

c) During the feast, people are attentive to who provided the largest turtles and who returned with small ones or none at all.

In short, the ability to return with a large turtle and offer it as a communal gift is an honest advertisement of physical strength and vigour as well as wealth. It is difficult to see how a physically unfit and resource-poor male could return from the hunt with a large turtle. The fact that the hunt is a signal rather than an economic necessity is indicated by the fact that this sort of hunting is too risky to be pursued except during public feasting. In other words the activity does not make much economic sense except as a device for young men to show off their virtues (Smith and Bliege Bird, 2000).

It is significant that Darwin chose to announce his theory of sexual selection in the same volume where he speculated on the evolution of humans (*Descent of Man and Selection in Relation to Sex*) and so it is fitting that it is to the evolution of our own species that we turn in the next chapter.

Summary

▶ At a superficial level, the mating behaviour of animals can be described in terms of species-characteristic mating systems. A deeper understanding is gained, however, by looking at the strategies pursued by individuals as they strive to maximise their reproductive success.

▶ Even where there is to be found considerable variance in the reproductive success between males and females, the sex ratio remains remarkably close to 1:1. The best ultimate explanation of this so far is that of Fisher, who suggested that natural selection gives rise to stabilising feedback pressures tending to maintain unity. Deviations from this sex ratio in some cultures have been observed as a result of culturally driven selective foeticide and infant neglect.

▶ Sexual selection results when individuals compete for mates. Competition within one sex is termed intrasexual selection, and typically gives rise to selection pressures (usually applying to males) that favour large size, specialised fighting equipment and endurance in struggles.

▶ Individuals of one sex also compete with each other to satisfy the requirements laid down by the other sex. An individual may require, for example, some demonstration or signal of genetic fitness or the ability to gather and provide resources. Selective pressure resulting from the choosiness of one sex about the other is studied under the heading of intersexual selection. Such pressure often gives rise to elaborate courtship displays or conspicuous features that may indicate resistance to parasites, or may possibly be the result of a positive feedback runaway process.

▶ The precise form that mating competition takes (such as which sex competes for the other) is related to the relative investments made by each sex and the ratio of fertile males to females. If females, for example, by virtue of their heavy investments in offspring or scarcity, act as reproductive bottlenecks for males, males will compete with males for access to females, and females can be expected to be discriminatory in their choice of mate.

▶ In cases where a female engages in multiple matings and thus carries the sperm of more than one male in her reproductive tract, competition between sperm from different males may occur. The theory of sperm competition is successful in explaining various aspects of animal sexuality, such as the high number of sperm produced by a male, the frequency of copulation and the existence of copulatory plugs.

Further reading

Alcock, J. (2013). *Animal Behaviour: An Evolutionary Approach.* 10th edition. Sunderland, MA, Sinauer Associates.

A good general book on evolution and animal behaviour.

Short, R. V. and E. Balaban (1994). *The Differences Between the Sexes.* Cambridge, Cambridge University Press.

A valuable series of specialist chapters by experts. Covers humans and non-humans.

Geary, D. C. (2010). *Male, Female: The Evolution of Human Sex Differences.* 2nd edition. Washington, DC, American Psychological Association.

Geary explains the principles of sexual selection and how these can be used to understand differences between males and females. Good discussion of the evidence for real cognitive differences between males and females.

Ridley, M. (1993). *The Red Queen.* London, Viking.

An enjoyable and well-written account of sexual selection theory and its application to humans.

Ryan, C. and C. Jetha (2010). *Sex at Dawn.* New York, NY, Harper.

Provocative and highly readable account of the evolutionary origins of human sexuality.

PART III

Human Evolution and its Consequences

The Evolution of the Hominins

Thus, from the war of nature, from famine and death, the most exalted object which we are capable of conceiving, namely, the production of the higher animals, directly follows. There is grandeur in this view of life, with its several powers, having been originally breathed by the Creator into a few forms or into one; and that, whilst this planet has gone cycling on according to the fixed law of gravity, from so simple a beginning endless forms most beautiful and most wonderful have been, and are being evolved.

(Darwin, 1859b, p. 491)

In this chapter we will be looking at the series of events that led to the evolution of our own species, one of Darwin's 'higher animals'. By investigating how the characteristics that define *Homo sapiens* arose, and the mode of life that led to the adaptations we now possess, we should be in a better position to understand our contemporary behaviour. This whole field is, of necessity, rich in speculation and conjecture, but the broad outlines of the story are reasonably clear. However, we can be sure that new fossil evidence will continue to be unearthed and there are still probably a few surprises in store. It will be helpful to start with a consideration of how we name and classify species.

5.1 Systematics

Systematics is the study of the diversity of life and the relationship between the different categories into which organisms can be grouped. A **taxon** (plural taxa) is a category of organisms in a classification system. The scheme that is universally accepted for naming a species was introduced by the 18th-century naturalist Carl Linnaeus. This is a hierarchical scheme in which the name of each species is a binomen (two parts) consisting of the genus plus the specific name. *Homo erectus* and *Homo neanderthalensis*, for example, are two species of hominins belonging to the **genus *Homo***. Above the level of genus lies the

taxon of 'family'. We might say therefore (based on morphological similarities) that humans belong to the family Hominidae, which also contains chimps and gorillas. Table 5.1 shows how this hierarchy can be extended upwards. It follows that we start from one species and rise upwards into groupings containing ever more species types. The kingdom Animalia in Table 5.1, for example, contains about one million different species from earthworms to humans. We should not be too fazed by the Latin names in this exercise. The groupings, although we hope they mirror objective realities, are also human inventions. To reduce the mystery of this, Table 5.1 also shows how we might classify a motor vehicle. We should not take the analogy too far, although it is tempting to note that just as members of a species can interbreed to produce fertile offspring, so cars on the bottom rung of the classification in Table 5.1 can usually swap parts and still remain functional.

5.1.1 How to classify humans and their relatives

When Linnaeus devised his system in the 18th century he had no notion of evolutionary change and classified organisms on the basis of physical similarities. Darwin himself argued that since species are similar because of genealogical descent then classification should be based on phylogenies. Surprisingly,

Table 5.1 Traditional taxonomy of the human species and a hypothetical classification of a motorised vehicle to illustrate the hierarchical principle of Linnaean classification

	Human	A motor vehicle
Kingdom	Animalia	Machines
Phylum	Chordata	Self-propelled
Class	Mammalia	Four wheel
Order	Primates	Petrol engine
Infraorder	Anthropoidea	Domestic motor vehicle
Superfamily	Hominoidea	Coupe
Family	Hominidae	Volvo
Genus	*Homo*	S60
Species	*Homo sapiens*	S60 GT

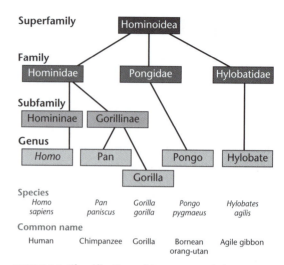

FIGURE 5.1 Classification of humans and the great apes based on phylogenetic information.
In this scheme humans, chimps and gorillas are based in the same family: Hominidae. Humans are then assigned to the subfamily Homininae, and chimps and gorillas to Gorillinae.

the debate about proper schemes of classification continues to this day. There are three major schools of thought about how classification should be done: phenetics, cladistics and evolutionary systematics. Phenetics stresses results of adaptations and hence morphological and anatomical similarities. Proponents of this school argue that it is based on real observable features. **Cladistics** (Greek *klados* = branch), on the other hand, only considers evidence relating to phylogenies (lines of descent) of organisms and is concerned to group them according to how they diverged from common ancestors (see Figure 5.1). The cladistic approach was developed initially by the German systematist Willi Hennig in 1950. Cladistics has some advantages over classical **taxonomy** in that by concentrating on branching points it relies less on morphological similarities, which may be open to subjective interpretation and weighting. This may seem a more objective system but the challenge then becomes how to infer what the branching times and sequences were. Evolutionary systematics is a scheme that combines elements of both.

In this book, we present the cladistic or phylogenetic system of classification as being the most valuable way of considering the relationship between humans and other species of apes. The term 'hominoids' will be taken to include humans, chimps, gorillas, orang-utans, gibbons and their ancestors to about 25 million years ago. **Hominids** include humans, chimps, gorillas and their ancestors to about 15 million years ago. The term **hominin** will be reserved for humans and their extinct ancestors after the lineage broke away from chimps some 7 million

years ago. We should note that there is still not universal agreement on this scheme (some would place chimps and humans in the subfamily Homininae; others have even suggested that chimps and humans should be assigned to the genus *Homo*).

As Table 5.1 shows, humans belong to the **order** Primates and there are about 200 living species of primate left. Their weights range from about 80 g in the case of the mouse lemur to gorillas at 150 kg. Most primates are **precocial** meaning that the young are born relatively mature and can soon look after themselves. Humans are the exception here and are **altricial**: the young are very immature and require a long period of nurture and protection. This turns out to be a crucial feature in human evolution. This tendency probably increased as the hominins evolved.

5.2 Origins of the hominins

5.2.1 Speciation and Earth history

In 1871 Darwin predicted that it was probably a species of ape living in Africa that was the common ancestor of modern humans and today's great apes. Advances in geology, climate reconstruction and genetics, coupled with the recovery of key fossils, lead us to believe that Darwin was almost certainly right. Many puzzles and problems remain but we now have enough information to describe the main outlines of our primate ancestry with some confidence. To put human evolution in an even broader time frame, Table 5.2 shows some key dates in the history of life on Earth.

5.2.2 Hominin speciation

Humans, chimpanzees and gorillas all originated in Africa and share some features in common, but all unambiguously belong to different species. The concept of a species has a fairly clear biological definition:

Table 5.2 Some key events in life history

Era	Epoch	Date, years before Common Era (BCE), or before present (BP)	Event
CENOZOIC	Holocene 10,000 BP	1859	Darwin's *Origin of Species*
		c. 2,600 BCE	Great Pyramid of Cheops at Giza built
		c. 3,000 BCE	Bronze begins to replace stone as a material for tools and weapons in Egypt and Western Asia
		c. 4,000 BCE	Agriculture reaches Western Europe
		c. 7,000–11,000 BCE	Beginning of Neolithic Age. People settle in villages and begin the domestication of plants and animals. Origins of agriculture
	Pleistocene 1.75 million BP	10,300 BP	End of last ice age
		c. 30,000–12,000 BP	Successive waves of migrant *Homo sapiens* enter North America
		c. 30,000 BP	Neanderthals become extinct
		c. 40,000 BP	Colonization of Australia by *Homo sapiens*
		c. 70,000 BP	Beginning of last ice age
		c. 72,000 BP	Movement of some humans out of Africa
		c. 74,000 BP	Large supervolcanic eruption in Toba, Sumatra, Indonesia. Six-year nuclear winter may have led to population of *Homo sapiens* reducing to 2,000 worldwide
		c. 180,000 BP	Emergence of modern *Homo sapiens* in Africa
		c. 1 million BP	*Homo erectus* migrates out of Africa
		c. 1.4 million BP	Use of fire by *Homo erectus*
	Pliocene 5.3 million BP	c. 1.8 million BP	Appearance of *Homo erectus* as a species in Africa
		c. 2.3 million BP	*Homo habilis* uses stone tools
		c. 3.6 million BP	Bipedal hominins (possibly **Australopithecines**) leave footprints in a trail of volcanic ash in Laetoli, Tanzania
	Miocene 23.8 million BP	c. 7 million BP	The last common ancestor of chimps and humans inhabits Africa
		c. 15 million BP	Beginning of the partial isolation of East Africa from the rest of Africa along the line of the Great Rift Valley
		c. 35 million BP	Ancestors of New World primates reach South America from their origin in Africa
		c. 65 million BP	Likely collision event between asteroid or comet and Earth leading to the extinction of the dinosaurs and new opportunities for mammals
		c. 65 million BP	Earliest primates
		c. 100 million BP	Common genetic ancestor of mice and humans
		c. 125 million BP	Earliest placental mammal: *Eomaia scansoria*. Probably resembled a modern dormouse
		c. 300 million BP	Origin of reptiles

Table 5.2 cont'd

Era	Epoch	Date, years before Common Era (BCE), or before present (BP)	Event
CENOZOIC		c. 480 million BP	Earliest jawed, bony fishes
		c. 600 million BP	Earliest multicellular organisms – probably resembled sponges
		c. 1,200 million BP	Beginning of sexual reproduction
		c. 1,300 million BP	Earliest plants
		c. 1,600 million BP	Blue-green photosynthetic algae appear
		c. 3,900 million BP	Beginning of life on Earth
		c. 4,500 million BP	Formation of the Earth
		c. 13,500 million BP	Best current estimate for origin of Universe

organisms are said to belong to the same species if they can interbreed to produce fertile offspring. Some different species, such as donkeys and horses, are closely related and produce viable offspring, but such offspring are inevitably infertile. But the fact that horses and donkeys are so similar is a reflection of the fact that they shared a common ancestor fairly recently in geological time; more recently than, say, a horse and a wolf. On the basis of viable offspring, present-day humans all belong to the same species called *Homo sapiens* (literally 'wise humans').

Through time, a single species can give rise to others through a process that Darwin and his followers called transmutation, which today is usually known as **speciation**. Speciation occurs when populations are reproductively isolated by geographical barriers such as islands or mountain ranges, isolated temporally by gradually breeding at different times, or separated in some other way. By gradual mutation and genetic drift, the two populations of what was once a single species reach a state in which gene flow between them ceases and cannot be revived by the removal of the barrier (Figure 5.2).

Where speciation is not quite complete, we often observe the existence of subspecies. Until recently there was debate over whether Neanderthal man and early *Homo sapiens* were really distinct species or whether they did, in fact, interbreed. Neanderthal man (with the possible exception of *Homo floresiensis*, see Section 5.3.2) was the last hominin species to die out, living in Europe as recently as 30,000 years ago and co-existing, at least in time, with *Homo sapiens*. Recent work has shown quite conclusively that Neanderthals did indeed interbreed with modern humans and that non-African modern humans typically contain about one hundred kilobases of Neanderthal DNA. Studies have shown that genes involved in keratin pigment formation are unusually rich in ancestral Neanderthal genes, suggesting that this interbreeding may have helped non-African *Homo sapiens* cope with a non-African environment (Sankararaman et al., 2014).

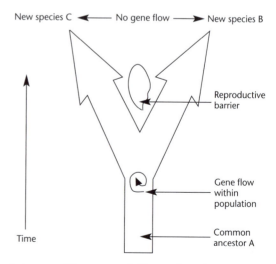

FIGURE 5.2 Diagrammatic illustration of speciation around a reproductive barrier.

5.3 Phylogeny of the Hominoidea

5.3.1 Branching sequences and dates

Early homins are gone, and there is now only one species left: ourselves, *Homo sapiens*. However, several species of great ape still exist today, and it is instructive to consider how closely we are related to them and when we last shared a common ancestor with each of them. To do this we need a phylogenetic tree detailing the evolution of the primates. There are two basic disciplines that can help us construct such a tree: comparative anatomy and molecular biology. Comparative anatomy examines the similarities between primates in terms of their basic body plan. Since humans more closely resemble chimpanzees than they do ring-tailed lemurs, it seems reasonable to suppose that we are more closely related to chimps than lemurs. By 'more closely related' we mean that we more recently shared a common ancestor. By such methods, humans and the four species of great ape (gorillas, chimpanzees, bonobos and orang-utans) were, by 1943, placed in the same superfamily called Hominoidea (Figure 5.1) where they still remain. Morphological evidence, however, tends to be qualitative, and it has proved difficult

to push the method further to establish unambiguously a branching order among the hominoids. More recent evidence from molecular biology has provided a clearer and more consistent picture of genetic similarities between the species of this taxon.

Genetic similarities between species can be ascertained by either examining DNA directly or looking at the structure of proteins that result from DNA expression and there is now a range of techniques that can be used to measure the degree of similarity between proteins or DNA from different species. When molecular studies began in the 1960s, the exciting news was that virtually every technique agreed in general terms with the broad conclusions from comparative anatomy. In terms of the amino acid sequences in a range of blood proteins, and hence the genes responsible, we are virtually identical to chimpanzees, slightly different from gibbons and different in many amino acids from the Old World monkeys.

Bearing in mind the promise and difficulties of molecular comparisons, we can now construct a phylogeny for the hominoids based on molecular clockwork calibrated using fossil evidence. Figure 5.3 brings together data from a number of studies.

The implications of Figure 5.3 are profound. The first point to note is that chimpanzees (and bonobos),

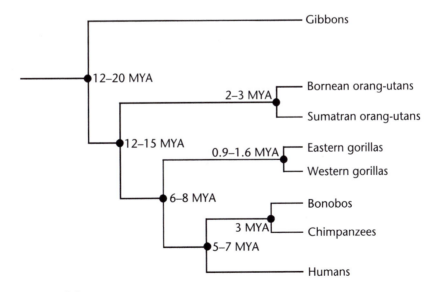

FIGURE 5.3 Phylogeny of the great apes.
MYA = millions of years ago.
SOURCE: Sequence constructed from various sources including Friday (1992) and Robson and Wood (2008).

not gorillas, are our closest relatives. In fact, chimpanzees are more related to us than they are to gorillas. We differ in our DNA from the chimps by just 1.6 per cent, the remaining 98.4 per cent being identical. Our haemoglobin, for example, is the same in every one of the 287 amino acid units as chimpanzee haemoglobin; in terms of haemoglobin, we are chimps. We must beware of making too much of the 1.6 per cent difference. Information coded on DNA is not a linear system; that 1.6 per cent could and has effected profound changes. A small change in DNA can bring about an enormous change in morphology through a change to the timing of the development of any particular feature. One obvious difference between chimps and ourselves is that we have a much larger brain; and even 2 million years ago, our ancestors had larger brains than modern-day chimps. Also, since DNA is a linear sequence of four bases, any given length will be at least 25 per cent similar to that of any organism (for example, a daffodil or a slug).

5.3.2 Early hominins

The first fossilised remains of humans were found in 1856 in the Neander valley near Dusseldorf. Such specimens quickly became called Neanderthal man, and for a time there was much speculation on whether they were ancestors of *Homo sapiens*. Modern molecular evidence suggests that they were cousins rather than ancestors of modern humans. They split away from the *Homo* lineage about 500,000 years ago and were a distinct species that died about 35,000 years ago (see Table 5.3).

It was in the 1920s that the first hominin fossils were found in Africa. In 1925 Raymond Dart described the species *Australopithecus africanus* based on a specimen called the Taung Child found in a cave at Taung in South Africa. The brain of this creature was small at about 410 cc, but the position of the foramen magnum (the hole in the base of the skull through which the end of the spinal cord enters the brain) suggested bipedal locomotion. It took a further 20 years for Dart's views on *A. africanus*'s bipedal stance to be taken seriously since the common assumption then was that a large brain preceded bipedalism. It was also widely assumed that Asia, not Africa, was the cradle of humankind.

An even older hominin was found by Donald Johanson and Tom Gray on 24 November 1974. The Beatles song 'Lucy in the Sky with Diamonds' was playing in their camp during the evening celebrations of the discovery; the name 'Lucy' stuck and she has since become one of the most famous of early hominin specimens. *A. afarensis* brain size is estimated at about 400 cc and is often said to be about the same size as that of a modern gorilla or chimp. We should note, however, that early Australopithecines were smaller than modern gorillas and that the ancestors of modern gorillas almost certainly had smaller brains 3 million years ago when they co-existed with *A. afarensis*. In short, brain expansion had already begun at the time of *A. afarensis*.

Since the 1920s numerous fossil remains have been unearthed and have enabled a whole raft of conjectures about the hominin lineage to be constructed. Figure 5.4 shows the current distribution of the African apes together with some find spots of early *Homo* species. Tables 5.3 and 5.4 provide some information on these discoveries in summary form, and Figure 5.5 shows a tentative branching sequence. The concept of EQ (a measure of

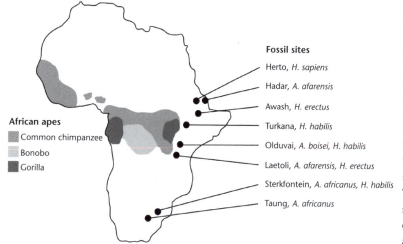

Fossil sites

- Herto, *H. sapiens*
- Hadar, *A. afarensis*
- Awash, *H. erectus*
- Turkana, *H. habilis*
- Olduvai, *A. boisei, H. habilis*
- Laetoli, *A. afarensis, H. erectus*
- Sterkfontein, *A. africanus, H. habilis*
- Taung, *A. africanus*

African apes
- ▇ Common chimpanzee
- ▇ Bonobo
- ▇ Gorilla

FIGURE 5.4 Current distribution of the African apes and find spots of early hominin species.

Table 5.3 Data on early hominins

Species	Time span (mya = million years ago; ya = years ago)	Mean cranial capacity (cc)	EQ calculated from observed volume/0.058 (body mass)$^{0.76}$ (see Martin, 1981)	Distribution	Other popular names and exemplars	Body size: height (m), mass (kg) Males	Females	Sexual dimorphism in body mass (males/females)
Australopithecus afarensis	4.0–2.9 mya	380–450	2.3		Lucy	1.51 44.6	1.05 29.3	1.52
Australopithecus africanus	3.2–2.5 mya	457	2.7	Eastern and possibly southern Africa	Taung child Mrs Ples	1.38 40.8	1.15 30.2	1.35
Homo naledi	Possibly 2.5 mya Firm date yet to be established	560 (males) 465 (females)	2.8	Several specimens found in the Rising Star cave complex 30 miles NW of Johannesburg in 2015		1.5 45		
Homo habilis	2.3–1.6 mya	552	3.3		Handyman	1.57 37	1.25 31.5	1.17
Homo ergaster	2.0–0.5 mya	854	3.5		The Nariokotome boy	63	52	1.21
Homo erectus	1.9 mya– 250,000 ya	1,016	4.1	Africa, Asia, Europe	Java man and Peking man	1.8 63	1.55 52	1.20
Homo neanderthalensis	150,000– 30,000 ya	1,512	5.7	Europe, Asia, Middle East	Neanderthal man	1.7 73.7	1.6 56.1	1.31
*Homo floresiensis**	?– 13,000 ya	380	2.5–4.6	Flores, eastern Indonesia	'Hobbit'	? ?	1.06 16–28.7	?
Homo heidelbergensis	600,000– 250,000 ya	1,198	4.6	Africa, Europe	Boxgrove man	1.8m 62	? ?	?
Homo sapiens (mean of several human groups)	200,000 ya– present	1,355	5.4	Africa, Asia, Europe, New World (in that order)	Cro-Magnon man (Western Europe *Homo sapiens*)	1.64 60.9	1.54 50	1.21

SOURCE: Information from various sources, see Collard (2002) and McHenry (1991).
* See Brown, P. et al. (2004); information on *Homo naledi* from Berger et al. (2015).

intelligence) shown in Table 5.3 is considered in the next chapter. New discoveries and reinterpretations will certainly add to this picture as the years go by.

By its very nature the naming of fossils provides ample opportunity for debate and controversy. In common with taxonomists in other branches of biology, there are the 'lumpers' who would rather group small variations found in fossils into one species, and the 'splitters' who use variations in new finds (such as slight changes in morphology) to argue for a new species. The splitters, for example, have argued (successfully it seems) that African *Homo erectus* should be called *Homo ergaster*. They would also

like *H. erectus* fossils found in Georgia to be called *Homo georgicus*.

5.4 The supremacy of *Homo sapiens*

5.4.1 Out of Africa or multiregionalism?

Over the period 2 million to 50,000 years ago several species in the *Homo* genus, such as *Homo erectus*, *Homo heidelbergensis* and *Homo sapiens,* migrated out of Africa and spread to Asia and/or Europe. But by 30,000 years ago (again with the possible exception of *H. floresiensis*) this taxonomic diversity had vanished, leaving one dominant species, *Homo sapiens*, now spread across east Asia, southeast Asia, Africa, Europe and poised to move into the New World. To explain this diminished diversity various models of speciation, replacement and genetic mingling (interbreeding) have been proposed. The debate is involved and complex but can be simplified into two main schools of thought. One, the **'Out of Africa' hypothesis**, suggests a single origin for *Homo sapiens* in Africa followed by an exodus that replaced other *Homo* species (Stringer and Andrews, 1988). The other, the **'multiregional model'**, argues for several regional populations of *Homo erectus* evolving simultaneously into *Homo sapiens* in different parts of the globe (Wolpoff et al., 1984). A variation on the multiregional model is the idea that the primary features distinguishing *Homo sapiens* arose in Africa but then spread to other populations of archaic *Homo* species by genetic mixing and population flow.

One interesting piece of evidence that points strongly towards the out of Africa model is that there is a strong correlation between the geographical distance apart of any two human populations and the molecular and genetic differences between these same two populations. This would suggest a single origin radiating across the world in time. Populations separated by a large geographical distance are also separated by time to arrive in those destinations and hence time for genetic changes to take place.

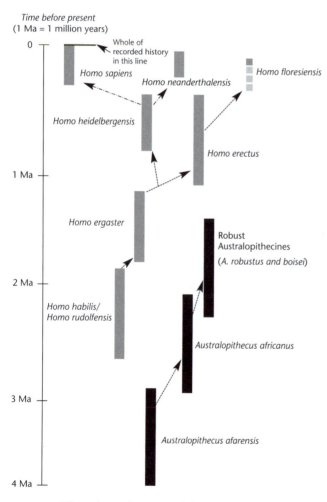

FIGURE 5.5 Time chart of some early hominids.
The exact lineages (that is, which species descended from which) remain uncertain and controversial.

Table 5.4 Summary of features of selected hominins

Species	Time range	Key features
Earliest hominins		
Ardipithecus ramidus	5.5–4.4 mya	The most primitive known hominin. Found in Ethiopia in 1994. Only fragments are known. Possibly bipedal although it was found in an environment that would have been wet and forested
Australopithecines		
Australopithecus afarensis	3.7–2.9 mya	Found in Ethiopia and Tanzania. Ape-like brain size but skeleton shows clear evidence of bipedalism. Lucy found at Hadar in 1974 by Donald Johanson. Best-known Australopithecine. Footprints also show bipedalism. High degree of sexual dimorphism. Toes and fingers curled suggesting forest habitation as well
Australopithecus africanus	3.5–2.5 mya	One of first fossils found to provide evidence an African origin for hominins. First found in South Africa at Taung by Raymond Dart in 1924. Smaller canines than chimps and *A. afarensis*; large back teeth suitable for grinding plant matter. Bipedal but also adaptations for climbing trees. Brain endocasts fail to show a Broca's area, the part of the brain associated with language. *Afarensis* and *africanus* known as gracile Australopoithecines because of slight and slender build
Robust australopithecines		
Australopithecus robustus and *Australopithecus boisei*	2.3–1.0 mya	Very similar species from South and East Africa. Show enlarged molars adapted to eating plant material (Leakey called one specimen of *boisei* 'nutcracker man'). Notice they co-existed with the *Homo* genus before becoming extinct. Some palaeontologists classify these in genus *Paranthropus* instead of *Australopithecus*
Transitional Forms		
Homo habilis/ rudolfensis	2.3–1.7 mya	Contemporary with australopithecines but larger brains and smaller teeth. *Homo habilis* first described by Louis Leakey in 1964. Found in Tanzania and shores of Lake Turkanya. Thought to be first hominins associated with tool making. Some doubt if these species are very different to Australopithecines, others merge *rudolfensis* and *habilis* into a single *Homo* species. Brain endocasts show a rudimentary bulge in Broca's area, suggesting onset of language
Archaic *Homo* species		
Homo erectus and *Homo ergaster*	1.9–0.10 mya	Found mainly in Asia. First discovered in Java in 1891. Very widespread, fossils found in South Africa, Tanzania, Kenya, Italy, Germany, India and China. Oldest fossils come from East Africa, species seems to have dispersed out of Africa and across the Old World as soon as species emerges resulting in geographically variable local populations. Some favour term *Homo erectus* for Asian specimens and *Homo ergaster* for African and Georgian fossils. Some argue that distribution of *erectus* and archaeological evidence suggest that *erectus* preyed on other animals. Tool use known from worked lavas and quartzites from Oiduvai Gorge at around 2.6 Ma. The Archeulean hand axe shows advance in sophistication and suggestion of a mental template. First Hominines found outside Africa. First use of fire and some evidence of systematic hunting. In these species dietary dependence on meat increases. Meat obtained from scavenging and possibly hunting. Brain size and pelvic measurements show that infants needed prolonged care

Table 5.4 cont'd

Species	Time range	Key features
Archaic *Homo sapiens*/ *Homo heidelbergensis*	0.60 mya– 250,000 ya	Archaic *Homo sapiens* was the appellation given to early humans that were thought to form the common stock for modern *Homo sapiens* and Neanderthals (*Homo sapiens neanderthalensis*). However, recent evidence that suggests that Neanderthals were a separate species (viz. *Homo neanderthalensis*) invalidates this term. Instead, the term *Homo heildelbergensis* is now used for this species. Found in Africa and Europe, showing signs of brain enlargement and strongly associated with hand axe technology
Homo neanderthalensis	200,000– 30,000 ya	Named after one of the earliest finds of a partial skeleton in the Neander Valley, Germany, in 1856. Now more than 275 individuals have been recovered from over seventy sites from Western Europe to Uzbekistan. Large-brained, thick-boned and large-bodied, showing adaptations to a cold glacial climate. Used fire and stone tools and buried dead. Last common ancestor with *Homo sapiens* probably about 600,000 years ago
Homo floresiensis	?–15,000 ya	Remarkable recently discovered (but controversial) species, identified from a skeleton of a female discovered in a cave in Liang Bua on the island of Flores in Indonesia. Named in 2004. Probably a dwarfed descendent of Javanese *Homo erectus*. Small in stature (approx. 1 m) with a small brain (350 cc). Possibly the smallest species of hominin
Modern humans		
Homo sapiens sapiens (*modern humans*)	0.15 mya– present	Modern humans appeared in Africa about 180,000–150,000 years ago. The 'Out of Africa' hypothesis (now the most favoured) suggests that this species then spread out of Africa to Europe and Asia between 100,000 and 60,000 years ago. Thought to have entered the New World about 12,000 years ago. Highly encephalised, sophisticated culture and technologies. Adult brain has capacity of about 1,350 cc. On this basis expected brain size for a child if *sapiens* was typical primate would be 725 cc. In fact the figure is 385 cc, showing that the human brain grows rapidly after birth

Evidence looking at **mitochondrial DNA** differences between individuals in a population points to sub-Saharan Africa as the region with the greatest diversity (see Figure 5.6).

Reviewing the recent genetic evidence, Stephen Oppenheimer (2012) concludes that modern human left Africa through a single exit route across the Red Sea near what is now called the Bab el-Mandeb Strait about 72,000 years ago (see Figure 5.7). Genetic diversity can also be used to calculate the size of the founding population for humans outside Africa. The figure has fairly wide margins but comes out between 1,000 and 20,000 breeding individuals in a total population of about 50,000. It has been suggested that one of the causes of this 'genetic bottleneck' may have been the Toba super eruption about 73,000 +/– 4,000 years ago – one of the largest known eruptions on Earth. The magnitude of the eruption is likely to have caused a volcanic winter lasting 6–10 years, severely reducing populations of many mammalian species (Williams et al., 2009, Figure 5.7 shows the dispersal of *Homo sapiens* across the globe).

What prompted the out of Africa diaspora for a group of *Homo sapiens* is a bit of a mystery and may never be established. It is not known whether this eruption preceded or post-dated the exodus of this population of humans leaving Africa. Genetic evidence tells us, however, that the departing group was fairly small since it carried with it only a small portion of the total African genetic diversity. This has effectively given rise to a founder effect across the globe but excluding Africa. For example, all indigenous, non-African Y chromosomes carry the single nucleotide polymorphism (SNP) mutation known as M168. In tracing back these non-African Y chromosomes to their most recent common ancestor we reach a man

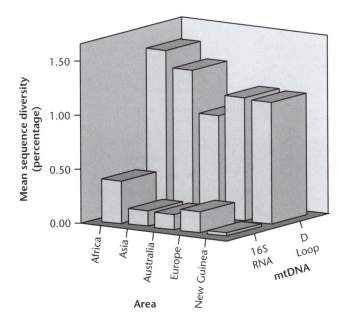

FIGURE 5.6 Chart showing molecular diversity amongst humans according to region.
Y axis shows diversity (measured as the mean pairwise divergence for individuals in a population) for two mitochondrial DNA alleles (D Loop, characteristically the area showing most variability, and the 16S region, an area usually showing the least variability). For both regions of the mitochondrial DNA, Africa shows the greatest degree of divergence, supporting the single origin out of Africa theory.
SOURCE: Data from Cann et al. (1987).

known as 'Eurasian Adam' – not to be confused with 'Y Chromosome Adam', the most recent male ancestor of all living men. Eurasian Adam probably lived in Africa about 50,000 years ago. His descendants moved across the mouth of the Red Sea to populate the rest of the planet (Underhill et al., 2001). Some of course, remained behind and the M168 mutation is found in places in Africa too. The fact that only a small group of *Homo sapiens* moved out points to the possibility of displacement by inter-group rivalry, although climate change such as initiated by the Toba eruption could be another factor.

The Americas were populated fairly recently, probably by a migration of small groups of Siberians across the land mass (Beringia) that would have then joined Alaska to Siberia. As the North American glaciers began to melt about 15,000 years ago so southward migration could begin in earnest. The original founding population was probably very small (possibly 100 or so individuals) since the genetic diversity of Native Americans is very low (there is almost a complete absence of blood type B, for example).

5.4.2 **Female exogamy**

Another feature of the life of *Homo sapiens* we can piece together is that females probably left their homes and encampments and went to live in the settlements of their male partner. This phenomenon is known as female **exogamy** and is still the common pattern in modern populations, and there is evidence from studies on mitochondrial DNA that this has been the condition of human bonding for tens of thousands of years. The useful aspect of mitochondrial DNA is that it is only passed on through the female line (Figure 5.8). But molecular studies show that the variation (between people) in mitochondrial DNA is similar to that of DNA found on the autosomal (that is, non-sex) chromosomes; so much so that the average local population contains about 80–85 per cent of all the variation in mitochondrial and autosomal DNA. In contrast to this, most local populations contain only about 36 per cent of the full genetic variation found on the Y chromosome (only passed down through the male line). This suggests that women have migrated more than men. Although each individual female's journey to her mate's home would (until recent times) have been a short one, over hundreds of generations this has had global effects on DNA distribution (Pennisi, 2001).

Some recent work suggests that female movement is more evident in non-foraging agriculturists and less so in hunter-gatherers. This is probably related to the fact that land ownership passes down

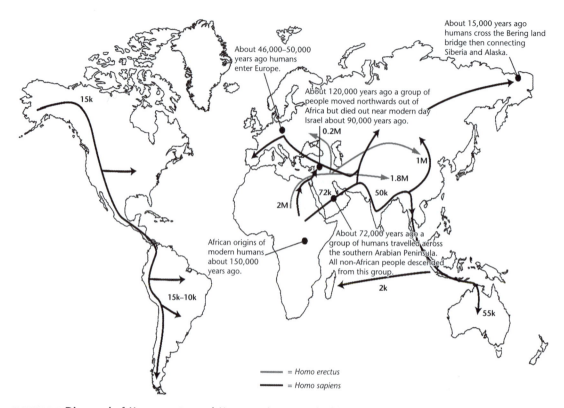

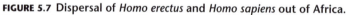

FIGURE 5.7 Dispersal of *Homo erectus* and *Homo sapiens* out of Africa.
The current consensus is that *Homo erectus* emerged in Africa around 2 million years ago and spread to regions of the Old World (India, China, Java, Spain) died out about 100,000 years ago. *Homo sapiens* originated in Africa about 150,000 years ago. An initial wave of *Homo sapiens* left Africa about 100,000 years ago and reached the Middle East but then died out about 70,000 years ago. One or possibly two more successful emigrations followed around 72,000 years ago, this time resulting in the colonisation of most of the world.
SOURCE: Data from Harcourt (2012); and Oppenheimer (2012), Figure 1.

the male line (hence males stay put on their land) whilst for hunter-gatherers land ownership does not exist (Marlowe, 2004). On top of this there are also a variety of local effects. Per Hage and Jeff Marck, for example, observe that through the Pacific Islands the movement of mitochondrial DNA is slow but that of Y chromosome markers much faster. This is consistent they argue with males dispersing widely through sea-faring but females tending to remain in or near the region of their birth (Hage and Marck, 2003).

5.4.3 The mitochondrial Eve hypothesis

The **mitochondrial Eve** concept is one of those scientific ideas that quickly permeates culture but is

prone to numerous misunderstandings. To understand the idea and its implications we need to consider the nature of mitochondrial DNA (mtDNA). Mitochondria are components (or organelles) of cells responsible for energy metabolism. There are hundreds of mitochondria in most human cells and mitochondria and mitochondrial DNA have some remarkable features and properties that make them very useful to study human evolution. Some of these key features are:

● Mitochondria contain their own genome: a circular molecule of DNA about 16,500 base-pairs long (compared to the 3,000 million base pairs in the nuclear genome). Many copies of this genome are found in each mitochondrion.

- The DNA of mitochondria codes for 37 genes and accumulates mutations about ten times faster than nuclear DNA. The DNA is haploid (that is, it has only one copy).
- Mitochondrial DNA is only inherited from the maternal line. This is because the mother supplies about 25,000 mitochondria in her ova but the mitochondria of male sperm (relatively few in number that supply the energy for sperm to swim) fail to penetrate the ovum (see Figure 5.8).
- mtDNA does not become mixed by sexual recombination as is the case with autosomal DNA. This has the effect of magnifying diversity and identifying bottlenecks in populations. Lineages and divergence times can be inferred from mutations in 'junk DNA' in the mitochondrial genome since these are not subject to natural selection.

The mitochondrial Eve hypothesis came to the fore in the 1980s in a ground-breaking paper by Cann, Stoneking and Wilson (1987). By examining the mitochondria of a sample of 147 modern humans they concluded that the mitochondrial DNA of all living people had descended from a single female who had lived in Africa about 200,000 years ago. The name 'mitochondrial Eve' quickly entered public consciousness. There are two common confusions:

firstly, that mitochondrial Eve was the only woman living at the time; secondly, that all existing humans were descended from this individual. To appreciate the true picture, imagine a line linking your mtDNA with your mother. The mtDNA you have in your cells was inherited from your mother; your mother inherited it from her mother and so on backwards into the depths of the past. If the line of each individual is compared it will be noticed that they start to converge in common female ancestors. This must happen of course since there are more people alive today than there were in the past so we all must be sharing numerous ancestors. If this mtDNA is pursued backwards along the female line of descent eventually all lines will converge on a single woman living about 200,000 years ago. There were of course many other women alive at this time, but Eve happens to be the only one with an unbroken line of matrilineal descent to the present day. Many other women at the time of Eve will also have descendents still alive today through a mixture of male and female lines.

There was some criticism of the original work by Cann et al. (only two of the 147 people sampled, for example, were sub-Saharan Africans). The work was repeated by Max Ingman and colleagues employing various refinements such as using a longer section of mtDNA and more Africans in the sample. They came

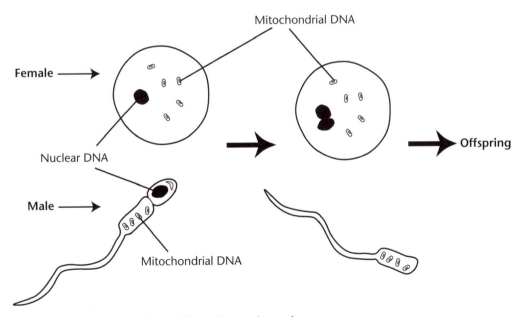

FIGURE 5.8 Mitochondrial DNA is inherited from the mother only.

to a similar conclusion that there was an African mitochondrial Eve that lived about 172,000 (+ or – 50,000) years ago (Ingman et al., 2000).

From recent studies on mtDNA it has also emerged that variability of mtDNA is low – only about one-tenth of that of chimps and gorillas. From such measurements Ruvolo et al. (1994) concluded that the two most different humans on Earth are less different in mtDNA terms than two lowland gorillas in the same forest in West Africa. This low variation in human populations implies that the human lineage faced a bottleneck at some time in the past. In addition, Africans display the greatest variation in mtDNA. This could imply that this is the oldest human population. Although there are still uncertainties about how to interpret the data and construct branching sequences, the balance of evidence points to the existence of a mitochondrial Eve who lived in Africa about 200,000 years ago before the exodus of *Homo sapiens* from that continent to others.

In conclusion then, the fossil and genetic evidence so far suggests the likelihood of *Homo sapiens* emerging as a result of speciation events in Africa about 150,000 years ago. The main issue for the future will probably be whether *Homo sapiens* replaced other species in the Old World completely, leaving them to go extinct, or whether there was some genetic admixture through interbreeding. This may be a difficult question to answer if the level of interbreeding was low but not zero, although strong evidence points to some interbreeding between non-African *Homo sapiens* and *Homo neanderthalensis*.

5.5 Hominin evolution and body size changes

5.5.1 Absolute body size

The vast majority of mammals, including most non-human primates, are considerably smaller than humans. It has long been recognised that, as mammals evolve, there is a tendency for them to grow in size. The Eocene (the geological period about 56 million to 34 million years BP) ancestor of the modern horse, for example, was only the size of a small dog about 40 million years ago. This trend, although it is not universal, is sometimes called Cope's law,

and humans, as mammals, have been part of this trend towards increasing size. Estimating the body mass of early hominins is fraught with difficulties, but there is general agreement on the broad picture: early Australopithecines were small, *Australopithecus africanus*, for example, probably weighing between 18 and 43 kg; body weight rose from *Homo habilis* to *Homo erectus*; Neanderthals were larger than modern humans; and the height and weight of *Homo sapiens* has declined slightly over the past 80,000 years (ignoring the effects of culture on diet and health since the Neolithic revolution). As noted earlier, *Homo floresiensis* is not part of this trend. Instead, it could be that this species reduced in size (as island species often do) in the absence of predators.

The causes of this gradual increase in body size are difficult to establish. It may be that a move from forests to a more terrestrial habitat relaxed the constraints on the size of arboreal species, or that in the more open environments that early hominins occupied, predation risks were greater and this selected for increased size (Foley, 1987). Whatever the causes, the ecological and evolutionary consequences were profound. One of the most significant effects relates to the fact that the metabolic rate of an animal rises with body weight in line with well-established physiological principles and according to the following equation:

$$M = KW^{0.75}$$

where: M = metabolic rate (energy used per unit of time)

W = body weight

K = some constant

The effect of this relationship is that, as the size of hominins grew, so the absolute requirement for the intake of calories grew, but the rate of input of calories per unit of body mass fell. This effect can easily be seen in mice and humans. Mice consume absolutely less food per day than a human, but a mouse will typically eat half its own body mass of food in one day and spend most of the day finding it. Humans eat only about one twentieth of their body mass in food each day and thankfully have plenty of time left over to read books on evolution. The consequence for hominins of the absolutely larger amount of food that had to be found as size grew was that their home range increased. This was not the only option available of course. Gorillas, in contrast, grew in size but became adapted to eating large quantities of low-calorie foodstuffs, such as leaves. This requires a large body to contain enough gut to digest the plant material, and this is essentially the strategy also pursued by large herbivores such as cattle. Chimps and early humans clearly opted for the high-calorie, mixed diet that required clever foraging to obtain.

Increases in size may in themselves have been a response to ecological factors, but it is important to note that such increases can have major effects on the lifestyle of an animal and, through complex feedback effects, force it to adapt further in other ways. As noted earlier, an increase in size means that animals incur an increase in absolute metabolic costs, costs that could be met by increasing the size of the foraging range. Large animals also have a smaller surface area to volume ratio than small ones, which in tropical climates would lead to problems of overheating and consequently a greater reliance on water. The upright stance of hominids may have been a response to this, since standing upright exposes less surface area to the warming rays of the sun. Body fur loss may have also have helped with temperature regulation. Larger animals take longer to mature sexually, so offspring

become expensive to produce and require longer periods of care. As a result, the hominins became increasingly K selected (see Chapter 8) and the kin group and larger social groups probably became important for care and protection (see Foley, 1987).

5.5.2 Sexual dimorphism

The altricial condition of human infants required a different social system for its support than the uni-male groups of our distant Australopithecine ancestors. As brain size grew, so infants became more dependent on parental care. Women would have used strategies to ensure that care was extracted from males. This would lead to the emergence of a more monogamous mating pattern since a single male could not provision many females. It is significant that the body size sexual dimorphism of hominids during the Australopithecine phase was such that males were sometimes 50 per cent larger than females. This dimorphism was probably driven by intrasexual selection (see Chapter 4) as males fought with males for access to multi-female groups. By the time of *Homo sapiens*, this difference in size had reduced to 10–20 per cent, signalling a move away from polygyny towards monogamy (Figure 5.9). Women probably ensured male care and provisioning for their offspring by the evolution of concealed ovulation. The continual sexual receptivity of the female and the low probability of conception per act of intercourse ensured that males remained attentive.

So our own species emerged about 150,000 years ago in Africa and has since come to dominate the globe. So much so that a new term – the Anthropocene – has entered the popular scientific literature to describe the geological period in which we now live. The next chapter examines the features of our species that make up the rather unusual hominin package.

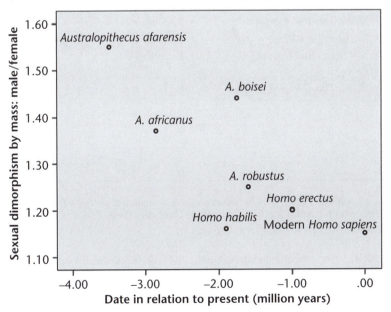

FIGURE 5.9 Sexual dimorphism in selected hominin species.
Dimorphism has gradually decreased over the last 4 million years suggesting a move away from polygyny towards milder polygyny or monogamy.
SOURCE: Data from Ruff (2002).

Summary

▶ According to phylogenetic classification, Humans belong to the subfamily Homininae (hence they are hominins); the family Hominidae (hence they are also, like chimps and gorillas, hominids) and the superfamily Hominoidea (hence, like chimps, gorillas, orang-utans and gibbons, they are hominoids). Hominins are humans and their ancestors lead back to the last common ancestor shared with chimpanzees.

▶ The hominins originated in East Africa and became adapted to an open savannah environment. A whole range of hominin species once existed but *Homo sapiens* are the last remaining.

▶ The most popular theory of the global dispersion of *Homo sapiens* is the 'Out of Africa model' which proposes that *Homo sapiens* originated in Africa about 150,000 years ago. A population left Africa around 72,000 years ago and gradually spread across the Old and then the New Worlds.

▶ The evolution of the hominins in the direction of *Homo sapiens* is associated with an increase in body size, reduced sexual dimorphism and increases in brain size.

Key Words

- Altricial
- Australopithecines
- Cladistics
- Exogamy
- Genus *Homo*
- Hominids
- Hominin
- Mitochondrial DNA
- Mitochondrial Eve
- Multiregional model
- Order
- 'Out of Africa' hypothesis
- Precocial
- Speciation
- Taxon
- Taxonomy

Further reading

Crawford, M. H. and B. C. Campbell (eds.) (2012). *Causes and Consequences of Human Migration: An Evolutionary Perspective.* Cambridge, UK, Cambridge University Press.
Explores latest research on ancient and recent human migrations.

Lewin, R. and R. Foley (2004). *Principles of Human Evolution.* 2nd edition. Oxford, UK, Blackwell Science.
Classic work on the hominin lineage.

Muehlenbein, M. P. (ed.) (2010). *Human Evolutionary Biology.* Cambridge, UK, Cambridge University Press.
Excellent set of articles about important contemporary issues in human evolution.

This Quintessence of Dust: The Hominin Package

What a piece of work is a man! How noble in reason, how infinite in faculty! In form and moving how express and admirable! In action how like an Angel! In apprehension how like a god! The beauty of the world! The paragon of animals! And yet to me, what is this quintessence of dust?
Shakespeare, *Hamlet*, Act II, sc. ii

As discussed in the previous chapter, humans are primates belonging to the superfamily Hominoidea, a taxon we share with other apes such as chimpanzees, gorillas, orang-utans and gibbons. Humans are a unique species, but so, by definition, are all species; yet humans have a number of characteristics, sometimes, but not necessarily, unique in themselves, which taken together identify us as a very unusual species. A definitive list of these characteristics is open to interpretation, but Table 6.1 gives a list of those often proposed.

In this chapter, space permits an examination of just a few of the attributes listed in Table 6.1: bipedalism, hair loss, brain size and language. Examinations of other features are scattered throughout the other chapters, especially Chapter 8.

6.1 Bipedalism

Since no other primate walks or runs with an upright gait, bipedalism represents one of our major defining characteristics. Bipedalism occurred before the rapid enlargement of the hominin brain and there are a variety of theories to account for its origins and adaptive advantages compared to quadrupedalism or occasional knuckle walking as practised by other primates. Habitual bipedality explains why humans have leg bones about 34 per cent longer than non-human apes relative to arm length: for bipedal locomotion leg length must be about 50 per cent of total height to be mechanically and hence energetically efficient. Table 6.2 shows a selection of popular theories to account

for the origins of bipedalism, although there is as yet no overall consensus on which theory is most likely.

6.2 Hair loss

Two notable features of *Homo sapiens* are hairlessness and bipedality and it has long been suspected that the two must be linked in some way (Wheeler, 1984). To be more precise, the human body is actually covered by hair with a similar follicle density to other primates: these hairs, however, are thin and short (so-called vellus hairs) and therefore of low visibility. Nevertheless, this means that of all the 193 species of primate only humans have this 'glabrous' appearance. There have been a variety of hypotheses to account for hair loss (or reduced coverage) in the hominin lineage and several are considered below (Rantala, 2007).

6.2.1 The cooling hypothesis

This hypothesis provides what is probably the most plausible account of hair loss and has been proposed a number of times in variant forms (for example, Morris, 1967; Mount and Mount, 1979). In a persuasive series of papers Peter Wheeler (1984) used mathematical models to examine the implications of heat loss and heat gain for a bipedal hairless primate. The basic argument was that as early hominins abandoned the cover of a forest canopy and began to spend more time in the open savannah so they faced

Table 6.1 Some defining characteristics of *Homo sapiens*

Feature	References
Lack of *baculum* or *os penis*. Humans differ from Old World primates and apes in the complete absence of a penile bone (*os penis*) (a feature they share with spider monkeys). One theory is that bipedalism left male genitalia open to injury; another is that the hydraulic erection required in the absence of a penis bone could serve as an honest signal of fitness.	Gilbert and Zevit (2001); Martin (2007)
Relative hairlessness. Humans have about the same density of hair follicles as other apes but clearly the hairs are short, thin and weak. Various theories have been proposed to account for this, including the cooling hypothesis and as a means to reduce ectoparasite load (see Section 6.2).	Pagel and Bodmer (2003); Ruxton and Wilkinson (2011)
Slow life histories. All great apes exhibit slow life histories but humans are even slower (see Chapter 8). Compared to other great apes humans live longer, have lower adult mortality, start reproducing later, have shorter inter-birth intervals and produce offspring of a higher altriciality. This complex mixture of characteristics is likely to be related to the advantages of cooperative breeding, rapid brain growth and the importance of cultural learning.	Robson and Wood (2008); Bogin (2010)
Large fatty neonates. Human babies are relatively large compared to offspring of other apes and are exceptionally fatty (see Section 8.3.1), carrying more body fat as a percentage of weight than any other primate neonate. It is likely that this fat serves as an energy store to supply a large and demanding brain.	Kuzawa (1998); Cunnane and Crawford (2003)
Prolonged postnatal brain growth. In most mammals brain growth slows down relative to body growth after birth. Humans have a remarkable period of rapid brain growth after birth, making neonates very dependent on parental care.	Martin (2007)
Heavy menstrual bleeding. Amongst mammals menstrual bleeding is mostly confined to Old World primates and a few shrews. But in humans menstrual bleeding is especially heavy. Currently there is no consensus on the adaptive function (if any) of this phenomenon (see Section 8.3).	Strassman (1996); Martin (2007)
Largest brain. Humans have the largest brain of any primate, and, when scaled for body mass, the largest brain of any animal. Given that brain tissue is energetically expensive, that brain size grew rapidly during hominin evolution, and that our brain size is perhaps the single most important attribute that separates us from other species, there are many theories to explain the evolutionary function of brain growth (encephalisation).	Walker et al. (2006); Herculano-Houzel (2009)
Brain lateralization. Compared to other primates the human **cerebral cortex** has a highly asymmetrical distribution of functions. Most humans (90 per cent), for example, are right handed and most language functions exist in the left hemisphere.	Sherwood, Subiaul and Zawidzki (2008); Alba (2010)
Language. The evolution of human language is one of the most controversial topics in anthropology. Whereas other animals have communication systems, humans are unique in their ability to construct immediate and symbolic meaning from sounds, gestures and images.	Dunbar (1996b); Sherwood, Subiaul and Zawidzki (2008)
Bipedalism. Other animals can walk on two legs (primates temporarily, kangaroos, birds) but the type of upright walking practised by humans (so-called orthograde, using the pendulum motion of the limbs) is unique to the hominin lineage and is not found in any other extant or extinct taxa. Over the last 100 years there have been at least 30 hypotheses to account for the evolution and function of human bipedalism (see Section 6.1).	Niemitz (2010)

higher temperatures than those to which they had previously been adapted. In this scenario, the loss of body hair coupled with an upright stance gave several advantages. Bipedal locomotion meant that less solar radiation was absorbed during the middle of the day (a time when early hominins were assumed to forage and hunt) since the vertically exposed surface area of a bipedal ape is less than that of a quadrupedal one (Figure 6.1).

The loss of hair would then allow the evolution of sweat glands allowing evaporative cooling, particularly from upper body regions such as the head that would be exposed to higher air speeds (air speed rises with distance above the ground). For this process vellus hair might even help cooling since such hairs act like miniature wicks assisting the evaporation of sweat droplets.

But not everyone has been convinced by this story. Do Amaral (1996), for example, argued that

Table 6.2 A selection of theories devised to explain the origins of human bipedalism

Theory	Description	References
Technological manipulation	Darwin in *The Descent of Man* (1871) suggested that bipedalism freed the hands for a variety of manipulative tasks such as making tools and weapons. Bipedalism frees two hands to make tools and fashion objects. One problem is that bipedalism appeared at least 1.5 million years before stone tools. Another problem is that primates do not assume a bipedal stance when they manipulate objects.	Darwin (1871)
Throwing of stones	Bipedalism helped early humans throw stones effectively at predators and prey. One problem is that bipedalism predates stone tool production (which may have been used as missiles) and also before the rapid brain growth in the *Homo* genus – and brain power would have been required for the fine hand–eye coordination required in accurate throwing.	Fifer (1987); Young (2003)
Thermoregulatory	Standing upright reduced the thermal load on hominins in tropical climates. Walking upright would reduce exposure by one third of that experienced walking on all fours (Wheeler, 1991). More recent work suggests that the most important source of heat that needed to be dissipated came from internal metabolic energy and not incident sunlight. Another problem is that this idea works for a savannah scenario with high incident radiation and yet many now believe that bipedality first evolved in a more forested environment.	Wheeler (1984); Wheeler (1991); Ruff (2002)
Carrying objects and infants	As the dietary resources of hominoids in the late Miocene became scarce and more dispersed, so an efficient mode of travel carrying objects over large distances became advantageous. This idea was extended to suggest that the carrying of infants during bipedal locomotion was energetically efficient. Watson et al. (2008), however, estimated the energetic burden of women carrying a child at the hip (i.e. without tools such as slings) and found it to be very large. Moreover, other Old World monkeys such as baboons seem to have solved the problem of carrying infants efficiently without bipedal locomotion. In essence, carrying a child whilst walking upright only becomes efficient after the evolution of bipedalism and the brain growth needed to devise support mechanisms. It cannot, therefore, be a precursor to bipedalism. There is more support for the idea that bipedal posture may have initially helped in food gathering and then freed the hands to enable food to be carried over large distances.	Rodman and McHenry (1980); Zihlman (1989); Watson et al. (2008); Niemitz, (2010)
Long-distance running	Humans are very efficient at running at a medium speed (jogging) even in the middle of a hot day. So bipedal running efficiency may have helped catch prey. One problem is that early hominins were probably vegetarian.	Bramble and Lieberman (2004)
Efficiency of walking in savannah compared to arboreal habitat	Lynne Isbell and Truman Young have suggested that increases in the ancestral human group size necessitated travel over large distances to find enough food to support larger groups and this selected for an energetically efficient walking gait.	Isbell and Young (1996)
The orthograde scrambling hypothesis	This hypothesis draws upon the behaviour of orang-utans who often can be observed walking in an upright posture (on the ground but more usually in the upper canopy) using branches to support their weight. The argument is that this type of locomotion represents an ancestral form which was retained and then modified in later hominin evolution. The problem with this idea is that the pongo clade split from the human/gorilla/chimp clade some 12–15 million years ago leaving many millions of years of separate evolution between humans and orangs.	Thorpe and Crompton (2006)

Table 6.2 cont'd

Theory	Description	References
The Amphibian Generalist Theory	One of the more recent theories suggests wading behaviour drove the evolution of bipedalism. In wading, the weight of the body is reduced and an upright posture is almost obligatory in deeper water. The suggestion is that during the late Miocence in Africa there were many patchwork areas of forest containing miles of rivers and banks of lakes and streams containing valuable food resources. Proponents of this hypothesis such as Carsten Niemitz (2010) suggest that wading may have been the original selective force but then once bipedalism was tentatively established the advantages suggested by the other theories may have consolidated the transition.	Niemitz (2010)

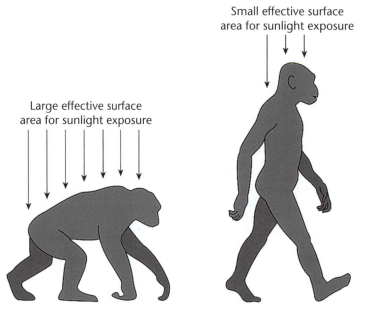

FIGURE 6.1 An explanation of the advantage of bipedalism in the middle of the day.
With the sun nearly directly overhead in the middle of the day in tropical regions bipedal primates would absorb less energy from the sun. The reasoning is probably flawed, however, since body hair greatly reduces the external energy actually absorbed.

over the course of 24 hours a naked skin was at a disadvantage compared to a hairy skin. This follows from the fact that hair is a good thermal insulator so on a hot day whilst the surface of the fur becomes hot it quickly loses heat by radiation and convection rather than by conduction down to the skin and so into the body. From this perspective a hominid evolving in a hot open environment should actually increase rather than decrease the density of its body hair. Such a trend is seen in various subspecies of savannah monkeys (such as vervet monkeys – *Cercopithecus aethiops pygerythrus*) which inhabit south and east Africa (a similar region to where humans are thought to have originated), are active during the day and yet have a denser coat of fur than most forest-dwelling primates (Mahoney, 1980). Furthermore, exposure of naked skin to sunlight risks skin damage whilst at night the lack of a coating of fur means that heat and hence energy resources are lost from the body.

Clearly to resolve these opposing arguments the balance of various factors must be considered using robust mathematical models. One key factor insufficiently considered by early theorists is the fact that physical activity, such as walking or running, itself generates waste heat that must be dissipated if humans are to avoid hyperthermia. More recently Ruxton and Wilkinson (2011) have extended Wheeler's earlier work and have developed a plausible mathematical model considering the heat balance of bipedal and quadrupedal apes walking at various times of the day in a hot open savannah environment. In one typical model they assumed the ape to be moving at about 1.2m/second on a hot cloudless day that reaches a maximum temperature of 40°C.

The model revealed that there was little variation across the day in the heat to be lost by a hairy biped or quadruped; this is because the main factor governing the heat balance is the metabolic heat generated through activity (about 200 watts), which does not change under the assumption of walking at all times. Even in the middle of the day there is relatively little heat gain from the environment since fur is such a good insulator: the outer part rises in temperature but then loses heat by radiation and convection, minimising the flow of heat down into the body. But the model demonstrated that for a hairless biped or quadruped the heat load varied considerably across the day since the thermal buffering provided by fur had been lost. Early in the day the heat load is negative meaning that the primate is actually losing body heat; by the middle of the day quadrupedal primates need to lose more heat than bipedal ones due to the larger surface area exposed to incident radiation.

The authors argue that the best interpretation of this model is that bipedalism was probably not selected for thermoregulatory reasons: hairless bipedal and hairless quadrupedal organisms face a similar pattern of heat load variation which must be dissipated through sweating. But the model, they argue, is consistent with the idea that hairlessness may have evolved to cope with heat loads once bipedalism had evolved for other reasons. When sweating mechanisms evolved to modern levels of efficiency, walking in the middle of the day became feasible for hairless bipeds (so long as they had a supply of water) but not for hairy ones where the heat generated is greater than that that can be lost through sweating. If this model is accurate it suggests that bipedalism evolved first (for non-thermoregulatory reasons) then loss of hair coupled with the evolution of more sweat glands occurred as early hominins adapted to walking during the middle part of the day.

Fossil evidence suggests that bipedalism evolved about 4 million years ago with the Australopithecines. Pointers to the timing of loss of hair are hard to come by (hair loss obviously leaves no trace in the mineral fossil record) but studies on the evolutionary history of hominin skin and fur parasites suggest loss of hair at about 3 million years ago (Weiss, 2009). Another analysis based on the examination of silent mutations at the melanocortin I receptor locus of genes involved in skin and hair pigmentation suggests that humans have been hairless for at least 1.2 million years (Rogers, 2004).

6.2.2 Hairlessness and hunting

For many years it was thought that hairlessness was an adaptive advantage for the hunting activities of early man. Vegetarian primates (like Lucy) so the argument runs, did not need to chase their food so cooling would not be a big problem; but, as Desmond Morris (1967) postulated in his bestselling popular science book *The Naked Ape*, carnivorous primates would overheat during hunting unless they lost their body hair. The model of Ruxton and Wilkinson above shows that even at a walking speed of 1.2m/second (approximately 2.9 miles per hour – a relaxed walking pace) hairlessness is favoured for thermoregulatory reasons. But we could suppose that running during hunting accelerated this trend. One big problem with the hunting hypothesis is that it is usually assumed that males did more daytime chasing and hunting of prey than women, suggesting that we ought to observe some sexual dimorphism in body hair with men having less pronounced body hair than women. Yet in fact the opposite is the case: irrespective of the tendency of many Western women to depilate, females do have less pronounced body hair than males.

6.2.3 Hairlessness, ectoparasites and sexual selection

Cooling through sweating may be the front-running theory in attempts to account for hairlessness but this does not rule out other selective pressures pushing in the same direction. One interesting idea, first suggested by Belt (1874) then rejected by Darwin (1899) but recently revived by Rantala (1999) and Pagel and Bodmer (2003), concerns the role of ectoparasites (lice, fleas, ticks and so on) that plague the human body. The central argument, in its revised form, is that once humans became group hunting primates and had developed adaptive strategies such as tool making, control of fire and food sharing, so they tended to live in relatively stable home bases. It was these home bases and camps, however, which also provided favourable breeding grounds for human parasites. In support of this idea is the fact that out of 193 species of primate only humans harbour fleas (Rantala, 1999), most likely because fleas need a constantly inhabited den or lair to complete their life cycle (they have a larval stage that feeds on matter outside of the body). The idea then is that the loss of body hair made life more difficult for these parasites and enhanced human survival; it would also have enabled humans to inspect each other for signs of infestation. Such parasites, we should remember, were not merely a source of inconvenient itching and irritation: fleas and ticks can carry a range of lethal viral and bacterial diseases that still threaten public health today (Cutler et al., 2010).

One advantage of this hypothesis is that it may contribute towards an understanding of the apparent male preference for females with low levels of body hair. Various studies have shown that young Western females regularly remove leg and underarm hair and offer reasons related to femininity and attractiveness to account for this (Tiggeman and Lewis, 2004; Tiggeman and Hodgson, 2008). Any inspection of a pharmacist or a drug store will also show that there is a mass market for hair-removal products aimed primarily (although not exclusively) at women. This may of course be a passing cultural phenomenon (after all, the fashion to grow or remove the male beard comes and goes for a variety of religious and cultural reasons); or it may be an exaggeration of an already sexually dimorphic trait. This raises the question as to why women should have less body hair in the first place and why men find this attractive. One possibility, suggested by Rantala (2007), is that if early hominin females spent more time in the home camp compared to males who were out hunting, then the loss of hair in females would have been adaptively more advantageous in reducing ectoparasite infestation. This ecological advantage then became hitched to the even more powerful force of sexual selection, since males who preferred hairless females had higher reproductive success; in addition hairlessness becomes an indicator of femininity in itself. With this reasoning in mind Prokop et al. (2012) predicted that women living in areas with a high prevalence of pathogens would be less likely to show a preference for male body hair compared to women living in areas of low pathogen density. When a study was conducted on female university students in Turkey (high pathogen density) and Slovenia (low pathogen density) no difference was found, and both groups of women were recorded as disliking male chest hair. This casts some doubt on the linkage between hairlessness, ectoparasites and mate choice criteria; but the sample population (students living in countries with access to medical facilities and subject to other associations of human body hair) may not be the ideal group to study.

Box 6.1 Of lice and men: dating the origin of clothing

Hairlessness might be a fine solution for early hominins roaming and running over open territory in the tropics during daylight hours, but during the night and before sunrise humans would have faced the problem of a net loss of body heat. In addition, as soon as those hairless bipedals *Homo sapiens* moved out of equatorial Africa, lower ambient temperatures would have made the problem more acute. The problem was solved, of course, by the invention of clothing, a cultural rather than physiological adaptation. Dating the innovation of clothing is not easy. There are archaeological remains associated with clothing manufacture such as bone needles which can be dated to about 40,000 years ago. There are also flint tools known as 'scrapers' which are thought to have been used to scrape flesh and tissue off animal hides which are much older. One ingenious approach to answering this question has come from the analysis of human body and hair lice. The human head louse (*Pediculus humanus captis*) and the human body louse (*Pediculus humanus corporis*) are strict obligate human ectoparasites – meaning that they only thrive by feeding on the living tissues of a single species (us). As indicated by the subspecies nomenclature, the head louse lives only on the scalp; the body louse, however, lives on the body but lays its eggs in clothing and needs clothing to complete its life cycle. It a reasonable assumption then that these two species, morphologically identical and differing only in habitat, diverged when humans began to wear clothing. Ralf Kittler et al. (2003) at the Max Planck Institute in Germany realised that if the date of this initial divergence could be dated (using random changes to DNA bases as a form of molecular clock) then it would give an approximate date for the origin of clothing. To do this they obtained nuclear and mitochondrial DNA sequences from 40 head and body lice collected from humans at various geographical locations. For calibration of these diverging molecular clocks they used sequences from the chimpanzee louse (*Pediculus schaeffi*),

making the reasonable assumption that when humans and chimps diverged about 5.5 million years ago then the louse would have co-speciated into the host-specific *P. humanus* and *P. schaeffi*. Based on this molecular analysis they concluded that the latter speciation event of human lice into *P. humanus corporis* and *P. humanus capitis* originated about 72,000 ± 42,000 years ago. Despite the relatively wide margins, this figure neatly coincides with current estimates for the human migration out of Africa, suggesting that coping with a cooler climate was indeed a driving force behind the adoption of clothing (Figure 6.2).

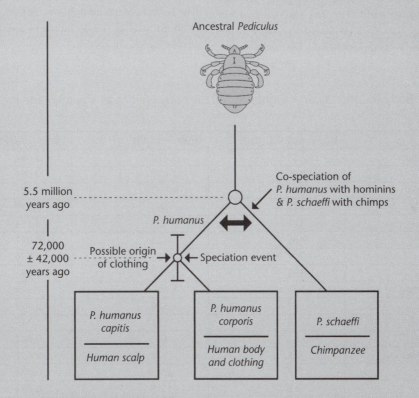

FIGURE 6.2 The speciation of lice.
When humans diverged from chimps about 5.5 million years ago it is suggested that the ancestral louse (Pediculus) split into two species *P. humanus* and *P. schaeffi* living on human and chimpanzee hosts respectively. *P. humanus* then speciated about 72,000 years ago into two human types: *P. humanus corporis*, living on the body and in clothing, and *P. humanus capitis*, living on the scalp. This gives a rough date for the origin of clothing at about 72,000 years ago.

6.3 Brain size

6.3.1 What makes humans so special?

Is there an area of the brain unique to humans? In the middle of the 19th century (c. 1858), Richard Owen, an anatomist who vigorously opposed the application of Darwinism to humans, thought that there was. He claimed that humans have a special structure in the brain called the 'hippocampus minor' that is not found in apes. This, he argued, was clear evidence that we could not have descended from the apes; here was the seat of human distinctiveness. His hopes for a special status for humans were, however, short-lived: in 1863, a few years after Owen announced his thoughts, Darwin's 'bulldog' champion, Thomas Henry Huxley, rushed to the fray and conclusively demonstrated that apes possessed the same structure that Owen had identified. The debate was parodied in popular culture. In Charles Kingsley's *Water Babies*, published in 1863, for example, there is much talk of 'hippopotamus majors' (see Gross, 1993).

Since the time of Owen and Huxley, there have been numerous attempts to establish which features of the human brain, if any, confer upon humans their unique qualities. It is tempting to think that we simply have bigger brains than other mammals, but even a cursory examination of the evidence rules this out. Elephants have brains four times the size of our own, and there are species of whale with brains five times larger than the average human brain. We should expect this of course – larger bodies need larger brains to operate them. The next step would be to compare the relative size of brains among mammals (that is, the ratio of brain mass to body mass). The results are unedifying: we are now outclassed by such animals as rhesus monkeys and the common mole (Table 6.3).

Table 6.3 Body weights, brain weights and brain/body weight as a percentage for selected animals

Animal	Body weight (kg)	Brain weight (g)	Brain/body (%)
African elephant	6,654	5,712	0.086
Pilot whale	3,178	2,670	0.084
Asian elephant	2,547	4,603	0.181
Giraffe	529	680	0.129
Horse	521	655	0.126
Cow	465	423	0.091
Gorilla	207	406	0.196
Pig	192	180	0.094
Bottlenose dolphin	154	1,600	1.039
Human	**62**	**1,320**	**2.129**
Sheep	55.5	175	0.315
Chimpanzee	52.2	440	0.844
Rhesus monkey	6.8	179	2.632
Cat	3.3	25.6	0.776
Rabbit	2.5	12.1	0.484
Mountain beaver	1.35	8.1	0.600
Guinea pig	1.04	5.5	0.529
Rat	0.28	1.9	0.679
Mole	0.122	3.0	2.459
Mouse	0.023	0.4	1.739
SOURCE: various sources including Jerison (1973); and Sacher (1959).			

6.3.2 Allometry

If we wish to find some basis for our special status we can find some reassurance in the phenomenon of **allometry**. As an organism increases in size, there is no reason to expect the dimensions of its parts, such as limbs or internal organs, to increase in proportion to mass or volume. If we simply magnified a mouse to the size of an elephant, its legs would still be thinner in proportion to its body than those of an elephant. This happens in primates too: the bones of large primates are thicker, relatively speaking, than the bones of smaller primates and there are sound mechanical reasons for this.

One of the fundamentals of allometry is that one variable Y (such as brain size, gestation period or lactation period) can be related to a more fundamental one X (usually body size) by an equation of the form:

$$Y = CX^k$$

where C and k are constants. The scaling variable, a, can be obtained from the gradient of a Log Y v. Log X graph since:

$$Log\ Y = kLog\ X + Log\ C.$$

If we apply this to the scaling of brain size with body mass then:

$$Brain\ size = C\ (body\ size)^k$$
$$or\ brain\ size = C\ (W)^k$$

where C and k are constants and W is the body weight (in grammes) of the organism (equation 1).

The constant C represents the brain weight of a hypothetical adult animal weighing 1 g. The constant k indicates how the brain scales with increasing body size and seems to depend upon the taxonomic group in question. Much of the pioneering work in developing these equations was carried out by Jerison (1973), who concluded that, for the entire class of mammals, k was about 0.67 and C about 0.12. There is much discussion about the precise values for these constants, and even within primate groups k varies from 0.66 to 0.88. Martin (1981) revised Jerison's work and concluded that k = 0.76 and C = 0.058.

If we plot a graph of brain size against body weight for mammals on linear scales, a curve results, showing that brain size grows more slowly than body size (Figure 6.3).

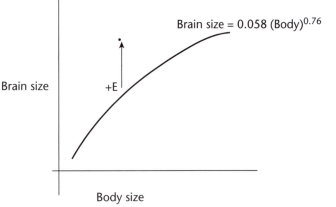

FIGURE 6.3 Growth of brain size in relation to body size for mammals.
An animal occupying a point above the line is said to be encephalised; that is, it has a brain larger than expected for an animal of its body mass. This could be the result of a relative growth in brain size (positive encephalisation, +E).

If we take log values of both sides of equation 1 with the constants for mammals inserted, then:

$$Predicted\ weight = 0.058\ (W)^{0.76}$$
$$log\ (brain\ size) = 0.76\ log\ (W) + log\ (0.058)$$

Thus, a plot of the logs of brain and body size (Figure 6.4) should give a straight line of slope 0.76 and an intercept of –1.24 (i.e. log 0.058).

Figure 6.4 starts to give an indication of what makes humans so special: we lie well above the allometric line seen for other mammals (the deviation above the allometric line may seem small but remember that the Y axis is a logarithmic scale where one unit represents a tenfold difference). If we insert a value of 65 kg as a typical body mass for humans into equation 1, our brains should weigh about 264 g. The real figure is in fact nearly 1,300 g. Our brains are about five times larger than expected for

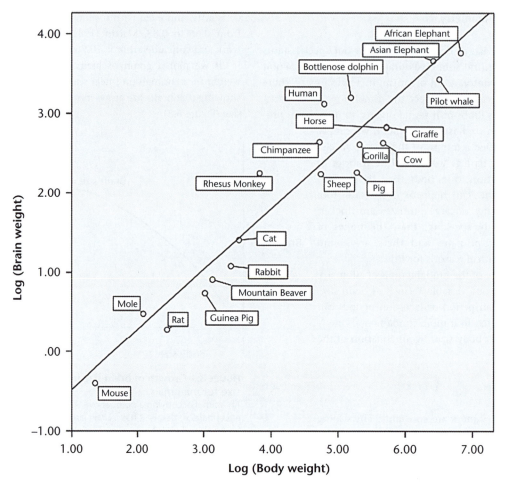

FIGURE 6.4 Logarithmic plot of brain size (g) against body size (g).
Organisms above the line have brains larger than expected from their body sizes. The line of best fit is drawn as:
Log (brain size) = 0.76 log (body weight) + log (0.058).
SOURCE: Data from Sacher (1959); and Jerison (1973).

a mammal of our size and about three times larger than that expected for a primate of our size.

6.3.3 Ancestral brains and encephalisation quotients

A reasonable estimation of the size of the brains of our early ancestors can be obtained by taking endocasts of the cranial cavity of fossil skulls. The construction of endocasts involves pouring some substance that sets inside the hollow cavity of a skull and measuring the volume of the cast that results. There is some debate over how to interpret the fine detail of these casts (such as evidence for folding), but there is consensus on the general trend: about 2 million years ago, the brains of hominids underwent a rapid expansion (Figure 6.5). Australopithecines possessed brains of a size to be expected from typical primates of their stature, but *Homo sapiens* now have brains about three times larger than a primate of equivalent body build. The departure of brain size from the allometric line is known as the **encephalisation quotient** (EQ).

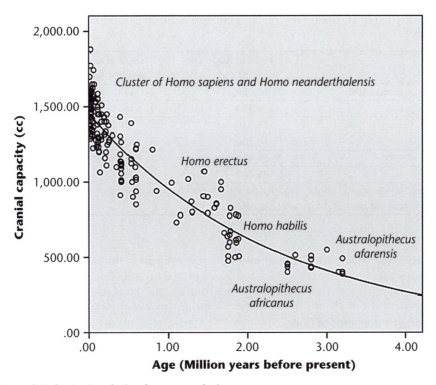

FIGURE 6.5 Growth in brain size during human evolution.
Cranial capacity and age of specimen for 175 fossil hominin skulls.

SOURCE: Data from Table S1 (supplementary information) in Shultz, Nelson and Dunbar (2012). The labels show clusters of specific species.

The encephalisation quotient (EQ) is defined as:

$$EQ = \frac{\text{Actual brain weight}}{\text{Brain weight predicted from allometric line}}$$

The interpretation of encephalisation remains controversial. Intelligence is likely to be far too complex to have a simple relationship with the EQ. This is illustrated by what Deacon (1997) has called the 'Chihuahua fallacy'. Small dogs such as chihuahua and Pekinese are highly encephalised; that is, they lie to the left of an allometric line of brain weight against body weight for carnivores. The reason is that they have been deliberately bred for smallness in body size, but since brain size is far less variable, the breeding programme that led to small dogs has left them with relatively larger brains. But reviewing the evidence, Deacon concludes that the high EQs of humans are not the result of negative somatisation.

In fact, the fossil record shows that hominid body size has been increasing over the past 4 million years and human brain size has grown rapidly in relation to increases in body size.

The calculation of EQs does, however, lead to some odd results. For example, imagine two species, one with a body weight of 5 g and the other of 50 kg, but both with brain volumes 50 per cent higher than expected from allometric scaling. The EQ of both is the same at 1.5, yet it seems unlikely that they are cognitively equal since the large brain has vastly more neural tissue.

It seems clear that EQ is a better measure of intelligence than absolute brain size. But it is probably not a perfect measure of intelligence; for criticisms of the EQ concept see Holloway (1996) and Roth and Dicke (2005).

Now we have established that there is something unique about the human brain, we need to examine

Table 6.4 Body weights, brain weights and encephalisation quotients (EQs) for selected apes and hominids

Species	Typical body weight (g)	Typical brain weight (g)	EQ
Pongo pygmaeus (orang-utan)	53,000	413	1.83
Gorilla gorilla (gorilla)	126,500	506	1.16
Pan troglodytes (common chimp)	36,350	410	2.42
Homo habilis	40,500	631	3.43
Homo erectus	58,600	826	3.39
Homo sapiens	65,000	1,250	4.74

* Calculated from: EQ = Actual brain weight ÷ Predicted brain weight from $0.058 (\text{Body weight})^{0.76}$ (Martin, 1981).

SOURCE: Data from Boaz and Almquist (1997); and Lewin (2005).

why such a risky organ should have evolved to the proportions that it has.

6.3.4 The energetic demands of brains

As humans, it is easy to take the value of high intelligence and large brains for granted and assume that they had self-evident survival value. It may seem obvious that natural selection would eventually deliver intelligent creatures capable of contemplating their own origins. But progress to large brains was never inevitable. Natural selection is not interested in brainpower per se – if small brains are sufficient for the purpose of genetic replication, all the better. There is a curious species of sea squirt (one of the tunicates or 'dead men's fingers'– an edible invertebrate that the French and Japanese consume avidly) that in its larval stage uses its brain to find a suitable rock to cling to; once found, there is no need for a brain and it is absorbed back into the body (the joke in university circles is that this is what happens to academics once they get tenure). It also absorbs its primitive spine and rudimentary eyes.

The problem in explaining encephalisation among the primates is that brains are very energetically expensive. An adult human brain accounts for only about 2 per cent of body mass but consumes about 20 per cent of all the energy ingested in the form of food. The idea that we only use part of our brains is one of those urban myths: brains are made of expensive tissue; natural selection would not continue to fuel brains if there was not some payback. As we rest, the power output of the brain is somewhere between 16 and 20 watts. Unfortunately for thinkers who want to lose weight, this percentage and power output value hardly alters whether we think a lot or little.

To explain the rise in brain volume in the human lineage it is useful to look in a more general way at how the considerable energy demands of a large brain might conceivably be met. There are two basic strategies that a hominin with a primate and mammalian body plan could follow: increase energy inputs relative to body size overall and/or reduce energy allocation to some other bodily function or organ (Figure 6.6).

If we examine first the possibility of an overall increase in energy input then at first glance it might seem that humans did not follow this route. The basic metabolic rates (BMRs) of young (65 kg) males and females are about 81 watts and 70 watts respectively and this is typical of what one would expect from the allometric scaling laws of body mass and BMR. But scaling laws have proven surprisingly controversial (for example, see White and Seymour, 2013); moreover, it is arguable that human body mass needs to be adjusted to take into account the fact that humans have an unusually large percentage of adipose fat compared to our nearest relatives. When this adjustment is done, humans end up with a larger BMR than expected for their body mass. The conclusion, if great apes are indicative of our last common joint ancestor, is that humans have derived in their evolution a larger brain volume, a higher percentage of body fat and a rise in net energy intake. This increase in energy could have come from any

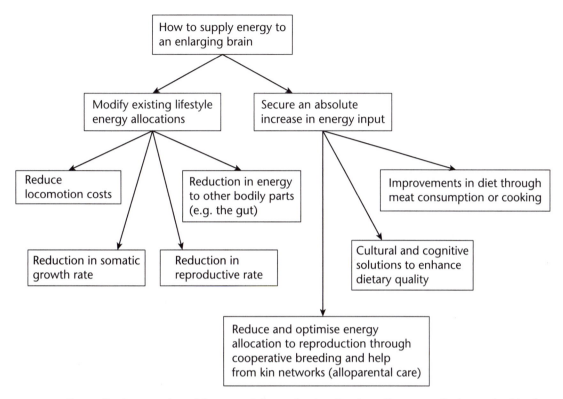

FIGURE 6.6 Generalised energy-based framework for understanding how the energetic demands of brains increasing through evolutionary time could be met.

SOURCE: Based on ideas in Navarrete et al. (2011).

combination of sources identified in Figure 6.6. Diet may have improved, for example, by an increased consumption of meat and energy-rich bone marrow, combined with the use of cooking to release more nutrients from raw food.

The potential importance of cooking is also illustrated by the linkage between diet and time budgets. An animal that travels along the trajectory of encephalisation by increasing overall energy intake also has to spend more time looking for and consuming food. Karina Fonseca-Azevedo and Suzana Herculano-Houzel (2012) have modelled how feeding time can be expected to vary in relation to body size and brain volume for primates. They conclude that there is a ceiling for primates such as orang-utans, gorillas and chimps of about 8 hrs per day feeding time. If humans ate raw food to support their larger brains we would have to spend more than 9 hrs per day feeding – an ecologically unrealistic figure. They argue that cooking may have been the key

to increasing calorific and nutritional inputs without exceeding ecological limits for feeding time.

Another possibility is that cooperative breeding and alloparental care (features ubiquitous in human societies but very rare among other animals) could have increased energy consumption above what would be possible for an isolated individual or single family unit (see Chapter 8). Finally, humans are renowned for eating an incredible variety of foodstuffs and the bio-cultural skills, based themselves on individual cognitive power, needed to exploit a wide range of foods would have ensured (at least up to the Neolithic Revolution) a steady supply of something to eat and so buffer the risk of severe food shortages.

In terms of phylogenetic alterations to energy allocations, there are also several possibilities. As noted earlier in this chapter, moving from a quadrapedal to a bipedal gait may have reduced the locomotive energy costs of early hominins beginning to move across an open, as opposed to an arboreal, landscape.

The other possibility is that hominins began to adopt a slower life history: reducing the rates of growth and reproduction and diverting energy to brain growth instead. Isler and Schaik (2012b) have argued that this tendency towards a slower life history but larger brain must eventually reach a limit (which they call the 'gray ceiling') whereby the risk of natural mortality to a slow-growing organism reduces any gains in reproductive rates from an increased lifespan. Above this limiting brain size the population ceases to be self-sustaining. They estimate that for a primate the upper limit of brain volume to be between 600 and 700 cc – a figure that is close to the value of today's great apes and the extinct Australopithecines. They postulate that humans broke through this barrier through cooperative breeding. The strong pair bonds between males and females, possibly emerging at the time of *Homo erectus* (c. 1.8 million years ago) would have encouraged meat sharing between mating pairs; this, coupled with help from kin (and perhaps especially grandmothers) would have enabled the inter-birth interval to be shortened (whilst still allowing energy to be spared for brain growth) and so ensure viable population growth rates.

This view is consistent with the fact that the real bottleneck in supplying energy to a demanding brain appears to be in periods of prenatal and early childhood growth. Prenatal brains require 60 per cent of basal metabolism and this high figure is maintained for several years after birth (Aiello et al., 2001). This obviously places a huge demand on the energy budgets of mother and infant. It could have been met by changes to social structures and the advents of more support from the father, grandmothering and the sharing of food between unrelated adults.

Alternatively, although reduction in gut size to facilitate brain growth seems to be an unlikely general characteristic in mammals and other primates (Navarrete et al., 2011) there remains the possibility that this was a unique feature of recent human evolution.

Such considerations lie behind the 'expensive-tissue hypothesis' proposed by Aiello and Wheeler (1995) in which they suggested that increases in the brain size of hominins must have been balanced by a reduction in the demands of other organs. Their argument was that since the overall metabolic rates for humans is as expected for mammals of our size, and since our brains are using much more energy than expected, some other organ or organs must have a reduced energy requirement. Figure 6.7 shows the contrast between actual and expected values in the mass of human organs.

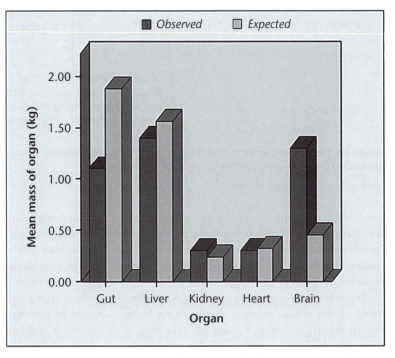

FIGURE 6.7 Expected and observed masses of human organs.
Expected masses are calculated for a typical 65 kg primate. Humans have guts smaller than expected and brains larger than expected.

SOURCE: Data from Aiello, L. C. and P. Wheeler (1995) 'The expensive-tissue hypothesis: the brain and the digestive system in human primate evolution,' *Current Anthropology* **36**: 199–221; Aiello, L. C., N. Bates and T. Joffe (2001) In defence of the expensive tissue hypothesis. In D. Falk and K. R. Gibson (eds.), *Evolutionary Anatomy of the Primate Cerebral Cortex.* Cambridge, Cambridge University Press.

Changes in the allocation of energy or possible net aggregate increases in overall energy inputs, as suggested in Figure 6.7, look like an important facet of human evolution. We should remember that there was nothing inevitable about this: the redirected metabolic energy could have been channelled to other organs to increase reproductive output. The fact that it seems to have been channelled to the brain shows that the brain must have been under strong selective pressure at this time. Energy changes allowed the brain to develop in response to these pressures. So what were these forces placing a premium on brain size?

6.3.5 Theories of brain enlargement

It is interesting to note that for virtually every feature that makes us distinctive as a species, such as bipedalism, hair loss, culture, language, concealed ovulation and, the topic of this section, the evolution of high intelligence, there are multiple theories and no universal consensus on what were the key selective pressures. Table 6.5 gives a selection of some of the more popular theories that have been advanced to account for brain enlargement in the *Homo* genus. These theories are considered in more detail below.

6.3.6 Comparing social and ecological theories

All single-factor models seem to face anomalies and problems. This suggests a multifactoral approach may be better. In the case of social complexity, for example, although chimpanzees have similar social systems to capuchin monkeys, the successful capuchin monkey only has to outsmart other capuchins, not chimpanzees. The immediate ancestors of chimps were already more encephalised that capuchins, perhaps because of ecological driving forces, and so the chimp line started on the road to increased brain size from a different beginning point. We should not expect, therefore, equalities in encephalisation in relation to social systems between taxa.

Teasing out the relative contribution of social, ecological and climatic factors in the evolutionary expansion of the human brain is going to prove difficult. These three categories, of course, need not be mutually exclusive: in harsh environments, there may be social competition to access limited environmental resources, requiring that the brain solve both ecological and social problems to thrive. The relative influence of these factors may also alter with time. Enhanced cognitive ability could be initially selected

Table 6.5 Selected theories of brain enlargement

Theory	Brief details	References
Dietary complexity	Humans adopted complex diets that required enhanced cognitive power to supply.	Milton (1988); Kaplan and Gangestad (2004); Previc (2009)
Tool use	Tool use and development acted as a stimulus to brain growth as the advantages of ever more complex tools were obtained.	Oakley (1959); Wynn (1988)
Ballistics	Throwing objects accurately would have once had great survival advantage, but this act requires sophisticated cognitive coordination.	Calvin (1982)
Sexual selection	Human intelligence is far higher than needed for ecological survival and bears all the signs of a sexually selected trait. Intelligence may serve as a costly honest signal valued by both males and females. Female choice may be the driver of male intelligence but females also need to be intelligent to appreciate displays (through art, language, music, and so on) of male intelligence.	Miller (2000)
Genomic imprinting	Advantages to mothers of well-developed neocortex and advantages to fathers of a more instinct-driven limbic system set up an escalating genomic conflict that increases the size of both components.	Badcock (2013)

Table 6.5 cont'd

Theory	Brief details	References
Machiavellian intelligence and the social brain hypothesis	Due to external ecological factors human group size increased in the *Homo* lineage. This set up tensions in the group to maintain and monitor relationships, this then exerted a pressure selecting for cognitive power to keep track of relationships, assess intentions of others, and engage in manipulation and deceit. A related idea is that of social exchange: humans needed high intelligence to interact and reap benefits of reciprocal altruism and non-zero sum games. This required the detection of cheats.	Byrne and Whiten (1988); Cosmides et al. (2010); Dunbar (1993); Dunbar and Shultz (2007)
Climate changes	Over the last 4 million years – a crucial period in hominin evolution – the Earth's climate has undergone several drastic changes. In the last 750,000 years the climate has fluctuated very rapidly. This selected for enhanced cognitive power to cope with changes in temperature, vegetation growth and resource availability.	Ash and Gallup (2007); Shultz and Maslin (2013)

for in the face of ecological and environmental problems (food and shelter) which once solved give rise to new social problems requiring further brain enlargement to tackle (Holloway, 1975).

A number of studies have been done that attempt to assess the relative contribution of ecological, dietary and social factors in the evolutionary enlargement of the human brain. Both the need to learn complex foraging skills and the need to learn how to function in complex groups would tend to increase the duration of the human juvenile period. By using a series of multiple regression models Walker et al. (2006) tried to estimate the relative importance of these and other factors in the evolution of primate brain sizes and the length of the juvenile period. They collected comparative data on such things as female age at first reproduction, brain weight, lifespan, body size, home range, group size and dietary type for 67 species of primate and performed a variety of regression analyses. They found that three crucial parameters of primate life – length of juvenile period, relative brain size and neocortex brain ratio – were driven by three different sets of socioecological factors. Length of juvenile period was most strongly correlated with diet and lifespan; relative brain size (a measure of overall brain size relative to body size) correlated most strongly with home range; whilst neocortex brain ration (a measure of the size of the neocortex relative to the rest of the brain) correlated most strongly with group size. Overall the authors

concluded that group size (and so by implication social complexity) is the best predictor of neocortical expansion and higher executive-to-brain stem ratios across all primates (see Table 6.6). But for the typical slow life histories experienced by primates – small litter size, long juvenile period and long lifespans – there was no evidence for a single prime mover. Rather it was suggested that there is a complex mixture of social and ecological pressures acting at variable intensities on different primate groups.

Table 6.6 A multiple regression analysis of the effects of various factors on neocortex ratios

Factor	Beta co-efficient[#]	P	Sig. or not sig.?
Body size	0.267	0.013	Sig.
Group size	0.405	0.001	Sig.
Lifespan	0.285	0.007	Sig.
Home range	0.086	0.412	Not Sig.
Fruit and seed in diet	0.067	0.547	Not Sig.

[#] The beta coefficient is a measure of how strongly the factor identified affects the dependant variable (here the neocortex ratio). So, for example, here the biggest effect on neocortex ratio comes from group size where one standard deviation (SD) change in group size will bring about a 0.405 SD change in neocortex ratio.

SOURCE: Data from Walker et al., 2006a, Table 3, p. 484.

In another study to analyse the relative strength of these different pressures, Bailey and Geary (2009) collected data on the cranial capacity of 175 fossilised hominin skulls together with associated measures (that is, in relation to where the fossil specimen was found) of population density, latitude, parasite prevalence, mean temperatures and temperature variability. The idea was to compare the relative influences of ecological factors (as indicated by latitude, parasite prevalence, temperature and climate) and social factors (as indicated by population density). The authors found that all parameters measured – latitude, population density, parasite prevalence, mean temperature and temperature variation – were significant predictors of cranial capacity (see Table 6.7).

As Table 6.7 shows, higher latitudes and temperature variation are both associated with higher cranial capacity (the correlation is positive) – results in keeping with the work of Ash and Gallup noted earlier. The negative correlation with parasite prevalence is interpreted as the availability of metabolic resources to support brain growth in areas of low parasite threat where the immune system does not command high resource allocation. Once the relative contribution of these factors was determined using multivariate analysis tools, it emerged that the primary predictor of cranial capacity was population density, and hence by inference social competition. The study did, however, make a number of questionable assumptions. Temperature variation at the time of the fossil find and the time during which the hominin lived, for example, was estimated from contemporary weather reports which may not accurately reflect ancestral climates. Similarly, population density was estimated from the frequency of fossil crania found in the area, yet the probability of a skull surviving may vary from area to area adding uncertainty to this measure.

Table 6.7 Correlation coefficients of factors influencing cranial capacity based on a sample of 175 crania (from Bailey and Geary (2009))

Factor	Correlation coefficient
Latitude	0.61
Population density	0.79
Mean temperature	−0.41
Temperature variation	0.30
Parasite prevalence	−0.47

A synergistic approach

Studies tend to show that social, ecological and climatic factors all seem to be involved in driving encephalisation. Perhaps the best conclusion so far is that multiple factors acted in synergy. Figure 6.8 shows how social and ecological factors could act in synergy to select for, or reduce the constraints on, brain growth.

6.4 Language

The application of evolutionary theory to the origins of human language has always been controversial. The arguments in the 19th century were so heated that, in 1866, one scientific society, the Societé de Linguistique de Paris, banned all further communications to it concerning the history of language. Over 145 years later, there are still wide-ranging disagreements over such basic issues as the probable timing of the start of human language and even whether language is a product of natural selection or is merely some emergent property of an increase in brain size. There are those who maintain that language first appeared in the Upper Palaeolithic period about 35,000 years ago, and those who suggest that language arrived with the appearance of *Homo erectus* about 2 million years ago. There is only room here to glance briefly at some of the main arguments in these debates.

6.4.1 Natural selection and the evolution of language

The leading exponent of language as a product of natural selection is probably Steven Pinker, a linguist at the Massachusetts Institute of Technology in the United States. Pinker advances a number of arguments tending to suggest that language has been the outcome of a selective force (Pinker, 1994). In summary, these are:

● Some people are born with a condition in which they make grammatical errors of speech. These disorders are inherited (see also Gopnik et al., 1996).

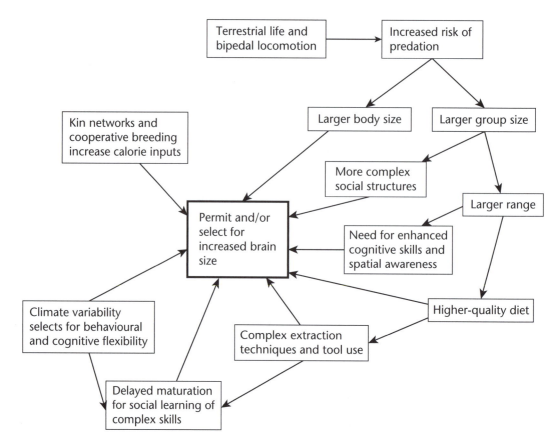

FIGURE 6.8 Conjectured relationship between social and ecological factors acting synergistically to exert a selective pressure on brain size in hominins.

- Language is associated, although not in a simple way, with certain physical areas of the brain, such as Wernicke's and Broca's areas (see below).
- Complex features of an organism that have been naturally selected, such as the eye of a mammal or the wing of a bird, bear signs of apparent design for specialised functions. Pinker argues that language bears these same types of design feature.
- Children acquire language incredibly quickly. Parents provide children with only complete sentences and not rules, but children nevertheless infer rules from these and apply them automatically.
- The human vocal tract has been physically tailored to meet the needs of speech. Specifically, humans, unlike chimps, have a larynx low in the throat, a feature that allows humans to produce a greatly expanded range of sounds compared with a chimp.

- The workings of the human ear indicate that auditory perception is specialised in a manner ideal for decoding speech.
- If languages were products of culture, we would expect some correlation between the level of sophistication of a culture and the grammatical complexity of its language, yet no such correlation is found and even the language of hunter-gatherers is grammatically complex.

The problem that all evolutionary accounts of the origin and development of a complex structure such as an eye or a wing or language have to face is the task of explaining the adaptive advantage of all the intermediate stages. Assuming that an eye, a wing or language did not suddenly fully emerge – and the probability of all the mutations appearing at once would truly be infinitesimal – we must ask what the advantage is of 5 per cent of a characteristic? The charge often levelled against adaptationist accounts of language is that 5 per cent of a language would be useless, and in any case the first language mutant would have no one to talk to. Pinker and Bloom respond to this by arguing that the early language-prone genetic variants would probably be close kin and could thus benefit from sharing language (Pinker and Bloom, 1990). Also, as is well known, 'pidgin' languages, the language of children and non-fluent tourists, show that a continuum of language skills and attributes can exist that still have use. Other animals also show the value of even a limited vocabulary. Vervet monkeys, for example, have alarm calls that distinguish between leopards, eagles and snakes. If language is an adaptation then it can be expected to be associated with specific areas of the brain, and this indeed is what is found. In the second half of the 19th century there were two milestones in understanding the neural basis of language with the discovery of two areas of the brain directly linked with language: Broca's area and Wernicke's area, named after their discoverers Pierre Paul Broca (1824–80) and Karl Wernicke (1848–1905). Both areas are found in the left hemisphere: if you place your finger on your head just above your left temple, Broca's area lies just beneath the skull; by placing your finger just above and behind your left ear you are likely to be pointing to Wernicke's area.

6.4.2 Dating the origin of human language: anatomical evidence

Predictably, if there is debate about whether language is an instinctive evolved trait or a cultural achievement, there will be debate about when it began. If we agree with Pinker that language is a feature of human biology, it must have begun no later than about 200,000 years ago when *Homo sapiens* appeared, and could have been around in earlier hominid species. If we take the cultural achievement view, we must place the origin of language with the origin of symbolic culture, generally around 35,000 years ago. At this point anatomical evidence may be useful.

A sudden expansion in brain size among our ancestors began about 2 million years ago with the first member of our genus, *Homo habilis*. By about

1 million years ago, *Homo erectus* had a brain capacity of about 1,100 cm^3, which is not far from our own of 1,300 cm^3. If language coincided with sudden brain expansion, this points to *Homo erectus* and *Homo habilis* as having had language capabilities. The problem is of course that we do not know whether there is a minimum brain size for language to start. If the minimum were 1,300 cm^3, only early *Homo sapiens* at 200,000 years ago could have had language. We know that language is associated with certain areas of the brain such as Broca's area and Wernicke's area. It has been claimed that an examination of endocasts of fossil skulls shows evidence of the existence of a Broca's area in the cranium of a *Homo habilis* specimen dated at about 2 million years old (Falk, 1983; Leakey, 1994). There have also been attempts to link asymmetries in fossil crania with the origin of language. Language in most people is associated with the left hemisphere, and, partly as a consequence of packing language circuitry into one half of the brain, the left hemisphere is slightly larger. Holloway (1983) suggests that this too can be detected in a *Homo habilis* specimen. Objections to this have included the fact that even specimens of Australopithecines show brain asymmetries, and the attribution of language to this small-brained ancestor seems improbable.

The most compelling anatomical evidence, however, comes from an examination of the throats of humans, apes and our ancestors. The voice box of humans is called the larynx and contains a cartilage, called the Adam's apple, which bulges from the neck. If you feel your Adam's apple, you will notice that it is some distance down from your mouth. Humans have their larynx low in their throats, which allows the space for a large sound chamber, the pharynx, to exist above the vocal cords. In all other primates, the larynx is set high in the throat, which restricts the range of sounds that can be made but does at least allow the animal to breathe and drink at the same time. Humans are in fact born with their larynx high in the throat, which helps the infant to suckle while breathing. After 18 months, the larynx begins to move down the throat, finally reaching the adult position when the child is about 14.

In adult humans, the tongue forms the front wall of the pharangeal cavity, so movement of the tongue enables the size of the pharynx to be altered. In chimps and infants, the higher position of the larynx means that the tongue is unable to control the size of the air chamber immediately above it. The differences between humans and primates in the position of the larynx strongly suggest that we have evolved vocal apparatus to convey a wide range of sounds.

It would seem at first glance an easy task to determine the position of the larynx in ancestral fossils and thus work out when the enlarged speech-facilitating pharynx was established. The problem is that all this vocal apparatus is composed of soft tissue, which does not fossilise. Clues can, however, be found in the bottom of the skull, the basicranium, which in humans is arched as a result of our specialised vocal equipment and quite unlike the flat shape of other mammals. Following this approach, Laitman (1984) found that a fully fledged basicranium emerged about 350,000 years ago. The earliest *Homo erectus* specimen of about 2 million years ago has a slightly curved basicranium typical of the larynx position of a modern 6-year-old boy.

Evolution provided our species with a series of adaptations (bipedalism, hair loss, a large brain, language and so on) to cope with the various environments and ways of life encountered during our history. But it also provided us with the ability to adapt at the individual phenotypic level to conditions around us. These differing types of ongoing adaptations and the timescales over which they operate form the subject of the next chapter.

Summary

🔺 Humans are an unusual species. Some of the characteristics commonly listed as underpinning our distinctiveness are: lack of penis bone; reduced hair thickness; a slow life history; large fatty babies; prolonged brain growth after birth; heavy menstrual bleeding; a large, highly lateralised brain; language; and bipedalism.

🔺 The idea that bipedality reduced thermal heat stress due to a lower effective surface area for incident radiation now seems flawed. The main heat stress on early humans would have come from metabolic energy, and it is likely that this was dissipated by reduced hair covering and the evolution of profuse sweat glands.

🔺 Recent work on parasites places the origin of clothing at around 72,000 +/– 42,000 years ago.

🔺 Possibly the most adequate measure of human brain size is the encephalisation quotient, which measures the ratio of actual brain size to the size of brain expected for an organism of a given size.

🔺 There is currently considerable debate about the relative role of social and ecological factors in accounting for the enlargement of brain size in the hominin lineage.

🔺 A theory currently receiving much attention is the idea of climatic factors causing ecological variability and selecting for a large hominin brain to learn ever-changing coping strategies. The real reason for rapid brain growth in the hominins may be a synergistic mix of factors and influences.

Key Words

- **Allometry**
- **Cerebral cortex**
- **Encephalisation quotient**
- **Machiavellian intelligence**

Further reading

Byrne, R. (1995). *The Thinking Ape*. Oxford, Oxford University Press.
A clear exposition of the Machiavellian intelligence hypothesis.

Deacon, T. (1997). *The Symbolic Species*. London, Penguin.
An evolutionary account of the growth of human brains that stresses the importance of the coevolution of language and the brain. Deacon argues that the ability of the mind to construct and hold symbols is key.

Dunbar, R. I. M. (1996). *Grooming, Gossip and the Evolution of Language*. London, Faber and Faber.
A readable and popular account of the evolution of brain size and its possible causes.

Grove, M. (2012). 'Orbital dynamics, environmental heterogeneity, and the evolution of the human brain.' *Intelligence* **40**(5): 404–18.
Paper that stresses the importance of the Earth's orbital cycles in influencing climate and brain growth.

Jones, S., M. Robert and D. Pilbeam (eds.) (1992). *The Cambridge Encyclopedia of Human Evolution*. Cambridge, Cambridge University Press.
Numerous experts have contributed to this book. Thorough and well illustrated with plates and diagrams. Excellent for comparing human and primate evolution.

Rizzolatti, G., L. Fogassi and V. Gallese (2006). 'Mirrors in the Mind.' *Scientific American* **295**(5): 30–7.
Explains the discovery of the mirror neuron.

Shultz, S. and M. Maslin (2013). 'Early human speciation, brain expansion and dispersal influenced by African climate pulses.' *PloS one* **8**(10): e76750.

Shultz, S., E. Nelson and R. I. Dunbar (2012). 'Hominin cognitive evolution: Identifying patterns and processes in the fossil and archaeological record.' *Philosophical Transactions of the Royal Society B: Biological Sciences* **367**(1599): 2130–40.
Paper that examines the role of climate variability and concludes that social factors are more important.

Adaptations and Developmental Plasticity

Adaptations and Evolved Design

Every organized natural body, in the provisions which it contains for its sustentation and propagation, testifies a care on the part of the Creator expressly directed to these purposes.

(William Paley, *Natural Theology*, 1836, p. 486)

Slow though the process of selection may be … I can see no limit to the amount of change, to the beauty and infinite complexity of the co-adaptations between all organic beings, one with another and with their physical conditions of life, which may be effected in the long course of time by nature's power of selection.

(Darwin, *Origin of Species*, 1859b, p. 109)

7.1 Teleonomy

One of the central problems of biology is the phenomenon of adaptation: how organisms and parts of organisms come to look as if they have been designed for a purpose. Prior to Darwin, many commentators on the natural world, such as William Paley, quoted above, struck by the exquisite match between the shape, appearance and behaviour of living things, employed teleological language: things were designed for a purpose, and even nature itself was conceived as having goals and purposes. Consequently, so the argument ran, just as there can be no artefacts without an artificer, so evidence of design implied the existence of a Creator.

Despite the fact that we now have a naturalistic account of how the form of living things is so suited to their function, it has proved difficult to eradicate teleological language, and many biologists make copious use of inverted commas to signal a non-literal interpretation of purposefulness in their writing or speech. It is very easy to suggest, for example, that a plant grows a flower *in order to* attract pollinators. Taken literally it implies that flowers have some sort of conscious plan, such that, knowing the preferences of insects, they have designed colourful, odorous and nectar-rich extensions. But to say plants grow flowers because over a long period of time those that didn't failed to reproduce as successfully as those that quite by chance threw up appendages that attracted insects sounds terribly pedantic and long-winded. The dilemma is such that a new term, '**teleonomy**', has arisen to cater for the need to use purposeful language within a scientific paradigm that denies any ultimate purpose. Teleonomy expresses the idea that adaptations can arise by purely naturalistic causes, such as natural and sexual selection, which serve goal-directed functions in the absence of any prior purpose.

It is also a problem for science educators since it seems that teleological-style explanations of biological phenomena represent a default position in human cognition. This may be linked to the tendency for humans to see agency at work (see also Chapter 9). Deborah Kelemen and Evelyn Rosset (2009) looked at the tendency among children and adults to resort to, or accept as correct, teleological-type explanations (such as 'the Earth has an ozone layer to protect it from UV light', 'earthworms tunnel underground to aerate the soil', or 'bees frequent flowers to aid pollination') when forced to make decisions under time pressures. Their study showed that even after completing college-level science courses adults easily flipped over to teleological explanations – a tendency

111

they called 'promiscuous **teleology**'. Interestingly, no link between belief in God and the acceptance of teleological ideas was observed.

In evolutionary studies of human behaviour the concept of adaptation is obviously crucial. The problem is that the term has a variety of different meanings and usages. In their early manifestations, sociobiology and evolutionary psychology tended to focus on the behaviour and cognition of adults to demonstrate how these promote (or once promoted) fitness-related goals. As discussed in Chapter 2, human behavioural ecologists make the phenotypic gambit that contemporary humans will still behave in an adaptive way; that is adjust their behaviour to secure reproductive fitness. Within a wider evolutionary framework, physical anthropologists might talk about the human body as showing adaptations to a particular mode of life. We also have the more common everyday usage of the term with the suggestion that humans can gradually become adapted or acclimatised to local conditions. This chapter aims to clarify what adaptations are, how they arise and how they are inherited.

One way of thinking about the process of adapting and acquiring adaptations is along a time continuum. Figure 7.1 shows how this can be conceptualised. At one end of the spectrum, acting over seconds, hours or days, humans have inbuilt feedback mechanisms that respond to environmental variables (both external such as temperature and internal such as blood glucose levels) that allow rapid adaptation to changing conditions to maintain homeostasis. At the other extreme, acting over hundreds and thousands of years, we have the processes of natural and sexual selection determining the fate of genes and acting on gene frequencies as they tumble down the generations. At the short end of the spectrum we have adaptations that allow individuals to cope with changing conditions; at the long end we have the process of speciation and the development of species-typical characteristics.

The primary concern of this chapter is to explore adaptations that lie between these two extremes. Over slightly shorter timescales than those giving rise to human universals, and reflecting the fact that humans have diversified over a range of ecosystems, we have local population-level adaptations to climate and regional ecologies that now contribute

to genetic diversity. Examples here include dietary adaptations, such as lactose tolerance and the ability to digest starch (examined in Chapter 20), as well as skin colour and body shape examined in this chapter. Recently, however, there has been much interest in adaptations occurring during shorter time periods, most notably during individual development, and the possible role of transgenerational epigenetic modifications.

7.2 Types of adaptations

7.2.1 Short-term reversible changes: homeostasis, acclimation, habituation and acclimatisation

Homeostasis means 'steady state' and refers to processes whereby animals achieve a relatively constant internal environment. Humans, for example, maintain glucose levels in the bloodstream at about 70–100mg of glucose per 100ml of blood; blood pH at 7.4+/–0.1; and a body temperature that deviates little from 37^0C. This is achieved through hormonal feedback mechanisms, and, in the case of temperature, sweating and behavioural changes (such as shivering). **Acclimation** is a closely related term but one usually reserved for one-off responses to experimentally induced environmental stressors, such as placing humans in a heat chamber.

Habituation refers to the gradual reduction of the response due to repeated stimulation. This may not always be adaptive in the sense that the organism is any better off: habituation to a loud noise or pain, for example, can still be damaging. Strictly, habituation that is not simply the result of motor or sensory fatigue can be adaptive when it reduces the frequency or intensity of a response that would otherwise be costly. An example would be prairie dogs giving an alarm call in the presence of humans studying them. Once they realise that the humans are of no threat the alarm call frequency reduces and prairie dogs save their energy.

The adaptive function of **acclimatisation** is much more obvious. Acclimatisation refers to physical and psychological changes in the organism that take place over an extended period of time (in humans, days, months or years) in response to environmental

Mechanism	Homeostasis, acclimation, habituation	Acclimatisation	Developmental acclimatisation and plasticity	Transgenerational epigenetic effects	Natural and sexual selection	
					Population-level genetic diversity (recent)	Human universals (ancient)
Process	Short-term changes to maintain bodily stability. Uses gene-based physiological mechanisms common to all humans. Examples include release of insulin to regulate blood glucose, sweating to keep cool, shivering to generate heat	Temporary adaptations to changes in environmental conditions. Typical examples: increases in red blood cell count when moving to higher altitudes; increases in muscle size following exercise; release of melanin in relation to UV exposure. Changes usually reversible	Phenotypic changes to organism during growth and development to enable it to cope with different predicted social and physical environments and lifestyles. Examples include early activation of sweat glands in relation to climate; *in utero* influences on life history strategy (see Chapter 8). Changes usually irreversible	Modifications in histone proteins leading to silencing or expression of genetic information. The environment experienced by a parent may cause epigenetic modification of DNA which is inherited by offspring. This could ensure a match between gene expression and selective pressures. Example is effect of diet through the generations (see Section 7.2.3)	Differences between populations due to genetic adaptation to local conditions and ecologies. Examples include skin colour and body shape	Selection over a long period of time of species-typical characteristics. Examples: bipedality, hair loss, universal cognitive mechanisms for processing information. A universal system of emotions. In general, species-typical characteristics
Timescale	Seconds to days	Days, months, years	Months and years	One or more generations	Many generations: hundreds and thousands of years	Many generations: tens of thousands of years

TIME SCALE

FIGURE 7.1 Human adaptability.
Towards the left of the table we have evolved physiological systems to enable individuals to adapt to rapid change in internal and external environments. Moving towards the right we observe longer-term, often irreversible, changes in phenotypes, gene imprinting and finally gene frequencies.

stresses. People moving to higher altitude, for example, gradually increase the oxygen-carrying capacity of blood. The mechanism is that in response to the initial drop in blood oxygen levels the kidneys increase the secretion of a hormone called erythropoietin which stimulates the production of red blood cells which in turn enhances oxygen absorption and transport. If people then move to lower altitudes the red blood cell count will remain high for a few weeks (a phenomenon often taken advantage of by athletes in training) before returning to normal levels. Other examples include the phenomenon whereby the salt content of sweat reduces as people acclimatise to hot climates; and the acquisition of a sun tan when people are exposed to high levels of sunshine.

7.2.2 Developmental acclimatisation and developmental plasticity

Acclimatisation during the developmental period of a human's life can lead to irreversible changes and this falls under the heading developmental acclimatisation. Examples include the increase in chest size when a person grows up at high altitudes (Frisancho, 1993). These changes can also be culturally induced. In the late 19th century it was fashionable for European and American women to have narrow waists. To achieve this, tight corsets were worn as a girl was growing, this had the effect of deforming the lower rib cage. Chinese foot binding, now illegal, provides another example. It was a common practice among some working-class families in the city of Guangzhou (south China) in the late 19th century to bind the feet of the eldest daughter in the hope that this would make her attractive to a rich man. The result would be tiny pointed feet that were considered aesthetically pleasing. Sweating again provides another instructive example. At birth, all humans have a similar number of sweat glands, somewhere between two and five million. Initially, however, they don't function. Instead, over a period of about three years a proportion of sweat glands are programmed to start working; the hotter the environment the more glands that are brought into service.

The phenomenon of developmental plasticity is now recognised as a key process in evolutionary adaptation since it allows organisms to finely adjust their phenotypes towards optimal fitness in a range of environments. An important point is that this process can itself be the subject of higher-level natural selection and so a variety of plasticities may have been selected in response to the range of ancestral environments encountered.

The way a phenotypic trait emerges in response to different environments is called the reaction norm and often the trait shows a continuous range. Occasionally we find not a continuous variation but a discontinuous set of different outcomes. In many reptiles, for example, the sex of an egg is determined by temperature. Generally, shaping of phenotypes by environmental cues tends to occur early in the life of an organism and the whole process is subject to a critical window. A dramatic case is seen in the case of the masculinisation of the male body in humans which is determined by foetal exposure to testosterone. Another example is found in the honey bee where a female could become a queen or a worker bee depending on the degree of nutrition supplied in the larval stage. This emergence of distinct morphs is called polyphenism.

In the case of humans some profound and interesting work has recently demonstrated how adult health is related to early developmental conditions. Two contrasting ideas here are the thrifty genotype and thrifty phenotype hypotheses.

The thrifty genotype

The **thrifty genotype** hypothesis was an idea used to explain the prevalence of type II diabetes in populations that were assumed to have been exposed to famine sometime in their past. Populations examined so far in this way include Native Americans, South Pacific Islanders, sub-Saharan Africans, Inuit, and Aboriginal Australians (Eaton, Konner, Shostak, 1988). The basic idea was initially proposed by the geneticist James Neel (1962) who saw it as a solution to the problem that diabetes is a debilitating illness with a known genetic component yet these damaging genes still persist. Neel suggested that a gene that promoted a rapid insulin trigger such that glucose was efficiently converted into fat in times of food abundance would carry an advantage since in times of food scarcity such individuals would have fat reserves to draw upon. Problems then arise with modern diets since with a constant abundance of food insulin levels remain high (due to the rapid thrifty genome-based insulin trigger) resulting in reduced sensitivity to insulin, raised glucose levels and hence type II diabetes.

If there were such a set of 'thrifty' genes that were selected for in populations exposed to famine then it would be an example of local genetic diversity whereby this feature is more prevalent in some populations than others. However, plausible though this hypothesis sounds, academic opinion now seems to rule it out as a major factor in the explanation of diabetes and obesity. These are a series of difficulties that challenge the whole concept. Firstly, the frequency of type II diabetes has increased among Europeans that have probably not been exposed to periodic bouts of famine; secondly, recent studies on the genetic loci known to be associated with diabetes and obesity show no signs that such loci have been subject to natural selection (Southam et al., 2009); furthermore, studies on contemporary hunter-gatherers show that even in times of plenty obesity is rare. If hunter-gatherers do not put on excessive weight in times of food abundance it is hard to see how these thrifty alleles, that are supposed to help during times of famine, can be there at all. Also, it seems very likely that Pleistocene populations of hunter-gatherers did not face periodic bouts of famine. Famine was a characteristic of post-Neolithic societies where reliance on a few crops could, as a result of variable climatic conditions and periodic crop failure, occasionally bring about a massive reduction in food supply (see Chapter 19).

The thrifty phenotype

Although the thrifty genotype idea has met with scepticism, the **thrifty phenotype** hypothesis has enjoyed more explanatory success. A major boost to this hypothesis comes from the work of David Barker (2007) on the incidence of coronary heart disease in the UK. Using data from the UK over the period 1911 to 1979, Barker noted that adult lifestyle had only a limited ability to predict the likelihood of coronary heart disease. He did, however, find a very strong correlation between low birth weights and (for the same cohort) the risk of heart disease later in life. He concluded that the best way to interpret this was in terms of developmental plasticity (he also called this perspective the 'developmental origins theory'). The essential idea is that the foetus is programmed *in utero* to develop according to the environment it is likely to encounter. Hence a maternal environment of poor nutrition (and hence low birth weight) programmes the foetus to develop in a compensatory way, such as increased insulin resistance, high blood levels of fatty acids and reduced growth, as a strategy to make the most of the limited energy resources experienced *in utero* and expected in the future. However, if the growing child later encounters abundant food resources, this 'thrifty phenotype', conditioned to expect poor nutrition, will tend to put on weight and have an increased risk of type II diabetes and heart disease (Figure 7.2).

The plausibility of this model depends upon selection for such a mechanism over generations

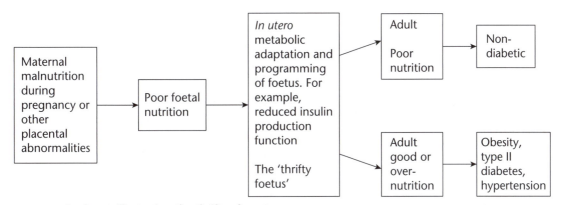

FIGURE 7.2 A schema illustrating the thrifty phenotype.
SOURCE: Based on ideas in Frisancho, A. R. (2010). The study of human adaptation. In M. P. Muehlenbein (ed.), *Human Evolutionary Biology*. Cambridge University Press (adapted with permission); and Walker et al. (2012).

where predictions about future environments were accurate; in other words, where foetal experiences did on average reliably predict future environments and hence the inherent developmental plasticity did confer real fitness advantages.

Since the work of Barker, hundreds of others studies have produced similar findings, consistently relating lower birth weight to a host of adult health problems, including: hypertension, insulin resistance and diabetes, high-risk patterns of fat deposition and increased risk of cardiovascular disorder (See Kuzawa and Quinn, 2009).

So how and why does the body have the ability to modify its development in response to early life conditions? The obvious answer that early experiences predict later conditions (which is probably in part correct) faces the problem that a few years of experience in early infancy may not be a reliable guide to many years of adult life decades into the future. One ingenious idea is that it is the early timing (*in utero*) of the phase of the foetus's sensitivity to programming that actually helps to shelter it against local and immediate conditions and variations. As the foetus develops the mother's physiology buffers it against short-term (days, weeks, months) variations in the local environment. But at the same time the adjustment that the mother's body has already made to a lifetime of biological experiences, as reflected in such parameters as her blood pressure, the nutrients she passes to the foetus, the quality of milk she provides for the neonate, could perhaps more reliably signal the sort of environment the growing child is likely to encounter. The mother has, as it were, already integrated information over a number of years to arrive at a more considered prediction of the future than the foetus or neonate could hope to make (Figure 7.3).

Evidence in favour of this model comes from studies on birth weights and the nutritional status of the mother. Birth weights tend to be lighter in populations where nutrition has been restricted over several generations. Obviously one might expect poor nutrition for the mother to result in a reduced birth weight of the child. Yet nutritional supplements given to pregnant women have little if any effect on birth weight. It is as if long-term environmental conditions have more impact than short-term dietary variations (see Kramer, 2000; and Kuzawa, 2005). Another set of evidence pointing to the reality of developmental plasticity in terms of early life influences *in utero* with the possibility of transgenerational effects comes from the Dutch 'Hunger Winter' of 1944–5 (Box 7.1).

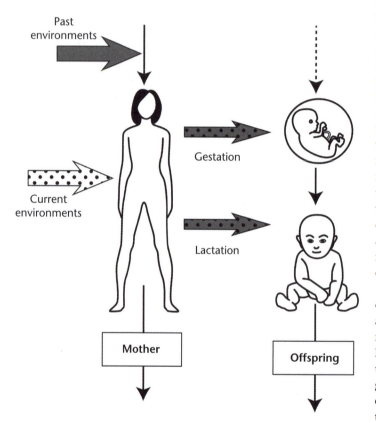

FIGURE 7.3 Possible adaptation of offspring to ecological information collected and transmitted by the mother.
The mother is able to pass on information about both past and current environments through gestation and lactation. Adapting to the cumulative lifetime experiences of the mother is likely to be a more reliable predictor of future environments for a developing foetus, neonate and infant than only responding to local and immediate information.
SOURCE: Based on ideas in Kuzawa and Quinn (2009).

Box 7.1 The Dutch Hunger Winter 1944–5

A tragic episode in World War II provided a set of data that was used to test theories about developmental plasticity and life history programming. Following the Normandy invasion of 1944 Allied troops advanced across occupied Europe, but were halted in September of that year by the failure of Operation Market Garden to capture the bridge over the Rhine at Arnhem (the subject of the film *A Bridge too Far*). To help the allies the Dutch resistance had organised a railway strike to hamper the movement of German troops. In a savage act of reprisal, the Germans then placed an embargo on food transport to the Netherlands. This, coupled with a severe winter that year, effectively imposed starvation conditions on the civilian population. Food supplies were restored by the Allies in May 1945 but by this time over 20,000 people had died.

During this terrible time medical care was still given to pregnant women by the civic authorities and careful records kept of the progress of newborn babies. Follow-up studies on the adult life of these children enabled a comparison to be made between the health of those who were *in utero* before, during and after the famine. Consistent with the findings of Barker and his colleagues on UK populations (Barker, 2007), babies born small due to maternal under nutrition during pregnancy developed high blood pressure and glucose intolerance later in life (Roseboom et al., 2011). A similar study showed that those exposed to famine early in gestation had a much greater incidence of coronary heart disease later in life. When children born during the famine grew up to reproduce themselves they also tended to produce children smaller in weight than average, suggesting some sort of epigenetic influence (Painter et al., 2008). Females who were prenatally exposed to famine, later in life had more children, more twins and started reproducing earlier compared to unexposed women. This seems to suggest that prenatal exposure to undernourishment led to a faster life history strategy. More recently, there have also been reports of a premature decline in cognitive function of those exposed to famine during their gestation (de Rooij et al., 2010).

These Dutch famine studies have also provided the first direct evidence of prenatal epigenetic programming. Heijmans et al. (2008) showed that individuals who were exposed to famine very shortly after conception had less DNA methylation of the imprinted insulin-like growth factor 2 gene (IGF2) compared to same-sex unexposed siblings even 60 years later. Loss of imprinting suggests a developmental epigenetic adjustment to allow IGF2 to be synthesised and hence growth promoted.

Humans have managed to settle in a wide range of ecosystems partly helped by permanent genetic changes that characterise local populations, and partly by the response of our various mechanisms of acclimatisation and adaptive phenotype plasticity. But without doubt one of the main reasons behind the fact that we have colonised most of the globe is our ability to devise cultural solutions to environmental problems. Fire, clothing, architecture, central heating and air conditioning are all examples of how we have managed to surround ourselves with our own microclimate more in keeping with our African ancestry. This might be called a form of cultural developmental acclimatisation and is a subject considered in more depth in Chapter 21.

7.2.3 Epigenetics, imprinting and transgenerational effects

Imprinting

As discussed above, studies on malnourished embryos and young children have provided evidence in favour of the various developmental hypotheses: the idea that early life conditions programme the phenotype for later development. There is also mounting evidence, as we will shortly examine, that such effects can even persist through generations. Yet, as noted in Chapter 3, one of the central tenets of modern biology is that acquired characteristics cannot be inherited; information flows from genotype to phenotype but not the other way. In this view there is simply no way that Lamarckian inheritance can work. But the recent work on the thrifty phenotype and the various developmental hypotheses discussed in the previous section points to the possibility that some changes caused by the effect of the environment on one generation can be passed to succeeding generations (grandchildren and great grandchildren) even though the underlying base sequence code on the DNA has not altered. This type of transfer of information falls under the heading of '**epigenetics**' (in Greek επι means 'over' or 'above'), which is now a burgeoning and potentially transformative academic field of study.

The key to understanding how such a mechanism might work lies in the idea of genetic switching. Biologists have known for generations that since nearly all cells in the human body contain the same genetic information then genes must be switched on or off to enable cells to behave according to their location: so that cells in the liver carry out liver-related functions and cells in the heart carry out heart-related functions. It has also been known since the 1980s that genes can be 'imprinted'; that is, a given type of gene can be switched off or on depending if it is inherited from the mother or father (see Barton et al. (1984) for a pioneering paper on imprinting in mouse development). These epigenetic effects, originally established in the germ line, then remain *in situ* in all somatic cells in the body. Two key mechanisms seem to be involved: methylation of the DNA bases so they become inactive; and modifications to the histone proteins around which DNA winds. The underlying genetic code is unchanged.

Sex-specific imprinting and birth weight

Genomic imprinting is a relatively new field of study. The traditional view within Mendelian genetics was that with the exception of the sex chromosomes (XX in females and XY in males), the chromosomes inherited from each parent were functionally equivalent. Genes could of course be dominant or recessive, but there was no reason to suppose that the operation of normal functional genes could depend upon their parental origin. Indeed, considering the process of inheritance in diploid organisms, it would seem highly unlikely that the fortunes of any gene should depend upon which parent is transmitting it, since as a result of meiotic recombination a gene transmitted by the mother in one generation could have descended to her from either her father or mother. Similarly, she could pass it to both daughters and sons. We now know, however, that genomic imprinting proves an exception to this picture.

Sex-specific genomic imprinting entails silencing the operation of a gene (imprinting) when it is inherited from one parent but not the other. Why this happens may be related to asymmetries in parental care and differences in genetic relatedness between mothers and fathers in relation to their children (the so called 'parental conflict hypothesis'). In this view, paternal genes have more to gain by promoting the rapid growth of the embryo since any future offspring may not be those of the same father. In contrast, mothers will be more desirous of conserving resources for their own survival and their future offspring, which they have an equal interest in. Hence a mother, it is argued, will 'imprint' her own growth genes to reduce the rate of embryonic growth. A more plausible alternative suggestion is that the imprinting of both paternally and maternally derived genes is the under control of the mother in the zygotic stage of development of the embryo (Keverne and Curley, 2008).

If the imprinting of parental genes is related to asymmetries in parental care then we would not expect to find much imprinting, if any, in taxa where parental care is at a minimum after fertilisation; that is, where the biological burden to the mother does not increase significantly after the zygote is formed. Consistent with this expectation, so far there is little evidence for imprinting in fish, amphibians and reptiles (Killian et al., 2000). In contrast, genomic imprinting is widespread in placental mammals where resource requirements after fertilisation are high and are delivered by the mother via a placenta. In birds and monotremes (a class of oviparous, that is, egg-laying, mammals such as the platypus), resource allocation after fertilisation is also low and again imprinting is not found in these taxa (see Bartolemei and Ferguson-Smith, 2011).

Sex-specific imprinting carries the risk of haploinsufficiency. If a diploid organism, following imprinting, only has one functional copy of a gene and this is compromised in some way, then a crucial gene product may not be manufactured leading to a diseased state. There are still numerous unknowns associated with imprinting but the fact that it is maintained in the face of this selection pressure against it suggests it has important adaptive benefits.

Further support for this phenomenon comes from correlation studies on human birth weights. When examining the birth weights of half-brothers and half-sisters it is notable that there is a higher correlation between half-siblings who share the same mother than those who share the same father. In addition, birth size correlates more strongly with the birth weight of the mother than the birth weight of the father (See Table 7.1).

This greater influence of maternal-line factors than paternal-line factors on birth weight is now a well-established principle that applies to humans and many other mammalian species. In contrast, adult height reached is equally correlated with the height of the mother and father with correlations lying between 0.4 and 0.6 (see Luo et al., 1998). It looks, therefore, as if foetal growth but not post-birth development has been wrested from paternal control. This of course makes good adaptive sense. The mother's body is probably a better source of information about environmental conditions, nutrient availability, her ability to support a growing foetus, and,

of course, the size of her birth canal, than the genetic and epigenetic information a father can provide.

Epigenetics – a resurgence of Lamarckism?

When imprinting was discovered it was generally assumed that such physical imprints preventing gene expression were removed in each generation and that the genome was reset during meiosis. Indeed, this is a requirement if genes are to be imprinted according to the sex of the parent and imprinting must begin anew in each generation during spermatogenesis for paternal imprinting and oogenesis for maternal imprinting. Early signs that epigenetic effects might persist through generations came from some remarkable work by the Swedish medical researcher Lars Olov Bygren and colleagues (2001) on the health records of people living in Norrbotten in northern Sweden. This area lies inside the Arctic Circle and was so isolated in the 19th century that if the harvest was bad people starved. Climatic variations also influenced crop yields. The years 1800, 1812, 1821, 1836 and 1856, for example, were times of total crop failure; whilst in 1801, 1822, 1828, 1844 and 1863 the land produced an abundant surplus of food. One consequence of this dramatic variability in food supply was that people alternated between near starvation and over-consumption. To examine the intergenerational effects of these consumption patterns on health, Bygren's group analysed a random sample of 99 individuals born in the Överkalix parish of Norrbotten and examined their health records in relation to the food supplies available to their parents and their grandparents. The results were surprising: boys

Table 7.1 Correlation of birth weight with other parameters.
Studies show that birth weight is more correlated with maternal relationships than paternal.

Relationship	Correlation of birth weights (r)	Sample number in study	Reference
Maternal half-sibs	0.581	30	Gluckman and Hanson (2005)
Paternal half-sibs	0.102	168	Gluckman and Hanson (2005)
First cousins through mother and her sister	0.135	554	Gluckman and Hanson (2005)
First cousins through father and his brother	0.015	288	Gluckman and Hanson (2005)
Child's and mother's birth weight	0.226	67,795	Magnus et al. (2001)
Child's and father's birth weight	0.126	67,795	Magnus et al. (2001)

approaching puberty who went from normal eating to gluttony in a single season left grandsons who died an average of six years earlier than grandsons of boys who had eaten sparingly due to a poor harvest. In later work on a similar data set, Kaati et al. (2002) looked at deaths due to cardiovascular disease and diabetes mellitus. They found that if food was not readily available during puberty of the father then cardiovascular mortality was low; but if the paternal grandfather was exposed to a surfeit of food during puberty then mortality for his grandchildren from diabetes was high. In a subsequent paper the research group identified the male line as a prime candidate for a sex-specific transgenerational effect of food supply experienced during the early puberty of grandparents on the mortality rates of later generations (Kaati et al., 2007).

In short, Bygren, Kaati and co-workers had found clear evidence for a transgenerational response to environmental stress whose effects were passed down at least two generations and yet did not involve changes to the genetic code. This ground-breaking work pointed to the existence of an epigenome – some sort of marker that affected gene expression and was transmitted along with DNA in gametes. Since these early findings there are now at least 100 well-established examples of transgenerational inheritance in such taxa as prokaryotes, plants and animals (Jablonka and Raz, 2009).

Bygren was initially unclear as to what mechanism could be at work and wondered if similar responses to stress could be observed in other populations. He then met Marcus Pembrey, a geneticist working at University College London. Together they worked on data from the Avon Longitudinal Study of Parents and Children (ALSPAC), a study of the health of parents and their offspring living in or around the city of Bristol in the UK. The children in the study, who were born in 1991 and 1992, have been measured every year since for a whole variety of physical traits. They found that fathers who started to smoke before the age of 11 – a time when boys are just entering puberty and when it might be expected that gamete formation would be subject to external stresses – produced sons who, by the age of 9, had BMIs significantly higher than other boys. This effect was found down the male line only (Pembrey et al., 2006). The same study group also examined the Swedish data on people from Norrbotten and found that the impact of a grandfather's nutritional status on the mortality risks of grandchildren was confined to grandsons and that of grandmothers to granddaughters (see Table 7.2).

The sex-specific nature of these epigenetic, transgenerational effects led the authors to suggest that they are associated with the X and Y sex chromosomes:

> We conclude that sex-specific, male line transgenerated responses exist in humans and hypothesise that these transmissions are mediated by the sex chromosomes, X and Y. (Pembrey et al., 2006, p. 159)

Since Pembrey and Bygren's study on the effects of smoking, a whole range of transgenerational effects have been noted in the paternal line, including effects of drugs, alcohol and toxins (see Curley et al., 2011, for a review). Subsequent research has pointed to two main ways in which DNA can be

Table 7.2 Mortality risk ratios (RR) for female and male subjects in relation to their grandparents' food availability during their slow growth period.
Mortality risk ratio (RR) is observed mortality rate divided by expected mortality rate for that group. RR values departing significantly from 1 are shown in bold. High values of RR imply a higher than expected risk of death.

	Good food availability for grandparents		Poor food availability for grandparents	
	Granddaughters RR and P value	Grandsons RR and P value	Granddaughters RR and P value	Grandsons RR and P value
Paternal grandmother	*2.13 (p = 0.001)*	1.02 (p = 0.93)	0.72 (p = 0.12)	1.23 (p = 0.27)
Paternal grandfather	0.81 (p = 0.32)	*1.67 (p = 0.009)*	1.17 (p = 0.43)	*0.65 (p = 0.025)*
p = significance				

imprinted: chemical modification of cytosine bases and the clustering of proteins around active genes (Box 7.2).

In the case of non-human animals, one of the most astonishing recent studies comes from the work of Dias and Ressler (2014) who have reported that odour-fear conditioning can be passed down several generations. They gathered a group of mice (F0 generation) and conditioned them to associate fear (through administering an electric shock) with the odour of acetophenone. What is remarkable is that the F1 and F2 generation (that is, offspring of F0 and their offspring in turn) also showed the behavioural response to this odour but not to other odours. The authors suggest that the mechanism may involve CpG hypomethylation in the Olfr151 gene carried by sperm from the males. Even more recently Yehuda et al. (2015) looked at the cytosine methylation of a specific gene (coding for a binding protein called FKBP5) in Holocaust survivors and their adult offspring. The group found that, compared to controls, methylation levels were different in Holocaust victims and their offspring. The researchers suggest this is the first recorded example of transgenerational epigenetic transmission of the effects of trauma.

Box 7.2 Epigenetic effects: mechanisms and functions

Epigenetic influences appear to work in two ways. One is the influence of histones, the protein 'spools' around which DNA winds, on gene expression; the other is via methylation, which is the chemical modification of cytosine (the letter C in the four base pairs: A, G, C, and T). It is clusters of DNA wound round histones that form chromatin. DNA can only be expressed when it is unwound from these histone proteins and comes into contact with RNA polymerase. The histone process is not well understood but it is possible that modified histones (caused by changes to the amino acids that make up the protein chains of which they are formed) can be carried into a new cell along with the copied DNA and then serve as a template for further histones. The well-established existence of prions (proteins that act as infectious agents) is evidence that a specific protein state (for example, a prion) can induce in cells conversion of native proteins to this same state. Methylation converts cytosine to 5-methylcytosine; this modified base still acts much like cytosine in that it pairs with guanine. But it is known that areas of DNA that are densely methylated have a lower transcriptional activity. In other words, methylated genes are muted or turned off altogether. Interestingly, the sites where methylation tends to be found are in the regulatory regions of the gene upstream from the site where transcription starts. Moreover, these methylation patterns are stable and survive cell division and it seems cellular differentiation relies on this type of epigenetic modification. (Methylation may pass down through the generations by a process whereby an enzyme, such as DNA methyltransferase, when it encounters a methylated strand of DNA proceeds to methylate the other half.)

The evidence suggests that methylation is more stable than histone modification. When cytosine (C) bases are methylated they are converted to 5-methylcytosine and this alters the 3D structure of the DNA strand as well as attracting binding proteins and histones. The net effect of this accumulation of methyl groups, proteins and histones around a specific site is that the chromatin adopts a tighter conformation which restricts access of transcriptional factors and RNA polymerase resulting in a silencing of the gene (Gluckman et al., 2009). These epigenetic changes are not usually localised on the gene itself but rather on a gene promoter 'upstream' of the gene. It is now realised that epigenetic modification is involved in at least four important functions: sex-specific imprinting, cell differentiation, developmental plasticity and X chromosome inactivation in females (Figure 7.4).

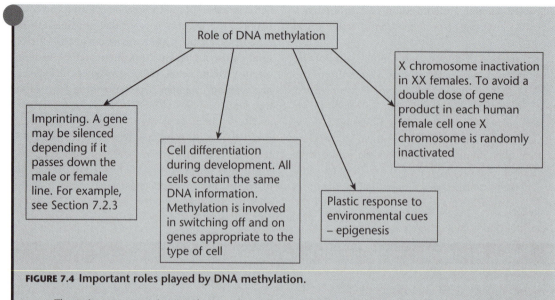

FIGURE 7.4 Important roles played by DNA methylation.

There is now a project underway to map the human epigenome. It is steered by the International Human Epigenetic Consortium (IHEC) launched in Paris in January 2010 (see http://ihec-epigenomes.net/). It will be a huge undertaking compared to the human genome project. The genome of an individual is the same in each type of tissue but the marks that constitute the epigenome are different in each cell and change during a person's lifetime.

Epigenetic effects and paternal care: rats and humans

The problem with investigating epigenetic effects in humans is the availability of accurate longitudinal data. Something may be learned from experiments on short-lived mammals such as rats and mice, however. One approach has been to exploit natural variations in maternal care in Long–Evans rats (an outbred strain of *Rattus norvegicus* developed by Long and Evans in 1915). Shortly after giving birth, female rats lick and groom (LG) their offspring. The frequency of grooming is an individual characteristic but varies widely between individuals, enabling females to be labelled as high, medium or low in their LG behaviour. Now the LG behaviour of pups when they mature and give birth themselves is highly correlated with that of their mother. This is not a simple genetic variation, however, since female pups born to high LG mothers and then raised by low LG mothers go on to exhibit low LG behaviour to their own pups; and pups of low LG mothers fostered by high LG mothers go on to give high levels of LG to their pups. The neurobiology of this phenomenon has been investigated by a number of people and the effects are complicated (see Champagne, 2008, for a review). In essence, the evidence points to the strong possibility that low maternal care influences the expression of a gene involved in the neurobiology of maternal behaviour and the stress response, and this can be passed down several generations.

Could this discovery have any bearing on human parental behaviour? It has been known for several decades that about 20–30 per cent of abused infants are likely to grow up to become abusers themselves, and that about 70 per cent of abusive parents were themselves abused in childhood (Chapman and Scott, 2001). A similar effect is found with attachment (see Section 8.3.3) in that the level of a mother's attachment to her own mother correlates positively with her own infant's attachment (Benoit and Parker, 1994). Furthermore there is now evidence emerging that the experience of abuse in infancy can lead to epigenetic modifications in the human brain (McGowan, 2009).

Are identical twins epigenetically distinct?

Identical twins are of obvious interest to researchers since they share an identical genome. Yet these monozygotic (MZ) twins are also known to demonstrate phenotypical differences, particularly in their susceptibility to diseases such as schizophrenia and bipolar disorder (Cardno et al., 2002). The usual temptation is to account for these in terms of different environmental experiences, but are there epigenetic differences between such twins as well? Causes of these examples of 'phenotypic discordance' are poorly understood but it is possible that epigenetic differences may be involved (Fraga et al., 2005; Poulsen et al., 2007). To assess the degree of epigenetic difference between MZ twins, Mario Fraga et al. (2005) recruited 80 Spanish volunteers who were Caucasian twins. They found that patterns of epigenetic modification on the genome of MZ twins diverged as the twins grew older (see Figure 7.5). As well as age effects, spending a longer time apart also increased the epigenetic distance between twins. Causes of these differences are not fully understood but could include lifestyle habits (smoking, diet, levels of exercise) or a process of epigenetic drift. In future, epigenetics might have a role to play in explaining how similar genotypes give rise to very different phenotypes (Wong et al., 2005).

Differences may also appear in females due to unbalanced X chromosome inactivation. Males possess one X chromosome (derived from the mother) and females two (one from each parent). To avoid an excessive dose of proteins coded for on each X chromosome, early in the development of female embryos one X chromosome in each cell is randomly inactivated. Throughout the whole female body this means that roughly half the X chromosomes derived from the father and half from the mother are inactivated. This process also has the useful advantage of masking any recessive genes derived from just one parent. Suppose, for example, that a female receives an X chromosome from her father which carries a recessive gene for colour blindness. Half of these genes will be inactivated as will half of those versions of the normal dominant colour vision genes acquired from the mother, but there will be enough normal copies of the vision genes left over to ensure normal vision. If, however, the activation is skewed towards one parent or other a recessive trait may manifest itself. In the case of identical female twins, if the fertilised egg splits soon after conception and before this skewed X chromosome inactivation takes place (so that it happens to one twin and not the other) then supposedly genetically identical twins can end up showing phenotypic discordance. Cases so far reported showing this effect include Hunter syndrome, haemophilia, colour blindness, Lesch–Nyhan syndrome and Duchenne muscular dystrophy (Winchester et al., 1992; Bennett et al., 2008; Jorgensen et al., 1992; De Gregoria et al., 2005; Richards et al., 1990).

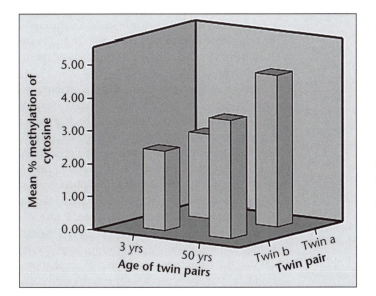

FIGURE 7.5 Methylation of DNA in pairs of twins according to age. At 3 years of age twin pairs have a similar degree of DNA methylation. At 50 years of age, not only is the degree of methylation higher but also the difference between any two twins is also higher.

SOURCE: Data from Fraga et al. (2005), Figure 1C, p. 10606.

Transgenerational epigenetic inheritance and the theory of evolution

The existence of the epigenome offers a potential way of explaining how organisms as complex as humans can manage with only about 25,000 genes – about the same number as found in many fish and mice (even the humble nematode worm (*Caenorhabditis elegans*) at 1 mm in length has a genome containing about 20,000 genes). Various analogies have been proposed: one is that the genome is like the computer hardware and the epigenome the software, another is that the genome is a book and the epigenome decides how to read it and what bits should be read. In either case the epigenome introduces a whole new layer of complexity.

One crucial question, of course, is whether the facts of transgenerational epigenetic inheritance compromise the modern synthesis (or 'central dogma') that formed the bedrock of biology for most of the 20th century (see Pigliucci, 2007). Marcus Pembrey, for example, controversially referred to himself as a 'neo-Lamarckian'. The answer is probably no, even though epigenetics may turn out to be an important addition to knowledge of the way the genome works. At one level, epigenesis is an example of phenotype plasticity. The human genotype of 25,000 genes is incredibly similar across different human populations yet people are capable of living in a wide range of ecological conditions. They do this by the various adaptive processes discussed in this chapter and so epigenetics can be conceptualised as one of many ways in which a developmental response is achieved.

This key question – whether the epigenetic response is adaptive – is still hotly debated. If transgenerational epigenetics (TGE) does serve adaptive ends then its role may be understandable in terms of the timescale of fitness-relevant environmental changes. For changes over short time periods then DNA encoded regulatory and facultative response systems (acclimatisation) may suffice. So, for example, exercise will produce stronger muscles (as opposed to wearing them out as in a mechanical system) since this is a sensible response to a perceived need. Over much longer timescales (such as climate change over several million years that led to the emergence of the hominins in east Africa) then natural selection will alter gene frequencies and drive

speciation – allowing genes to persist in new environments through species change itself. Superimposed on top of these systems is brain-mediated learning and culture. An individual can learn new tricks in its lifetime to cope with short-term changes, whilst collective social learning can generate cultural solutions to common problems (such as the science of clothmaking to keep warm as hominins moved into ever-colder regions of the globe).

Within this perspective, possibly TGE has itself evolved to enable an organism to cope with intermediate timescale changes. If the environment changes very quickly then passing on an epigenetically acquired response may be pointless, natural selection will not have time to change gene frequencies, and individual or social learning may solve the problem more effectively (hence the selective pressure towards brain enlargement discussed in Chapter 6). Change over a much longer timescale will allow DNA to respond through natural selection. But with intermediate timescale changes the probability of an organism facing an environment similar to its parents or grandparents may be high enough to select for a fitness-enhancing transgenerational epigenetic system. Probably the best interpretation of epigenetics is that it does not challenge neo-Darwinism but holds that these epigenetic effects are, as David Haig said, 'switches evolved by natural selection to enable genes to exhibit conditional behaviours' (Haig, 2007, p. 420). In a way epigenetic processes are functionally equivalent to learning systems, that is adaptations (such as the neural hardware and software in the case of learning) that enable a capture and transmission of valuable ecological information (Scott-Phillips et al., 2011).

There is mounting evidence pointing to transgenerational epigenetic effects in a whole range of organisms. What is open to doubt, however, is whether this phenomenon is adaptive. Some examples of epigenetic inheritance can be interpreted in adaptive terms, such as the differential methylation of maternal and paternal chromosomes in the phenomenon of imprinting. But as noted earlier, these effects only last one generation. Research has tended to focus on establishing the effect, itself a radical enough finding, rather than searching for functional explanations. The significance, therefore, of this phenomenon to evolutionary psychology remains uncertain.

The potential of early life experiences to influence later life in an adaptive way is further explored in the next chapter.

7.2.4 Universal adaptations and adaptations to local ecologies

Universal human adaptations, features possessed by all normally functioning humans, are genetically based and were shaped by natural selection to solve fitness-related problems. They lie at the right end of the spectrum shown in Figure 7.1. Examples at the physical level include skeletal adaptations to bipedality, and physiological adaptations for temperature control (such as sweating); at the psychological level, examples include mechanisms for processing sensory data, the recognition of kin, motivational systems, mechanisms for evaluating potential mates and, in short, the whole species-typical architecture of the human brain. Taken together the effect of these adaptations on behaviour is what we mean by human nature. Traditionally, evolutionary psychology has been most interested in these human universals since they serve as an essential platform from which to examine the functional (in the Tinbergen sense) basis of human behaviour. Typically, for these adaptations there will be little genetic variance within and between human populations around the globe. In the context of the examples given earlier, all humans need a skeleton to support bipedal gait, systems for temperature control, mechanisms to recognise kin and process sensory information, and motivational and affective systems for choosing mates. Much of this book, of course, is about these human universals.

But about 72,000 years ago a small group of humans moved out of Africa and spread across the globe, taking up residence in a wide variety of climatic zones. Consequently, it might be expected that there will be genetic differences between human groups corresponding to populations adapting to their local ecologies (Harcourt, 2012). On this important topic in human evolution Darwin got it completely wrong. He thought that physical differences between different groups around the world could not be accounted for by adaptation to the environment. He said, for example:

If, however, we look to the races of man as distributed over the world, we must infer that their characteristic differences cannot be accounted for by the direct action of different conditions of life, even after exposure to them for an enormous period of time. (Darwin 1871, p. 246)

To be fair to Darwin, this observation appears in a chapter where he is anxious to show that all humans belong to the same species, on which count he was right. But there is now abundant evidence that humans do show genetic adaptations to local conditions and that for populations outside of Africa they have acquired these since the exodus from this continent some 72,000 years ago. A whole range of these have been recorded such as: lactose tolerance; skin colour variation; body shape variation; cranial shape variation; the incidence of genetic diseases (where a genetic disorder might be more prevalent in some populations if it confers a local advantage – see Chapter 18); and adaptations to living at high altitudes. In this section we will look at two examples of local adaptations that have led to, and so are based on, genetic diversity: body size and shape, and skin colour.

Body size and shape variations

The African diaspora meant that some humans exposed themselves to a wide variety of ecosystems and climatic challenges. One example of the adaptive process is the relationship between climate and physical size and shape. There are two well-established and widely supported rules that inform the study of the relationship between bodily morphology and ambient temperature; these are known as Bergmann's rule (1847) and Allen's Rule (1877). Both of these are explained in Box 7.3.

Pioneering work on the application of these rules to human populations was carried out by D. F. Roberts (1953). Roberts examined a sample of 116 males from 10 different geographical areas and found considerable support for the application of Bergmann's and Allen's rules. Forty five years later Roberts' analysis was repeated by Katzmarzyk and Leonard (1998) on a slightly larger population (223 males and 198 females) but from the same geographical areas.

Box 7.3 Rules that impact on the shape of animals: Bergmann's rule and Allen's rule

Bergmann's rule

This rule was first proposed by the German biologist Carl Bergmann in 1847. He originally formulated it in terms of species and populations stating that in a widely distributed taxonomic group (for example, a genus) smaller species tend to be found in warmer environments and larger species in cooler environments. Today the rule is more commonly applied to explain the variation in size of individuals within a population or species group. The rule derives from the fact that warm-blooded animals such as mammals will lose heat to their surroundings at a rate proportional to their surface area (SA). But the rate of heat loss per unit of cellular mass (which indicates the temperature stress on the whole body) is proportional to the ratio surface area/volume. If we imagine for a moment a spherical animal then $SA = 4 \pi r^2$; and volume $= 4/3 \pi r^3$. Hence heat loss per unit mass of tissue is proportional to $4 \pi r^2 / 4/3 \pi r^3$ or $1/3r$. In short, the greater the linear size of an animal (r) the lower the rate of heat loss per unit mass of tissue. Since the heat generated by an animal is proportional to its cellular mass, then in cool climates large animals (high r) will conserve heat better than small animals. This rule explains why there is a physical size minimum for warm-blooded animals. The smallest warm-blooded vertebrate is probably the bee hummingbird (*Mellisuga*

helenae), which weighs about 1.8 grams. The high SA/vol. ratio is possible since this bird survives on a diet of energy-rich (and hence highly calorific) nectar. Any smaller than this and warm-blooded animals simply could not keep their bodies warm.

There are, however, some exceptions to Bergmann's rule and Harcourt (2012) suggests that a better term than rule would be 'effect'. Despite this, in a recent study Meiri and Dayan found that looking at 94 species of bird, 75 per cent of them showed the Bergmann effect, and for the 149 species of mammal they examined 65 per cent showed the effect. It appears then that Bergmann's rule or effect is a fairly valid ecological generalisation for birds and mammals (Meiri and Dayan, 2003).

Allen's rule

Allen's rule was proposed by the American zoologist and ornithologist Joel Asaph Allen (1838–1921). Allen argued that animals in colder climates will tend to have shorter and more rounded body parts, whilst animals in warmer climes will tend to have longer and more slender limbs. The rule follows from the fact that for a given mass long thin shapes will have a larger SA/vol. ratio compared to shorter, more rounded shapes. Therefore, shorter, more rounded shapes will conserve heat better in cold climates.

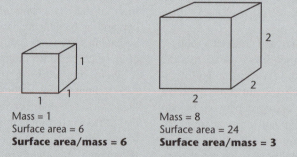

Mass = 1
Surface area = 6
Surface area/mass = 6

Mass = 8
Surface area = 24
Surface area/mass = 3

FIGURE 7.6 Illustration of relationship between SA/mass ratios for objects of changing size. The two cubes have a density of 1 g/cc. An increase by a factor of two in linear size from the small to the larger cube has brought about a reduction by a factor of two in the SA/mass ratio.

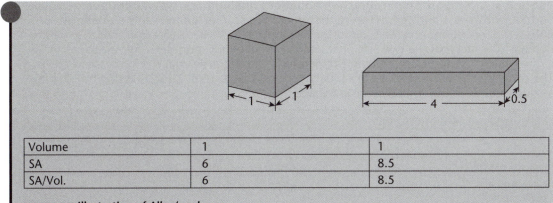

Volume	1	1
SA	6	8.5
SA/Vol.	6	8.5

FIGURE 7.7 Illustration of Allen's rule.
Both shapes have the same volume and mass, yet the long thin shape has a higher SA/Vol. ratio and so will lose heat faster to a cooler environment.

A comparison of the two sets of findings is instructive. The work of both sets of researchers shows Bergmann's and Allen's rules in operation: people living in hotter climates do tend to be lighter in weight and more slender in bodily proportions than those living in cooler climates: as local temperatures rise then the body mass of indigenous peoples tends to fall whilst their surface area/body mass (a measure of slenderness) tends to rise (see Figure 7.8).

What is interesting about the more recent work of Katzmarzyk and Leonard, however, is that the size of both the correlation coefficients linking body weight and surface area to volume ratios to temperature and the regression coefficients of the two regression lines were smaller than those found by Roberts. In other words, there was more variability in the data, and the effect of temperature on body size and proportions seems to be having a diminished effect 45 years after Roberts' initial study (see Table 7.3).

The authors attribute these changes to two main factors: firstly, secular trends in body size over the 45 years after Roberts' work, particularly those changes resulting from improved nutrition in tropical regions tending to increase overall body size; secondly, the effect of improved and more widespread technological means to mitigate exposure to extreme temperatures during development. This latter point points to the possibility of some developmental plasticity in the growth of human body size and proportions.

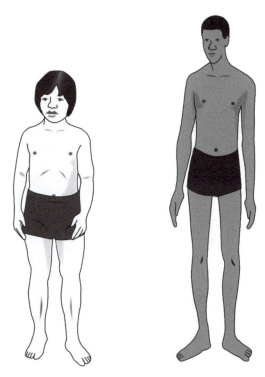

A) Arctic body proportions (Inuit)

B) Hot climate body proportions (Sudanese)

FIGURE 7.8 Typical body shapes of humans living in two different climatic zones.
Humans in hot climates tend to have long slender limbs since this gives a higher SA/volume ratio and hence allows greater heat loss.

Table 7.3 A comparison of regression equations linking body weight and temperature, and SA/body mass and temperature, between two studies on human populations. Both sets of equations provide support for the application of Bergmann's and Allen's rules to humans. The fact that the two sets of equations are different suggests some secular changes in the effect of temperature on bodily dimensions had occurred over the intervening 45 years. The correlation coefficients have fallen between 1953 and 1998 suggesting that other factors are increasingly affecting body size and shape.

Equation: Body weight = k Temp. + C				
	k value (regression coefficient)	C value	r (correlation coefficient)	Significance
Roberts (1953)	−0.55	65.8	−0.59	p<0.001
Katzmarzyk and Leonard (1998)	−0.26	66.86	−0.27	p<0.001
Equation SA/Body mass = k Temp. + C				
	k value (regression coefficient)	C value	r (correlation co-efficient)	Significance
Roberts (1953)	1.15	267.55	0.59	p<0.001
Katzmarzyk and Leonard (1998)	0.49	267.00	0.29	p<0.001
SOURCE: Roberts (1953); and Katzmarzyk and Leonard (1998).				

Skin colour

Skin colour is primarily determined by the amount of melanin in skin cells. It is likely that the earliest hominins had a skin resembling that of a modern-day chimpanzee; that is, white or lightly pigmented skin covered with dark hair. This dark hair reduces environmental heat gain in hot climates. It does this even if the fur is black since the hair will absorb short wavelength radiation before it reaches the skin, and reradiate away heat as infrared radiation. This hairy but unmelanised skin was probably the primitive (that is, ancestral) condition for all primates. Even today humans have about the same number of hair follicles per square inch as a chimpanzee but our hairs are finer and less developed. But as activity levels in open sunlit areas increased, concomitant with humans adopting bipedalism and walking and foraging over long distances during daylight hours, so this fur-based protection proved to be inadequate and overheating became a problem. The evolutionary response was the development of an efficient set of sweat glands and a decrease in body hair. This then presented the problem of protecting the skin from harmful UV radiation. This protection was achieved by the evolution of melanin-producing cells (melanocytes) of which you have (irrespective of your skin colour) about 1,000–2,000 per square mm in your skin.

Even a cursory glance at the complexion of indigenous peoples shows that melanin content varies around the globe, with darker skin in equatorial regions and lighter skin at higher latitudes (north or south). Most authorities agree that these variations must reflect some sort of biological adaptation to some aspect of the environment. But, perhaps surprisingly, from here on there are a variety of hypotheses about exactly what selective factors are at work. Melanin clearly blocks light penetration but what problems are solved by this? Today we are conscious of the carcinogenic effects of excessive sunlight but whether skin cancer (which appears later in life) was a real factor in reducing reproductive fitness is questionable. One suggestion is that ultraviolet radiation (UVR) damages sweat glands and therefore disrupts temperature regulation (Roberts and Kahlon, 1976).

A more convincing scenario has come from an examination of levels of UVR around the globe in relation to the skin colour and absorbance properties of melanin itself. Melanin has two main effects in the skin: firstly, it blocks light penetration, particularly at 545 nm (1nm = 1×10^{-9} m); secondly, it absorbs toxic chemicals produced by photochemical reactions in the skin. These two effects combined mean that melanin is particularly effective at protecting against damage to vitamin B9 (folic acid, sometimes called folate) in blood vessels just under the skin. Jablonski and Chaplin (2000) suggest that it is this protection of folate that had led to the adaptive variation of skin around the world. Two lines of evidence strongly support this. Firstly, folate is an extremely important vitamin involved in a huge array of essential physiological functions (such as synthesis of DNA, cell division and growth) and low levels of folic acid are known to be implicated in embryonic defects and reduced male fertility; secondly, the correlation between skin reflectance and UVR is highest for reflectance measured at 545 nm and it is this wavelength that represents the absorption maximum of oxyhaemoglobin. This suggests that an important function of skin pigmentation is to protect the content of subcutaneous blood vessels in the skin (Table 7.4).

The correlations are impressive but we now need to explain why skin pigmentation, which evolved in the tropics to prevent the destruction of folic acid and damage to sweat glands, should have evolved towards lighter tones as humans moved north and south. Why not keep the dark skin? The answer is probably related to the fact that certain wavelengths of UVR (specifically UVB radiation at 280–315 nm) are required for vitamin D synthesis. As one moves north or south out of the tropical regions the intensity of UVB falls off particularly rapidly, partly due to the increased depth of atmosphere that light has to pass through at oblique angles and partly due to the ozone layer. Humans moving northwards retaining deeply pigmented skin would have suffered from vitamin D deficiencies unless they adopted a diet rich in vitamin D. This latter point is nicely illustrated by the Eskimo-Ammassalimiut peoples of the North American and north-east Asian Artic, who have darker skin pigmentation that would be expected from their high latitude and low level of UVB in their environment (see Figure 7.9). Their relatively dark skin is explained by several factors. One is that the main diet of these people – fish, caribou and marine mammals such as seals – provides rich sources of vitamin D. In fact they inhabit an environment of such low UVB intensities that life would be impossible there without dietary sources of vitamin D. Another factor may be the fairly recent (in evolutionary terms) arrival of these people from their Asian homelands at lower latitudes, resulting in an evolutionary lag between an adaptively optimal pigmentation and a new environment. Finally, there may have been a selective pressure to maintain a fairly dark skin pigmentation as protection against UVA radiation incident onto the skin both from direct irradiation and from reflection off snow and ice (Jablonski, 2004).

Hence skin colour has to be seen as a product of two opposing selective forces: melanisation to protect the skin and prevent the photolysis of folic acid, and de-melanisation to facilitate UVB penetration and so facilitate vitamin D synthesis (Figure 7.10). In long-standing populations it is likely that these opposing tendencies have been resolved by natural selection to give an optimal skin tone.

Table 7.4 Correlation coefficients between skin reflectance and UVMED, and skin reflectance and latitude.
UVMED represents the minimal dose needed to cause reddening of the skin and can be calculated around the world using NASA data. Reflectance is the proportion of light reflected from the skin at specific wavelengths. Both correlation coefficients are high. The correlation between reflectance and latitude is high and positive since people living at high latitudes (north or south) have highly reflective (that is, lightly pigmented) skin. The correlation between reflectance and UVMED is high and negative suggesting that skin of low reflectance (that is, darkly pigmented) requires a higher UVMED (that is, can withstand a higher dose of UV before reddening is observed).

	UVMED (Ultraviolet Minimal Erythemal Dose)	Latitude
Reflectance at 545 nm	–0.964	+ 0.957
SOURCE: Data from Jablonski and Chaplin (2000), Table 4, p. 72.		

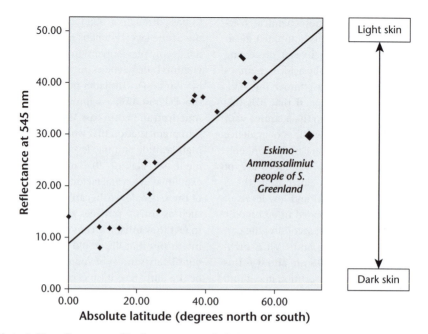

FIGURE 7.9 Plot of skin reflectance of indigenous people living at various latitudes against absolute latitude (degrees either north or south).
The strong positive correlation supports the idea that some component of latitude (probably UVR) has been a prime determinant of skin tone. A high reflectance implies a pale skin. The point noticeably outside the trend is from the Eskimo-Ammassalimiut people of southern Greenland, who have skin darker than one would expect from their latitude. Their case is discussed in the text. Data on skin reflectance from Jablonski and Chaplin (2000); latitudes from standard world atlas.

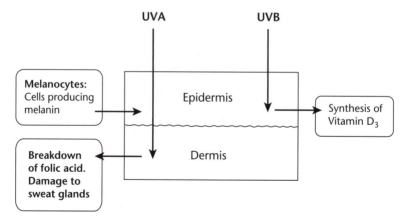

FIGURE 7.10 In melanin production and hence skin tone natural selection has achieved a delicate balance between the positive and negative effects of UV radiation.
If UVA penetrates to the dermis layer it breaks down valuable folic acid and can damage sweat glands. Hence, melanocytes produce melanin to prevent this. However, penetration of UVB to the dermis is required to enable the synthesis of vitamin D. If vitamin D is readily available in the diet then more melanin can be tolerated to reduce UVA damage.

Table 7.5 Reflectance values from male and female skin.
Females have consistently lighter skin than males.

Filter wavelength (nm)	Mean reflectance for females	Mean reflectance for males	p value from two-sample 't' test
425	19.20	16.88	$p < 0.001$
545	23.93	22.78	$p < 0.01$
685	47.20	45.09	$p < 0.001$
p = significance			
SOURCE: Data from Jablonski and Chaplin (2000), Table 3, p. 72.			

Sex differences in skin pigmentation

It has been noticed for several decades that human females are noticeably lighter in skin tone than human males (for example, see Robins, 1991). A plausible adaptive reason for this is likely to relate to the extra calcium needs of females during pregnancy and lactation. A lighter skin promotes a higher synthesis of vitamin D compounds. Jablovski and Chaplin (2000) compared the reflectance values of the skin of males and females from a variety of indigenous populations. Reflectance at three different wavelengths was found to be significantly higher for women than for men (see Table 7.5).

The sexual dimorphism in human skin colour is complicated by the effects of sexual selection since in many cultures men are reported to prefer women with a lighter skin tone (for example, see Dixson et al., 2007). It is possible that a feature that began as a product of natural selection was reinforced by sexual selection implying that pigmentation in human females might represent a complex balance between sexual selection, photo-protection and vitamin D synthesis. It also suggests the possibility of more work looking at sexual dimorphism in skin tone where vitamin D synthesis is not a priority in areas where diets are already rich in vitamin D.

7.3 The problem of individual differences: heritability and genetic diversity

As noted earlier, evolutionary psychology has tended to focus on human universals to the neglect of explaining individual differences. The idea that human genetic diversity is trivial or that observed behavioural variation is just noise in the universal psychological machinery is, for a variety of reasons, no longer a tenable positon. Firstly, human geneticists have demonstrated the existence of genetic differences within and between populations that have arisen as a result of natural selection. Secondly, as David Buss (2009) points out, in some contexts it is only individual differences that really matter. The obvious example is mate choice: we don't just fall for and mate with anyone that satisfies the criteria of being of the same species. Rather, difference-defining attributes such as agreeableness, intelligence, dependability, wealth and so on are paramount. Thirdly, a whole branch of psychology, sometimes called differential psychology, has been concerned with individual differences and has shown how many personality characteristics have strong heritabilities (>50 per cent) and are stable throughout a person's life (Plomin et al., 1999). This strong heritability of personality differences sits uneasily with the idea that differences between people is just noise in the system that natural selection is reducing as it favours optimum genes. An urgent task for all evolutionary accounts of human behaviour is to develop models that can incorporate individual differences within a broader framework of our species-typical psychology.

As Figure 7.11 shows there are a variety of approaches that have been postulated to solve this problem. The two broad categories are those models that assume some underlying genetic variation (and hence high heritability of personality traits) and models based on a more universalist conception of a human nature which is expressed differently in different circumstances and which can sometimes give the impression of high heritability of personality traits (for example, reactive heritability, see below).

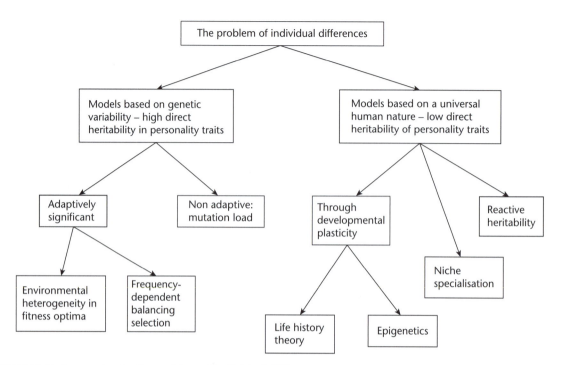

FIGURE 7.11 Approaches to the problem of individual differences.
Two basic approaches are the idea that there are significant genetic differences between people (left-hand side), or that differences arise due to a universal human nature with low genetic diversity developing in ways strongly influenced by the environment and epigenetic inheritance (right-hand side).

7.3.1 **Models assuming genetic variability**

a) Adaptively significant: environmental heterogeneity in fitness optima
    Life's problems do not always have a simple or single solution; furthermore, different environments will select for differing types of solutions. So long as humans have encountered different environments over an evolutionarily significant period of time so different gene-based solutions may persist in the gene pool. Some environments may favour, for example, extraversion and risk taking; while others may favour caution and rectitude. If humans have encountered both types of environments in their evolutionary history and have adapted to both then the human gene pool will contain genes relevant to both strategies, which might show up as personality differences between people in different environments, or (since humans have moved around a lot) differences between individuals now living in the same environment. This is sometimes referred to as

'environmental heterogeneity in fitness optima' (Buss, 2009).

b) Adaptively non-significant: mutation load
    All genes are susceptible to mutation. It has been estimated that about half of the 25,000 genes that humans carry find some sort of expression in the human brain. Hence the brain (and, correspondingly, behaviour) is a large target for the impact of inherited and newly arising mutations. Keller and Miller (2006) estimate that on average each person carries about 500 mutations that could affect brain function. A significant aspect of this mutation load is that due to statistical variation some will inherit more mutations than others: they estimate a standard deviation of at least 22 mutations affecting brain function. If these mutations affect personality and behaviour the result is possibly a continuum expressing itself as personality differences, with those having a very high mutation load of particular genes being classified as having a disorder (see Chapter 19).

c) Adaptively significant: frequency-dependent balancing selection

Another type of balancing selection is called frequency-dependent selection. This is where several different behavioural strategies can co-exist so long as their individual frequency in the population is not too high or low. This idea has been used to explain psychopathy in which a strategy of cheating and exploitation could be advantageous if it exists at a low frequency such that perpetrators can move to new groups where their reputation is unknown. It has also been used to explain handedness; being left handed could bring benefits in a population where most people are right handed (see Box 7.4).

Box 7.4 Handedness and frequency-dependent balancing selection

A puzzle awaiting a more complete evolutionary explanation is handedness in humans. A variety of studies suggest that right handedness is the most common state across a wide range of cultures, describing the habits of about 63–97 per cent of the human population, with true ambidextrousness being extremely rare (Faurie et al., 2005). There are essentially two questions that need answering here: 'Why are most humans right handed?' and 'Why is there a consistent minority of left handers in all populations?' It is easy enough to see why handedness should be lateralised. It is probably a good use of expensive neural tissue to have the neurons responsible for fine motor control primarily located in one side of the brain, resulting in one hand becoming more dextrous that the other, than to double up on brain functions and use up tissue on both sides of the brain.

There is no general consensus as to why it should be the right hand that is most commonly used by humans. Several theories have tried to link right handedness to the fact that language specialisation is predominantly located in the left hand side of the brain – the side that also controls the right hand (see also the ballistic theory of encephalisation discussed in Chapter 6). One problem with this is that it might suggest that left handers would process language in the right side of their brains, yet this does not seem to be entirely the case: about 95 per cent of right handers and 50 per cent of left handers process speech in the left hand side of the brain, and about 25 per cent of left handers use both sides of the brain equally (Santrock, 2008). Other theories focus on the idea that early humans may have cradled their infants with their left hand, allowing the head of the baby to rest near the mother's heartbeat, and this freed up the right hand for other uses (Hopkins et al., 1993). But such theories of right handedness usually fail to address the question as to why everyone is not right handed.

The problem of explaining left handedness is made acute by the fact that a variety of studies indicate that being left handed is associated with fitness costs. Compared to right handers, left handers tend to produce fewer offspring, have lower weight and height in adulthood, reach puberty later and have a lower life expectancy (Aggleton et al., 1993; Faurie et al., 2006). The reason for these costs is not clear. There are also a variety of proximate theories to account for left handedness, such as the effect of high prenatal testosterone (Geschwind, 1984), or even the presence of a recessive gene for left handedness (Annett, 1964).

The fact that left handedness is substantially heritable and has been around since the Upper Paleolithic (between 35,000–10,000 years ago) suggests it a polymorphic trait, possibly maintained by some frequency-dependent selective process (Faurie and Raymond, 2004). So the question arises: 'What possible advantages are there to being left handed?' The answer may lie in aggressive encounters between early humans.

In a world where right-handed men fight other right-handed men (and hence where there are adaptations to favour the right hand in this process), then being a left hander gains an advantage. But as the frequency of left handedness increases, and so left handers meet other left handers, so this advantage decreases. Other things being equal the relative frequency should even out at 1:1. If being left handed incurs other costs, however (and there is evidence as noted above to suggest this), then the final frequency might be biased towards right handedness, so that the fitness of both conditions becomes equal.

The costs themselves seem to have components (lower fertility, reduced life expectancy, delayed puberty) that are likely to operate in a variety of socioecological contexts. But the advantages carry more weight in societies where physical combat between males is intense. With this in mind, Charlotte Faurie and Michael Raymond investigated if the frequency of left handedness was positively correlated with measures of physical aggression in traditional societies. They used a variety of sources to establish the degree of left handedness in a culture, and used homicide rate as an indication of physical violence in that society (Faurie and Raymond, 2005). Figure 7.12 shows a plot of their findings, confirming that left handedness is indeed positively correlated with homicide rates, supporting what they call the 'fighting hypothesis' to explain the polymorphic nature of hand laterality.

The findings are very suggestive but not conclusive. It may be, for example, that some other factor, such as level of male testosterone, drives both handedness and the degree of violence in a society.

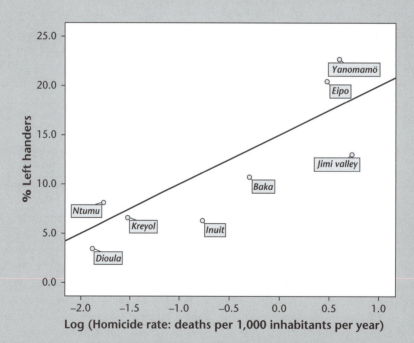

FIGURE 7.12 Frequency (%) of left handedness plotted against homicide rates for eight traditional cultures.
(Spearman rho = 0.83, p = 0.01 (two tailed)).

SOURCE: Data from Faurie and Raymond (2005), with permission.

But further support for the fighting hypothesis comes from an analysis of competitive sports. In interactive sports, such as fencing and tennis, where right-handed players hone their skills against other right-handed players, left-handed players might be at an advantage. In non-interactive sports, such as swimming, diving and cycling, this advantage disappears. One might expect, therefore, that interactive sports might have a higher frequency of left handers than non-interactive sports. A study looking into this was carried out by George Grouios and colleagues who looked at the characteristics of over 1,000 class A (that is, very good) athletes in Greece (Grouios et al., 2000). As expected, they found that interactive sports had a higher proportion of left handers than non-interactive ones; they also found, as is often reported, that left handedness is more common amongst men (Figure 7.13).

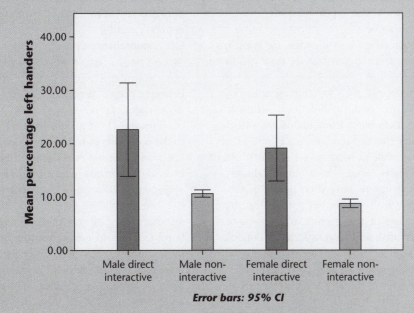

FIGURE 7.13 Frequency of left handedness according to gender and type of sporting activity. The difference between interactive sports (such as boxing, fencing and tennis) and non-interactive sports (such as diving, swimming, weight lifting and running) in the frequency of left handedness is statistically significant for both men ($\chi^2 = 11.4$, $p < 0.001$) and women ($\chi^2 = 10.2$, $p < 0.001$).
SOURCE: Data from Grouios et al. (2000), Table 3, p. 1278.

7.3.2 Models assuming low genetic variability

a) Reactive heritability
Differences between individuals could arise through what Tooby and Cosmides (1990a) call 'reactive heritability'. The idea here is that there might be several different solutions to an adaptive problem. To achieve a goal, for example, we might use aggression or cooperation. All humans might have the ability to call upon aggressive or cooperative strategies (stemming from a universal human nature) but the one chosen might depend upon an assessment of the context and, crucially for this perspective, one's own physical and mental qualities. Realising that one is strong and muscular compared to the group average, for

example, might incline you to adopt aggressive strategies. Now body build is strongly heritable, and so personality type might give the appearance of high heritability but only because it is riding on the back of this other heritable quality. Another potential example is the trait of extraversion, which has a heritability of about 50 per cent (Jang et al., 1996) and is often associated with both physical attractiveness and physical strength (Lukaszewski and Roney, 2011). A decision rule that says something like 'if you are above average in physical strength and attractiveness then adopt extrovert traits' would give rise to differences between people on a scale of extraversion that would also appear heritable.

b) Niche specialisation

Another potentially fruitful idea to explain human personality differences is that people adopt strategies and personality styles to pursue common biological goals by finding a behavioural niche that avoids competition with others. A classic example is Frank Sulloway's theory of birth order effects. Each child in a family has to develop a mode of being to get on with, and extract resources from, its parents. Sulloway argued that firstborns have their pick of potential niche personality types but tend to identify with the values of the parents and so appear more conservative and conventional in outlook than their younger siblings. Later-born children find this niche occupied and so behave differently. They tend to identify less with their parents' values and often experience domination

and bullying by older siblings. Sulloway argued that this led to them expressing greater empathy for weaker members of society, to resisting authority and being more questioning of the status quo. In short they were more rebellious than firstborns (Sulloway, 1996). Sulloway's ideas have prompted a wave of research on this topic with mixed results probably due to the large number of confounding variables and possibly the small nature of the effect anyway (see Healey and Ellis, 2007, for work supporting this hypothesis). Even if Sulloway's model turns out to be a poor predictor of family relationships it points to a type of explanation that could demonstrate how personality differences could arise without the need for genetic variation.

c) Life history theory and epigenetics

In Section 7.2 above we observed how developmental plasticity can give rise to differences between people depending on their early life experiences. Epigenetics provides one mechanism for this at the genetic level and even points to the possibility of transgenerational effects. The science of epigenetics is still in its infancy, however, and it remains to be seen just how adaptive epigenetic modification turns out to be. What is more established is the idea that energetic resources may be switched between different behavioural outcomes depending on what type of environment has been experienced and is to be expected in the future. This more established body of knowledge is called life history theory and is the subject of the next chapter.

Summary

- The fit between form and function and the phenomenon of adaptations can be explained by natural selection, although it has proven difficult to eradicate teleological language in describing this process.

- Humans adapt to their surroundings in a variety of ways. This can be conceived as a spectrum: from, at the one end, short-term individual acclimatisations to local changes that serve to maintain homeostasis or better enable individual survival in variable conditions, to, at the other end, long-term genetic changes which fall under the heading of natural selection. In the mid-ground of medium-term changes over one or a few generations there lies the important area of developmental plasticity and the new science of epigenetics – currently the subject of much research.

⮞ Epigenetics points to the possibility of the inheritance of acquired characteristics, something that Lamarck proposed and was dismissed from mainstream biology for most of the 20th century. Such changes, however, are not to the base codes of DNA but to modifying proteins and methylation patterns. There is some evidence for transgenerational effects but how widespread this is and whether is serves adaptive ends is still open to debate.

⮞ Since some groups of humans left Africa some 72,000 years ago we can see plenty of evidence of adaptations to local climates and ecologies. Two well-studied examples are skin-tone variation and bodily proportions.

⮞ A key problem in evolutionary accounts of human psychology is the explanation of individual but heritable differences. There are various models to address this. One group suggests that they represent real genetic differences and the product of past adaptations; another suggests they are effects of the developmental plasticity of a universal human nature.

Key Words

- Acclimation
- Acclimatisation
- Epigenetics
- Homeostasis
- Habituation
- Teleology
- Teleonomy
- Thrifty genotype
- Thrifty phenotype

Further reading

Armstrong, L. (2014). *Epigenetics*. New York, NY, Garland Science.

Looks at epigenetics from a molecular and biochemical perspective. Not specifically related to human behaviour but an excellent technical consideration of the current field.

Danchin, E. (2013). 'Avatars of information: towards an inclusive evolutionary synthesis.' *Trends in Ecology and Evolution* **28**(6), 351–8.

Interesting paper that suggests we need to revise the way we think about heredity in the light of findings from epigenetics and gene-culture co-evolutionary theory.

Buss, D. M. and P. H. Hawley (eds.) (2010). *The Evolution of Personality and Individual Differences*. Oxford, UK, Oxford University Press.

Thorough analysis of the current problem of explaining individual differences in an evolutionary framework. Well-structured series of articles by experts in their field.

Life History Theory

The woods decay, the woods decay and fall,
The vapours weep their burthen to the ground,
Man comes and tills the field and lies beneath,
And after many a summer dies the swan.
(Tennyson, *Tithonus*, 1859)

The fundamental processes of life – birth, growth, reproduction and death – are common to all animals, yet different species partake of these stages at different rates. There may even be variation in the timing of these events within a species according to local conditions and individual variation. Life history theory is the specialism addressing this phenomenon. The tragedy of Tennyson's Tithonus was that he was granted immortality but not eternal youth. As we will see in this chapter, eternal youth is not a Darwinian possibility; and ageing, senescence and death are also part of the purview of evolutionary theory.

8.1 Life history variables

Life history theory (LHT) is based on the idea that organisms must balance competing demands on their time and energy budgets. In this respect we can think of organisms as capturing energy from the environment through feeding and allocating it to three types of activity (Gadgil and Bossert, 1970):

- **Growth** – organisms must grow to a certain size so that they can fend for themselves and such that they are sexually mature and large enough to attract a mate. This investment impacts on future reproduction.
- **Repair and maintenance** – all organisms are daily buffeted by shocks and damage from their external and internal environments. Some energy

must therefore be allocated to coping with infections and repairing damaged tissue. This investment also impacts on future reproduction.
- **Reproduction** – all organisms must produce direct or indirect offspring. This investment, by definition, is concerned with current reproduction. In this category there are still decisions about the quantity and quality of offspring and the timing of reproduction.

It can be seen that these three allocations resolve to the problem of a trade-off between future and current reproduction possibilities. The allocation of energy into these three domains will require compromises and so it follows that activities that may seem maladaptive in the short term (such as delayed sexual maturity) may be perfectly sensible when viewed over the whole lifespan.

The natural world is remarkable for the huge diversity of ways in which living things organise these allocations. Some species, like birds, spread their breeding sequentially over a number of breeding seasons in their life; others, such as salmon and annual plants, have a single short breeding period and then die. Some species, like humans, have a few relatively large offspring, whilst others such as insects, oysters and many plants produce hundreds or thousands of very small offspring.

In seeking to explain this diversity, LHT takes an optimality approach and looks at how inclusive fitness can be maximised by the timing of growth, reproduction and behaviour across the life cycle. This can help an understanding of why life history strategies differ between species but it can also help explain how individuals within a species modify their behaviour according to local social and ecological conditions.

The allocation of resources between essential life functions such as growth, repair, survival and reproduction can be understood as responding to short-term, medium-term and long-term influences (Figure 8.1). Over the long term, natural selection will have operated to set, within margins, the timing of the main stages of life: infancy, childhood, juvenility, adolescence, adulthood, the menopause, senescence and death. This will give rise to a species-typical pattern, but may also result in differences (that is, genetic diversity) between human groups according to local ecologies. But it can also be expected that natural selection will have laid down the possibility of

developmental plasticity such that local information can be used to help fine tune these allocations in an adaptive manner. In the medium term this may occur through transgenerational epigenetic mechanisms (see Chapter 7). In the short term, local influences and early experiences in nutrition and parenting can act at the psychological or physiological level.

The relevance of LHT to an evolutionary understanding of human behaviour is that complex decisions have to be made about the allocation of resources and so it may be supposed we have adaptations to assist us in this task. This differential allocation in relation to the age of the organism does not require the existence of a centrally coordinated set of conscious calculations for its determination. Rather, **endocrine systems** have evolved to serve this role. In puberty, for example, the release of hormones initiates changes to energy budgets that can manifest effects over several years. In both males and females,

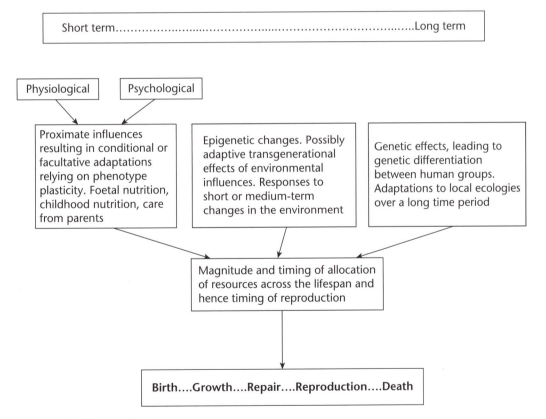

FIGURE 8.1 Life history allocations.
The way resources are allocated between growth, repair and reproduction will be influenced by long-term effects on the species as well as medium- and short-term effects on the individual and its immediate ancestors.

growth subsides as energy is allocated to reproductive functions and secondary characteristics. More specifically, in females such hormones lead to fat storage and regular menstrual cycling. In males, androgens lead to increased musculature and competitive mating displays. In both sexes, investments to the immune system are reduced. Once humans have reached middle age and have reproduced the investment in sexual display is less of a priority. Consistent with this is the finding that when men become fathers testosterone levels subside, enabling effort to be redirected from mating to fathering (Gray et al., 2002). Another very simple mechanism is illustrated by the fact that lactation suppresses ovulation. This makes adaptive sense since it would be unwise to invest further energy into reproduction whilst still breastfeeding a young infant. In the sections that follow we examine how some puzzles in relation to human life are addressed by LHT.

8.2 Quantity and quality, mating and parenting

K and r selection

Within the category 'energy directed to reproduction' (see above) there are still choices (either by the individual or in the phylogenetic line) to be made. Should an individual invest energy in a few high-value offspring or large numbers of lower-value offspring? Some species (for example, fish, frogs) are said to be 'r selected', that is they produce large numbers of energetically cheap offspring; others, such as humans and elephants, are said to be 'K selected' and produce few, but energetically expensive, offspring (Table 8.1).

The r–K spectrum is a reflection of two key decisions that animals have to make: firstly, whether to invest in current offspring or delay reproduction and invest in future offspring; and secondly, whether to make lots of low-quality offspring or fewer high-quality offspring. One of the main factors that will drive this decision-making will be external sources of risk and mortality. When adult mortality is high it becomes adaptive to shift resources to current offspring by mating earlier since waiting may mean the death of the adult and hence no reproduction at all. When juvenile mortality is high it probably pays to produce lower-quality offspring in the hope that a few will survive. More generally, high external risks suggest a preference for investing in reproduction rather than somatic effort and an investment biased towards mating and away from parenting. This latter point follows from the argument that in high-risk environments (with unpredictable and severe risk of mortality) increased parental effort will quickly cease to increase the fitness of offspring (Quinlan, 2007).

There arises the possibility that complex organisms that make day-to-day decisions about their lives can bias their reproductive allocation to a more r-selected or K-selected strategy in the direction of adaptive optima for local conditions. There has also been the suggestion that humans show developmental plasticity in that they are able to move across a short section of the r–K spectrum by adopting fast or slow life histories according to ecological contexts (Table 8.2). This idea is explored in the next section in relation to the timing of puberty.

Table 8.1 Some features of K-selected and r-selected organisms

	r selected	K selected
Climate	Variable and unpredictable	Constant and/or predictable
Population size	Variable over time. Often below carrying capacity of environment. Periodic recolonisation of habitat	Fairly constant and at equilibrium, near to carrying capacity of environment
Lifespan of organisms	Usually short, typically less than one year	Longer than one year
Reproduction	Production of many small offspring. Usually small body size and rapid growth to sexual maturity	Fewer but larger progeny. Delayed reproduction. Slower growth to sexual maturity
Mortality rates	High	Low
Example species	Mouse lemur, frogs, oysters	Gorillas, elephants, humans

Table 8.2 Slow and fast human life history strategies

Developmental feature	FASTER LIFE HISTORY STRATEGY (more 'r' selected)	SLOWER LIFE HISTORY STRATEGY (more 'K' selected)
Risks taken for potential gains	Higher	Lower
Rate of development	Faster	Slower
Onset of puberty	Earlier	Later
Biological ageing	Faster	Slower
First sexual experiences	Earlier	Later
Number of sexual partners	More	Fewer
Age of reproduction	Earlier	Later
Number of offspring	More	Fewer
Investment in each offspring	Lower	Higher
Immediate gratification	Take	Delay

SOURCE: Copyright © 2012 by the American Psychological Association. Adapted with permission. Ellis, B. J. et al. (2012). 'The evolutionary basis of risky adolescent behavior: implications for science, policy, and practice.' *Developmental Psychology* **48**(3): 598–623. Fig 1, p. 608.

8.3 The ages of man – a view from life history theory

All the world's a stage,
And all the men and women merely players;
They have their exits and their entrances,
And one man in his time plays many parts,
His acts being seven ages.
(Shakespeare, *As You Like It*)

Shakespeare's Jacques in *As You Like It* pictured the seven ages of a man's life as: the infant, the school boy, the lover, the soldier, the middle-aged 'Justice', old age and finally death and oblivion. These are not too far removed from the stages studied in contemporary anthropology. In the sections that follow we explore these stages in both men and women from an LHT perspective.

8.3.1 Birth and infancy

Neonatal adiposity – or why babies are born fat

At birth humans are born with more body fat than most other species (Figure 8.2). It was once the common view that this early layer of neonatal fat served some heat insulation function and was a compensation for the loss of hair in the hominin lineage (for example, see Prechtl, 1986). Yet the evidence

in favour of fat as a thermal insulator is very weak. Human populations living at cold high northerly latitudes may consume more fat in their diet as a source of energy but their bodies are not significantly higher in subcutaneous fat stores than humans living in temperate zones (Stini, 1981). A more convincing explanation for such high fat levels is given by Christopher Kuzawa (1998) who argues that body fat in neonates is there as a source of energy. As noted in Chapter 6 the human brain is energetically expensive tissue, consuming, in adults, 22 per cent of basal metabolic rate (BMR). But newborn babies devote around 70 per cent of BMR to support brain growth and maintenance during their first year of life. Moreover, this amount is not negotiable: it needs to be supported at this level to avoid serious damage. Adipose fat stores may, therefore, serve as an energy reservoir to ensure the brain is supplied with the calories it requires. Shortly after weaning, the child faces another drain on its energy reserves when he or she is exposed to pathogens and is no longer protected by the mother's immune system. Given that the immune system is 'naïve' in the sense that its needs infections to learn from and stimulate antibody production, infections in childhood are inevitable. Illnesses and their consequences such as such as diarrhoea and fever can place considerable energy demands on the metabolism of the child and hence require the mobilisation of fat reserves. Given

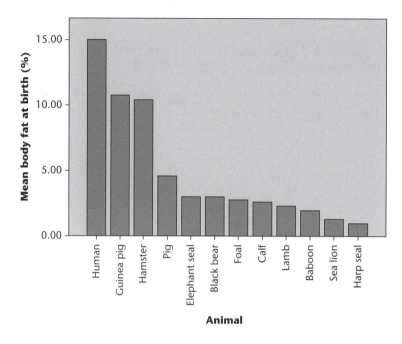

FIGURE 8.2 Percentage body fat at birth for selected mammals. Humans have one of the highest levels of neonatal body fat known. Even after allometric scaling humans have body fat levels at birth about 3.75 times that expected for a mammal of our body size.

SOURCE: Data from Kuzawa, C. W. (1998). 'Adipose tissue in human infancy and childhood: an evolutionary perspective.' *American Journal of Physical Anthropology* **107**(s 27), 177–209. Table 1, p. 181.

this twofold importance of body fat it is perhaps not surprising that babies continue to increase fat deposition (as a percentage of overall body weight) during the first year of life, after which it declines steadily.

8.3.2 Childhood and the inter-birth interval

Generally, compared to most other species, humans have a slow life history trajectory characterised by:

- Late reproduction
- Long lifespan
- Long period of reproductive immaturity
- Bi-parental care
- High investment in parental care and transmission of learning.

The odd feature of human life history, however, is that all of the above features would tend to predict low fertility. Yet compared to close primate relatives humans have a shorter inter-birth interval. Another puzzle is that humans also have two other periods of life history not shared by chimpanzees: childhood and adolescence (Figure 8.3). LHT may help to resolve these puzzles.

Inter-birth intervals

One important life history decision to be made in giving birth is the compromise between quantity and quality, and this decision reveals itself in the pattern of birth spacing. A long interval, for example, between births would imply that a woman is choosing to invest more in each offspring. The !Kung San people of Botswana have been the focus of a number of studies to investigate if birth spacing represents an adaptive strategy. Studies by Lee (1979) have shown that the birth interval for !Kung mothers was about four years and that an average woman bears 3.8 offspring in her lifetime. A factor that determines birth interval in this society is the harsh reality that in the dry season !Kung women have to travel several miles to collect their staple diet of mgongo nuts. Children under four years are carried by their mothers on these stressful foraging expeditions. Blurton Jones and Sibly (1978) developed models to see how the backload of women varied with inter-birth interval (IBI). Their model showed that as the IBI fell from ten years the backload remained fairly constant until at four years the predicted load rose sharply. In the model at least, an IBI of below four years would place a high strain on the mother and could jeopardise her health. From this simplified perspective it looks like

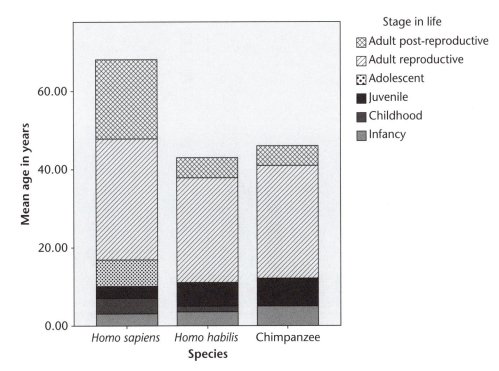

FIGURE 8.3 **Relative duration of life history stages for females of three hominid species.**
The estimates for chimpanzee and *Homo sapiens* are based on contemporary empirical evidence; data for *Homo habilis* is necessarily speculative and based on inferences from skeletal remains. The figure shows that the menopause is peculiar to modern humans and that increased female human longevity is largely a result of an extension to post-menopausal adult life rather than a significant extension of earlier life stages. Note also that chimpanzees have no period of childhood or adolescence as such (see text).

SOURCE: Data from Bogin, B. (1999). *Patterns of Human Growth* (Vol. 23). Cambridge University Press. Fig. 4.9, p. 185, and Fig 4.13, p. 219; and Robson, S. L. and B. Wood (2008). 'Hominin life history: reconstruction and evolution.' *Journal of Anatomy* **212**(4), 394–425. Table 2, p. 398.

four years may be the optimal IBI for !Kung mothers. There is an ongoing debate on this subject, however, and other factors may also be involved (see Pennington, 1992).

Childhood

Childhood is something distinctively human. Most mammals pass from infancy and dependency on nursing from the mother directly to a stage of independent feeding. But humans experience a distinctive period of about four years (roughly ages 3 to 7) of post-weaning dependence, otherwise known as childhood. Biologically, this period is characterised by the following features:

● Relatively small body size for age compared to growth curves of other primates.

● Fast-growing brain consuming at 5 years of age 44 per cent of resting metabolic energy needs – a figure far higher than any other primate.

● Immature dentition.

● Motor and cognitive immaturity.

● A mid growth spurt around 6–8 years.

According to Barry Bogin (2010), the evolution of childhood gave humans a reproductive advantage compared to primates of a similar size. For many mammals the termination of infancy is signalled when the first permanent molar tooth (M1) erupts. This is understandable given that the withdrawal of feeding by the mother at the end of infancy needs to coincide approximately with the arrival of dentition to enable the juvenile to process and consume its own food. Now this period of infancy, before the eruption of the first molar, is also for most mammals

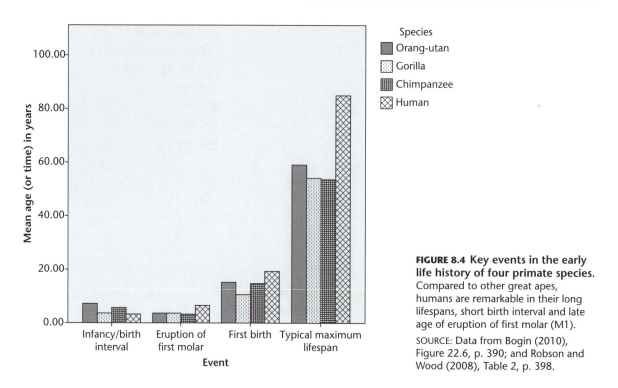

FIGURE 8.4 Key events in the early life history of four primate species. Compared to other great apes, humans are remarkable in their long lifespans, short birth interval and late age of eruption of first molar (M1).

SOURCE: Data from Bogin (2010), Figure 22.6, p. 390; and Robson and Wood (2008), Table 2, p. 398.

very closely aligned with the inter-birth interval: a large primate cannot energetically afford to be pregnant with another offspring whilst still lactating and providing expensive care to the current one.

But as Figure 8.4 shows, humans are an exception to this relationship. The birth interval for women in traditional societies is about three years and not the six years predicted by age of eruption of M1. Mothers achieve this by effectively outsourcing the needs of children for food and protection to other individuals – often relatives such as siblings and grandmothers of the child. Reiches et al. (2009) propose a 'pooled energy budget hypothesis' as a way of quantifying the combined resource allocations of a community (as opposed to an individual mother) for reproduction. The net effect of this pooling of energy is that humans can produce infants at about twice the rate of chimps and orangs. The help of grandmothers may be crucial here and this is explored later in this chapter.

Further work is probably needed to consolidate the definition of childhood as a uniquely human phenomenon and some anomalies remain. T. M. Smith et al. (2013), for example, studied wild chimpanzees in the Kibale National park in Uganda and did not find that eruption of M1 was associated with

cessation of weaning or the mothers returning to oestrus. Given the emphasis placed on the age of first molar eruption in the reconstruction from fossils of the life history of early hominins it will be important to better understand these relationships.

8.3.3 Adolescence, puberty and the timing of menarche

If adolescence is unique to humans (Figure 8.3) we need to find an evolutionary explanation for its occurrence. Bogin (2010) sees it as a stage when young adults practise life skills (see also Cameron and Bogin, 2012). A clue may lie in the mortality statistics of primates. In the case of yellow baboons, toque macaques and chimpanzees about 50 per cent of the firstborns die in infancy. But a closer study of wild baboons shows that the death rate falls with each successive offspring: 50 per cent for first born, 38 per cent for second and 25 per cent for third (Altmann, 2001). The difference is probably due to knowledge and skills gained by the mother with each birth and her increasing ability to derive support from the wider social group (Altmann, 2003).

Applying these ideas to humans, Bogin suggests that during adolescence girls learn child-rearing skills by observing their mothers and helping to raise their own siblings. This results in a higher survivability of human infants. Figures for infant mortality amongst traditional human groups are very variable. In a survey of 28 forager cultures Frank Marlowe found that mortality ranged from 10–40 per cent with a median of 21 per cent, a figure obviously higher than developed countries but still lower than chimpanzees in the wild (Marlowe, 2005).

The value of adolescent males is perhaps less obvious. The puzzle is exemplified by the fact that boys typically begin producing sperm between 13–14 years of age, yet across a wide range of cultures few males actually father children until they reach 20 years of age (Hill and Kaplan, 1988). The answer again probably lies, at least in part, in the time needed for young adults to practise and acquire sociosexual skills. Marlowe also has noted that a higher mean male contribution to the diet of a group from adult and adolescent males does serve to lower weaning time and increase fertility rate, although it does not reduce the infant mortality rate (Marlowe, 2005). The distribution of male effort between mating and parenting is probably influenced by testosterone.

Testosterone and the distribution of life history allocations

It is well known that the costs of reproduction are divided unequally between men and women. Compared to the physiological investment in reproduction by females, the energetic costs of spermatogenesis are fairly trivial and only account for about 1 per cent of BMR in human males (Elia, 1992). In addition, sperm quality and quantity vary little with the energetic expenditure of males, suggesting little trade-off between spermatogenesis and other pursuits (Bagatell and Bremner, 1990). However, males invest in reproduction in other more energetically demanding ways; specifically, competing for mates and then provisioning and protecting mates and offspring. It is here that testosterone and other androgens come into play since testosterone is known to promote muscle anabolism (that is, build-up) and to divert energy to building up cortical (that is, the cortex or hard outer shell) bone density and increasing the number of red blood cells.

This testosterone-induced diversion of metabolic energy means that less is available for routine maintenance and other bodily functions. In addition, testosterone is also associated with a range of negative effects such as increased likelihood of injury due to greater risk-taking activity and increased aggression, and the suppression of the immune function (Muehlenbein and Bribiescas, 2005). There is also the likelihood that testosterone increases the risk of prostate cancer; the links are not totally clear but certainly reducing testosterone is a standard treatment for prostate cancer (Soronen et al., 2004).

Men also have to make a trade-off between parenting effort and reproductive effort. Direct evidence in support of this comes from the work of Mascaro et al. (2013). This group measured the testosterone levels and testes volume of 70 male volunteers who were also biological fathers. These measures of reproductive effort were then correlated against questionnaire-derived measures of parenting effort. They found a significant negative correlation between both testes volume and plasma testosterone levels when mapped against parenting effort. In other words men with large testes and high testosterone levels (signifying high investment into reproduction) seem to invest less into parenting.

Given the role of testosterone in shifting between risks and body growth, it is conceivable that levels of testosterone might vary between individuals and groups in a functionally adaptive way. It is plausible, for example, that in environments with a low risk of pathogens (such as at high altitudes) testosterone levels might be higher than areas where pathogen pressure is more intense since the suppression of the immune system which testosterone causes will be less debilitating where pathogen risks are low. Also, in regions of high resource availability testosterone levels might also rise above the average since the diversion of metabolic energy to muscles and bones incurs a smaller risk of an energy deficit elsewhere than if resources were scarce.

Some of the most thorough studies on testosterone and life history trade-offs in non-humans have been conducted on a songbird species called the dark-eyed junco (*Junco hyematis*), a greyish sparrow-like species found across North America. If males of these birds are given a testosterone implant then they perform more courtship displays, have more extra pair copulations and have a larger home range than non-implanted males. However, this same implanted group provided less parental care, had a more suppressed immune system and a lower survival rate compared to unaltered junco males (Reed et al., 2006). Similar findings have been observed in other bird species (Ketterson et al., 1999).

Michael Muehlenbein and Richard Bribiescas suggest that this type of adaptive and conditional expression of testosterone might explain why levels of testosterone tend to be lower for males living risky resource-poor environments, such as the Ache of Paraguay and the !Kung San of Namibia, compared to males living in the resource rich environments of Europe and the USA (Muehlenbein and Bribiescas, 2005).

More recently, Lassek and Gaulin (2009) looked at the costs and benefits of testosterone by examining data from the third USA National Health and Nutrition Examination Survey (NHANES III) conducted over the years 1988–94. The data set provides information about the nutritional intake (and hence energy expenditure) and overall muscularity of over 5,000 men aged 18–59. This was supplemented by interview-derived information about the sexual success (indicated by age at first intercourse and number of sexual partners) of the men. Testosterone was not measured directly but various studies have shown that muscle mass is related to testosterone levels (for example, Griggs et al., 1989) and that the sexually dimorphic greater muscle mass of men develops primarily at puberty as testosterone levels rise. The data set also gave information about the efficiency of the immune system of the men as indicated by C-reactive protein (CRP) levels and white blood cell count (WBCC). The higher these two values the better the immune system. The two researchers hypothesised that the benefits of high muscularity (and by

inference high testosterone) might be manifested by a positive correlation of muscularity with number of sexual partners and a negative correlation with age at first intercourse. They also reasoned that the negative effects of testosterone would be indicated by a positive correlation with energy intake and a negative correlation with immunocompetence as measured by CRP and WBCC levels. Table 8.3 shows a summary of some of their key data in terms of the regression coefficients of muscularity (LMV and FFM) with measures of sexual success, energy intake and immunocompetence. The table shows that the main predictions were supported and reinforces the idea that increasing muscle mass increases sexual success but carries costs in the need for increased energy input and a weaker immune system.

This work on the costs and benefits of testosterone suggests that the level attained in an individual or a population might be the result of a balance between natural and sexual selection. Sexual selection – in the 'intra' sense that enhanced muscularity increases competitiveness among males, and in the 'inter' sense that females are attracted by muscular males and other signs of testosterone – drives levels of testosterone upwards. Natural selection, however, driven by the advantages of low energy intake, reduced risk and a good immune system, serves to drive the levels downwards. It is feasible that the ability of testosterone to shift resources from survival functions to reproductive functions in this way means that it is an ideal mediator in how an organism might use social and ecological information to adjust life history strategies. Figure 8.5 gives a schematic view of how this might operate. It now appears very likely that some sort of mechanism like this exists but the precise details have yet to be formulated (see Hau, 2007, for a review).

Table 8.3 Regression coefficients of measures of muscularity with measures of sexual success, energy intake and immune system response. CRP = C-reactive protein; WBCC = white blood cell count. Both CRP and WBCC are indicators of immune system efficiency.

	Numbers of sexual partners	Age at first intercourse	Energy intake	CRP	WBCC
Limb muscle volume (LMV)	0.079 ($p<0.001$)	−0.046 ($p<0.01$)	–	–	–
Fat-free mass (FFM)	–	–	0.246 ($p<0.0001$)	−0.061 ($p<0.01$)	−0.160 ($p<0.0001$)
p = significance					
SOURCE: Data taken from Lassek and Gaulin (2009), Tables 3 and 4, p. 326.					

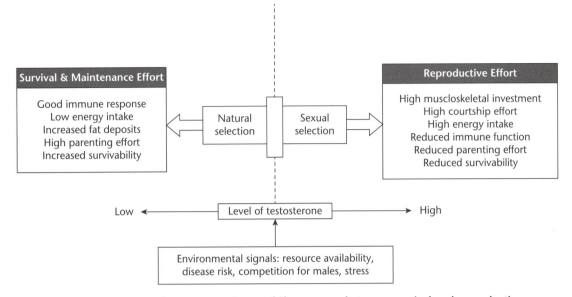

FIGURE 8.5 **Testosterone as a hormone serving to shift resources between survival and reproduction.**
Testosterone may serve as a proximate mechanism to serve life history strategies. Environmental signals can be used to predict the relative benefits of investing in survival or reproduction and testosterone shifts resources between these two life history functions. In addition, sexual selection in males may tend to increase testosterone levels to promote mating, whilst natural selection acts in the other direction to promote survivability.

Menarche, puberty and nutrition

The onset of reproductive maturity is a crucial time for animals and in the case of humans there is mounting evidence that the timing of puberty is influenced by a range of factors and that it may respond in adaptive ways to local conditions both between and within different human groups and populations.

Menarche is defined as the time of the first menstrual bleeding in human females and it signals the onset of puberty and fertility. It varies within and between cultures but in developed countries it is typically between 12 and 13 years of age. Strangely for such a vivid phenomenon, the function of menstruation itself is not well understood (Box 8.1).

Within LHT the age of menarche can be viewed as a 'decision' on when to begin reproduction so as to maximise some aspect of reproductive fitness. If this is the case, then it might be expected that this age is subject to alterations according to how cues from local contexts might be interpreted as indicators of reproductive opportunities and survival probabilities.

Box 8.1 The function of menstruation

Menstruation itself is something of a problem for evolutionary biology. There are essentially two phenomena at work: firstly, the cyclical regeneration of the lining of the uterus (called the endometrium), which if implantation and pregnancy do not occur is shed with the menses or reabsorbed; and secondly, vaginal bleeding. The first phenomenon is universal in mammals, whilst the second (bleeding) is mostly confined to Old World primates and a few shrew species. Menstruation usually refers to bleeding and the problem becomes why this wasteful process occurs in some mammals but not others. One suggestion is that it is a signal of fertility. But several genera of primates (for example, baboons and chimpanzees) show both sexual swellings (that is, ovulation is not concealed) and menstruation. In such cases it is hard to maintain that menstruation is a sign of fertility when other signals are revealing enough. In 1993 Marge Profet proposed the ingenious idea that menstruation served to flush out sperm-borne pathogens from the uterus. Profet then reasoned that suppression of the menses, by oral contraceptives for example, could compromise the body's natural defences. However, the theory has not stood up well to the evidence. Several studies have shown that menstruation exacerbates infection. Furthermore, menstruation would seem to be a very poor defence against pathogens when sexual intercourse in pre-industrial societies would have been quite frequent at those times when menstruation was absent: pregnancy, lactational suppression of ovulation and during the post-menopausal years (Strassmann, 1996b). In addition, menstruation is not noticeably more copious in groups that mate in a polygynadrous fashion.

Another view is that it is energetically more frugal to remove the endometrium each cycle than to maintain expensive tissues in a state of 'implantation alert' for a few days each month. Beverley Strassmann supports this theory and argues that vaginal bleeding is a side effect of this process (Strassmann, 1996b). At present no real consensus exists for the explanation of menstruation.

Two nutritional factors thought to influence the onset of puberty in females are the diet of the mother while the child was *in utero* and the quality of early childhood nutrition. In the early 1970s Rose Frisch and Roger Revelle (1970, 1971) published a series of papers arguing that early menarche was associated with high body weight; this became known as the Frisch 'critical weight–menarche hypothesis'. As well as purporting to explain the cross-sectional variation in the age of the onset of puberty, this hypothesis had the advantage of offering an explanation for the longitudinal secular trend towards earlier menarche in countries that had experienced rising standards of living and better nutrition over a long period of time. Over the last 100 years the age of menarche in developed countries, for example, has fallen considerably from about 17 to 12.5 (Gluckman and Hanson, 2006). Sadly this rather neat hypothesis did not stand up to lasting scrutiny and the situation appears more complex with a possible role for endocrine-disrupting chemicals from the environment (Parent et al., 2003).

However, the timing of reproduction will have a major impact on the reproductive output of any female and so it would be surprising if humans had not adapted to respond conditionally to environmental information to optimise reproduction in anticipated environments. The variety of subtle mechanisms at work are only just now being unravelled (Sloboda et al., 2007).

Despite the complex causation of declining age at puberty, it is now well established that energy input deficiency during late childhood, caused, for example, by excessive exercise or the condition anorexia nervosa, can delay puberty. A life history interpretation of this is that it makes adaptive sense to delay puberty (and hence reproduction) when energy supplies are restricted since an energetically demanding pregnancy and motherhood under these conditions might jeopardise both the mother's and the child's survival prospects. Conversely, despite the problems with the Frisch hypothesis in explaining long-term population trends, under conditions of good nutrition it might enhance reproductive fitness to bring forward puberty allowing the mother to benefit from a longer reproductive life.

In contrast to these effects of late childhood nutrition, prenatal exposure to good or poor nutrition experienced by the mother seems to have effects in the opposite direction. That is, children of low birth weight have accelerated menarche compared to those of a high birth weight. One interpretation of this is that a poor early environment is interpreted by the organism (still in a state of developmental plasticity) as an indicator of a harsher later environment and a reduced lifespan. In this case it is advantageous to bring puberty forward to a younger age to ensure reproduction (Sloboda et al., 2009).

Puberty, body stature and environmental risk

LHT also addresses the linkage between litter size, body size and lifespan amongst mammals. Mice are smaller than elephants, they have a higher relative metabolic rate and live shorter lives. As small mammals mice are frequently predated and so it makes sense for adults to produce large litters at frequent intervals to ensure that a few make it to adulthood. The sheer size of elephants offers protection against predation and so breeding pairs need only produce a few offspring in the knowledge that some will survive to grow and reproduce themselves.

As noted earlier, LHT has been successful at explaining these between-species variations in the differing allocations of energy to growth survival and reproduction. But we can also apply this reasoning to within-species variations, and the variation in human body size provides an instructive example. Some human groups have small stature and are referred to as pygmies. Pygmy stature is defined as an average male height of less than 155 cm. There are pygmy populations in African tropical forests – the best known being the Aka, the Efé and the Mbuti – but others are populations of people showing pygmy-like stature outside of Africa. Various explanations have been forthcoming to account for the small body size of these people, including the idea that it is an adaptation to avoid starvation in low productivity environments (Mann et al., 1987).

FIGURE 8.6 A photograph from the 1920s showing a Caucasian male with an African pygmy family.
One interpretation of the short stature of pygmies is that that they lie towards the fast end of the life history spectrum.
SOURCE: *Colliers New Encyclopaedia* (1921) p. 58.

Yet pygmy stature is found in warm humid environments and cool dry ones, and food shortages do not always give rise to short stature in human groups. A life history perspective might be a useful tool to approach this problem. Based on a careful analysis of demographic data, Migliano et al. (2007) concluded that it was the high levels of childhood mortality (as a consequence of the parasites and pathogens in their environment) that drove the life history trade-off towards early fertility and reproduction and reduced stature. Pygmies, therefore, occupy the 'fast' end of the life history spectrum: they reach sexual maturity earlier, and with a smaller body size, than most humans, but also die earlier. This makes adaptive sense, since to delay sexual maturity and reproduction in environments where mortality is high risks the chance of not reproducing at all. It looks likely that this pattern is genetic rather than a result of poor nutrition since the growth rates of pygmy children are surprisingly fast, even exceeding those of USA children in some cases (Perry and Dominy, 2009). Perry and Dominy reasoned that if the life history perspective were correct then larger body stature would be associated with increased longevity. They collected data on a number of pygmy tribes. The results offer modest support for this hypothesis; but, as the authors note, there are many confounding factors such as the fact that poor nutrition would reduce stature and life expectancy and give an impression of a correlation between these latter two variables. In addition, mortality rates are extremely difficult to establish for these peoples.

This suggested relationship between mortality risk, body size, growth rates and age of reproduction was also investigated by Walker et al. (2006b) who examined anthropological data on 22 small-scale

societies (including farmers and foragers). Among the various findings of the group it was noted that the age of menarche was positively correlated with the probability of survival of girls to the age of 15. This is consistent with the idea that high mortality tends to promote earlier menarche, presumably as a way of ensuring reproduction before death in high-risk societies (see Figure 8.7).

Attachment and the timing of menarche

Compared to most other mammals humans have fairly slow life histories, but since humans have faced (and still face) a range of challenging environments that can vary drastically (especially outside of the tropics) even between generations it would be unwise to commit to a single rigid strategy. With this in mind, numerous workers have looked to early

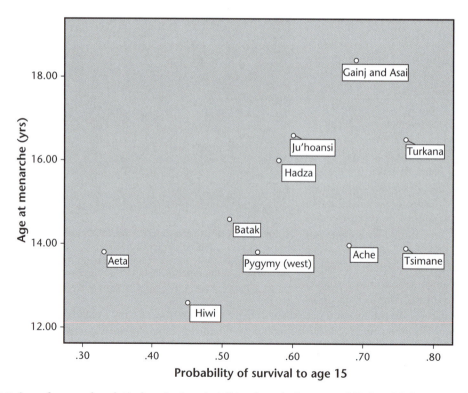

FIGURE 8.7 Age of menarche plotted against probability of survival to age of 15 for girls in ten small-scale societies.
In cultures with a high mortality risk and a low probability of survival into adulthood it makes adaptive sense to bring forward menarche.
SOURCE: Plotted from data in Walker et al. (2006b), Table 2, p. 300.

childhood as a period when information about local environments and interpersonal relationships could be used to drive the personality of the growing child towards behaviour patterns that are best adapted to predicted future ecological and social conditions. A key set of ideas here is **attachment theory**, pioneered by John Bowlby. In the 1930s, Bowlby worked as a psychiatrist at a centre for troubled youngsters in London. Here he was struck by the blank emotional response of many of the youths he met. Then he noticed a pattern: children who showed a lack of emotional response, and typically were in trouble with the authorities, tended to be those who had been separated from their mothers at an early age.

At the time of Bowlby's original insights, the study of human emotions and relationships was dominated by Freudian psychoanalysis. The study of non-human animals, particularly in America, was also heavily influenced by the behaviourism of Watson, Skinner and their followers. Bowlby felt that both Freudianism and behaviourism were not adequate to the task of explaining his observations on attachment. Fortuitously, Bowlby then came across the work of two men, the Cambridge comparative psychologist Robert Hinde and the Austrian ethologist Konrad Lorenz. Both were sensitive to the importance of evolutionary theory in explaining animal behaviour. This was just the encouragement that Bowlby needed and he went on to formulate his attachment theory within an evolutionary framework.

Bowlby (1969) proposed that all primates are born pre-programmed to form close emotional ties with nearby adults such as mothers. He argued that the way a growing child conceived of social relationships depended on this early infant experience. According to Bowlby, the attachment bond had to form before the age of 2½ years or else it would be difficult for an attachment to form after this. Should attachment not take place then permanent emotional damage would result leading to problems with relationships, sadness and depression later in life (Bowlby, 1980). Bowlby also suggested that the process of forming an attachment was adaptive for both parents and offspring. Babies, he argued, are born programmed to attach to the adult that gives them care early in life; this of course is usually the parent who provides the baby with a source of protection and nurture. He suggested that just as the newborn baby is attracted to the smiles and face of its mother so the faces of babies and their voices serve to stimulate and release the nurturing response of the mother.

Bowlby's work has had an enormous impact on thinking about the relationship between parents and children in the first few years of life. The central tenets of attachment theory have entered mainstream psychological thinking and the consensus seems to be that attachment has an important role to play in the long-term mental health of the growing child. Where the theory is weak is in explaining why children differ so much in their resilience to neglect.

The first systematic attempt to incorporate attachment theory into a life history perspective was made by Belsky et al. (1991). They drew upon the idea that a child's relationship with its primary caregiver led to an internal working model of beliefs and expectations about the world, people and relationships. The group argued that children who were securely attached to a reliable and empathetic caregiver used this as an indication that in future people can be trusted and the world would be relatively stable. Children who experienced rejections and neglect from an unsympathetic caregiver would become insecurely attached and draw from this the expectation that people cannot be trusted, relationships would not last and resources are unpredictable. They argued that these two extremes would align with two different styles of reproduction in later life (Figure 8.8).

A similar set of ideas was proposed by Chisholm (1996) who suggested that cultures with high mortality rates and unpredictable resources should push optimal mating strategies towards early reproduction, insecure attachments and high levels of sociosexuality. In lower-stress cultures with more abundant resources Chisholm's model predicts later reproduction and lower sexuality. The logic of both these sets of ideas is that if environments are harsh, unpredictable, stressful, and resources scare, then it pays not to adopt a high investment strategy since such investments may be lost or impossible to sustain. Better instead to produce more offspring earlier and with lower levels of investment in the assumption that some will survive.

There has been a great deal of research supporting the Belsky–Chisholm model. It has been found, for example, that girls from father-absent homes reach puberty before girls living in families with

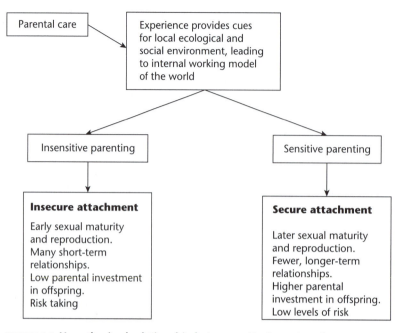

FIGURE 8.8 Hypothesised relationship between attachment and reproductive styles.
The suggestion is that styles of parenting are used as a predictor of future life chances by the growing child who then shifts allocations between mating and parenting in adaptive ways.

3. Women experiencing early stress would be more likely to show insecure attachments as adults.
4. Early stress would be reflected in an expectation of a shorter lifespan.

Note that hypothesis 4 refers to the expectation of lifespan rather than the actual lifespan. This is important since it is the subjective life expectancy that may influence or be related in some way to reproductive behaviour. The data from the group of 100 women was controlled for income and education. The researchers found that the level of early stress was correlated with all four variables: age of menarche (negative correlation), age of first birth (negative correlation), expected lifespan (negative correlation) and levels of insecure attachment (positive correlation). Amongst older mothers, for example, 82 per cent showed a secure attachment style compared to only 58.3 per cent for younger mothers. Figure 8.9 shows a sample of their data using two groups: younger mothers and older mothers.

The results support a life history perspective on the timing of reproduction and the central idea that risky environments promote early reproduction. What is not clear, and now needs elucidating, is how childhood experiences of stress are actually communicated to these other functions such as menarche and attachment relationships. This will be the goal of psychoneuroendocrinology.

More recently, Daniel Nettle et al. (2013a) tested Belsky's idea that adverse early life experiences (such as low parental investment and family stress) would promote accelerated reproduction strategies (as evidenced by, for example, earlier menarche and pregnancy) by examining longitudinal data from the UK's National Child Development Study (NCDS). The NCDS is an ongoing project recording extensive medical and sociological data on all children born in the UK between 3 and 9 March 1958. So far

their biological fathers (Surbey, 1990; Wierson, Long and Forehand, 1993). Ellis and Garber (2000) investigated the effect of maternal psychopathology and the arrival of a stepfather (or mother's boyfriend) on pubertal timing for 87 American girls. Consistent with the Belsky–Chisholm model they found that girls who were exposed to father absence and familial stress reached puberty earlier than girls from stable homes. A significant part of this study was also the fact that the mother's mental health was also implicated in early puberty.

In another study, James Chisholm and colleagues investigated the effects of stress on the life histories of a sample of 100 women. The basic premise was that a stressful, risky and uncertain environment would tend to promote early reproduction (Chisholm et al., 2005). From this life history perspective the authors drew up four hypotheses:

1. Early stress would correlate with early menarche.
2. Early stress would be associated with an early age of first giving birth.

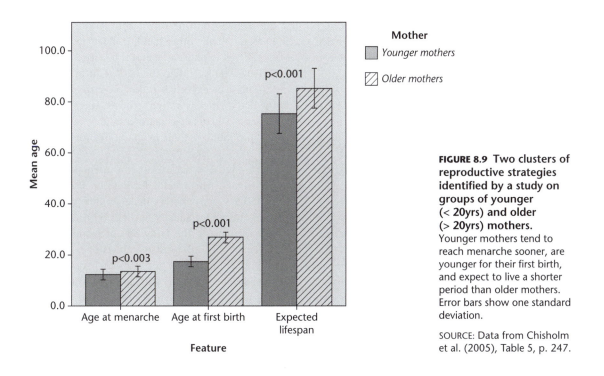

FIGURE 8.9 Two clusters of reproductive strategies identified by a study on groups of younger (< 20yrs) and older (> 20yrs) mothers. Younger mothers tend to reach menarche sooner, are younger for their first birth, and expect to live a shorter period than older mothers. Error bars show one standard deviation.

SOURCE: Data from Chisholm et al. (2005), Table 5, p. 247.

data exists for various parameters at birth and ages 7, 11, 16 and 23 years. The key prediction made by Nettle's group was that women who became mothers at a young age (that is, had teenage pregnancies) would also be shown to have poor growth early in life, rapid weight gain in middle childhood, early cessation of growth and early puberty compared to older mothers. The advantage of this data set is that it allows socio-economic status to be controlled for – an important requirement since it is well known that early pregnancy is also associated with low social status. In a number of important dimensions the group did indeed find an association between fast life history indicators and those women who had experienced early pregnancies (defined as <20 years old). Early childbearing was also associated with significantly higher levels of behavioural disturbance (for example, depression and hostility to adults) at ages 7 and 11.

The work is interesting in that it offers a fresh perspective on teenage pregnancy – something often thought to be a cause for concern. Teenage mothers and the control group did not differ in their exposure to sex education. Indeed young mothers also tended to state an earlier ideal age for having children

compared to the control group. It looks like teenage motherhood is not a consequence of mistakes caused by poor knowledge of contraception but part of a longer life history trajectory. Improving conditions at birth might, therefore, possibly be more effective at reducing teenage pregnancies than more sex education.

Attachment theory and life history theory have both been very successful at generating hypotheses and predictions. One nagging doubt has to be whether this mechanism to enable a growing child to predict the future state of the world would really have been that useful. One might ask why a child should use experience of its parents to generalise about all other people; given that the human brain is large and subtle, we might expect that children could form different models for different people. Finally, given that even in the Palaeolithic era conditions may have changed over a period of say 20 years, it seems that using the experience of childhood to devise a reproductive strategy 20 years in the future is relying on shaky foundations. There also needs to be more work done on whether these putative adaptive mechanisms actually work. So, for example, do children who have an insecure attachment and grow up in a risky and unpredictable world (that is, their predictions from childhood were correct) actually do better in these conditions than those who had a secure attachment but later found themselves in a more hostile world than expected?

One answer to the problem of predicting the future comes from the work of Kanazawa (2001). He argues that father absence is used as an indication of polygyny – either serial or simultaneous. In serial polygyny the divorce rate will be high and children will experience father absence; in the case of simultaneous polygyny fathers will also have less time to devote to each wife and child compared to monogamy. Now in cultures where either type of polygyny prevails it will be advantageous for girls to reach puberty early. Since girls tend to mature before boys, a girl aged 12 in a monogamous culture will have a low chance of mating since sexually mature men will all be paired up and 12-year-old boys will not be sexually mature. But in a polygynous culture, a girl just reaching puberty could become another wife of an already married (or recently divorced) high-status male. In testing these ideas Kanazawa did indeed find two things in support. Firstly, the age difference at marriage between men and women tends to be greater in polygynous societies than monogamous ones: men on average are 4.5 years older than women in polygynous cultures compared to 3 years older in monogamous ones. Secondly, the degree of polygyny in a culture does seem to be negatively correlated with the age of marriage (and by implication menarche) – that is, the more polygynous the culture the lower the age of menarche.

8.3.4 Old age, the menopause and the function of grandparents

The evolutionary purpose of the declining years in humans is still something of a mystery. It may be that the period of life after 60 years is merely a non-functional period of decline that simply takes years to complete. The alternative is that even this period has been shaped by natural selection to yield fitness benefits. As Hill (1993) notes, these two alternatives make different predictions. If old age is simply a time of somatic collapse then all faculties could be expected to decay roughly synchronously – there is no point in maintaining a healthy heart, for example,

if the liver is shutting down and cannot detoxify the blood. On the other hand, if older people have been selected to play a role in, for instance, skill and knowledge transfer, then the decline in mental faculties might be expected to be slower than the rest of the body. There is indeed some evidence that this is the case. Age-related neuropathology in humans is less pronounced than in macaques for example (Finch and Sapolsky, 1999).

At around the age of 50, the **menopause** sets in for women (the average for developed countries being 50.5 years) whereas male fertility continues beyond this and gradually declines with age rather than experiencing the fairly abrupt cessation seen with female fertility. Most mammals of both sexes, including chimpanzees and gorillas, experience a gradual decline in fertility with age, totally unlike the menopause of human females. So why the sudden shutdown experienced by human females?

One hypothesis about the menopause is that it is an artefact of an unnaturally elongated female lifestyle. According to this view, most early humans died by the age of 50 or 55 and so there was no problematic post-reproductive period of life in females that needed explaining. In contrast, the 'grandmother hypothesis' proposes that even in early human groups females enjoyed a significant period of post-reproductive life and that this contributed to their individual lifetime fitness. In favour of the second functional hypothesis we do find that many contemporary hunter-gathering women live into their 60s and 70s. Furthermore, if reproductive cessation was part of a general process of wear and tear we ought to expect a similar decline in function of other aspects of human physiology starting around 50 for females. Figure 8.10 shows how fertility decline compares with the age-related performance of other functions and demonstrates that fertility decline does indeed occur well before that of other functions, pointing to a purposeful period of post-reproductive life.

One possible explanation for the menopause may lie in the impact of the mortality risks of human childbirth. At birth, the human infant is enormous compared with the offspring of gorillas: 7-pound baby humans emerge from 100-pound mothers compared with 4-pound gorilla babies emerging from 200-pound mothers. Consequently, on a relative scale, the risk of death to the mother during childbirth is relatively

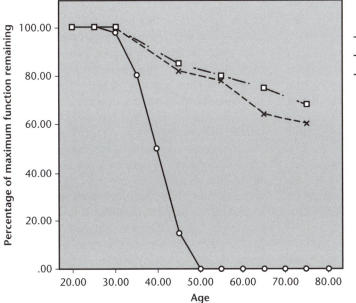

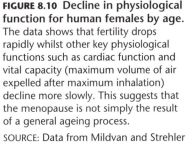

FIGURE 8.10 Decline in physiological function for human females by age. The data shows that fertility drops rapidly whilst other key physiological functions such as cardiac function and vital capacity (maximum volume of air expelled after maximum inhalation) decline more slowly. This suggests that the menopause is not simply the result of a general ageing process.

SOURCE: Data from Mildvan and Strehler (1960); Wood (1990); and Hill and Hurtado (1991).

high for humans. Because human infants are altricial they remain dependent on parental (especially maternal) care for a long period of time. For a woman who already has children, every extra child, while increasing her reproductive success, involves a gamble with the risk that she may not survive to look after the children. Now, the risks of death in childbirth increase with age, and there comes a point at which the extra unit of reproductive success of another child is exactly balanced by the extra risk to her existing reproductive achievement. Beyond this point (at which an economist might say the marginal benefit equals the marginal cost), it is not worth proceeding. To protect the mother's prior investment in her children, natural selection may have instigated the menopausal shutdown in human fertility. Since childbirth carries no risk to the father, and fathers can always increase their reproductive success with other partners, men did not evolve the menopause (Diamond, 1991). Given the difference in parental investment in the early years of a child's life, we can see that children born to old men but young women have a better chance of survival than those born to old women.

If reproductive shutdown at the menopause occurs to protect a mother's investment, it could be supposed that females should live just long enough after the menopause to protect and raise their children to independence. In fact, women tend to live longer than is strictly necessary. Among other mammals, a mother will spend only about 10 per cent of her life after her last birth, whereas human females can live nearly a third of their lives after their last child. This has led to much speculation about the evolutionary function of grandmothers. But grandmothers aside, it is a reasonable hypothesis to suppose that women who bear children late in life and survive childbirth subsequently live longer. Voland and Engel (1989) claim to have found support for this hypothesis. Modern medical care will probably iron out any effect in modern cultures, so they examined the records of 811 women born between 1700 and 1750 in a rural district of Germany and found that life expectancy did increase with the age of the last child. In the same study, these authors confirmed that childbirth on the whole decreases life expectancy since married childless women tended to live longer than married mothers. However, if women became mothers, their life expectancy was increased significantly by a late age of the last child.

More recently, in a study of the health records of three groups of settlers in Quebec and Utah in the 19th and 20th centuries, Gagnon et al. (2009) found that as age of mother when she last gave birth increased, so did the length of her post-reproductive survival. They found that for every year added to the age of the woman at the birth of her last child her mortality ratio decreased by somewhere between 2.2 per cent and 3.6 per cent, an effect remarkably consistent between the three populations. The finding that age of last birth increases longevity fits with an adaptive explanation but there are others. It may be that a late age of motherhood is a general sign of slow ageing, or that pregnancy and late age of giving birth promotes healthy behaviour.

This same group also found that the more children a mother had (increasing parity) the lower her life expectancy (Gagnon et al., 2009). One explanation of this latter finding comes from what is sometimes called the 'disposable soma' theory, which suggests that resources allocated to reproduction are not then available for maintenance and survival; hence increasing parity reduces lifespan. In the study of the three populations the effect was remarkably consistent and an excess mortality of between 1.6 per cent and 3.2 per cent was observed for each extra child. The data is also consistent, however, with the 'antagonistic pleiotropy' theory (see Chapter 18) whereby mutations having a late onset reduction in lifespan might be selected for if they promote early fertility.

But we still have to explain the fact that although menopause begins at around 50, women live another 20 years or so – longer than seems necessary to make sure their last offspring has survived to reach sexual maturity. The conjecture to explain this continued longevity is called the 'grandmother hypothesis' and it suggests that women can still achieve fitness gains late in life by diverting investment to grandchildren. It is a plausible idea but one that is difficult to test. There are other arguments, however, that suggest that post-reproductive women are advancing their genetic fitness. As Hill and Hurtado (1997) point out, if elderly women had no reproductive value then natural selection would be powerless to resist the accumulation of recessive alleles that would lead to senescence and death. In which case we would expect women to die sooner than men since men are reproductively active after 50. This is patently not the case, and the fact that women live slightly longer than men suggests that fitness gains are still being gained by elderly women. Since this is not through direct offspring (they are beyond childbearing) they must be having a positive effect on kin.

Blurton Jones et al. (1999) have also suggested that if the grandmother hypothesis is correct then women should have shorter inter-birth intervals than predicted from life history theory applied to individual women. The reasoning here is that if grandmothers are helping their family then a mother can raise more offspring than simply predicted by considering her own time constraints and access to resources. Some support for this comes from Sear et al. (2000) who conducted a study on nearly 2,000 children in Gambia. They found that children with living grandmothers had better nutritional status and higher survival chances (the study used longitudinal data spanning 25 years) than children without. Interestingly, the effect was confined to maternal grandmothers and not paternal grandmothers. This is consistent with the notion of discriminatory grandparental solicitude in relation to paternity certainty (see Chapter 11).

Another line of reasoning links the menopause to the relatively short human inter-birth interval noted earlier. As noted above, in hunter-gathering populations the birth interval is about 3 years (see Figure 8.4), which although high compared to post-Neolithic pre-modern farming populations, is low compared to chimpanzees (4–5 years) and orangutans (approximately 8 years). One explanation of this has been that the strong male–female pair bond elicited considerable help from males. However, work from anthropologists has often demonstrated that the calories provided by hunting (an activity primarily carried out by men) are much fewer than those provided by women gatherers (Hawkes et al., 2001). Might the presence of grandmothers explain both the phenomenon of the menopause and that of the short birth interval? To address this question Rebecca Sear and Ruth Mace carried out a meta-review of the literature to establish the impact of a relative on the welfare of a child by recording cases where loss of kin (fathers, mothers, grandmothers and so on) had an effect on the survival of a child. From this they were able to infer how kin impacted positively or negatively on child survival (Sear and Mace, 2008). As expected, mothers in all studies had a positive impact on child survival. But what is less expected and perhaps more interesting is that both maternal and paternal grandmothers are more commonly reported having a positive effect on child survival than fathers. The review by Sear and Mace provides additional support for the grandmother hypothesis. The positive impact of grandparents and older siblings also supports the idea explored earlier that help to the mother from immediate kin enables a reduction in the expected birth interval. It does not entirely account for the menopause since the end of reproduction may occur for other reasons and then grandmothers are selected to help grandchildren. One concerning feature of this whole area is that so far attempts to build mathematical models to show that a woman can compensate for loss of fertility at menopause (around 50 years of age) by investing in the fitness of children and grandchildren have not been conclusive (Grainger and Beise, 2004).

8.3.5 Ageing, senescence and death

The inevitability of death

In the light of LHT, ageing, senescence and death – surely candidates for suboptimality if there ever were any – begin to make sense. Ageing and senescence should not be confused. Ageing is inevitable and simply refers to the passing of time and increases in chronological age. Senescence, on the other hand, refers to the age-related physical degeneration that occurs in all sexually reproducing species.

So why, we may ask, has natural selection not engineered an organism that lives forever and produces copies of itself (or spreads its genes by sexual reproduction) at regular intervals for evermore? Richard Law (1979) called such an organism a 'Darwinian demon' in that it would outcompete all other life forms and monopolise life on Earth. To understand why, after 3.5 billion years of evolution, no such organism, thankfully, has evolved we need to appreciate how a finite intake of metabolic energy has to be distributed according to the trade-offs between growth, repair and reproduction.

In very broad and simple terms, an organism that aimed to live forever would have to allocate an enormous quota of energy to maintenance and repair at the expense of reproduction. A rival organism that

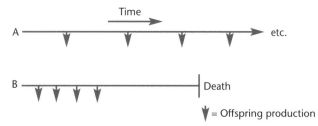

FIGURE 8.11 Alternative growth, reproduction and repair strategies.
Organism A delays sexual maturity and opts for high levels of investment in repair and maintenance. Organism B breeds more rapidly but at the expense of an earlier death. Over time Organism B will leave more offspring and, other things being equal, become the norm. In this way, death and senescence are inevitable.

allocated more energy to reproduction early in life, at the expense of later senescence and death, would quickly begin to leave more offspring (children, grandchildren and so on). Such rapidly multiplying shorter-lived offspring would quickly fill the ecological niche and become the norm (Figure 8.11).

One consequence of this logic is that natural selection is also largely blind to diseases and degenerative processes that take effect after the period of reproduction. Williams (1957) even postulated the possible existence of genes that exhibit antagonistic pleiotropy. Pleiotropic genes (genes that code for more than one effect) would become antagonistic if they had a positive effect on fertility in youth and a negative effect later in life. They would be favoured even though they were detrimental to the later life of the organism that had already reproduced. In essence then, post-reproductive death is natural selection's answer to the trade-off conflict between reproduction and repair: bodies are disposable, temporary vehicles.

How long should organisms live?

An immortal Darwinian demon might not exist but one could expect natural selection to maximise reproductive success over a lifetime. But what determines a lifetime?

There have been two basic approaches to this question: the 'rate of living theory' (and related hypotheses), and an evolutionary optimisation approach strongly rooted in LHT.

The rate of living theory

The rate of living theory of senescence was first proposed around 1908 by the German physiologist Max Rubner (1854–1932). He was struck by the fact that animals with a high metabolic rate died sooner than those with a low one – a case of 'live fast, die young'. Updated, the idea is that damage to cells results from the by-products and processes of metabolism and animals are at their limit of what they can repair. Hence the theory makes a very definite prediction: life expectancy should be negatively correlated with metabolic rate. For example, a mouse weighing about 40 g uses about 29 kJ of energy per day, equivalent to a metabolic rate of 961 kJ/kg/day; whilst a human weighing about 70 kg uses about 9680 kJ per day, a metabolic rate of 138 kJ/kg/day. The metabolic rate (per kg of tissue) of mice is higher than that of humans; mice live about 2 years and humans about 70. So far so good, but from the theory it follows that all species should therefore expend about the same amount of energy per gramme of tissue per lifetime. This was tested by Austad and Fisher (1991) for 164 mammalian species. The energy spent per lifetime values varied considerably, however, from 39 kcals/g/lifetime for an elephant to 1,102 kcals/g/lifetime for a bat, a result, therefore, that failed to support the rate of living theory of ageing.

Evolutionary optimisation

Evolutionary approaches to ageing are, theoretically at least, more plausible and satisfying. There are several approaches. One is the idea that there could be deleterious mutations that easily become fixed in the gene pool since their effects occur after a significant number of offspring have already been produced; these will be largely invisible to the effects of natural selection and operate as it were in a post-reproductive 'selection shadow'. It is possible that there are pleiotropic genes that enhance reproduction early in life but cause senescence and decay later. Another view is that since an organism constantly faces the risk of death by an accident or predator throughout its life it will only be a matter of time before fatality strikes, in which case it may be advantageous to shift the bulk of reproduction to early in life before death

makes it too late – the so-called disposable soma idea discussed earlier.

Given these possibilities we might expect to find some connection between longevity and the timing of reproduction. Early reproductive effort, for example, might be associated with a short lifespan since energy has been directed to reproduction at the cost of maintenance; also, early reproduction is a way of avoiding the fitness costs of the effects of late-onset damaging genes. There could be, in effect, a facultative trade-off across variable environments between the timing of reproduction and longevity. We might expect that living in a risky environment, where death is an ever-present threat, should be associated with earlier reproduction. This was exactly what was noted earlier in the context of age of menarche and probability of survival to age 15 (see Figure 8.7). These ideas have proved quite rewarding for the understanding of non-human animals. Austad (1993), for example, compared two populations of Virginia opossums: one heavily predated, living in a risky environment with a high ecological mortality; the other living on an island free from mammalian predators and enjoying a lower level of ecological stress. As expected, the island opossums lived longer; but, more significantly, the island females had smaller litters each season – they were slowing down the rate of reproduction. But overall this approach has not worked so well for human studies and the balance here between fecundity, reproduction and senescence seems much more complicated (see Helle et al., 2005).

8.4 LHT, human behavioural ecology and the demographic transition

A special challenge for human behavioural ecology (predicated as it is on measuring real fitness gains through offspring numbers) is the explanation of the **demographic transition** (Figure 8.12). How can humans still be regarded as fitness maximisers by choosing to have fewer children when presumably (since the transition is associated with the growing wealth of a culture) they could afford more? Data on reproductive rates and birth spacing for foraging

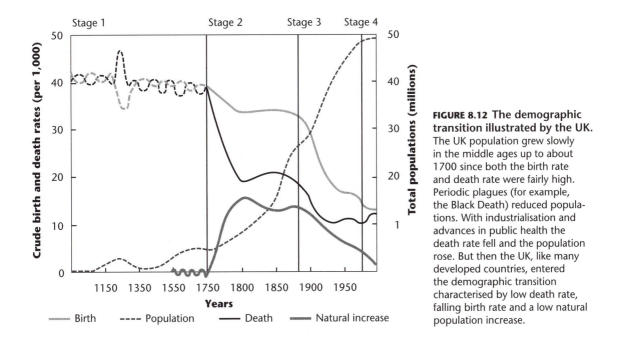

FIGURE 8.12 The demographic transition illustrated by the UK. The UK population grew slowly in the middle ages up to about 1700 since both the birth rate and death rate were fairly high. Periodic plagues (for example, the Black Death) reduced populations. With industrialisation and advances in public health the death rate fell and the population rose. But then the UK, like many developed countries, entered the demographic transition characterised by low death rate, falling birth rate and a low natural population increase.

Table 8.4 Birth intervals, reproductive rates and other variables for women in foraging, rural and modern societies.

These figures will vary widely between cultures but this table is illustrative. Modern women reach menarche much earlier than their forebears and also spend much less of their overall life lactating. Some of these parameters are almost certainly related to lifestyle and environmental factors such as nutrient supply and health care. But clearly modern women are making the decision to delay the age when they first conceive and have fewer children than their ancestors. Trying to identify adaptive reasons for these decisions is a difficult task.

Culture	Age at menarche	Age at marriage	Total time spent lactating	Typical number of offspring	Age of menopause	Birth interval
Hunter-gatherers	15	16	20 years	5	40	3
Rural/pre-modern	14	18	20 years	10	45	1.5
Modern/urban	12–13	24	5 years	2	50	4
SOURCE: Data from Zihlman, A. L. (1982). *The Human Evolution Colouring Book*. Harper Resource, Section 6.9.						

societies, rural societies and modern societies reveal considerable differences in the life history decisions made by women (Table 8.4). Modern adults in the developed world tend to begin reproduction later, have fewer children, experience earlier menarche, have later menopause and spend less time breastfeeding than their early industrial and hunter-gathering ancestors.

A life history theory framing of this problem would start by seeing parents making a decision to increase investment in each child, or even a single child, compared to a lower investment per child in a larger number of children. The rationale for this might be that in complex societies children can only grow to be effective and competitive adults, and hence attract mates, if they have undergone extensive training to acquire skills useful in such cultures. If this is the case then children who have received higher investment (but fewer brothers and sisters as a consequence) should in the long term do better (leave more offspring than children from larger families who received less investment). We might also expect higher-status adults to eventually leave more offspring than low earners. But studies exploring such ideas have met with numerous problems. Kaplan et al. (1995), for example, looked at the number of third-generation descendants (that is, grandchildren) of men in Albuquerque, New Mexico, and found it to be highest amongst men who initially had the most children. Indeed, many studies seem to indicate that whereas higher fertility is associated with lower educational and economic status of offspring, the reduced earning capacity of such offspring does not reduce their fertility. Furthermore, although some studies show that in pre-industrial societies there is a positive correlation between status and resources and reproductive success (Barkow, 1989; Voland, 1990), studies on low fertility societies, such as those associated with modernity, reveal no such correlation or even a negative one (Kaplan et al., 1995).

It is here then that there has been a proliferation of ingenious theories to account for these anomalies within life history theory and the human behavioural ecology approach. Within evolutionary psychology the problem disappears since this whole approach suggests that current actions and lifestyle practices are not necessarily fitness maximising. We are driven to have sex, which is a natural adaptation, but in figuring out contraception we have thwarted what were once the inevitable consequences. A variety of suggestions have been forthcoming:

- Perusse (1993) suggested that men's psychology is adapted to securing resources since in the past this reliably increased mating opportunities and hence, in the absence of contraception, number of offspring. Today male psychology still dictates wealth pursuit but without the same ensuing reproductive gains.
- Similarly Irons (1983) and Turke (1989) have suggested that humans are evolved to track cultural success as a proxy for what enhances fitness and

the costs of achieving such success in some modern cultures (for example, the time demands of a successful career) impact negatively on the number of children.

- Lancaster (1997) has suggested that parents need to increase investment in children at the expense of number of children to enhance the competitiveness of their children in the marriage market.
- Kaplan and Lancaster (1999) developed a model that emphasises the importance of skill acquisition by offspring. As noted earlier, human life history is such that children mature slowly to acquire the complex skills needed to function in the ecological niche to which we are adapted. According to this model, parental psychology may be such as to detect the relationship between investment in offspring and the return on this investment (that is, the income or proxies for income) in the offspring. As societies modernise, therefore, and increasingly complex skills are required to function successfully so the resources that parents need to invest in offspring (schooling and so on) increase and can only be met by a lower overall fertility. In response contraceptive technologies are sought. It follows if this model is accurate that contraception is the outcome not cause of forces tending to lower fertility.
- Turke (1989) has suggested that it is the development of the nuclear family and the break up of extended family networks that follow modernisation that has reduced fertility. With modernisation comes social and physical mobility, this in turn breaks down networks of kin support for couples rearing children. In effect the burden of children on the parents (and especially the mother) rises and it becomes optimal to have fewer of them.

Despite these ideas the problem of the demographic transition for human behavioural ecology (HBE) is acute. The idea, well supported in the animal literature, that low fertility represents a life history trade-off between quality and quantity in a competitive environment faces problems. In a study of the fertility of 14,000 Swedes over the years 1915–2009, Goodman et al. (2012) found that low fertility and high socioeconomic status did enhance the socioeconomic success of descendants. However, high fertility of parents was not associated with reduced survival or reduced reproductive success of their descendants – contradicting the theoretical expectation from quality–quantity trade-offs. And, crucially, low fertility and high socioeconomic status did not predict long-term reproductive success. Hence it seems unlikely that the low fertility we see in developed countries is a strategy to maximise long-term reproductive success in a competitive environment where complex skills expensively acquired enhance social standing and thence reproductive success. Whatever the cause of low fertility in the developed world it does not really fit easily in the adaptive optimality models of HBE. Even Nettle et al. (2013b), otherwise enthusiastic proponents of HBE, suggest that the demographic transition points to 'some kind of maladaptation or mismatch between the conditions under which decision-making mechanisms evolved and those under which they are now operating' (p. 1038).

This phenomenon of 'mismatch' also forms part of the subject of the next chapter where we examine how even our very processes of reasoning are not foolproof and are prone to errors and biases in interesting ways.

Summary

- Life history theory (LHT) is concerned with how organisms make decisions about the allocation of energy into different essential functions over the lifespan. This time-orientated perspective is needed to help make sense of behaviour that can only be appreciated as adaptive over a longer perspective.

- The way life history allocations are made reflects the phylogenetic experience of a species and as such has been shaped by natural selection. But there is evidence that the precise calibration of life history decisions is also open to medium- and short-term influences. Epigenetic mechanisms, for example, may track environments in adaptive ways. Similarly, humans may use the experience of their early lives *in utero*, infancy or childhood to influence life history decisions such as the timing of reproduction.

- LHT has the potential to explain between-population differences in such things as age of menarche and stature. Early menarche and reduced stature could be the result of evolved fast life history strategies in hostile environments.

- The function of the menopause is an evolutionary puzzle but it looks likely that the answer lies in the function of grandmothers and their ability to help in raising their grandchildren.

- The demographic transition is a serious problem for human behavioural ecology since it suggests contemporary humans in wealthy societies are not maximising their reproductive fitness.

Key Words

- **Attachment theory**
- **Demographic transition**
- **Endocrine systems**
- **K and r selection**
- **Neonatal adiposity**
- **Menarche**
- **Menopause**

Further reading

Hawkes, K. and R. R. Paine (eds.) (2006). *The Evolution of Human Life History*. Oxford, UK, James Currey Publishers.

A well-structured edited collection of articles with many useful figures and data sets.

Bogin, B. (1999). *Patterns of Human Growth*. Cambridge, UK, Cambridge University Press.

A thorough and comprehensive book that tackles human development from a variety of perspectives: historical, philosophical, cultural and of course primarily biological.

Bribiescas, R. G. (2006). *Men: Evolutionary and Life History*. Cambridge, MA, Harvard University Press.

Life history perspective on men. Looks at growth, puberty, mortality, criminality and sexuality.

PART

V

Cognition and Emotion

Cognition and Modularity

Plato … says in Phaedo that our 'imaginary ideas' arise from the pre-existence of the soul, are not derivable from experience – read monkeys for pre-existence.
(Darwin, 1838, quoted in Gruber, 1974, p. 324)

9.1 The modular mind

9.1.1 Epistemological dilemmas: rationalism or empiricism?

In *The Phaedo*, Plato advances the notion that the mind is born pre-equipped with ideas or ways of structuring experience. Plato was developing the ideas of his mentor Socrates (469–399 BC) who thought that knowledge was essentially a recollection of universal truths that the soul understood in a previous life. Both Plato and Socrates believed in the doctrine of innate ideas – we are born knowing certain things. This position is sometimes called rationalism (to distinguish it from empiricism) or nativism. The doctrine of innate ideas has had a long history (Table 9.1).

In contrast to this rationalist approach stands the empiricist tradition. The English philosopher John Locke, for example, argued that the mind started out as a formless mass and was given structure only by sensory impressions. In a note to himself, Darwin dismissed Locke in the same way as he dispatched Plato's pre-existing soul: 'He who understand baboon would do more toward metaphysics than Locke' (Darwin, 1838, quoted in Gruber, 1974, p. 324).

The debate between rationalism and empiricism is a long-standing one in philosophy (see Kenny,

1986) and the details do not overly concern us here. But Darwin was probably right: the brain at birth is not a formless heap of tissue, nor does it carry a recollection of eternal verities associated with an immortal soul. The brain enters the world already structured by the effects of a few million years of natural selection having acted upon our primate and hominin ancestors. It is therefore born ready shaped, but also eager for experience, for calibration and fine-tuning to the world in which it finds itself. Eibl-Eibesfeldt expressed this succinctly:

> The ability to reconstruct a real world from sensory data presupposes a knowledge about this world. This knowledge is based in part on individual experience and, in part, on the achievements of data processing mechanisms, which we inherited as part of phylogenetic adaptations. Knowledge about the world in the latter instance was acquired during the course of evolution. It is, so to speak, a priori – prior to all individual experience – but certainly not prior to all experience.
> (Eibl-Eibesfeldt, 1989, p. 6)

Evolutionary epistemology

The term for this view of knowledge is **evolutionary epistemology** and was coined by Donald Campbell

165

Table 9.1 Empiricism and rationalism in European thought

Rationalist tradition (certain ideas are innate)	Empiricist tradition (the mind forms knowledge only from experience)	Evolutionary epistemology (a rational–empirical hybrid: our mind is born structured by the effects of selection on our ancestors)
Socrates (469–399 BC)	John Locke (1637–1704)	Herbert Spencer (1820–1903)
Plato (428–348 BC)	George Berkeley (1685–1753)	Charles Darwin (1809–82)
Descartes (1596–1650)	David Hume (1711–76)	William James (1842–1910)
Spinoza (1632–77)		
Leibniz (1646–1716)		
Immauel Kant (1724–1804)		

(1974). Somewhat confusingly, the term can have three distinct meanings corresponding to three levels of analysis: phylogenetic, ontogenetic and cultural. It is the phylogenetic level we are concerned with here – the idea that during evolution the mental circuitry of a species becomes attuned to certain inbuilt ways of grasping experience so that we now come to possess specific **cognitive adaptations**. The ontogenetic level is the idea that the process of variation and selection goes on in the individual mind as it matures, a sort of selection of the fittest concepts or neural connections as the organism interacts with its world. At the cultural level, we have the idea that processes and metaphors from evolutionary biology can be applied to the development of ideas and scientific theories in culture as a whole. Karl Popper, for example, was a selectionist in this sense. His view in *Conjectures and Refutations* was that scientific ideas that are falsified are rejected and by analogy become extinct, leaving the fittest to remain in the canon of science (Popper, 1963). More recently both Daniel Dennett and Richard Dawkins have explored the issue – the latter coining the phrase 'universal Darwinism' to describe the attempt to apply the idea of natural selection to phenomena outside speciation and phylogeny. This idea is examined in Chapter 20. It is important to realise that the three approaches are analogous but distinct: the truth of any one does not depend on any other – although there have been attempts to link all three. The phylogenetic approach is often linked to the modular view of the mind – the notion that the evolved biological substrate of cognitive activity resides in specialised and dedicated areas of the brain called modules.

Innate biases in learning

Evidence for the existence of domain-specific learning biases in humans came in the 1980s from experiments by Alan Leslie on very young children. Leslie showed images of moving objects to young children (Leslie, 1982, 1984). When an unrealistic event was simulated, such as a ball (A) moving towards another ball (B), stopping before its reached B and then B moving off, children as young as 6 months showed more interest in this event than realistic ones involving collisions. Similarly, Renee Baillargeon and her colleagues have shown that female infants at 18 weeks show surprise when physically impossible displays are presented to them – such as an image sequence showing a lower block removed from a pile of blocks leaving the upper ones poised in mid air (Baillargeon et al., 1985; Baillargeon, 1987; Baillargeon and DeVos, 1991). Results like these are consistent with the idea that the human mind is preformed to some degree and that it arrives with an a priori intuitive physics. It is possible that children have or are programmed to develop a 'theory of body'.

9.1.2 The localisation of brain function

At its simplest interpretation the notion of the modular brain is simply the unobjectionable view that different components of the brain deal with different specific functions. Since the nineteenth century, studies on people with damage through disease or injury to specific parts of the brain have revealed a whole array of specialised functions. Even rather abstract faculties such as the ability to make plans and enact behaviour to further them seem localised in specific brain regions.

In more recent times the use of the term '**modularity**' as applied to the mind was initiated by the linguist and philosopher Jerry Fodor. In his now classic monograph *The Modularity of Mind* (1983) Fodor argued that many perceptual and cognitive processes such as vision, hearing and talking were organised and performed with dedicated modules or 'input systems'. Fodor argued that these modules worked independently of each other and dealt with their own designated inputs from the environment; they were, in other words, 'domain specific' (Fodor, 1983). The important point about Fodor's concept of perceptual modules is that they operate quickly and are relatively immune from interference by memory, experience and reflection. This is illustrated by the intriguing nature of

optical illusions (see Box 9.1). In optical illusions, for example, the observer still sees the illusion (i.e. the inaccurate conclusion about the image) even when measurements are made and we become conscious and rationally aware of our error. Knowledge and understanding cannot override the processing of the image.

This may look as if the human perceptual system is not well calibrated if it is prone to such errors. However, relatively hard-wired modules like this make good adaptive sense if they are fast. Often humans need a quick response to a threat or danger and the odd mistake is worth making so long as the interpretation errs on the side of caution. Fodor was insistent, however, that higher-level functions such as reasoning were not modular. These central cognitive processes, he argued, are slow and non-mandatory – we can choose, for example, whether to think about an intellectual problem but we cannot choose

---

Box 9.1 Two optical illusions

Edward H. Adelson

The Adelson chequered square.
Probably the most uncanny optical illusion in existence. Square A is exactly the same shade as square B.

SOURCE: Created by Adrian Pickstone based on an original by Edward H. Adelson. Used with permission from Wikipedia and Prof. E. Adelson (personal communication).

The white rectangle.
The shade of white in the rectangle is exactly the same as the background, yet it probably appears whiter or more distinct, as if there were an object there.

In fact these illusions demonstrate the success of the human optical system rather than its failure. The visual system did not evolve to be an objective light meter but rather to create meaning from the image information and so perceive the nature of the objects in view.

not to see the illusions shown in Box 9.1. But just as the modules that comprise our visual system are prone to illusions, there is increasing evidence that our very rationality is subject to its own set of illusions, errors and biases in cognition.

9.2 Problems with rational thought

9.2.1 The problem of optimisation

Quite understandably, humans as a species are proud of their powers of rational thought. So much so that it was natural that when mathematicians began to model human behaviour they tended to assume humans would behave in ways that were optimal from a rational point of view. By the 1960s the idea of rational optimisation had entered such diverse fields as economics, psychology and animal behaviour. What was common to this endeavour was the idea that the behavioural agents under study (be they human or other animals) acted to maximise returns to themselves in the contexts they were operating in. In economics, for example, humans were expected to maximise net benefits, profits or utility; in psychology cognition was supposed to be directed towards optimal decision-making; and in behavioural ecology animals were supposed to behave so as to optimise reproductive fitness or, in the case of optimal foraging theory, calorie intake.

The problem with this approach is that it tended to assume almost unlimited cognitive power and access to huge amounts of information to take into account. Gigerenzer (2001) refers to these models as 'unbounded rationality' and argues that many of the underlying assumptions are unrealistic. The realisation in these disciplines that all the relevant information is rarely available led to 'optimisation under constraint' models. Imagine an animal A (human if you like) trying to make a decision about a prospective sexual partner B. If A is trying to decide if B is suitable she has to factor into her calculations an estimate of the worth of B (looks, resources, health status, social status, age, future prospects) and her estimation of her own worth. If A concludes that her own reproductive value (a measure, for the sake of argument, of all the relevant facts suitably weighted and averaged) is much

greater than that of B she may decide to reject B and search for something better. But the decision to reject also has to take into account the probability of something better than B showing up and also the costs of further searching. The latter consists of time, resources and opportunity costs – instead of searching around for Mr Right she could have settled down and started to fulfil her Darwinian imperatives with Mr B. In addition, the whole process of making a decision, weighing costs and benefits demands time. Suddenly the whole process is entangled in complexity and leads us to question whether such computations actually take place or whether the organism uses a quicker route.

As optimality models faced these challenges, the work of Amos Tversky and the Nobel Laureate Daniel Kahnemann in the 1970s and 1980s demonstrated that humans are prone to errors of reasoning and that such errors were very revealing about the sort of mental strategies that people actually use to solve problems under conditions of uncertainty.

9.2.2 Errors of thought and reasoning: heuristics and cognitive illusions

Errors in probabilistic reasoning

In an experiment that has become a classic in psychology, L. G. Humphreys (1939) studied the ability of people to predict when one of two light bulbs in front of them would light up. The bulbs were actually programmed to light up randomly but such that one lit up for 80 per cent of the time and the other for 20 per cent of the time. People quickly grasped this frequency and were then asked to score as highly as they could in predicting which bulb would be the next to light up. Subjects typically distributed their predictions in the same ratio as the observed frequencies, so they predicted the lower frequency bulb 20 per cent of the time and the higher one 80 per cent of the time. Although this sounds intuitively correct it is not the optimal strategy for maximising the number of correct predictions. This can be seen with a few calculations. Since the process is random, placing 80 per cent of your predictions on a bulb that only lights 80 per cent of the time gives a hit rate of 80 per cent × 80 per cent = 64 per cent. To this we can add the return from placing 20 per cent of your predictions on a bulb that only lights 20 per cent of the time = 20 per cent × 20 per cent = 4 per cent. This gives a total success rate of 68 per cent. A moment's thought shows it to be better to have simply placed all your bets on the 80 per cent bulb, giving a return rate of 80 per cent. Why people do not do this has been the subject of much research. One line of enquiry following the work of Humphreys considered the reward from actually acting rather than being passive.

There are many ideas as to why people exercise suboptimal rationality. Various robust patterns have emerged with corresponding labels such as 'risk aversion' and 'framing effects'. In the case of the bulbs it may be a tendency of humans to desire to use knowledge and insight once they have gained it rather than remain passive. Barash (2003) suggests that this is why many people invest in the stock market by selecting specific stocks (or paying someone to do so) despite the fact that tracker funds (baskets of shares that simply follow the leading shares on the stock market) outperform the vast majority of managed funds most of the time. Perhaps each believes that they can outperform the average or that some knowledge must surely be of use and preferable to no knowledge.

Another effect is called 'source dependence' and again seems to illustrate the preference for using established knowledge over the uncertain. A good illustration of this is the 'Ellesberg jar'. Test yourself by answering the following question quickly.

Imagine a jar containing 100 balls, 50 red and 50 white; and another jar also containing red and white balls numbering 100 balls in total but in an unknown proportion. You will be given $100 to correctly guess the colour of a ball pulled out at random from one of the two jars. Which jar would you prefer to take your chances on? Most people opt for the first jar even though the chances are 50/50 for either jar.

In the 1970s a large body of evidence had accumulated that suggested that ordinary people were rather poor at solving problems that required probabilistic reasoning. Rather poor, that is, compared to professional mathematicians who could solve such problems at their leisure. More worryingly, even professionals such as physicians were not perfect. When medics were given the task of estimating the probability that a patient had a disease based on a scenario where a given test threw up false positives and negatives, their answers were consistently way out (see Box 9.2). Many of these errors that people

---

Box 9.2 The base-rate fallacy

This fallacy has received a lot of attention; possibly, in part at least, because otherwise intelligent and supposedly numerate people, such as medics, who should know better, perform quite badly. One version of this test is phrased as follows:

> If a test to detect a disease whose prevalence is 1/1000 has a false positive rate of 5 per cent, what is the chance that a person found to have a positive result actually has the disease, assuming that you know nothing about the person's symptoms or signs?

In one study 60 students and staff at Harvard Medical School attempted this problem. Nearly half judged the answer to be 0.95 (that is, 95 per cent chance) (Casscells et al., 1978). The average response was 0.56 and only 18 per cent gave 0.02 (Gigerenzer, 1994). The correct answer by Baysean reasoning is 0.02. This is not just of academic interest, of course, since many tests for diseases do not have a 100 per cent correct hit rate, and if you test positive, thinking you have a 95 per cent chance of the disease when the real answer may be 2 per cent can have dire consequences.

Baysean reasoning

The origins of a rule for inferring the likelihood of a hypothesis being correct from certain types of data is attributed to the Reverend Thomas Bayes (1702–61). Although Bayes' original formulation is slightly different a modern expression of the false positive-type problem is given here.

Let   H   =   a person with the disease
        H' =   a person without the disease
        X   =   a person who tests positive
Then p(H) =   baseline probability that some-
                one picked randomly will have
                the disease = 0.001 (that is,
                1/1,000)
  p(X|H') =   probability of 'X' given H'. In
                this case that someone will test
                positive without the disease =
                0.05 (that is, 5 per cent).
    p(H') =   probability that someone picked
                at random does not have the dis-
                ease = 0.999 (that is, 1 – 0.001)

p(X|H) =   probability of 'X' given 'H'. In this
              case the probability that someone
              will test positive if they have the dis-
              ease = 1
p(H|X) =   probability that person will have
              the disease if testing positive

According to Baysean reasoning:

$$p(H|X) = \frac{p(H)p(X|H)}{p(H)p(X|H) + p(H')p(X|H')}$$

Substituting the values above gives:
  p(H|X) = 0.001/0.05095 = 0.02
          (rounded to 2 decimal places)

succumb to have been categorised with names such as the 'conjunction fallacy', the 'base-rate fallacy' and 'overconfidence bias'. Thanks to pioneering work by Kahneman and Tversky, the study of such 'cognitive illusions' is now an important part of mainstream psychology (Kahneman, Slovic and Tversky, 1982). One typical example is the 'the conjunction bias'.

The conjunction bias

The following problem is typical of that used by Kahneman and Tversky (Tversky and Kahneman, 1983)

> Linda is 31 years old, single, outspoken and very bright. She majored in philosophy. As a student, she was deeply concerned with issues of discrimination and social justice, and also participated in antinuclear demonstrations.
>
> Which of the two alternatives is more probable?:
> a) Linda is a bank teller (T)
> b) Linda is a bank teller and is active in the feminist movement (T and F)

Of the participants in Tversky and Kahneman's original study, 85 per cent chose b). This conclusion is a fallacy since the probability of two events (T and F) must be less than one of its constituents. For example, if you throw two dice then the probability of a six

and a four is less than just one being a six. Rationally, it is much more likely that Linda is a bank teller than a bank teller and active in the feminist movement. For b) to be true Linda has to be a bank teller anyway and it is conceivable that she may not be a feminist. Tversky and Kahneman explained this effect, which has since become widely reported in the literature of a number of disciplines, by positing a 'representativeness heuristic'. Hence, because Linda was described as if she was a feminist and this fitted one part of the description, this biased our better judgement.

9.3 Cognitive illusions and the adapted mind

So is the human brain, after some 2 million years in the hominid lineage, a rather poor device at calculating probabilities? We can take some comfort in the fact that we can (thanks to schooling in Baysean reasoning) eventually work out the right answer. But why is it such an effort fraught with pitfalls, even for educated people? Also, if some survival advantage were to be attached to this sort of reasoning then coming up with an answer of 95 per cent (half the respondents in the base-rate fallacy experiment) when the actual answer is 2 per cent suggests a catastrophic mismatch between representation and actuality.

Gigerenzer attacks these so-called fallacies by arguing that questions based on single events are inappropriate to test probabilistic reasoning (Gigerenzer, 2000, p. 246). Another explanation of the effect may be that in both cases the mind is led astray by assumptions that it makes rather than a cognitive error as such. In the bank teller case, for example, it could be that the question is framed so as to suggest Linda is already a bank teller and then we have to estimate if she is also a feminist.

Faced with the phenomena of cognitive illusions, the evolutionary psychologist would probably start to work from two premises:

a) The human mind is well adapted to solving problems, but these problems are the sort encountered in the EEA not those of abstract mathematics.
b) If problems of the sort that generate cognitive illusions could be recast into more ecologically and socially relevant formats then performance should improve.

Applying these premises we can revisit the fallacies described above.

9.3.1 Bounded rationality and adaptive thinking

At the time when notions of unbounded rationality and optimisation under constraints were being propounded, Herbert Simon advanced the competing idea of '**bounded rationality**' (Simon, 1956). Although Simon's and subsequent definitions were not always precise, the concept was destined to capture a more realistic account of how real people make decisions which, although not strictly irrational, are nevertheless not optimising because of the limitations of cognitive power or the incompleteness of information. The concept has been developed and applied in recent years with some success by Gerd Gigerenzer (2001). Gigerenzer compares human reason to an adaptive toolbox. The function of this toolbox is to enable the organism to achieve its proximal goals such as finding a mate, avoiding predators, obtaining food, achieving status in a group and so on. The crucial point is that Gigerenzer does not assume that the strategies in the toolbox seek to optimise. The reasons for this assumption are that

the number of decisions that an organism has to make is so huge that it would require immense and unrealistic computing power to deal with them all. Furthermore, given the absence of complete information, assumptions have to be made that could generate large errors if optimisation were always sought. In describing this toolbox Gigerenzer borrows a phrase from Wimsalt in his description of nature: 'a backwoods mechanic and used parts dealer' (quoted in Gigerenzer, 2001, p. 30).

The analogy is that a backwoods mechanic will have neither a general purpose tool, nor a complete set of tools for all cars, nor a complete set of spare parts. Such a mechanic will have to make do with whatever is available for the problem at hand, improvising and adjusting accordingly. If the mind contains, metaphorically speaking, such a toolbox, there are some features that we may expect to be present. They are:

- Psychological plausibility. Models will be based on what cognitive capacities humans have rather than assumptions of Divine Intelligence or pure logic, against which humans will inevitably fall short.
- Domain specificity. The rules and devices of the toolbox will be specialised.
- Ecological (and social) rationality. The rationality of these domain-specific heuristics is not to be judged against the canons of pure logic. In this respect it should come as no surprise if such rules as consistency and optimality are broken. Rather, the success of bounded rationality should be judged by its performance in social and ecological contexts. Humans in particular have to make decisions that are fair and ethical (in the sense of conforming to a group ethos) and that can be defended within a group.

The toolbox will contain **heuristics**; that is, strategies for generating a solution to a problem. Such strategies may consist of simple and economical rules to guide decision-making. Gigerenzer gives two examples: search rules and stopping rules.

Search rules. Searching can be thought of as looking amongst alternatives for cues to evaluate the choice on offer. Search rules can be illustrated by the so-called 'secretary problem' much investigated by statisticians (see Todd and Miller, 1999). Suppose you have to hire a secretary from 100 applicants. You interview in a random order and either accept or reject. A candidate that is rejected cannot be returned to for hiring. The dilemma of course is this: if you choose too early you may pick someone of poorer quality than the mean and poorer than many that would follow; if you pick too late it becomes increasingly likely that the few left may not contain someone suitable. Mathematical demonstration shows that the best solution is to interview a certain proportion, remember the best of them and choose the next who is better than that. The optimum number to interview is $1/e$ (or 37 per cent) where e is the natural number 2.718. This is not foolproof but this 37 per cent rule will give the best result about 37 per cent of the time.

Stopping rules. Searches eventually have to stop. A very simple combination of a search and stop rule might be sample a dozen and take the best. Stopping rules need not be cognitive; an emotional system could perform the task. In this respect love and affection for a sexual partner or children might serve as emotional stopping rules, obliging the organism to stay with its mate and offspring and forestall any

further calculations of whether better options lie elsewhere. Such a system could, over a whole life, be highly fitness enhancing compared with the constant abandonment of husbands, wives and children as better options present themselves. As another example of the need for a stopping rule consider the problem of finding the best deal on a new car. Suppose that there is some price variation for the same model and that the actual price is only to be ascertained by visiting widely dispersed showrooms of dealers and haggling with the salespeople. One solution to the problem of when to stop looking is to make a 'reservation price' and decide when the marginal costs (time effort, travel costs) of further searching equals the marginal expected improvement over the best seen so far. When it starts to cost $300 of travel effort to find a car $300 cheaper than the best seen so far, common sense (ecological rationality) says stop.

Herbert Simon termed the type of examples and solutions given above 'satisficing' (Figure 9.1), which he defined as 'using experience to construct an expectation of how good a solution we might reasonably achieve, and halting a search as soon as a solution is reached that meets the expectation' (Simon, 1990, p. 9). These search rules look simple but are they enough for real-life situations which throw up a much greater array of parameters?

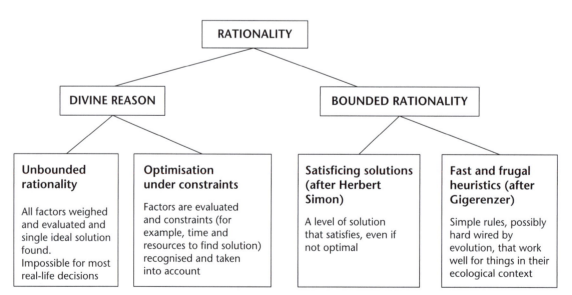

FIGURE 9.1 Some alternative concepts of rationality.

In mate choice, for example, the organism has to weigh up the quality of mates it encounters, its own quality, search and opportunity costs (the biological clocks of males and females tick away whilst decisions are delayed), and reactions of potential partners to advances. The crucial point to emphasise again is that the mind does not solve the secretary problem or the cost–benefit analysis of marginal costs mathematically and formally each time. Natural selection solved such problems for us, not by giving us an onboard omnipotent computer but by providing a set of short, fast and simple rules that work reasonably some of the time.

The research programme of Gigerenzer's group consists of three approaches. Firstly, to try to identify likely heuristics that could work; secondly, testing these heuristics in artificial and real-world environments; thirdly; ascertaining if people actually use them. Gigerenzer's concept of heuristics must be distinguished from other such uses of this term in the psychological literature. The 'heuristics and biases' research programme of Tversky and Kahneman, for example, has tended to suggest that heuristics are mentally sloppy procedures that usually lead to fallacies (see below). For Gigerenzer's research programme, however, heuristics are treated more positively as ways 'the human mind can take advantage of the structure of information in the environment to arrive at reasonable conclusions' (Gigerenzer et al., 1999, p. 28). In other words, they are rather efficient devices serving the immediate needs of the organism:

> The function of heuristics is not to be coherent. Rather, their function is to make reasonable, adaptive inferences about the real social and physical world given limited time and knowledge. (Gigerenzer et al., 1999, p. 22)

Two examples of potential heuristics that humans may be inherently disposed to employ are the prestige bias and the conformist bias:

Copy prestigious members of a group. Much research suggests that humans everywhere display a propensity to copy prestigious individuals – a process sometimes known as 'prestige-biased transmission' (see Henrich et al., 1999). Cues for prestige can easily be gleaned from displays of wealth or attention given to individuals by others. In gatherings, for example, high-status individuals are listened to more attentively, interrupted less and generally treated with deference. The adaptive function of this behaviour is fairly obvious. By copying successful individuals the copier obtains valuable information about behaviour patterns that have led to success. This form of learning can be very time efficient. Acquiring hunting skills by copying the best hunter is probably more efficient than learning through individual trial and error. Such a mechanism could explain the almost self-perpetuating phenomenon of fame. In contemporary Western cultures the media spotlight often seems to be trained continually on people for no other reason than that they are themselves often in the media.

Conform to the mode. Running parallel to prestige-biased transmission, conformist transmission is a heuristic that says preferentially copy the most common behaviour in a population. The adaptive value of this strategy is also clear: the most common behaviour must have brought some success and cannot be too fitness reducing otherwise the group would have died out or abandoned the behaviour. Imagine you are hungry, walking across a strange landscape and come across a group of people consuming purple fruits but assiduously avoiding the red ones. Which fruits would you eat? Significantly, and as expected, theoretical modelling shows that individuals do rely heavily on this heuristic when cues from the environment are ambiguous or uncertain (Henrich and Boyd, 1998). Numerous studies of social learning in species such as pigeons, rats and guppies suggest that these animals adopt a 'do-what-the-majority-do' strategy (see Laland et al., 1996).

9.4 Biases and fallacies revisited

9.4.1 The conjunction fallacy revisited

Gigerenzer has argued that if we replace the Linda bank teller question with one based on frequencies then the conjunction fallacy should disappear. In this modification the question can be rephrased as follows:

The are 100 persons who fit Linda's description (see p. 170).

How many of them are:
a) bank tellers
b) bank tellers and active in the feminist movement?

Hertwig and Gigerenzer performed such studies and found that the percentage of subjects committing the conjunction fallacy (that is, T and F) fell from the 80 per cent level observed in Tversky and Kahneman's work to between 10 and 20 per cent (Hertwig and Gigerenzer, 1999).

9.4.2 The base-rate fallacy revisited

Part of the problem with the way the disease test question is presented to subjects is that it is unclear what to do with the base rate. To answer correctly the

mind first needs to assume that the person to whom the medical test is administered is randomly selected from the population where the prevalence is 1/1000. It is obvious that those giving the answer as 0.95 are not doing this.

When the problem is rephrased in a frequency format the effect on performance is dramatic. Table 9.2 shows how two probability problems were rephrased in frequency terms by two different research groups. Figure 9.2 shows the improvement in performance when the problems are rephrased.

The overall conclusion from this approach is that the biases and fallacies identified by cognitive psychologists, and some of the heuristics invoked to explain them, may be an artefact of the experimental procedures. An evolutionary view of the mind predicts that it will be good at solving problems when expressed in frequency terms rather than the fractional probabilities used by statisticians. The idea that

Table 9.2 Alternate ways of expressing a probability problem

Baysean probabilistic expression of problem	Problem recast in frequency format
Disease test problem 'If a test to detect a disease whose prevalence is 1/1000 has a false positive rate of 5 percent, what is the chance that a person found to have a positive result actually has the disease, assuming that you know nothing about the person's symptoms or signs?' (Casscells et al., 1978, p. 999)	'One out of 1000 Americans has disease X. A test has developed to detect when a person has disease X. Every time the test is given to a person who has the disease, the test comes out positive. But sometimes the test also comes out positive when it is given to a person who is completely healthy. Specifically, out of every 1000 people who are perfectly healthy, 50 of them test positive for the disease.' 'Imagine that we have assembled a random sample of 1000 Americans. They were selected by a lottery. Those who conducted the lottery had no information about the health status of any of these people. How many people who test positive for the disease will actually have the disease?' (Cosmides and Tooby, 1996, p. 24)
Mammography problem 'The probability of breast cancer is 1 per cent for women at age 40 who participate in routine screening. If a woman has breast cancer, the probability is 80 per cent that she will get a positive mammography. If a woman does not have breast cancer, the probability is 9.6 per cent that she will also get a positive mammography.' 'A woman in this age group had a positive mammography in a routine screening. What is the probability that she actually has breast cancer? ___ per cent.'	'10 out of every 1,000 women at age 40 who participate in routine screening have breast cancer. 8 of every 10 women with breast cancer will get a positive mammography.' '95 out of every 990 women without breast cancer will also get a positive mammography. 'Here is a new representative sample of women at age 40 who got a positive mammography in routine screening. How many of these women do you expect to actually have breast cancer?' ___ out of ___ (Gigerenzer and Hoffrage, 1995, p. 9)

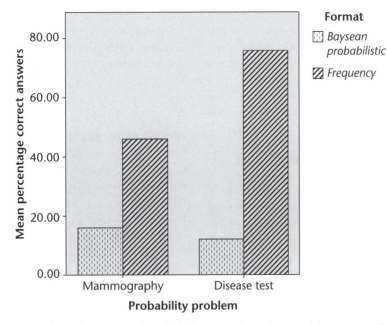

FIGURE 9.2 Improvement in performance of probability reasoning when problems are rephrased in a more 'mind-friendly' frequency format.

SOURCE: Data on mammography test from Gigerenzer and Hoffrage (1995); data on disease test from Cosmides and Tooby (1996).

'two people out of ten that ate this meat were sick the next day' is more in keeping with the grain of the mind than 'it is estimated that the probability of sickness following the consumption of this meat is 0.2'.

We have seen that heuristics and biases can sometimes lead to 'cognitive illusions' so-called in analogy with visual illusions where the mind 'sees' something different to what is the case. The analogy is quite instructive. If we consider again the optical system it becomes apparent that it was not designed to faithfully present an objective picture of what is 'out there', but rather to present the mind with a useful interpretation. David Marr (1982) added a new impetus to the whole field of vision science when he argued that the visual systems made compensations to perception to enhance the usefulness of the knowledge to be gained. In Box 9.1, for example, we see a rectangle when none is there and are convinced the shade of white of the rectangle is different to the background since this is a reasonable assumption. In the natural world it is highly unlikely that we would come across circles with precise slices missing as shown there. As Workman and Reader point out, applying this type of illusion to

an ecological context, seeing a tiger partially hidden behind long grass as a series of puzzling tiger-coloured strips is less useful than seeing the whole tiger (Workman and Reader, 2004). But the obvious advantage of making such assumptions can sometimes lead to errors, and to this we now turn.

9.5 Making decisions amid uncertainty: error management theory

Over the last few decades there has been a remarkable convergence of ideas about the way humans make decisions amid uncertainty. One important contribution comes from Martie Haselton and David Buss (2000) who have developed a framework for integrating whole classes of such biases which they call error management theory (EMT). The basic approach of EMT can be illustrated by imagining yourself walking alone at night through a dark area of woodland and hearing the snapping of a twig. There are four basic possibilities linking physical reality to your mental beliefs about the cause of the sound, which can be understood in terms of the traditional statistical nomenclature of type I and type II errors. If you believe it to be the result of some agent (for example, a hostile predator), while in reality it is a natural phenomenon (such as a twig falling from a tree), you are committing a false positive or type I error. If you assume it is a natural phenomenon but in reality it is caused by a hostile agent you are falling prey (probably in more ways than one) to a type II error or making a false negative. Now the costs consequent to these errors may be different in each case. Committing a type I error may leave you feeling a bit silly and you may waste calories running away or mobilising adrenalin; but committing a type II error could leave you dead. Being biased in decision-making towards type II errors, that is being slightly over-anxious and mildly paranoid (seeing agency at work where there is none), might, therefore, in the long run make evolutionary sense (Haselton and Nettle, 2006). To recast the analogy in terms of bears and boulders, if perception is an imperfect system it is better for your health to see some boulders as bears than some bears as boulders.

Figure 9.3 shows how this can be expressed as a contingency table. To add generality to this model we could say that we are making a hypothesis (H_1) about the world and accepting or rejecting the hypothesis as correct. In the formal language of hypothesis testing we are trying to decide between two hypotheses:

H_0: The sound is a harmless natural phenomenon (for example, a twig falling from a tree)

H_1: The sound of the twig is an indication of the presence of a hostile agent

The bias in cognition that might be suggested by the scenario shown in Figure 9.3 depends entirely on the asymmetry of costs and benefits associated with false belief. In the example shown, the costs of type II errors (mistaking a bear for a boulder or concluding that a snapping twig is just a natural phenomenon) greatly exceed those of type I errors (mistaking a boulder for a bear or a harmless snapping of a twig for a predator) so according to EMT we will be biased towards making type I errors (that is, accept H_1 even when the evidence is weak).

Within the whole statistical process of the testing of hypotheses central to many branches of the life sciences it is generally considered that type I (false positive) errors (that is, rejecting the null

	Reality	
Your beliefs	H_1 is true. Sound caused by hostile agent	H_1 is false and H_0 true. Natural phenomenon
Believe H_1 is true. Assume agent	Accurate. True positive 	False positive, type I error 
Believe H_1 is false. Assume natural phenomenon	False negative, type II error 	Accurate. True negative 

FIGURE 9.3 Type II and type I errors of judgement.
The consequences of type II and type I errors may be different, leading to a bias in human cognition towards committing the least costly error.

hypothesis and accepting the research hypothesis as correct when in fact it is false) are more serious and costly than type II errors (rejecting a perfectly good research hypothesis). Hence statistical testing procedures and associated probability estimates are biased towards avoiding type I errors but of necessity increasing the risk of type II errors. A typical procedure is to state the hypothesis, collect the data, calculate a test statistic (for example, X^2, student's 't') and then assess the probability (P) of obtaining that value of the statistic if the research hypothesis were false. When P is below some threshold (often 0.05) we can say this is low enough to accept the fact that the hypothesis is probably not false and is true (or at least supported). We have thereby in effect reduced the chance of a type I error to 5 per cent.

In terms of human psychology, EMT predicts that when the costs of the two types of error (false positive (FP) and false negative (FN)) are different then decision-making (given that it cannot often be spot-on and is made amid uncertainty and background noise) will be biased towards a higher likelihood of committing the least costly error.

In the section that follows we will examine how EMT can help make sense of a variety of cognitive biases, errors of reasoning and behavioural tendencies (putting aside for the moment the distinction between these categories) reported in the literature. The structure follows Figure 9.4.

9.5.1 Biases that offer protection

All the examples in this section exhibit an asymmetry such that false positive (FP) costs are much smaller than false negative (FN) costs (FP<<FN) and hence suggest a bias towards accepting FP type I errors. They accord with the maxim 'better safe than sorry'.

a) Moving objects
    If an object making a noise approaches, the intensity of the sound rises as the object nears the observer; conversely as it recedes the intensity of the sound falls. Humans can use this effect to estimate the speed of approach and departure of an object. Experiments conducted by Neuhoff (2001) using volunteers listening to sounds emitted by speakers on moving cables showed that subjects overestimated the speed of approach and the proximity of an approaching object. This bias makes adaptive sense in so far as the cost of an FP (being ready too early for an approaching object) is small compared to the cost of an FN – underestimating the speed and hence being unprepared for the arrival of a predator or hostile assailant. In later work Neuhoff et al. (2009) even found sex differences in this effect with the bias found to be greater in women than men. They were able to rule this out as a product of

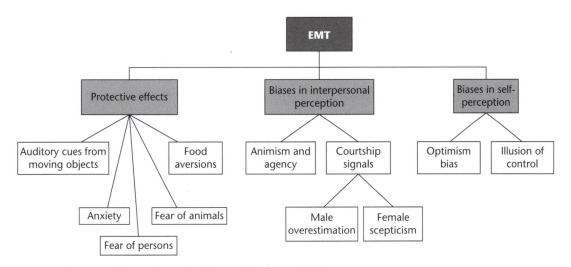

FIGURE 9.4 An organising schema for the application of EMT.
SOURCE: Based on Haselton and Nettle (2006).

poorer audio processing by women generally and concluded instead that the slower running speed and strength in women meant that the adaptive bias in perception was even stronger (women need a longer time to run away or prepare to take action).

b) Anxiety

Smoke detectors work by being biased towards FP errors – they go off when in reality the house is not burning but there is a harmless source of smoke (for example, burning toast). Randolf Nesse (2005) argued that many human defence systems might work like this giving rise in some cases to excessive anxiety (see Chapter 18).

c) Fear of animals

Experimental work illustrating this category also falls under the heading of 'preparedness theory'. Evidence suggests that humans are quick to acquire a fear of such animals as snakes and spiders and in experimental situations overestimate the association between images of snakes and spiders and an induced mild electric shock. This is in keeping with EMT: in ancestral environments it would be better to overestimate the threat of an animal than to risk serious injury or death from venomous snakes and spiders. This bias would suggest that faced with ambiguous images humans would more often interpret images of sticks (say) as snakes than snakes as sticks.

d) Fear of persons

A parallel argument to the risks from dangerous animals can be made to examine the threats posed by potentially dangerous humans. Indeed, given that humans are large animals it is possible that the greatest violent existential threat to early humans was from other humans. Unlike in relation to other animals, however (where FP costs are low), inferring danger and hostility from a fellow group member where none is present could incur high FP costs: lost opportunities for cooperation, for example. But FP costs due to encountering someone in another group are likely to be low (fewer future opportunities for mutualism) with FN costs potentially higher. In support of this idea there is some evidence that people think that their own ethnic or racial group is more benevolent and less hostile than other outgroups (Brewer, 1979; Quillian and Pager, 2001).

The fear of ill people is logically similar although the potential source of harm different. Communal panic over contagious diseases is fairly easy to induce. There are studies showing how on the basis of very little evidence people assume someone is contagious (Faulkner et al., 2004), and how persistent stigmas can be concerning physical afflictions (Bishop et al., 1991). Faulkner et al. also suggest that fear of disease transmission may underlie xenophobia (Faulkner et al., 2004).

e) Food aversions

Aversions to certain types of food may be a protective reaction against toxins and microorganisms. It has been suggested that one reason why children avoid leaves and vegetables and are notoriously picky about food is that children's livers are immature and less able to detoxify foods compared to adults (Cashdan, 1998). Similarly, food aversions by pregnant women may be a way of avoiding potentially harmful foodstuffs (see Chapter 18 and Fessler, 2002). The 'disgust' response is instructive in this context and may be an intuitive form of microbiology – best to avoid even a tiny amount of something which is suspected to contain harmful microorganisms. Otherwise palatable foods presented in disgusting shapes (for example, chocolate mousse in the shape of dog faeces) are refused by people even though rationally the food is demonstrably harmless (Rozin and Fallon, 1987). In EMT terms the subjects err on the side of an FP once the disgust response is activated since even a small dose of such a contaminated substance could be very harmful.

9.5.2 Biases in interpersonal perception

a) Animism

In what has become a classic experiment, Fritz Heider and Mary-Ann Simmel (1944) asked subjects to view images of geometric shapes (a big triangle, a little triangle and a small circle) moving about and interacting with each other on a screen (this can be viewed at www.youtube.com/watch?v=n9TWwG4SFWQ, last accessed 29 January 2016). When they asked observers what they saw, replies were always couched in terms of agency and intention such as the little triangle being in love with the small circle but the large triangle behaving aggressively. The implication here is that the human mind seems to have an agency or animism bias as a default setting. In terms of EMT it may have been advantageous to see agency where there was none (FP) than to miss an occasion where agency was present (FN). Agents (that is, humans and other animals with a sense of purpose and their own interests) often have high fitness implications for humans and having a low threshold to infer their presence may have been adaptively useful. Guthrie (2001) extends this notion and suggests that it may underpin the almost universal tendency of humans to hold religious beliefs.

b) Courtship signals

When men and women looking for romance interact each has to make a whole cascade of difficult judgements and adjust behaviour accordingly. One obvious decision to make and signal is whether they are at all interested in the other as a potential romantic partner. They will also need to detect if the other person is similarly interested. It is a situation ripe with ambiguity and the potential for misunderstanding, the very stuff of tragi-comedy. Haselton and Buss (2000) used EMT to predict sex differences in the interpretation of mating signals.

Male overestimation of sexual intent

From a male perspective, thinking that a female is showing sexual interest when she is not is committing an FP error, incurring the possible costs of embarrassment and disappointment when the truth dawns. But an FN error (assuming there is no sexual interest when there is) is potentially much more injurious to a male's reproductive success since it involves a missed mating opportunity. The prediction made, therefore, by Haselton and Buss was that males will be biased to avoid FN errors and to make FP-type errors.

A variety of empirical studies have been conducted that do indeed provide support for this effect. Frank Saal, Catherine Johnson and Nancy Webster (1989), for example, conducted several studies involving male and female subjects viewing video material of real-life brief encounters between men and women. In all studies the male viewers reported observing more sexual interest on behalf of the women than did the female viewers. Studies based on asking people about their real-life experiences outside the laboratory paint a similar picture. Martie Haselton (2003) asked 102 female and 114 male undergraduates about experiences where a member of the opposite sex erroneously interpreted encounters in terms of sexual intent. She found that women reported significantly more episodes of FP committed by men (men assumed sexual interest when none was present) than FN errors. Men reported that women made slightly more FP than FN errors but the difference was not significant at the 0.05 level (see Figure 9.5). There are other findings in the literature also supporting this hypothesis (see Farris et al., 2008, for a review).

Such findings are of obvious academic and theoretical interest in furthering the scientific study of human behaviour; but they also have practical implications. Haselton cites the example of the Safeway supermarket policy of 'service with a smile' where employees were encouraged to make eye contact with customers and smile. The policy was phased into branches in the mid-1990s but soon faced a barrage of complaints from female employees who complained that male customers interpreted this as a flirtatious sign resulting in numerous cases of sexual harassment (see Mendell and Bigness, 1998). Such studies may throw further light on contexts where inappropriate behaviour might be inadvertently promoted and so inform policy.

Female scepticism about male intentions

Many of the examples discussed previously show human reasoning and/or behaviour to err on the side of FPs. But it could be the other way round, depending on the situation and the way the initial supposition or hypothesis is framed. Consider the case of an ancestral female (that is, before the days of contraception) trying to decide if a male will be willing to make a significant post-reproductive commitment to her and any offspring if she agrees to mate with him. Let the hypothesis now be:

H1: 'The man's intentions are genuine and he will stay and provide care and support'

Here the asymmetry in costs is different to those of male perception of sexual intent. The costs of FP

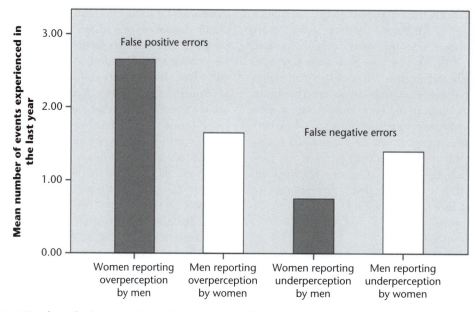

FIGURE 9.5 Number of misperceptions relating to sexual intent made by men (as reported by women) and women (as reported by men).
Overperception: reporters' friendliness was interpreted as sexual interest. Underperception: reporters' sexual interest was interpreted as friendliness. Men seem much more likely to make false positive errors inferring sexual intent where there is none than false negative errors (p<0.001). Men report that women make slightly more FP errors than FN errors but the effect is not significant (p>0.05).
SOURCE: Data from Haselton, 2003, Figure 2, p. 40.

(she mates with the man, bears his child and then he clears off) are conceivably much higher than FN (she loses a mating opportunity with someone who is caring and interested in her). After all, given the biological asymmetry in reproductive rates there will be no shortage of future willing males. Furthermore, bearing a child in traditional societies without male support has been shown to more than double the risk of the mother's offspring dying young (Hurtado and Hill, 1992). Having a child also decreases the residual reproductive value of the female from the perspective of future males. So, in short, if the relative costs of desertion compared to lost mating opportunities are high it is adaptive to make more FN errors than FP errors.

The balance of risks in this scenario may underlie what Haselton and Nettle (2006) call 'commitment scepticism'. In support of this effect several studies do seem to show that women underestimate the level of commitment a male genuinely intends and is trying to convey. Haselton and Buss (2000), for example, issued questionnaires to 217 undergraduates asking questions about the interpretation of various statements. They found that women thought the statements describing attention given to women by men indicated less commitment than did the male subjects. In contrast male and female subjects did not differ in their interpretation of women's behaviours. Such studies, however, are very indirect, using a narrow age range and responses to questionnaires. It is desirable that this effect is explored in more real-life contexts.

9.5.3 Biases about self

a) The optimism bias

Up until the late 1980s conventional psychological wisdom suggested that a mentally healthy person had accurate views about the world and themselves, and those who had distorted views about reality were either vulnerable or mentally unwell. Taken to extremes this is obviously still true (there is something wrong with a person who thinks he is Napoleon Bonaparte). But in 1988 there appeared a landmark paper by Shelley Taylor and Jonathan Brown that reviewed a whole range of evidence on self-perception and concluded that 'the mentally healthy person appears to have the enviable capacity to distort reality in a direction that enhances self-esteem, maintains beliefs in personal efficacy and promotes an optimistic view of the future' (Taylor and Brown, 1988, p. 201). The authors did explore the possible adaptive basis of these illusions but a more formal treatment was given by Daniel Nettle (2004a). The conclusions of Taylor and Brown can be interpreted in terms of EMT as another example of the 'smoke detector principle': FP errors may be worth making if their cost is low and the cost of FN errors high. In this case, if we start with the linked set of suppositions that 'the future is rosy and I am a capable person that can control events to my advantage', then the costs of a FN (the pessimist's view) might lead to a fitness-damaging passivity. The costs of FP would be actions taken that do not succeed. But if the benefits of success are high (such as a desirable mate or a rewarding job) then an optimistically biased self-belief system might just tilt the balance in competitive situations and so be better than a perfectly accurate one. By analogy, poor football and baseball teams sometimes beat better ones (so long as they have self-belief and think they can win), even though on paper the prospects look unlikely. This effect has now been well documented and is sometimes called the 'optimism bias'. There is now a mass of evidence showing that we underestimate our chances of being in a car accident, overestimate our driving ability, underestimate the chance of getting divorced, expect to live longer than is statistically likely and think our children will be more talented than they usually are. It is now one of the most robust and well-documented biases in psychology and behavioural economics (Sharot, 2011).

b) Behavioural superstitions and the illusion of control

Many otherwise sane and rational people suffer from superstitions, beliefs and behavioural tics that seem bizarre. Actors, for example, avoid saying the name of Shakespeare's play *Macbeth* and instead refer to the 'Scottish play'. Wishing someone 'good luck' before a performance is also a no-no, and instead 'break a leg' is the preferred alternative. Sportsmen

and sportswomen are also notorious for following superstition-based routines. We may think these examples as distinctly human foibles but in 1948 the behaviourist B. F. Skinner published a paper called '"Superstition" in the pigeon' (Skinner, 1948). Skinner fed captive pigeons automatically and in a regular manner totally unrelated to anything that the bird did. Despite the perfectly predictable and mechanical nature of food delivery, the pigeons developed rituals (such as swinging their heads or walking in a circle) in the mistaken belief (according to Skinner's interpretation) that such actions caused the food to appear. Skinner called these behaviours superstitions; interpreted another way the pigeons were suffering from the illusion that they controlled events. There is now a growing body of evidence that humans also suffer from this illusion of control (see Rudski, 2001, 2004). The propensity of humans to resort to superstitions and rituals seems to vary with the level of rewards. Jeffrey Rudski and Ashleigh Edwards (2007) asked college students to report on their use of charms, rituals and superstitious routines in relation to tasks of varying difficulty and different levels of importance. They found that students were more likely to rely on superstitions when they thought that they were poorly prepared for the task or felt that their own level of skill was low; and as the importance of the task increased so did the reliance on superstition. The authors argue that such behaviour provided participants with a feeling (however illusory) of control over the situation and was instrumental in keeping at bay 'learned helplessness' and the inertia people experience if they really do have no control over outcomes.

Applying EMT to this we might suggest that in ancestral environments it really was important to establish rules and patterns that did involve prediction and control. Finding these where in actuality they were absent (not factually supported) would not have mattered much so long as the costs of FP were low. Imagine a hunter who touches the point of a spear before each throw at an animal in the belief that this helps guide its path. The cost of this exercise, futile in mechanical terms, would be low compared to the benefits of rules that did occasionally pay off and the rewards that followed from believing that the environment did have patterns and regularities that could be understood and controlled; 'where ignorance is bliss, 'tis folly to be wise' as the poet Thomas Gray said.

We have seen that according to EMT it sometimes pays to have a bias towards perpetrating FP errors and in other cases FN errors depending on the balance of costs and benefits and of course how the conjecture about the situation is formulated. One general principle that guides EMT is that overall costs will be minimised. But costs and rewards also have a cultural dimension and some biases, such as positive illusions about the self, could be expected to be shaped by the cultural context since culture gives value to specific skills and abilities. So what might be valued in Western individualist cultures might not be applauded so loudly in Eastern collectivist cultures (Haselton and Nettle, 2006). One of the most useful aspects of EMT is its ability to generate new predictions. All one needs is a series of problems that would have impacted on fitness over evolutionary time, where the degree of uncertainty in the information available is such that quick and accurate solutions are precluded and where the two types of error in relation to a conjecture about what might be the case give rise to stable asymmetrical costs. Haselton and Nettle suggest the following areas as likely candidates for these conditions and hence further work: mate guarding and sensitivity to sexual interest in a partner by an outsider, dangerous animals (other than spiders and snakes), disease and contamination.

EMT may turn out to be an important tool in psychology with widespread application since for any decision-making domain the likelihood that the costs of FP and FN errors are exactly the same (i.e. zero asymmetry) is very small. Hence such fitness-maximising biases (as opposed to truth-seeking mechanisms) might be extremely common in the behaviour of humans and other animals. EMT also has the advantage of supplying functionality to many of the cognitive biases unearthed by mainstream cognitive psychology (Johnson et al., 2013). One urgent task for EMT is to disentangle cognitive from behavioural biases. Given the logic of the cost asymmetries it would pay an organism to develop both behavioural and cognitive biases. In other words, to both behave in a biased way without thinking and in other cases to cognitively reach a biased solution which then informs action. Against this sanguine (possibly optimistically biased?) outlook for EMT must also be balanced the fact that many asymmetries in costs may have shifted and reversed over time, not allowing (outside a few examples related to stable biological differences between the sexes) many such biases to emerge. But this in itself might guide a new prediction: in contexts influenced by shifting and unpredictable factors perhaps other cognitive mechanisms (resilient to bias) are recruited for the task. There is room here to explore the linkage with the various ecologically based theories of intelligence discussed in Sections 9.3 and 9.4.

9.6 Module plasticity

Virtually all neuroscientists now agree that there are significant gene-driven constraints on the way brains gather and process experience. The issue that polarises current debates is the degree of plasticity available to the organisation of the neo-cortex. At the low-plasticity end of the spectrum we have the massive modularity hypothesis with the suggestion of high internal constraints (Tooby and Cosmides, 1992; Caruthers, 2006). At the other end of the spectrum we have the idea that the neo-cortex is subject to only a few inherent constraints but is shaped by developmental experience (Finlay et al., 2001).

It is, of course, possible, even likely, that both systems are present to some degree. Among different species the balance of these two systems (domain specific and domain general modified by learning) will probably depend upon the result of selective forces having acted upon the costs and benefits of each system in relation to changing environments and local ecologies. So, for example, environments that are invariant over long periods of time, and where incoming information is stable and predictable across generations, might be expected to promote relatively invariant cognitive systems. Such 'quick fixes' would provide rapid solutions to local problems. These invariant and fairly constrained systems respond fast and frugally to information input. They give quick answers even at the cost of accuracy (for example, optical illusions). In this category belong the fast and frugal heuristics investigated by Gigerenzer (2000) and Kahneman (2011) (see above). In this type of neural hardware and software we might find systems for coping with input's three major problem-related areas: self and others, living things, and the behaviour of objects. The fast and rapid heuristics with which we process information from these domains is often referred to as folk psychology, folk biology and folk physics respectively (see Chapter 2). Conversely, ecologies that are unstable and of low predictability might promote more flexible learning systems. Table 9.3 shows how these two extremes can be conceptualised in cost–benefit terms.

Recognition of faces may be an example of both systems at work. It is important for humans to recognise faces and so identify parents, partners, friends and enemies. Yet the basic features of primate faces (two eyes above one nose and one mouth) have been stable for millions of years. It might be expected then that humans share with other primates a fairly constrained piece of facial recognition neural hardware. Beyond realising that the object in front is a face, however, there is then obviously a need for a more plastic system to recognise and remember individual faces. Geary and Huffman call this system of hard and flexible responses 'soft modularity' (Geary and Huffman, 2002). Geary and Huffman sum up their perspective as follows:

> constraint and plasticity are the evolutionary result of invariant and variant information patterns, respectively, that co-varied with survival or reproductive outcomes during the species' evolutionary history. (Geary and Huffman, 2002, p. 691)

9.7 Sex differences in cognition

It is interesting to speculate whether sex differences in cognition and related abilities in modern humans might be the result of ancestral selection pressures. If so, there might be an echo or vestige of such pressures revealed in the sort of tasks carried out by men and women in existing traditional societies – not because these people represent some sort of evolutionary throwback but because they can be expected to encounter similar ecological problems to those faced by our hunter-gathering ancestors and may solve these problems using inherited sex-specific skill sets and biases in various competences.

One important source of data comes from the ethnological studies of the American anthropologist George Peter Murdock (1897–1985). His wok culminated in the production of the Ethnographic Atlas – a database of 1167 societies published in instalments in the journal *Ethnology* between 1962 and 1980. Various researchers have used this database to look for robust patterns in the sex division of labour. Table 9.4 shows some typical findings.

In addition to knowledge about the sex division of labour in traditional societies, we have modern psychological knowledge about sex differences in competences possessed by humans living in modern societies. One person who did much to tease out

Table 9.3 Types of problem-solving and learning systems

Environment	Unstable, low in predictability	Invariant over long periods, stable and predictable features and regularities
Type of learning system favoured	High plasticity Low cognitive constraints	Low plasticity. High cognitive constraints
Costs of this system	Poor discrimination of relevant information. Slow to learn fitness-relevant patterns	Limited ability to adjust to unexpected changes in information patterns. May be inaccurate
Benefits of this system	Good ability to adjust to novel situations and unexpected information patterns	Fast accumulation and processing of fitness-relevant information
Examples	Symbolic culture and creation of language. Flexible rules of grammar allow expression of very diverse meanings and conclusions	Associative learning, for example, taste of food and sickness. Perception of depth, proportion and dimensionality (hence the problem of optical illusions). Folk physics, folk psychology and folk biology. Recognition of co-specifics. Homology across species expected here

Table 9.4 The division of labour by sex from 185 traditional societies in the Standard Cross-Cultural Sample.

The male index is calculated by taking measures of the proportion of males carrying out tasks in each culture and then calculating an average across all cultures. A low index (5–27 per cent) implies activity mostly carried out by females.

Exclusively or predominantly male activities (male index 93–100%)	Mostly male activities (male index 70–92.3%)	Swing activities – in some societies mostly carried out by men, in others mostly by women	Mostly female activities (male index 5.7–27.2%)
Hunting large land and aquatic animals	Butchering	Making fire	Fuel gathering
Smelting of ores and metalworking	Collecting wild honey	Preparing skins	Preparing drinks
Felling trees, working in wood	Clearing land	Gathering small land animals	Gathering and preparing vegetal foods
Making musical instruments	Tending large animals	Planting crops	Spinning
Boatbuilding	Making ropes and nets	Harvesting	Fetching water
Mining, quarrying and stone working		Milking	Cooking
		Manufacture of clothing	
SOURCE: Data from Murdock and Provost (1973).			

sex differences in cognition was the late Canadian psychologist Doreen Kimura (1933–2013). Table 9.5 shows a distillation of some of her findings concerning cognitive differences between the sexes.

So how might we connect the information about current sex differences in cognition with the sex-specific division of labour in traditional societies? The problem is partly that it is all too easy to describe a narrative that fits – in other words to spin a plausible set of Just So Stories. Throwing objects accurately is of obvious use in hunting, something predominantly practised by men. Successful hunting would also draw upon cognitive skills associated today with activities such as map reading, maze learning, wayfinding and the mental transformation of objects (see Box 9.3). In contrast, successful

Table 9.5 Some well-established differences in cognitive and related abilities between men and women

Problem-solving tasks where women excel over men	Problem-solving tasks where men excel over women
Tasks involving perceptual speed where subject must spot differences or match similar items	Representation of objects in space and mental rotation tasks (see Box 9.3)
Object location: recalling position of objects and whether any have moved	Directed motor skills: for example, throwing objects accurately
Ideational fluency: for example, listing objects of same colour	Water line determination (see Box 9.3)
Verbal fluency and verbal memory: for example, listing words beginning with same letter	Mathematical reasoning
Precision manual tasks involving fine motor coordination with hands and fingers	Navigation using sense of direction and distance
Navigation using landmarks	
Mathematical calculation tasks	
SOURCE: Kimura (1999).	

Box 9.3 Some exercises used to test for sex differences in cognition

The mental rotation test

Robert Shephard and Metzler (1971) originally investigated this phenomenon. Subjects have to decide which object on the right (A, B, C or D) can be formed by rotating the standard object on the left.

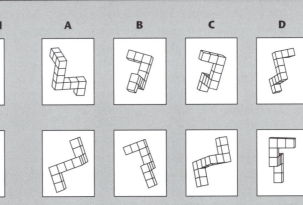

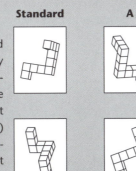

Memory location

Subjects are asked to look at each of the pictures below in turn and identify which objects have been moved.

SOURCE: Reproduced with permission from Silverman, I., Choi, J., & Peters, M. (2007). The hunter-gatherer theory of sex differences in spatial abilities: Data from 40 countries. *Archives of Sexual Behavior 36*(2), 261-268.

Water line

Subjects have to draw a line indicating where the water level would be.

foraging involves recognising and remembering the location of specific plants in a complex array of vegetation. Today such cognitive skills are likely to be associated with recalling the spatial distribution of objects. In today's environment then it is predicted that men should be better at this first set of skills and women better at the second. A number of researchers have investigated this with a series of tests that have become standard (see Box 9.3). But gathering also requires navigational skill and a sense of direction and location. Furthermore, how can we explain the value of higher perceptual speed and acuter discrimination in women compared to men?

Irwin Silverman and Marian Eals (1992) have been prominent in advancing the idea that sex-related cognitive differences may be attributable to differing selective pressure on males and females (as a result of a division of labour) in hominin evolution, and tested males and females on mental rotation, space relations and object location. In later work, Silverman et al. (2000) examined the performance of males and females on the water line test and also an outdoor wayfinding task. For the latter, subjects were taken to a dense wooded area and their performance on finding their route back and understanding of their position at various times were measured. Males were found to significantly outperform females on this wayfinding task. A digest of their results for the paper-based cognitive tests is shown in Table 9.6.

As the data shows, males outperformed females and females outperformed males in skills of proposed relevance to hunting and gathering respectively. This could be evidence for sex differences in domain-specific competencies in male and female brains, but as the authors point out there are other interpretations. Spatial skills, for example, are dependent on the larger domain of allocentric perception competencies – that is, the ability of subjects to detach themselves from ingrained egocentric perceptual modes. There could still be sex differences in this, but the case for the relationship of differing levels of allocentric perception to hunter-gathering skills differentiated by sex is weaker. It could also be that sex differences in early experience with toys involving straight lines and angles can help the development of mental rotation skills. Furthermore, young males tend to have a larger home range right from childhood (Gaulin and Hoffman, 1988) and this early experience could foster better wayfaring skills or allocentric abilities in general.

Despite these qualifications the findings seem to be robust. In a later study Silverman et al. (2007) examined data on male and female abilities in mental rotation and object location tasks using data from 40 countries and 7 ethnic groups. The data set consisted of responses from over 200,000 individuals gathered as part of a larger BBC internet survey on sex differences in behaviour. Broadly in keeping with predictions, they found that men scored better than women on mental rotation tasks in all 40 countries; while women did better than men on object location tasks in 35 out of 40 countries.

Table 9.6 Summary of results from two separate studies on sex differences in cognitive tests.

Silverman and Eals (1992)	Males	Females	Performance	Significance of difference
Object location[1]	83	134	F>M	p<0.01
Silverman et al. (2000)	**Males**	**Females**	**Performance**	**Significance of difference**
Water line determination[2]	9.91	7.68	M>F	p<0.001
Mental rotation[3]	25.07	16.50	M>F	p<0.001
Notes:				
(1) Points awarded for correct identification of objects added to picture in set time				
(2) Numbers show number of correct responses in 3 minutes to 12 similar water line tests				
(3) Numbers show correct responses to 24 tests taken in 10 minutes with one point deducted for each incorrect identification				
SOURCE: Data from Silverman and Eals (1992); and Silverman et al. (2000).				

Gathering skills in a real ecological context

The idea that the superior object location memory shown by women compared to men relates to ancestral divisions of labour (with men hunting and women gathering) is called the gathering hypothesis. It is a plausible conjecture but testing it in a real ecological setting, amongst people who actually do hunt and gather, has proven difficult. Hence it is not known if differences in object location memory also correlates with differences in gathering capability. A step towards addressing this gap in our knowledge was made by Elizabeth Cashdan et al. (2012) who looked at sex differences in spatial cognition among Hadza foragers. On the water level task (see Box 9.3) men were better than women at identifying the correct answer, although more women than men gave answers that were heading in the right direction (Figure 9.6).

Water line picture	No. of males choosing	No. of females choosing	Answer is ...
	9	5	Incorrect
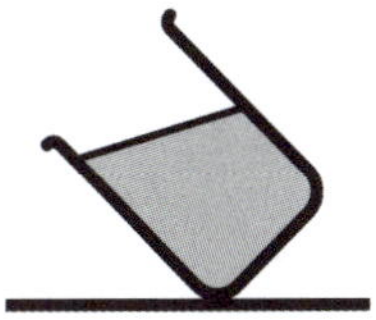	21	38	Incorrect but intermediate
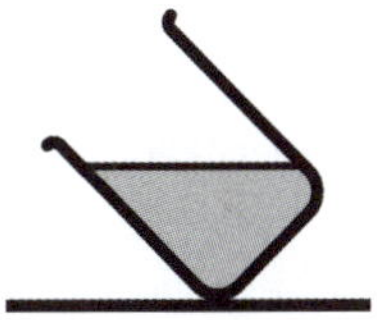	22	8	Correct
Number	52	51	

FIGURE 9.6 Response to the water level task by 51 female and 52 male Hadza foragers.
More males than females gave the correct answer, although more females than males were nearly correct. The study offers some support for ecologically relevant cognitive differences between the sexes.
SOURCE: Data from Cashdan et al. (2012).

Hadza subjects were also tested for object location memory using a series of cards printed on one side with images of plants and animals. Here the results were surprising: men tended to be better than women at object location memory; the skill of women in this task declined with age whereas that of men did not; finally, women who were nominated as good at gathering bush foods by other tribal members were not noticeably better than the group average at the object location task.

The early work of Silverman and Eals (1992) on this topic demonstrated superior object location memory in females. But Krasnow et al. (2011) point out that a static object location test is not necessarily the best proxy for a test of gathering skills. Instead they propose a gathering navigation theory which posits a female advantage in remembering objects in their absolute location in a way that assists navigation and not just an advantage in remembering relative location such as assessed by the original tests of Silverman and Eals. This group used subjects from Japan and the USA and devised more complex methods to assess male and female gathering-relevant memory and navigation skills. In all their studies they once again observed a female advantage in these tasks.

This study reveals that more work needs to be done before we can assume that laboratory-based cognitive competences map neatly onto the cognitive skills needed for a hunter-gathering lifestyle.

Humans rightly pride themselves on their cognitive powers – the ability to make decisions and derive conclusions form informational inputs. Yet this whole process only makes sense in the context of goals and desires (otherwise why bother?), features, that is, of our emotional life. The next chapter concerns the tricky subject of human emotion.

Summary

- Evolutionary epistemology views the mind as pre-structured by natural selection acting on our ancestors. We are born, therefore, with inbuilt biases of perception and a priori ways of thinking.

- Studies on perception, memory, cognition and decision-making have traditionally been the province of cognitive psychology. This was a discipline that grew up in the 1960s and tended to concentrate on proximate causes and hence the neural processes and decision-making algorithms that underlay behaviour and cognition. Evolutionary approaches look to ultimate levels of explanation and the sort of problems that the brain had to solve in the environment of evolutionary adaptation. Errors of reasoning and cognitive biases that may now appear suboptimal or even maladaptive may make sense when viewed as adaptations to an ancestral environment or as the result of trade-off in a mind that has to make compromises to arrive at fast and adaptively efficient (as opposed to logically accurate) solutions.

- The mind is subject to cognitive illusions. A fruitful way of understanding these is to see them as ecologically rational in the environment in which they evolved.

- Error management theory has proven to be a useful tool in explaining why people are subject to biases in reasoning and perception. The direction of a cognitive or perception bias can be predicted by this theory according to the relative costs of type I (false positive) and type II (false negative) errors.

- The anthropological literature shows that current hunter-gathering groups have distinctive patterns in the division of labour. The psychological literature shows cognitive differences in thinking skills between men and women. But testing the hypothesis that ancestral divisions of labour may have led to contemporary sex differences in cognition has proven to be difficult, although there is some supportive evidence.

Key Words

- **Bounded rationality**
- **Cognitive adaptations**
- **Evolutionary epistemology**
- **Heuristics**
- **Modularity**

Further reading

Barkow, J. H., L. Cosmides and J. Tooby (1995). *The Adapted Mind*. Oxford, Oxford University Press.

A highly influential work containing the manifesto of evolutionary psychology by Tooby and Cosmides. Other chapters of interest deal with cognitive adaptations for social exchange, the psychology of sex and language.

Gigerenzer, G. and R. Selten (2001). *Bounded Rationality: The Adaptive Toolbox*. Cambridge, MA, MIT Press.

A series of research papers based on a workshop held in 1999. The concept of the adaptive toolbox is here extended to cover reasoning, emotions and social norms.

Halpern, D. (2000). *Sex Differences in Cognitive Abilities*. Mahwah, NJ, Lawrence Erlbaum Associates.

Highly readable and broad-ranging analysis of the literature.

Hamilton, C. (2008). *Cognition and Sex Differences*. Basingstoke, UK, Palgrave Macmillan.

Tackles sex differences in cognition using evolutionary and sociocultural approaches. A careful examination of the research evidence.

Kimura, D. (1999). *Sex and Cognition*. Cambridge, MA, MIT Press.

Well-illustrated book that carefully documents sex differences in perception and cognition. Accessible and authoritative.

Kurzban, R. (2012). *Why Everyone (Else) is a Hypocrite: Evolution and the Modular Mind*. Princeton University Press.

A readable and interesting application of the idea of modularity to how the brain thinks, why we make mistakes and how hypocrisy often lies at the root of our decision-making.

Emotions

Reason is, and ought to be the slave of the passions, and can never pretend to any other office than to serve and obey them.
(David Hume, *A Treatise of Human Nature*, 1739, p. 460)

One is tempted to define man as a rational animal who always loses his temper when he is called upon to act in accordance with the dictates of reason.
(Oscar Wilde, *The Critic as Artist*, 1891)

The relationship between emotion and reason is one that has long troubled philosophers, psychologists and more recently neuroscientists. One of the most long-standing metaphors in cultural history has been that of master and slave, with the dangerous and brutish impulses of our emotional life guided by the masterly exercise of reason. Until recently, a common stance was to distrust emotions. The behaviourist Skinner in his utopian novel *Walden Two*, for example, wrote: 'We all know that emotions are useless and bad for our peace of mind and our blood pressure' (p. 92).

Sometimes the pendulum has swung the other way, but this dualistic view of emotions and the intellect still survives even when emotions are accorded higher praise. One of the effects of the Romantic movement in art and literature (beginning about 1790) was to suggest that Enlightenment rationality had for too long repressed the emotions and that they should be allowed to flourish, containing as they do sources of deep natural wisdom, authenticity and creativity. Indeed, within the whole Western tradition of thought emotion and reason seem to have been cast as polar opposites compelled to orbit each other like some binary star system, each periodically coming to the fore in cultural approval: the head and the heart, the Apollonian and the Dionysian, Enlightenment rationalism and Romantic emotivism. The quotation from Hume at the start of this chapter is so surprising because it reverses a whole tradition of

thinking on the relative positions of reason and emotion. Yet modern research is beginning to suggest that Hume may have been right; moreover, a sharp dividing line between emotion and reason is not that easy to find.

So what are emotions? This is a vexed and tricky question still unresolved by over 100 years of scientific research. For a start, it is not obvious what the observable phenomena are. Does the psychologist concentrate on inner feelings, measurable physiological states, facial expressions or the whole process of stimulus, context and response in which emotions are somehow buried and live?

A common approach (see Eysenk, 2004) is to suggest that emotions are associated with three types of responses:

1. Behavioural: such as a prototypic form of expression (usually facial).
2. Physiological: a pattern of consistent autonomic changes.
3. Verbal and/or cognitive self-report: a distinct subjective state.

So, for example, in the case of fear, a subject's eyes may widen, their lips part and eyebrows rise (behavioural); the heart beats faster and sweating may follow (physiological); finally, the subject describes their feelings as terrified or afraid (verbal self-reporting).

The problem is that these three components do not always show high concordance; that is, they are only weakly correlated and are not expressed to the same degree in each individual or between individuals. This is partly to be expected since each subsystem serves different functions and its activation and expression may depend upon the fine details of the context. One function of the behavioural system, for example, is probably to communicate the emotion to others; the physiological system serves to prime the body for a specific course of action such as run, fight or remain motionless; finally, the internal self-reporting can motivate the individual to change a course of action, continue it or more generally draw upon the emotional feeling to make appropriate choices.

10.1 Some early theories of emotions: Darwin, James and Lange

Darwin

Many histories of psychological theories of the emotions start with William James's well-known essay 'What is an emotion?' published in 1884. But James's ideas were a development and modification of those of Darwin who explored the same issue in 1872 in his *The Expression of the Emotions in Man and Animals*. Darwin tackled several important questions: why emotions were expressed, why their associated facial expressions took the forms they did (for example, why a smile was a raising of the corners of the mouth and not a drooping) and whether emotions were universally expressed and recognised by all races of people.

For Darwin, facial expressions were universal, innate and inherited from our primate ancestors. In his reasoning Darwin tended to assume that the universality of emotions implied that they were innate. Logically, of course, this is not a conclusive argument. A universal trait could be the result of humans socially learning the same solution to the same type of problem found everywhere.

After some initial interest by Darwin's followers, the long-term influence of *The Expression* was delayed and animal behaviourists and psychologists only really began to take inspiration from Darwin's approach in the 1970s. This is somewhat strange since in its day the book was a bestseller, and by then Darwin was a famous scientist.

The James–Lange theory

Following Darwin, the next major theory was advanced independently by William James (1842–1910) in 1884 and by the Danish scientist Carl Lange in 1885. It became known as the **James–Lange theory** although James was the main proponent. James was particularly interested in emotions that have accompanying bodily reactions. In essence, his theory reverses what was then (and probably still is) the common sense or popular view that we experience an emotion and our body responds accordingly. So, for example, we perceive a threat, experience fear and our heartbeat quickens because we are afraid. James (1884, p. 189) partially inverted this view and suggested:

> [T]he bodily changes follow directly the PERCEPTION of the existing fact, and that our feelings of the same changes as they occur IS the emotion.

A more simple formulation, in keeping with the views of James, is that it is not 'I see a bear, I feel afraid, I run away', but rather 'I see a bear, I run away, I feel afraid'. The James–Lange theory is represented diagrammatically in Figure 10.1.

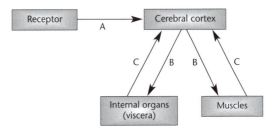

FIGURE 10.1 The James–Lange theory of emotions. The sequence of events according to this theory is A, B, C. In this view an emotion is an afferent feedback from disturbed organs and muscles.

Subsequent research indicated that reality was not so simple or straightforward as James supposed. Walter Cannon, for example, showed in 1927 that people who have suffered a severance between their viscera and the **central nervous system** (as in spinal damage) still experience emotions. Other problems were also found, such as the absence of a specific pattern of autonomic response that corresponds to each emotion and the fact that blocking signals from the autonomic system with drugs does not prevent emotions from being felt.

In the light of this, other theories emerged such as 'arousal-interpretation theory' and 'appraisal theory' both of which stressed the importance of the cognitive processing and appraisal of the information gathered by the receptors. After all we might run from a wild bear in a wood but not from one behind bars in a zoo. Nevertheless James's theory was invaluable in pointing to the possible role of physiology in the experience of emotions. The eminent neuroscientist Antonio Damasio (2000) thinks that James hit upon a mechanism that is still essential to the understanding of emotions but that he did not develop it fully or adequately consider the role of cognitive evaluation.

10.2 Are emotions adaptations?

In the wider psychological literature on emotions this is still a vexed question. From the viewpoint of evolutionary biology it is worth re-examining what constitutes an adaptation and how we can tell if any trait is the result of one. At a very simple level, since adaptations result from a long process of natural selection sculpting phenotypes to survive and reproduce, we could say that a trait that is adaptive looks like it has been designed for a purpose (see Chapter 7). Extending this we can start to draw up a list of distinguishing criteria. Three very basic ones are:

1. The trait or behaviour enhances reproductive success. Whereas there may be some behaviours that do this that are not adaptations (for example, donating sperm to sperm banks) adaptations must now or have once done this otherwise natural selection could not have created them.
2. Adaptations are traits that are heritable. If we are focusing on biological (as opposed to cultural) adaptation then natural selection can only work if the features of a trait are inherited through genetic means.
3. The trait will usually be a modification of an earlier form. Natural selection needs something to work from – it cannot design features *de novo*. So, for example, the legs of tetrapods evolved from fins, human bipedalism from quadrupedalism,

the mammalian eye evolved from light sensitive cells, and the fins of cetaceans from the four limbs of their land-dwelling tetrapod ancestors that returned to the water. In the case of emotions we might expect a complicated emotional system to evolve from simple innate responses such as fear.

At a more detailed level biologists (for example, Williams, 1966; Dawkins, 1986) have suggested a variety of criteria to identify an adaptation. These include:

*   Precision: adaptations should map onto specific causes and consequences. Emotions should be based on specific, physiological structures and mechanisms and initiate specific consequences
*   Reliability and constancy: adaptations should reliably develop in all normally functioning humans and operate in a similar manner across populations. There should be a strong concurrence between emotions and outcomes
*   Complexity: if a feature is complex it is unlikely to have arisen by chance
*   Economy and efficiency: if adaptations have resulted from a long period of natural selection then they should now achieve their function efficiently and with economy of time and resources. If emotions are adaptations, for example, they should achieve their intended outcomes more efficiently and quickly than learned responses
*   Functionality: adaptations should manifestly help further survival and reproductive goals, although genome lag may make this difficult to establish for humans in contemporary environments.

Figure 10.2 is an attempt to synthesise some of these points.

It has proven quite difficult to apply these criteria to the operation of emotions, but in their review of the evidence Lench et al. (2015) conclude that the emotions of sadness, anxiety and anger, at least, do show signs that they fulfil the precision requirement in that they are reliably associated with neuro physiological mechanisms and phenotypical expressions. They also concluded that they showed constancy and reliability in that changes in physiology and behaviour co-occur with these emotions in a

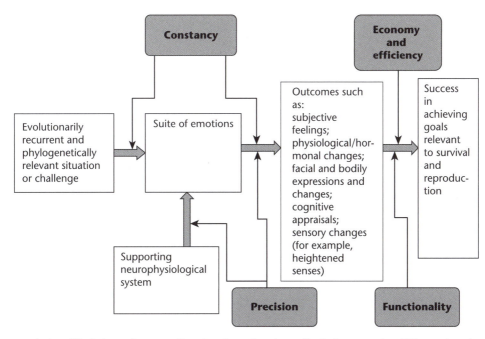

FIGURE 10.2 A simplified view of an emotional system showing criteria for assessing if the system is an adaptation.
If emotions are adaptations then their outcomes should show clear functionality, this should be achieved with economy and efficiency, there should be precision in their construction from neurological systems, and they should fire with constancy in relation to stimuli and challenges and in their effect on subjective feelings and physiological changes.

coherent manner. Economy and efficiency, however, have been harder to demonstrate. One of the most important attributes of an adaptation is that it should show functionality and this crucial question is considered in the next section.

10.3 The functionality of emotions

10.3.1 The case for functionality

The difficulty of research into emotions (and the variety of approaches in psychology) is illustrated by the fact that there is as yet no consensus on exactly how many emotions there are, and which emotions should be regarded as primary and which secondary (that is, made up from primary emotions). The behaviourist Watson thought that a child is born with only three types of basic emotion: fear, rage and love. In his early work, Paul Ekman (1973) thought there were six basic emotions that corresponded with six types of facial expression: happiness, surprise,

anger, sadness, fear and disgust. Johnson-Laird and Oatkey (1992) suggested just five: all those of Ekman minus surprise.

Despite uncertainties about how many emotions we can actually experience, the fact that emotions are likely to serve some adaptive functions should be clear from an examination of some of their characteristics:

● Emotions elicit or are associated with autonomic and endocrine responses.
● Emotions motivate action.
● Emotions are capable of being communicated and so could allow social bonding.
● Pleasant emotions are positively reinforcing.
● Emotions can affect the cognitive evaluation of events and memories.
● Emotions can bias the storage of memories in specific directions.

The functionality of positive emotions is easy to see: it is understandable that natural selection should give us warm feelings for things that are beneficial to our reproductive interests such as good food, a secure environment, sex, supportive friends. Understanding why we behave in ways that communicate (often unintentionally) our emotional state is trickier. Steven Pinker (1997) suggests that there are sound evolutionary reasons for honest displays of an emotional state and that this helps explain why emotional displays are often linked to parts of human physiology that are not normally under conscious control such as blushing, flushing, blanching, sweating and trembling. So, for example, blushing in public out of shame for some misdemeanour becomes a sure sign to others that the individual recognises that he or she has done something wrong and is experiencing the remorseful prick of conscience. This could be an honest signal that the person will be minded to behave more responsibly in the future. Darwin himself seems to have appreciated the importance of emotions as honest signals when he wrote:

> The movements of expression give vividness and energy to our spoken word. They reveal the thoughts and intentions of others more truly than do words, which may be falsified. (1872, p. 359)

10.3.2 Evidence of functionality

So far we have speculated that emotions serve functional needs and are adaptations. There are three types of evidence that could potentially lend more credibility to this claim:

1. **Homology**. The idea that the expression of emotions may be shared with our nearest relatives among the primates.
2. **Universality**. If humans in widely different cultures show the same facial displays of emotion and conceive of emotions in similar ways this lends support for their pan-human biological origins.
3. **Neurophysiological correlates**. If specific parts of the brain can be identified as involved in the processing of emotions then the presence of this hardware, likely to be shaped by natural selection, is strong evidence for their genetic and hence functional basis.

We will now examine each of these areas of evidence in turn.

10.3.3 Homology

Darwin was one of the first to suggest that there were strong similarities between the sounds and expressions associated with emotions between humans and other primates. As he observed:

> We may confidently believe that laughter, as a sign of pleasure or enjoyment, was practised by our progenitors long before they deserved to be called human; for very many kinds of monkeys, when pleased utter a reiterated sound, clearly analogous to our laughter, often accompanied by vibratory movements of their jaws or hips. (Darwin, 1872, p. 356)

Darwin went on to speculate that some emotions such as laughter may be shared with our immediate primate relatives since they were present in our common ancestors but others such as blushing may have been derived in the human line after it split from other primates and therefore are more recent.

Modern research on homologies between the facial expressions of humans and other primates is still at a fairly early stage. As an illustration we might consider smiling and laughing. In humans they are used in similar contexts and it is tempting to regard smiling as a diminutive form of laughing. The nearest correlates to smiling and laughing in chimps are the 'silent bared teeth display' (SBT) and the 'relaxed open mouth display' (ROM) respectively. But Van Hooff (1972) found that SBT and ROM were displayed by chimpanzees in different behavioural contexts, with ROM associated with play and SBT connected with affinity or friendship. Van Hooff also speculated that the primate prototype of the human smile was the prosimian or monkey 'grin' which originally functioned as a fear expression. With the evolution of the apes the expression developed into three forms: SBT type 1, signalling submission; SBT type 2, signalling fear and affection; and SBT type 3, signalling affection. In humans, he argued, the first function of fear or submission (SBT1) had been lost and the fear function of the second (SBT2) dropped, leaving us with two expressions – smiling and laughing – both signalling affection and play (see Figure 10.3). Preuschoft (1992) built upon this work and by looking at the behaviour before and after the display of these expressions concluded that in Barbary macaques (*Macaca sylvanus*) the SBT display was associated with submission by the sender. Consistent with this, Waller and Dunbar (2005) found that SBT and ROM in chimpanzees were found in dissimilar contexts. In short they found that rates of affinitive behaviour increased following SBT, suggesting that SBT is a signal of affinity, whereas ROM was observed primarily during play, and play between two individuals was increased in length when ROM was bidirectional. Rates of affinitive behaviour also increased after ROM and this might suggest that both displays assist social bonding and could explain why the two displays have converged in humans.

Looking at Figure 10.3, we might think that the chimp faces do not closely resemble human smiles or laughter. We need to realise however that the facial muscles of chimps are likely to be stronger than those of humans since for the last 2 million years hominins have used tools and later fire to help them extract and process food.

Preuschoft (2000) suggests that the apparent contrast between the smile of humans (where the corners of the mouth rise upwards) and the SBT of non-human primates such as chimps may simply be the result of homologous muscles stretched over a differently shaped muzzle (Preuschoft, 2000). Further insights into the ape–human phylogeny of facial expressions are likely to come from a precise analysis of the muscle groups involved (see Chevalier-Skolnikoff, 1973; Schmidt and Cohn, 2001; and Vick et al., 2006).

10.3.4 Universality

The crucial question in this area is whether our experience of emotions and their expression by gestures and facial movements are culturally determined (the social constructionist view) or universal. Put simply, is the raising of the mouth in the corners (a smile) a sign of happiness or friendliness in all cultures or are there some culture where a smile might mean 'I am disgusted' or 'I am angry' or 'I am afraid'? As we noted, Darwin, for a variety of reasons, was keen to insist upon the universality of facial expressions. In *The Expression of the Emotions* he made an important contribution by taking photographs of ordinary people and actors staging emotional gestures to show that they are readily recognised by others. He then used his extensive network of correspondents overseas to show that similar emotions were displayed by people untouched by Western civilization.

In recent times the true heir to this tradition has been Paul Ekman. Ekman realised that testing for the universality and instinctive nature of emotional expression and experience poses numerous experimental problems. One is to disentangle the effects on people of mass culture, with its supply of ready-made images of emotions and their expression, from what might naturally develop. Another is the problem of translating terms for emotions between languages. To overcome this, in the 1960s Ekman went about interviewing and testing people in a wide variety of cultures. One of his studies concerned the Fore people of the New Guinean highlands. At the time these

Description of face and context	SBT1 'Silent grin' Submission 	SBT2 'Silent grin' Appeasement	SBT3 'Silent grin' Affection	ROM Play face Play
Human homologue and context	?	Smile Happiness, joy		Laugh Playfulness, happiness

SBT = silent bared teeth   ROM = relaxed open mouth

FIGURE 10.3 Chimpanzee–human homologues in facial expressions.
SBT2 and SBT3 are likely to have evolved into the human smile. The ROM may be a homologue for the human laugh. It is not clear if SBT1 simply faded out in the human lineage or has transformed into another expression.

SOURCE: Diagrams based on a series of photographs and descriptions in Van Hooff (1971). By permission of Graphic Design Dept., University of Chester.

people were still living a Stone Age lifestyle without a written language and relatively impervious to the global reach of Western culture. Many of them had never even seen their own faces in a mirror. In an ingenious experimental design, subjects attached an emotional label to a picture by identifying the face of a character in a story such as that of a man unexpectedly attacked by a wild pig. In this example, his subjects identified the character as one that we would say was displaying fear. Ekman then asked these people to act out the emotions of such characters as he filmed them. When the film was shown back in the USA volunteers easily identified the intended emotion. Later, such studies were extended by Ekman and others to include a total of 21 countries. The results were highly significant: despite the numerous differences between such cultures in terms of language, economic development and religious values, there was overwhelming agreement about faces that showed happiness, sadness and disgust (Ekman, 1973). Eventually, as we have seen, Ekman suggested that there were six basic emotions found in all cultures: fear, anger, disgust, sadness, joy and surprise.

From such studies it increasingly begins to look as if our emotional life (at least in the way it is expressed facially) is part of a standard development programme. Since this programme develops consistently in all cultures, it would seem reasonable to suppose that it has been shaped by natural selection. Further evidence comes from the work of Tracy and Matsumoto (2008) who found that judo athletes who were blind from birth showed facial expressions of pride when they won a match and shame when they lost. The expressions were rated from photographs presented to research assistants who were ignorant of the study's goals. Since the athletes could not have learnt these expressions through visual modelling on others it suggests an innate biological mechanism.

However, in arguing for the pan-cultural nature of emotions we need to distinguish between the expression of the emotion (that is, the response output) and the stimulus that elicited it (the input). Work by Ekman and others has indeed shown that there may be a set of universal cross-cultural emotions (such as joy, anger, fear, sadness, surprise and disgust) but the circumstances that are associated with each expression may depend upon the individual's life history and hence the experience of the cultural value attached to different situations and emotions. In other words the outputs may be pan-cultural but the input system may be flexibly acquired. The work of the behaviourists (although misguided in so many respects) did seem to demonstrate that it is highly unlikely that we have an innate 'fear a large-fanged animal' or 'fear a snake' reaction, which then prompts the display of fear. It looks more likely that we have innate emotional responses called fear or sadness but the cues for these are socially learned. Babies are not born with a 'fear of a snake' response but they quickly learn it and, more importantly, learn it from watching the reactions of adults (Klinnert, Campos et al., 1982). This system looks like a good compromise between the need to adapt to a changing world (where threats vary from generation to generation) and the need to avoid the time-wasting process of learning afresh through trial and error. Having an innate fear response that alerts the system to a state of readiness coupled with its linkage to an input that is learnt through experience and watching what adults fear is perhaps the best of both worlds.

This cultural dimension was also captured by Ekman's work. We know from common experience that it is possible to have an emotion and either suppress its facial manifestation or allow its full expression. Ekman suggests that there are 'display rules' that may govern this decision process and that such rules may be socially conditioned. In one experiment he filmed American and Japanese students as they were shown both neutral and stress-inducing films. When the experimenter was absent the facial responses were broadly similar; but when the experimenter entered the room and asked the subjects about their emotions as the film was played again the Japanese subjects suppressed their facial expressions. Ekman interpreted this as the operation of the Japanese cultural display rule of not showing negative emotions in the presence of an authority figure. However, Ekman did observe 'leakage'; in other words the emotion was never completely suppressed, suggesting that there are two types of circuit. By 1973, then, Ekman had concluded:

> Our neuro-cultural theory postulates a facial affect program, located within the nervous system of all human beings, linking particular facial muscular movements with particular emotions. It offers alternative

nonexclusive explanations of the movement of the facial muscles. Our theory holds that the elicitors, the particular events which activate the affect program, are in largest part socially learned and culturally variable … but that the facial muscular movement which will occur for a particular emotion (if not interfered with by display rules) is dictated by this affect program and is universal. (Ekman, 1973, p. 220)

Facial expression programme (FEP) or behavioural ecology theory (BET)?

Ekman's idea that certain facial expressions are driven by specific emotions which are universal is called the facial expression programme (FEP). If the FEP is correct then basic emotions and facial expressions should reliably co-occur. The classic example of this is the Duchenne smile named after Duchenne de Boulogne, a French neurologist whose photographs of induced facial expressions featured in Darwin's *Expression of the Emotions* (1872). This smile involves two groups of muscles which raise the corners of the mouth and the cheeks and is an expression long thought to be associated with happiness. Paul Ekman, for example, used this type of smile as an example of a natural and involuntary smile (compared to a forced and controlled smile) since the set of muscles controlling the cheeks and corners of eyes (orbicularis oculi) are difficult to control.

But since the 1990s the FEP has met with increasing scepticism. Surprisingly, perhaps, the evidence linking emotions and facial expressions is mixed. In their review of naturalistic (that is, real, non-laboratory settings) studies on emotions and faces Fernández Dols and Crivelli (2013) noted that numerous studies have discovered only a weak correlation between, for example, the Duchenne smile and happiness. In a more recent study, Crivelli et al. (2015) showed that smiling was better predicted by the interaction with others than an underlying emotional state. Judo fighters who had just won a medal (and were obviously happy) tended to smile far more when interacting with others and being seen than when alone and unobserved. A proponent of the FEP might of course argue that this shows the modifying effect of display rules. But a similar picture emerges from looking at laboratory-based studies where an emotion is stimulated and then correlations sought

with facial expressions. In their review of the topic, Reisenzein et al. (2013) found that only amusement showed robust correlations with the appropriate facial expression. It is possible, however, that a laboratory situation does not elicit an emotion of sufficient intensity or authenticity to provoke the corresponding display; or that, once again, inhibitions arise because of the context and display rules.

The alternative theory to explain facial expressions is called the behavioural ecology theory (BET). This suggests that facial expressions are used as signalling tools. It may not be to the advantage of the signaller, for example, to accurately advertise an emotional state to another as would be the case if facial expressions were involuntarily associated with an interior emotional state. In this model, a smile could have a variety of functions such as the marking of social status, sexual invitation or the willingness to cooperate and share resources. This view draws upon Fridlund's influential *Human Facial Expression: An Evolutionary View* (1994) where he argues that it is wrong to think of facial expressions as simply the signs of emotion leaking out into the interpersonal world. Faces serve to communicate behavioural intentions and their expression will be influenced by their audience. In Fridlund's view, there is no one-to-one correspondence between emotion and facial expression because the same emotion can be accompanied by different social intentions. Anger, for example, could be associated with or followed by aggressive action, withdrawal or cold planning to take revenge. The so-called angry face might simply be one expression of one motive – in this case the intention to be aggressive. In general, Fridlund's theory is superior to that of Ekman in explaining why facial expressions vary according to the social context, although it has yet to yield convincing evidence as to the direction, duration and intensity of modifying effects and where they might be expected.

One problem with the BET approach, however, is that people make emotional faces even when alone. In fact it is possible that people cry more frequently, for example, when alone than when in company. We also are capable of smiling and expressing sorrow on our faces even when no one is looking. So if facial expression is about communication this poses a problem. In general there is not yet a single satisfactory theory or model that accounts for the full complexity of evidence linking facial expressions, social motives and emotions (for a review see Parkinson, 2005).

10.3.5 Neurophysiological correlates

Until recently, research into physiological activity associated with specific emotional states tended to focus on the behaviour of the **autonomic nervous system** (or ANS: the sensory and motor **neurons** that control the operation of the viscera such as organs involved in cardiovascular, digestive and respiratory functions) and the endrocrine system (glands that secrete hormones into the bloodstream). This was partly because such studies were easier to perform than trying to probe the brain directly, and partly because it was obvious that changes in such things as heart rate and hormone levels were involved in some way with emotional responses. The problem that bedevilled this research was that these very systems were also involved in the routine housekeeping functions of the body (such as energy metabolism, tissue repair and homeostasis generally) and so did not offer ideal pointers to the working of the emotional system.

One more direct method has been to examine patients who have suffered damage to specific areas of the brain and try to correlate changes in emotional and cognitive functioning with the areas damaged. One famous often-quoted case of this is the experience of a US railway foreman called Phineas Gage. In 1848 Gage placed blasting powder into a hole drilled into a boulder; as he tamped down the explosive charge with an iron bar it exploded sending the bar through his left cheek and up through the front of the brain. The bar landed about 100 metres away. Amazingly, Gage survived but was a changed man. Whereas previously he was a thoughtful and industrious person, he now became childish and thoughtless in his behaviour. Later analysis of the damage caused showed that the iron bar obliterated much of the **orbitofrontal cortex** of the man's brain. This tragic incident was one of the first of many clues that leads neuroanatomists to think that this part of the brain is responsible for integrating signals from the emotional centres below (especially the **amygdala**) and sensory information presented to it by the sensory systems and other parts of the cortex.

More recent research has been able to look at the brain under more controlled conditions. There have been two main approaches:

1. Neuro-imaging techniques. Developments in technology have given neuroscientists a whole array of techniques to study brain activity. In general these techniques allow researchers to identify which parts of the brain are active under specific circumstances and so, in principle, enable the determination of when and where cognitive processes are occurring. The two primary techniques here are:
   a) positron emission tomography (PET), which involves the detection of positrons given off by some atomic isotopes and enables a spatial resolution of brain activity down to 3–4 mm
   b) magnetic resonance imaging (MRI and functional MRI), which relies upon the excitement of atoms in the brain by radio waves and the detection of any magnetic changes produced using a large magnet that surrounds the head of the patient under test.
2. Use of drugs. It is well known that drugs produce changes in behaviour and feeling. The problem with research in this area is that drugs may affect several areas of the brain and it is not always clear, therefore, which affected area is responsible for or associated with a behavioural or mood change.

Using these techniques has vastly improved our knowledge of how emotional states are related to the activation of different areas of the brain. The next section considers some elementary findings about emotions and brain structure.

10.4 Emotions and some specific functions

The preceding material has provided evidence that emotions are strong candidates for adaptations serving fitness-enhancing functions: they are universal, are shared with other primates and are associated with specialised hardware centres in the brain. In this section we examine the way functionality may be achieved in specific cases.

10.4.1 Emotion, commitment and decision-making

We saw in the last chapter that unbounded rationality is unavailable as a decision-making tool and that instead humans often use fast and frugal heuristics to solve everyday problems. With this in mind it is worth exploring the notion that emotions aid decision-making. The economist Robert Frank (1988) is a strong advocate of this position. Frank illustrates his approach with the examples of love and guilt. The function of love, he conjectures, is to reinforce the pair bond and so provide a stable unit for cooperative child-rearing. In this view love serves as a commitment device to enable both sexes to reap the long-term rewards of a stable relationship and resist or ignore the temptations from other members of the opposite sex. If this were the case then, as Gonzaga and colleagues predicted, feelings of love towards a partner should tend to suppress sexual desire for other attractive potential mates. On the other hand, feelings of sexual desire only towards a partner should be less effective than feelings of love at suppressing desire for others. Gonzaga et al. (2008) placed subjects in these roles by asking them, for example, to write about a partner whilst trying to suppress thoughts of another. Their predictions were confirmed: feelings of love suppressed thoughts of other attractive potential mates better than feelings of sexual desire.

Findings in support of the motivational function of guilt come from studies on ultimatum games. In these games volunteers are assigned the role of proposer or respondent. The precise details and rules can be varied, but in typical experiments the proposers are given a sum of money (for example, $10) and told to decide upon a split between themselves and the respondent. The split is the decision of the proposer but if the respondent refuses the allocation then neither party receives any money. Typically, grossly unfair splits (for example, $9 for proposer and $1 for respondent) are refused, even though economic rationality would suggest that the respondent is better off with one dollar than none. A study by Ketelaar and Au (2003) found that those proposers who experienced guilt after an unfair divide tended to reverse their behaviour and become more generous a week later. In contrast, few of those who benefited from an unfair divide but experienced no guilt reversed their behaviour. This tends to suggest that guilt can impact upon behaviour, possibly to take more account of longer-term benefits from cooperation and mutualism.

10.4.2 Emotions as superordinate cognitive programs

Tooby and Cosmides (2005) (true to their fondness for computer metaphors) view emotions as types of superordinate programs that exist to orchestrate other subordinate programs to achieve the best response to any given situation. Such programs are required, they suggest, because the mind has a whole cluster of domain-specific programs that need activating in a correct sequence otherwise conflicts might occur. This view may explain why emotions have proved so difficult to isolate and study selectively, since as Tooby and Cosmides (2005, p. 53) explain:

> An emotion is not reducible to any one category of effects, such as effects on physiology, behavioural inclinations, cognitive appraisals, or feeling states, because it involves evolved instructions for all of them together.

To illustrate their reasoning, Tooby and Cosmides consider fear. The 'front end' of this superordinate program is to detect a fearful situation. Consider, for example, walking alone at night and suppose elementary perceptual cues indicate noises consistent with being followed or stalked. The emotion program called fear is activated, which directs a whole cascade of changes such as:

- Heightened perception and attention to other noises and sights. Threshold shifts such that less evidence is now needed to interpret other cues as threats.
- Goals and motivation change: safety now given a higher priority over other goals such as hunger, thirst, finding mates.
- Communication processes initiated. Depending on the context this could be a cry or facial expression of fear.
- Physiological changes occur: blood leaves the digestive tract, adrenalin released, heart rate changes (up if flight reaction, lower if freeze reaction).
- Behavioural decision rules enacted: hide, flee, self-defence, remain still.

In this whole framework there are other emotion programs whose function is not to direct short-term behavioural responses as in the case of fear, but rather to periodically cause revaluation and recalibration of one's own estimation in relation to others. Guilt is probably an example of the latter, which serves to recalibrate the inclination to distribute resources between self and others, such as when favours should be returned. Depression is possibly another example of a condition designed to force a rethink on strategies that are not working and trigger the depressive state as a means of curtailing their pursuit (see Chapter 19). Table 10.1 shows how some specific emotions may be linked with evolutionary (fitness-related) functions.

Table 10.1 Selected research on some postulated functions of emotions

Emotion	Possible function	References
Fear	Primes the body to take action. In panic and agoraphobia, for example, blood supply to the muscles is increased and the mind becomes focused on finding escape routes. Fears that seem unreasonable and disproportionate are called phobias. Phobias may represent ancestral hard-wired developmental tendencies to fear fitness-relevant objects and organisms.	Marks and Nesse (1994) Nesse (2005b) Ohman and Mineka (2001) Seligman (1971)
Anger	May provoke extreme acts of destruction. Even spiteful acts may signal to another the costs of provoking anger in this and future situations. Anger may therefore be costly to the actor in the short run but in the longer run it may be adaptive in modifying future behaviour of others – a 'zero tolerance' approach. Anger may serve to punish cheaters in prisoners' dilemma-type situations.	Fehr and Gaechter (2002)
Sadness and depression	Weakens our motivation to continue with present course of action. May be a means of telling the organism to stop current strategies and conserve resources since they are unrewarding. Could also be a signal that help is required	Hagen (1999) Watson and Andrews (2002) Price et al. (1997)
Jealousy	Forces us to be alert to signs of deception by partners. May activate aggressive behaviours to force the defecting partner back to original relationship or deter partner and new mate from continuing in the new relationship.	Buss et al. (1992) Harris (2003) Haselton et al. (2005) Daly, Wilson and Weghorst (1982)
Love	Love for family members obviously can increase inclusive fitness. Passionate love can further reproductive goals by cementing the pair bond long enough to help raise children.	Frank (1988) Ketelaar and Goodie (1998) Nesse (2001)
Disgust	Many believe that disgust is a universal emotion but the targets of disgust are socially determined. In a nutshell disgust is often the fear of ingesting an undesirable substance. Children acquire a sense of what is disgusting as they grow. In this sense they are using their parents and adults as guides to what is edible and what is not. Disgust is often elicited even when something unpleasant has merely come into contact with something edible. For example, many people will refuse to drink orange juice from a sterilised bed pan or eat chocolate in the shape of dog faeces. As Pinker observes 'Disgust is intuitive microbiology'. Sexual disgust (the thought of mating with a close relative) may serve to inhibit inbreeding (see Chapter 21).	Pinker (1997) Fessler and Navarrete (2004); Lieberman, Tooby and Cosmides (2003)
Guilt	A remorseful feeling that follows self-awareness of having been unfair. May serve to drive more cooperative or generous behaviour in future encounters.	Ketelaar and Goodie (1998); Ketelaar and Au (2003)

10.4.3 **Resolving the paradox of emotions**

As Haselton and Ketelaar (2006) note, emotions do seem to pose a paradox. Decades of research confirm that they are probably universal features of human nature; they are found in our primate relatives; and they are correlated with distinct areas of brain activity. If, as seems likely, they are adaptations it would be odd if they evolved to disrupt decision-making and cloud our judgements on important matters. Yet this is often the experience of all of us when we struggle to remain objective, dispassionate and wrestle with emotions we perceive to be irrational, disturbing and unruly. This is the paradox of emotions often explored in the arts such as the dynamic between the hybrid human–Vulcan Star Trek character Dr Spock and his commander Captain Kirk: clear-sighted logic versus a messy mixture of logic and passion.

The way round this paradox is perhaps twofold. Firstly, we should realise that our emotional system was designed for conditions very different to modern environments. Consequently the system may appear suboptimal or irrational in the context of modern lifestyles. This is a constant refrain within evolutionary psychology and we will return to it in Chapter 18. Secondly, evolution operates to maximise our reproductive fitness and not our experience of well-being (although the latter can sometimes be co-opted to serve the former). It may be then that saddling organisms with unpleasant emotions such as an easily activated rapid fear response, a constant level of anxiety, or periodic bouts of jealousy, makes adaptive sense. Moreover, as conscious complex organisms we have our subjective personal goals (contentment, security, peace of mind and so on) that are different to the dictates of our emotions. Possibly then, the heart of this paradox lies in the tripartite tension between the pull of ancestral emotions, personal goals and modern environments.

Such reasoning may also help to explain why happiness is so hard to obtain, for it would be unlikely that natural selection would have designed a state called happiness that could be experienced for any great length of time. Temporary human happiness (contentment, satisfaction) is often linked with the achievement of goals that further reproductive fitness. For our hunter-gatherer ancestors this would be reliable sources of food, shelter from the elements, freedom from the risk of predators, a supportive social group, high status in the group, mating opportunities, healthy and loving children and so on. But consider the fate of someone who had all these but was still driven by the urge for more, who experienced some nagging discontentment that only more wives, or a better husband or a safer environment would satisfy. In some situations this might pay off and someone driven like this might leave more offspring than someone content with their lot. This could explain why happiness is a fleeting condition, why it correlates with wealth but only slightly, why, like the rainbow, it recedes as fast as we chase it. Thomas Jefferson was perhaps right in more ways than one when in drafting the American Declaration of Independence he referred to human rights as the 'preservation of life, and liberty, and the pursuit of happiness' – happiness being something we pursue rather than a stationary state.

One emotion found in virtually all humans is empathy and love towards our kith and kin. In biological language, humans display kin-directed altruism, and this is the subject of the next chapter.

Summary

- ▶ Darwin was one of the first to explore the idea that humans share emotional reactions with other animals. He suggested that human emotions were experienced universally.

- ▶ William James and Carl Lange argued that emotions may also be the experiences of physiological reactions that take place on our body that occur prior to any conscious evaluation of the emotion-inducing event.

- ▶ There is strong evidence that emotions are natural adaptations that served and serve specific fitness-related functions. Evidence comes from their homology with other animals, their universal expression in humans, and their localisation in specific bits of brain hardware.

▶ Emotions can influence reasoning and decision-making. Some of the simple heuristic rules explored in Chapter 9 may be affected by the emotional system.

▶ At first glance emotions seem to pose a paradox: why do we experience feelings that were presumably laid down by natural selection for our own benefit as disturbing and disruptive? The answer may lie in the fact that natural selection provided fitness-enhancing structures for a lifestyle different from today, and that enhancing fitness does not mean humans will remain in a state of happiness.

Key Words

- Amygdala
- Autonomic nervous system
- Central nervous system
- Homology
- James–Lange theory
- Neuron
- Orbitofrontal cortex

Further reading

Oatley, K. (2004). *Emotions. A Brief History*. Oxford, Blackwell.
Good survey of the history of emotions in psychology.

Damasio, A. (2000). *The Feeling of What Happens*. London, Vintage.
Lucid account of the work of Damasio and others on how emotions interact with physiology.

Nesse, R. M. and P. C. Ellsworth (2009). 'Evolution, emotions, and emotional disorders.' *American Psychologist* **64**(2), 129.
Useful introductory article on the evolutionary basis of emotions.

PART VI

Cooperation and Conflict

Altruism and Cooperation

He Ain't Heavy. He's My Brother.
Title of a column written by Roe Fulkerson for *Kiwanis* magazine (1924) and later a best-selling musical hit for The Hollies (1969).

In Chapter 3 we saw how William Hamilton and others expanded Darwin's original ideas on fitness into the concept of inclusive fitness – the idea that reproductive success can be measured by the propagation of genes into direct and indirect offspring. We also saw that this concept partly solved the riddle of altruism: 'selfishness' at the level of genes can give rise to altruistic acts by their vehicles if those acts benefit the genes themselves. Put very simply, **Hamilton's rule** suggests that the more closely we are related to another individual (the higher the r value) the more likely we are to be kind towards them. The first part of this chapter looks at the application of this idea to make sense of kin-directed human behaviour in traditional and modern cultures. It is obvious, however, that humans behave altruistically even to non-kin. Such behaviour calls for a separate set of concepts and other modes of explanation. The understanding of human non-kin-directed altruism has been greatly clarified through the use of concepts such as mutualism, **reciprocity** and game theory. Consequently the second half of this chapter is devoted to exploring these notions.

11.1 Conceptualising cooperation

So many words in science have a precise meaning different from the looser meaning of their everyday usage. In the main this does not matter too much, but in the case of the concept of altruism it has led to many unfortunate misunderstandings. The problem arises because any act that seems to imply that animals are not behaving selfishly has been called altruistic. In most cases, however, the acts are not altruistic at all because either the payoffs are not so obvious or are delayed. Similarly, the phase 'selfish gene' coined by Dawkins has led to the criticism that biologists are simply being anthropomorphic. To rescue these useful terms some clarification is sorely needed. Figure 11.1 shows a matrix illustrating the essential difference between altruism, mutualism, selfishness and spite. We can understand these terms as follows:

Selfishness, or parasitism

Parasites are the archetypal selfish organisms in that they benefit from a relationship with their host at the expense of the host itself. Often we find the genes of one organism manipulating the behaviour of the other. The cold virus, for example, not only invades your system and subverts its functions to produce copies of itself, but it also manipulates you into helping it spread further by persuading your lungs to expel an aerosol of droplets containing the virus at a great velocity. Dawkins referred to such effects as 'the extended phenotype'. In nest parasitism, the cuckoo manipulates the builder of the nest into giving aid to a completely different species. In these cases the 'altruism' of the donor is extracted by manipulation, and the donor gains nothing. We could view this as cuckoo genes reaching out beyond the cuckoo vehicle into the behaviour of the nest owner. Dawkins summed this up in his 'Central Theorem of the Extended Phenotype':

> An animal's behaviour tends to maximise the survival of the genes 'for' that behaviour whether or not those genes happen to be in the body of the particular animal performing it. (Dawkins, 1982, p. 248)

Mutualism

Mutualism is becoming the preferred term to symbiosis. We can distinguish two types of mutualistic behaviour: inter-specific (between two or more species) and intra-specific (within a single species). Some species form inter-specific mutualistic partnerships because individuals of each have specialised skills that can be used by the other. Aphids have highly specialised mouths for sucking sap from plants. In some species this is so effective that droplets of nutrient-rich liquid pass out of the rear end of the aphids undigested. Some species of ant take advantage of this by 'milking' the aphids in the same way as a farmer keeps a herd of cattle. The aphids are protected from their natural enemies by the ants, which look after their eggs, feed the young aphids and then carry them to the grazing area. The ants 'milk' the aphids by stroking their rears to stimulate the flow of sugar-rich fluid. Both sides gain: the ants could not extract sap so quickly without the aphids, the aphids are cosseted and protected from their natural predators by the ants.

In the case of intra-specific mutualism two or more individuals of the same species cooperate and each gains a net benefit. If two lionesses cooperate their chance of capturing a prey is probably more than twice that of each individual. Sharing half the meat then becomes a net benefit to each. In fact lionesses in a pride are related, as are the male lions that cooperate in taking over a pride, so cooperation is favoured by kin selection and mutualism.

Altruism

Kin-directed altruism was examined in Chapter 3 and further examples of altruism and reciprocal altruism are the main subjects of this chapter. But the problem with the definition suggested in Figure 11.1 is that, as Tooby and Cosmides (1996) pointed out, it does not give enough weight to the function of the behaviour. For example, an insect that is lured into the deadly embrace of an insectivorous plant, such as the Venus fly trap (*Dionaea muscipula*), which is then caught and digested by the plant, is certainly giving a benefit to the plant at a cost to itself. Yet this is not altruism in the sense that the insect possesses genes 'designed' to deliver benefits to the plant. A more precise definition therefore might be: *behaviour that has been naturally selected to benefit the fitness of the recipient organism at some cost to the donor*. This then focuses attention on how the behaviour could be naturally selected.

Spite

Spite has proved an elusive phenomenon to observe in the natural world. This is hardly surprising since it is difficult to see how genes for this behaviour could spread if they suffer a net cost, a cost moreover not outweighed by any inclusive fitness gains since, by definition, the recipient also suffers a cost. Indeed, if we consider Hamilton's equation ($rb > c$) then this can be simply phrased as an activity will flourish if the net benefits (rb, where here r can be regarded as the modifier of the value of the benefit) exceed the cost (c). If we think of benefit as gene frequency increase over a given period of time and cost as gene frequency decrease, then it is obvious that only when benefits exceed cost can genes for this behaviour flourish. Now in the case of spite we have a negative benefit and a positive cost – the initiator presumably expends some energy to inflict damage on another – and so it would seem impossible for spite to ever emerge. This explains why spite had been almost impossible to authenticate in animal societies. Indeed, what often seems to be spite often turns out to be good old selfishness. Male bower birds, for example, will sometimes go out of their way to wreck the carefully constructed bowers of other males. This may look like spite, but, as Hamilton argued, destroying a rival's bower reduces the reproductive success of competitors and leaves the field more open to the destructive male (Hamilton, 1970). But there are two possibilities proposed by Wilson and Hamilton. Wilson suggested that spite could emerge if a third party related to the initiator benefited from the costs suffered by the other two parties (Wilson, 1975). But strictly this could be seen as a form of indirect altruism. In contrast, Hamilton proposed the intriguing idea that spite could spread in the gene pool if r was negative. In the equation $rb > c$, if both b and r are negative

Recipient

		Gains	Loses
Initiator	**Gains**	Mutualism or reciprocity	Selfishness, for example parasitism
	Loses	Altruism	Spite

FIGURE 11.1 Matrix of relations defining mutualism, altruism, selfishness and spite.

then the condition for the spread of genes can once again be met (a negative × a negative gives a positive). The value of r can be negative if the recipient is less likely to share any gene of the initiator than the average for the general population (Hamilton, 1970). A spiteful gene can then flourish since it reduces the frequency of competing alleles in the gene pool. This system, however, requires some unlikely conditions: negative relatedness and very accurate kin discrimination. Such conditions are unlikely to be found in mammalian societies, but there is some evidence that they may be met in a few insect societies where Hamiltonian spite remains a distinct, albeit much debated, possibility (see Foster et al., 2001).

The presence of spite in humans probably requires a higher-level explanation. One possible explanation

is that spiteful behaviour brings maladaptive consequences for humans because it represents a miscalculation of the effects of certain courses of action. The threat of spite could bring rewards: 'if you don't do what I want we will both suffer' may sometimes work, but if the person threatened calls the bluff of the aggressor then it becomes maladaptive.

If we take a broader perspective we need to explain why all the behaviours in Figure 11.1 are found in nature. If they are products of natural selection then we also need to understand the roots of the behaviours in terms of the benefits obtained from enacting them. Adopting this approach we can see there are two classes of benefits: those that accrue directly to the player and those that are indirect in that they benefit inclusive fitness. Figure 11.2 shows

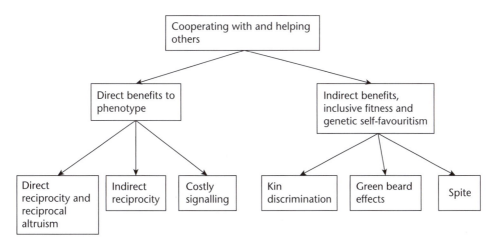

FIGURE 11.2 A schema for understanding cooperation and altruism.
Spite is included (counterintuitively) as it may serve inclusive fitness benefits if it is directed towards individuals with a lower than average population r value to the donor (thereby benefiting others to whom the donor is related).

a classification of cooperation using this perspective and the rest of this chapter explores the various components of altruism suggested in this figure.

11.2 Cooperation promoting inclusive fitness

11.2.1 Kin and parental certainty

Hamilton showed that kin-directed altruism is favoured between individuals who are closely related (that is, high r values). This raises the question of how humans can tell if another is related to them in a relationship such as brother, sister or offspring. The solution to sibling identification was probably twofold. One involved a simple rule of the form 'if you have grown up with someone from childhood in the same family they are probably a brother or a sister' (as we shall see in Chapter 17 one consequence of the application of this rule is probably to instigate a negative desire towards mating – that is, co-socialised children grow up not to find each other sexually attractive). Another solution may have used the sense of smell: there is increasing evidence that humans are able to react towards others using odour as a cue for genetic similarity and difference (see Chapter 17). In terms of offspring, however, there lies at the heart of family relationships an asymmetry in parental certainty, sometimes captured by the phrase 'mummy's babies, daddy's maybes'. Mothers can be sure of their parentage whilst fathers have to make do with the assumption that they are truly the father. Actual rates of non-paternity vary enormously between cultures and social groups. A thorough review on this topic was carried out by Anderson (2006) who estimated the rate of non-paternity amongst men who thought they were fathers to be about 1.9 per cent.

Given the asymmetries in parental certainty then it might be expected that men's investment in their offspring be influenced by any available cues of mate fidelity and the resemblance of the child to the father. A number of studies support these expectations (see Platek et al., 2003). A recent one was carried out by Apicella and Marlowe who conducted a questionnaire-based study of males with children. They asked the men questions about a) their investment in their children, b) the resemblance of children to

themselves, and c) the perceived faithfulness of their wife or partner. Men reported greater investment in their children if they perceived them having a greater resemblance to themselves. Men also claimed to place greater investment in their children if they perceived the children's mother as being more faithful (Apicella and Marlowe, 2004). The concept of parental certainty would also suggest that males will attend more closely to cues of resemblance than females (who are assured of their parental status). To test if males and females are affected differently by a physical resemblance to a child, Platek et al. (2004) took photographs of volunteer male and female subjects and blended their images with those of child faces creating 'self-morphs'. By this means the researchers were able to produce children's faces that resembled the subject. They then presented these faces, amongst other non-resembling faces, to the subjects. Subjects were asked a series of questions eliciting positive and negative reactions such as 'which child would you adopt?'; 'which child would you most resent paying child support to?' They found as predicted that males were more likely than females to select self-morphs in relation to positive questions. There was, however, no difference between males and females in response to negative questions.

11.2.2 Darwinian grandparenting

> Os filhos das minhas filhas, meus netos são.
> Os filhos dos meus filhos, podem ser ou não.

> My daughter's children, my grandchildren are
> My son's children, maybe or maybe not. (Old Portuguese proverb)

This old Portuguese proverb (it rhymes in Portuguese) carries more than a grain of truth. Within evolutionary theory the very existence of grandparents has been explained in terms of the advantages of investing in second-generation kin. Given the asymmetries in parental certainty between men and women we might expect differences in the investment of grandparents in their grandchildren according to levels of paternal and hence grandpaternal certainty. If we trace our ancestors backwards then every link through a male introduces some level of uncertainty.

As an illustration, a grandmother is certain that her grandchildren through her daughter are hers but less certain of her grandchildren through her son. A moment's thought should reveal that the order of certainty becomes MGM > MGF or PGM > PGF (See Figure 11.3).

The term 'investment' can be captured by a multitude of measures such as emotional closeness, frequency of contact, value of gifts, value of bequests and so on. In a study performed on 120 undergraduates, Todd DeKay (1995) found the following order in grandparental investment as evidenced by the measures of closeness, time spent together and gifts received: MGM > MGF > PGM > PGF. What is interesting is that maternal grandfathers (MGF) were consistently rated higher than paternal grandmothers

(PGM) on all measures even though the degree of grandparental certainty ostensibly looks to be the same (with one male link). DeKay explained this by suggesting that infidelity rates were higher in the parent's generation compared to the grandparent's generation. Hence there will be more uncertainty associated with the most recent male link.

A similar study was carried out by Harald Euler and Barbara Weitzel (1996) on German subjects. The sample size here was larger and 603 volunteers were identified who had all four grandparents alive until they reached the age of seven. The subjects then estimated the solicitude they received from each grandparent. Euler and Weitzel were able to rule out the possible effect of residential location on these estimations by showing that there was hardly any

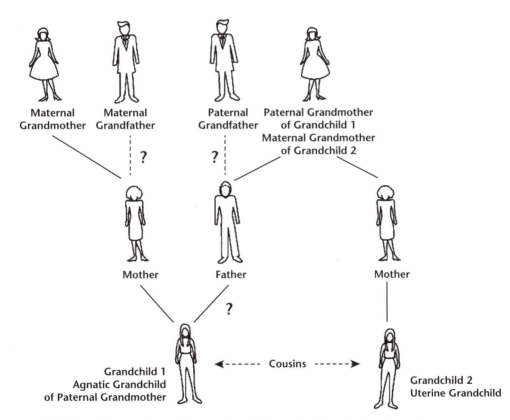

FIGURE 11.3 Differing degrees of grandparental certainty and options for investment.
A paternal grandfather (PGF) experiences two levels of descendent uncertainty, compared to one by maternal grandfathers (MGF) and paternal grandmothers (PGM). But paternal grandmothers may have a choice in investment towards agnatic (that is, via male line) grandchildren or uterine (via female line) grandchildren. This latter option could reduce investment from PGM to grandchild 1 if this alternative is available. The highest degree of certainty lies with the maternal grandmother (MGM).

difference between residential distance measures for the four types of grandparent.

However, there are other factors that influence **grandparental solicitude**. Salmon (1999), for example, found that birth order affected the quality of the relationship between grandparents and their grandchildren: specifically, children with a middleborn father or mother had less contact with their grandparents than those of first- or last-born parents. The type of culture is also important. Alexander Pashos (2000) found differences between rural and urban Greek communities in the relationship between grandparents and their grandchildren.

Pashos observed that overall rural Greek males rated grandparental solicitude higher than females did. He attributed this to the preference in this society for male offspring and hence grandsons. In rural Greece there is also a high frequency of **patrilocal** residence and so paternal grandparents generally reside closer to their grandchildren than maternal grandparents. This study illustrates the need to take into account societal factors other than kinship and **paternity certainty** in predicting family relationships.

Following this early work on grandparenting there are now a variety of hypotheses to account for biased grandparental investment (Danielsbacka et al., 2011). Table 11.1 shows a selection of the most common.

Several of the potential explanations detailed in Table 11.1 are not mutually exclusive, but do make some different predictions that can be tested. The 'sex effect' hypothesis is the suggestion that for a variety of biological and cultural reasons women tend to invest more resources in offspring than men. It has been argued, for example, that women in Western cultures are often socialised as 'kin keepers' – that is, they are encouraged to maintain family bonds and cement family relationships. Hence a preferential allocation of resources to a grandchild through the grandchild's mother may reflect the fact that women

Table 11.1 Five hypotheses addressing grandparents' behaviour towards grandchildren

Hypothesis	Discriminating grandparental solicitude	Preferential investment hypothesis	Sex effect hypothesis	Matrilateral hypothesis	Reciprocal exchange hypothesis
Main claims	Investment will be provided to grandchildren in order of genetic certainty	Investment will be biased according to genetic certainty and availability	For biological and cultural reasons women are more likely to provide childcare and care of grandchildren	Since women make more parental investment than men, grandparents can better enhance the prospects of their grandchildren by helping daughters and hence grandchildren by their daughters	Grandparents will help their children and hence grandchildren according to help they receive in return
Some predictions	The order of investment will be MGM > (MGF, PGM) > PGF. But MGF > PGM could be explained by generational differences in paternity certainty	Order of investment will be MGM > PGM > PGF > PGM if PGM have other uterine grandchildren. But expect PGM and PGF to be roughly equal if alternative uterine grandchildren not available for PGM	(MGM, PGM) > (MGF, PGF) that is, grandmothers always investing more than grandfathers	If PGM lack uterine grandchildren then still expect MGF to invest more than PGM	Investment will follow local cultural practices in caregiving and reciprocal exchanges

make stronger bonds to their parents (and children) than do men (Kaptijn et al., 2013).

This would explain any greater investment by maternal grandmothers compared to maternal grandfathers, but would not predict any difference between MGM and PGM. The matrilateral effect is the idea that since women have a prime responsibility for childcare, then grandparents will enhance the fitness of their grandchildren more by helping daughters than helping their sons. This predicts more investment by MGM compared to PGM but, contrary to the discriminating grandparental solicitude hypothesis, suggests MGF should invest more than PGM.

The reciprocal exchange hypothesis suggests that grandparents bias their support through their daughters in the expectation that daughters will later provide help and care when they (that is, the grandparents) become infirm. In an attempt to control for some of these confounding variables, Kaptijn et al. (2013) looked at grandparental generosity (measured by the willingness of the grandparents to help with childcare) towards grandchildren in two very different cultures: the Netherlands and China. They found that in the Netherlands children who gave emotional support to their parents (that is, grandparents of their own children) received help with childcare, whilst instrumental support (resources) had no effect. In China, emotional support had no impact on grandparental generosity but children who gave more instrumental support to grandparents received more help with childcare.

Interestingly, grandmothers in the Netherlands gave more support than grandfathers – something in agreement with paternity and grandpaternity confidence considerations. In China, however, a grandparental bias towards son's children was found, contrary to expectations from the paternity uncertainty hypothesis. This contrary finding may be explained by the fact that Chinese culture is predominantly patrilineal whilst Dutch culture encourages strong mother–daughter bonds. The importance of this finding is that even if paternity uncertainty effects do play a role in the relationship between grandparents and grandchildren (and this may still be the case at the individual family level in Chinese culture) higher-level cultural effects may predominate. More generally, if paternity confidence is quite

high (and the figure from Anderson (2006) quoted above of 1.9 percent of fathers who are not biological fathers is still after all a small percentage) then paternity uncertainty effects could easily be swamped by other factors – as we also saw in the case of Greek rural society.

By focusing on the options available to grandfathers and grandmothers, Laham et al. (2005) advanced a different explanation for why it is often observed that MGF > PGM. This they termed the 'preferential investment hypothesis'. A grandfather has two routes to invest in grandchildren involving one (through daughters) or two (through sons) degrees of uncertainty (see Figure 11.3). Hence a maternal grandfather does not have any more secure option for investment than through his daughter: he faces one degree of uncertainty through his daughter but any alternative through a son is worse in terms of genetic certainty. A paternal grandmother, however, has different options. She faces one degree of uncertainty through her son, but if she has a daughter this provides a more certain route to a genetic grandchild (zero degrees of uncertainty). Hence it is plausible that paternal grandmothers might bias her finite investment away from her son's children towards her 'uterine grandchildren' if she has a daughter. Interestingly this makes different predictions to the grandparental solicitude hypothesis: namely, that if a paternal grandmother does not have uterine grandchildren then her investment is predicted to be similar to a maternal grandfather; on the other hand, if a PGM also has uterine grandchildren then her investment in a grandchild (grandchild 1 in Figure 11.3) is predicted to be less than that of the grandchild's maternal grandfather due to the biasing effect of an alternative investment strategy.

These ideas can be tested. In one of the largest studies to date, Danielsbacka et al. (2011) used a data set of 33,281 Europeans collected as part of the Survey of Health, Ageing and Retirement in Europe project over the years 2006–7. The variable related to grandparental investment they examined was the probability that a grandparent would look after a grandchild in any one week, controlled for other potentially confounding variables such as distance, health, education and employment status.

As Figure 11.4 shows they found that MGM > MGF > PGM > PGF in agreement with the discriminating

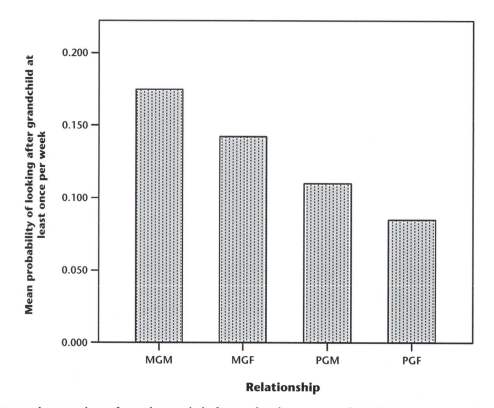

FIGURE 11.4 A comparison of grandparent help from a data base survey of 20,769 European grandparents. Investment follows the order of genetic certainty in agreement with the discriminating grandparental solicitude and the matrilateral effect hypothesis. The results are inconsistent with the sex effect hypothesis.

SOURCE: Data from Danielsbacka et al. (2011).

grandparental solicitude and the matrilineal hypothesis. But what if PGM with agnatic (that is, through the male line) grandchildren also had alternative uterine grandchildren? This investigation showed that in these cases PGM were observed to reduce their investment in agnatic grandchildren – presumably channelling investment through their daughters as an alternative more genetically secure line of transference. Figure 11.5 shows a comparison of MGF and PGM investment in two situations: one where the PGM has available only an agnatic route to grandchildren (that is, grandchild 1 in Figure 11.3) and one where the PGM has both agnatic and uterine routes to grandchildren (grandchildren 1 and 2 in Figure 11.3). In the first case investment from MGF and PGM is remarkably similar – consistent with the discriminating grandparental solicitude hypothesis, but contrary to the sex effect and matrilateral effect hypotheses. In the second condition the

investment of the PGM is lower (that is in grandchild 1, Figure 11.3), a reflection perhaps of the diversion of resources to an alternative grandchild in keeping with the preferential investment hypothesis but difficult to explain by the original and unmodified discriminating grandparental solicitude hypothesis.

11.2.3 Recognising kin

The studies above show how help and cooperation is biased towards kin. This tendency would logically suggest some reliable means of recognising kin and estimating degrees of kinship. Indeed, recognising kin should be important both for regulating altruism and for avoiding incest. Lieberman, Toobey and Cosmides propose the existence of 'kinship recognition' circuits (or a module) in the human brain that activates both these concerns and directs behaviour

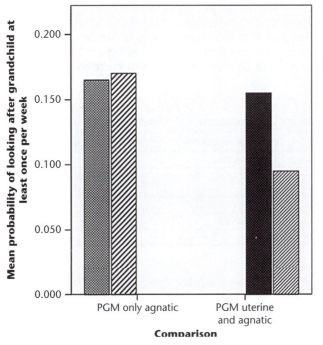

Relationship and investment from.

▨ MGF (only uterine)

▨ PGM (only agnatic available)

▨ MGF (only uterine)

▨ PGM (uterine and agnatic)

FIGURE 11.5 A comparison of investment from grandparents according to whether the PGM has only agnatic grandchildren (that is, no daughters) or both agnatic and uterine (that is, has a son and a daughter).

In the second case investment into the agnatic grandchild reduces, possibly as a consequence of preferential investment through the uterine route. MGF compared to PGM in only agnatic route p = 0.689, difference not significant; MGF compared to PGM in uterine and agnatic routes p < 0.001, difference significant.

SOURCE: Data from Danielsbacka et al. (2011), Figures 3 and 4, pp. 15 and 16.

accordingly (Lieberman, Tooby and Cosmides, 2007). They posit the operation of two primary mechanisms for estimating kinship. The first relies upon 'maternal perinatal association' (MPA), in other words, experiencing first-hand or observing close association between mother and child shortly after birth. The second suggested mechanism is the 'cumulative duration of co-residence summed over the full period they receive parental care' (p. 728). This second system is postulated to be needed by younger siblings who were not alive when older siblings were born and therefore cannot rely upon the MPA system. There is also the possibility of another system using additional cues such as odour and resemblance (see Figure 11.6).

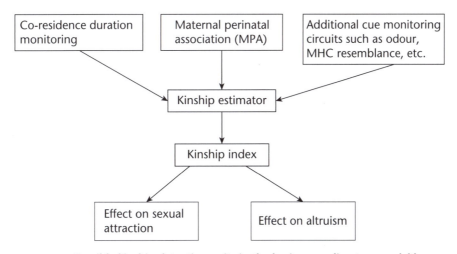

FIGURE 11.6 Possible kinship detection units in the brain according to a model by Lieberman, Tooby and Cosmides (2007).

Table 11.2 Data on revulsion expressed in contemplation of various acts
The two relevant for the purpose of illustrating kinship detection are brother and sister incest. The other two are added for comparison. The fact that the incest ratings correlate with years of co-residence when MPA is absent suggest two mechanisms as shown in Figure 11.6 with the dominant one being the MPA system (since years of co-residence does not correlate with rating when MPA is present).

Acts contemplated	Rating of acts (19 highest, 1 lowest)	Correlation of rating with co-residence and significance (p) values. NS = not significant	
		MPA absent	MPA present
A man killing his wife	14.92 ± 3.89	–0.06 NS	0.05 NS
Sex between a brother and sister (consensual)	11.53 ± 3.08	0.27 (p < 0.05)	–0.19 NS
Brother and sister marriage	10.65 ± 3.01	0.39 (p < 0.01)	0.16 NS
Speeding on the highway	2.60 ± 2.97	–0.02 NS	0.05 NS
SOURCE: Data re-tabulated from Lieberman, Tooby and Cosmides (2007).			

The authors issued questionnaires to over 600 subjects asking questions about their co-residence with siblings, their association with their mother, altruistic inclinations towards kin and the level of disgust evoked by the thought of sex with their siblings. The subjects rated disgust on an ordinal scale in relation to 18 other morally questionable practices, so that 1 would be rated the worst and 19 the least objectionable (see Table 11.2). From the responses the authors concluded that the MPA was the dominant system. Moreover, they found that when both MPA and co-residence experiences were available to the subjects the presence of the MPA effect eliminated the effect of co-residence. This fact the researchers take as evidence of the existence of some sort of intermediary 'Kinship estimator' and 'Kinship index' modules functioning between cues of kinship and the effects on behaviour (see Figure 11.6).

11.3 Cooperation promoting direct phenotype fitness benefits

According to Hamilton's equation, rb > c, altruism spreads because altruistic genes help others of the same type. But as this is played out, it is phenotypes helping other phenotypes that carry the same cooperation-promoting genes. But in many cases cooperation can also bring direct fitness benefits as found in mutualism (see Figures 11.1 and 11.2). In these cases it is cooperation that brings the rewards and so it is

not necessary that the cooperating phenotypes have high r values linking them. Cooperation will be a successful strategy, and cooperating genotypes will proliferate, so long as cooperators help other cooperators. Perhaps the clearest manifestation of this is **reciprocal altruism**.

11.3.1 Reciprocal altruism or time-delayed discrete mutualism

Acts can often appear altruistic when in fact we simply have not taken a sufficiently long-term view of the situation. Trivers (1971) was one of the first to argue that altruism could occur between unrelated individuals through a process he termed 'reciprocal altruism', which is in essence a more refined version of the maxim 'you scratch my back now and I'll scratch yours later'. We are really looking for genes that by cooperating with each other enhance their own survival and reproductive success through their own vehicles. In kin selection an individual may help another in the expectation that the helping gene is present in the recipient. In the case of reciprocal altruism, aid is given to another in the hope that it will be returned.

We can now see that there is a rather fine line between our definition of mutualism and that of reciprocal altruism. The most useful distinction is that mutualism involves a series of constant reliances. An extreme form of this would be lichen which are composed of an alga and fungus in an inextricable symbiosis. Similarly the bacteria in your gut that help you digest food are mutualistic in that the exchange of food products is virtually constant. We can think of reciprocal altruism as a sort of time-delayed mutualism. An exchange takes place in which it seems that the beneficiary gains and the initiator loses according to Figure 11.1. What we expect, however, is that in cases of reciprocal altruism the favour is returned at a later date. For direct reciprocal altruism we expect the favour to be returned by the individual who first received our help. Game theory (see below) has also been used to distinguish between reciprocal altruism and mutualism with the suggestion that rewards and punishments (for cheating) have different values.

Conditions for the existence of reciprocal altruism

We would expect the following conditions to be present if reciprocal altruism is to be found:

1. An animal performing an altruistic act must have a reasonable chance of meeting the recipient again to receive reciprocation. This would imply that the animals should be reasonably long-lived and live in stable groups to meet each other repeatedly.
2. Reciprocal altruists must be able to recognise each other and detect cheats who receive the benefit of altruism but give nothing back in turn. If defectors cannot be detected a group of reciprocal altruists would be extremely vulnerable to takeover by cheaters. Codes of membership for many human groups, such as the right accent and the right clothes, and the initiation rituals and signals of secret societies, could serve this function.
3. The ratio 'cost to donor/benefit to receiver' must be low. The higher this ratio the greater must be the certainty of reciprocation. This is sometimes called 'gains in trade' by economists – when something of low value to the donor is exchanged for something from the recipient that the donor values highly; in return the recipient receives something that they value highly and gives away something of low value for them.

Although humans spring to mind as obvious candidates, species practising reciprocal altruism need not be highly intelligent.

Examples of reciprocal altruism

One of the best-documented examples of reciprocal altruism concerns vampire bats (*Desmodus rotundus*). These were studied by Wilkinson (1984, 1990) who found that vampire bats, on returning to their roost, often regurgitate blood into the mouths of roostmates. Such bats live in stable groups of related and unrelated individuals. A blood meal is not always easy to find. On a typical night about 7 per cent of adults and 33 per cent of juveniles under two years old fail to find a meal. After about two or three days the bats reach starvation point. It might be thought that regurgitation is an example of kin selection and undoubtedly some of this is occurring. But the exponential decay of loss of body weight prior to starvation suggests that the conditions for reciprocal altruism could be present.

Figure 11.7 shows weight loss against time. In essence the time lost by the donor is less than the

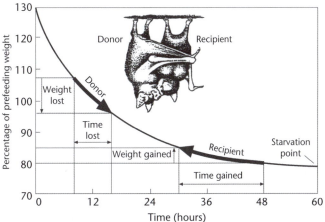

FIGURE 11.7 Food sharing in vampire bats shown in terms of a cost–benefit analysis.
In this diagram the donor loses about 12 per cent of its prefeeding body mass and about 6 hours of time before starvation. The recipient gains, however, 18 hours of time and 5 per cent of prefeeding weight. The fact that time gained is so much greater than time lost favours reciprocal altruism.
SOURCE: Adapted from Wilkinson, G. S. (1990). 'Food sharing in vampire bats.' *Scientific American* **262**: 76–82. By kind permission of Patricia Wynne.

time gained by the benefactor. The bats' mode of life also means they constantly encounter each other. Wilkinson conducted experiments whereby a group of bats was formed from two natural clusters. Nearly all of the bats were unrelated. They were fed nightly from plastic bottles. Each night one bat was removed at random and deprived of food. Wilkinson noticed that on returning to the cage it was fed by other bats from its original natural group. Reciprocal partnerships between pairs of bats were also noticed.

Recently some doubt has been cast on whether vampire bats really do exhibit reciprocal mutualism. It has been argued, for example, that apparent reciprocation could result as a by-product of kin-directed food sharing where recognition errors are made; or that it may result from harassment by hungry cospecifics (Clutton-Brock, 2009). Carter and Wilkinson (2013) responded to these criticisms by examining the behaviour of pairs of vampire bats held in captivity over a two-year period. They found little evidence for harassment effects. Instead, and perhaps surprisingly, they found that food previously received from another bat was the best predictor of food sharing and was a factor eight times more important than relatedness in predicting the donation of food.

Reciprocal altruism has also been documented in Gelada baboons. Dunbar (1980) found a positive correlation between support given by one female Gelada baboon to another and the likelihood that it will be returned. Evidence that reciprocal altruism may be at work in chimpanzee groups has been forthcoming from the work of de Waal (1997) on captive chimpanzees. De Waal found that if chimp A had groomed chimp B up to two hours before feeding then B was far more likely to share food with A than if it had not been groomed. Interestingly, B was equally likely to get food from A whether A had groomed it or not. De Waal's results suggest that grooming serves as a sort of service that is repaid later.

Generally, reciprocal altruism flourishes when there are differences in needs and different capabilities to meet them. This is why it is often observed between species (where these asymmetries are greater). So, as Dawkins points out, flowers need pollinating but since they can't fly they 'pay' bees with nectar to do it. Birds called honeyguides can find bees' nests but lack the strength to break into them.

Honey badgers (ratels) have the strength to break into nests but lack wings to find them. The solution? The honeyguides have a special flight that leads the ratels to the nest and the two species share the spoils (Dawkins, 2006).

But the most enthusiastic reciprocal altruists are humans. The very act of working for a company or a public body is an act of reciprocal altruism: we give our labour for a month in the expectation that we will be paid at the end. Without reciprocal altruism civilised life would be impossible. Reciprocal altruism is the basis for trade and barter in human societies. It is known that early humans often traded stone tools over great distances. Presumably they were not exchanged for other stone tools because there would be no point. Possibly, one group had access to raw materials, and the skills to knap stone-made tools, and exchanged them for skins or food with another group that had access to different resources and skills.

Reciprocal altruists, however, face one gigantic problem: how to ensure that a favour given to someone is likely to be returned at a later date. Such problems have been modelled in **game theory** and to this we now turn.

11.3.2 **Game theory and the prisoners' dilemma**

It is a safe enough bet that the early replicators were entirely selfish. From this primal state the step towards kin-directed altruism is a short one and it is easy to envisage how this could have occurred: care given to direct offspring could move to indirect offspring and all the benefits of inclusive fitness would follow. It is also easy to see how reciprocal altruism benefits both parties once it is established. A favour can be given to another individual at small cost to oneself on the understanding that the favour will be returned. There are conditions that need to be set for this such as the ability to recognise who donated the favour, a good chance of meeting that individual again and so on, but these are not unlikely conditions. The big problem, however, is to account for the origin of cooperative behaviour, given that in the very first interaction it would pay to act selfishly. The problem of the evolutionary emergence of cooperation is highly pertinent to humans. Humans spend a great deal of time cooperating and delaying immediate gratification for future rewards. In this section we will explore some models that show that cooperation may be a fitness-maximising strategy. If this is so then the simplistic linkage of nature–bad and culture–good is exploded. We may be caring and morally sensitive creatures by virtue of our biology.

The prisoners' dilemma.

In many real-life situations our best course of behaviour (in terms of bringing rewards) is often dependent upon the behaviour of others; but the problem is that we do not necessarily know how others will behave. Situations like this have been modelled using 'game theory' (Axelrod and Hamilton, 1981). A good starting point to investigate the moral basis of behaviour is a game known as the **prisoners' dilemma**.

The word 'prisoner' relates to one context where the logic of this game could apply. If two suspects are apprehended at the scene of one of a series of crimes one tactic the police could adopt is to separate the individuals and question them independently. If the overall evidence the police gathered is flimsy and a successful prosecution will rely upon a confession then each suspect may be offered the promise of a lighter sentence if they turn 'king's evidence' and inform (defect) on the other. If they both defect then they implicate each other and both receive a full jail sentence. If both cooperate and refuse to be tempted to defect, then given the lack of evidence each receives a smaller sentence for a minor part of the crime.

In the case of the crime suspects the punishment score is a high number of years, but in most games a high number indicates a better reward. Figure 11.8 shows a payoff matrix for a prisoners' dilemma scenario using values suggested by the political economist Robert Axelrod. The essential features are that cooperation brings a modest reward, defecting if someone else cooperates brings a greater reward, but both defecting brings the lowest reward of all.

The game of course is entirely hypothetical but it illustrates that 'rational behaviour' (in the sense of maximising returns to oneself) can result in the least favourable outcome; if both parties defect or inform, they are each worse off. They are in effect punished for failing to cooperate with each other. The dilemma then is whether to cooperate or defect. Defection will

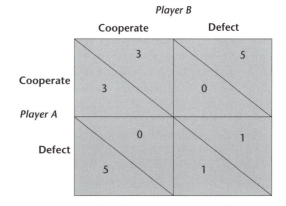

FIGURE 11.8 Prisoners' dilemma values for cooperation and defection.
Values are now shown for each player, the number on the left refers to Player A. The convention for such games became that R = reward, T = temptation (to defect), P = punishment, S = sucker's payoff. In the figure R = 3, T = 5, P = 1, S = 0.

bring the best rewards if the other cooperates, but each prisoner does not know what the other will do. It is difficult to see how cooperation could evolve in this system. One might imagine that suspects would confer before the crime and agree to both cooperate but this still begs the question of why cooperate and not defect. The game of prisoners' dilemma was first formalised in 1950 by Flood, Dresher and Tucker (Ridley, 1993). It is a situation that applies to many interactions in life. As Ridley points out, the gigantic trees in tropical rainforests are products of prisoners' dilemmas: if only they would cooperate and agree to, say, not grow over eight metres then all would be able to put more energy into reproduction and less into growing gigantic trunks to tower over their neighbours. But they can't. Human folly is also often the result of prisoners' dilemma situations. The arms race between the superpowers in the postwar period left both America and the former Soviet Union worse off. In all these cases the protagonists were locked into one form or another of the prisoners' dilemma. Another example from non-human animals concerns cleaner fish. These are small fish that swim inside the mouths of a larger fish and clean the teeth of the host and remove food debris, ectoparasites and so on. It is tempting for the cleaner to take a nip out of the larger fish and swim off – thereby gaining some useful calories; the temptation for the larger fish is to forego cleaning and swallow the cleaner.

In the language of game theory we could say that both players cooperating represents a Pareto optimum, which means that no player can increase their reward without making someone else worse off. This can be contrasted with a Nash equilibrium which exists when no player can do any better by changing their decisions unilaterally. At the heart of the prisoners' dilemma is the fact that the Pareto optimum is not a Nash equilibrium since each player does better by defecting no matter what the other player does.

The prisoners' dilemma problem has relevance to human behaviour in that many human interactions in the past must have taken this form. It might be expected then that the long evolution of humans and other animals should have given natural selection time to solve the problem. In essence we are looking for 'an evolutionarily stable strategy' – a strategy which if pursued in a population is resistant to displacement by an alternative strategy. Headway was made in relation to this problem when it was realised that social life is akin not to one chance encounter between individuals of the prisoners' dilemma form but to many repeated interactions. The problem then became to explain what strategy each should pursue if the game is played over and over again. It then transpires that defection is not always the best policy.

Tit for tat

When we are faced with such situations in life we have a wide range of options. We could play meek and mild and always cooperate. In the face of defection we would then 'turn the other cheek' and continue to cooperate to our ultimate detriment. Another option is to always play rough ('hawkish') and constantly

defect. To compare the strategies over time imagine a population of individuals that interact only once, then the strategy 'always cooperate' is easily displaced by a mutant 'always defect'. A single defector introduced into a population of cooperators will thrive; the population of cooperators will slump and become extinct because a cooperator will always lose on meeting a defector. Eventually the population will be composed of all defectors. As a group they will not do as well as the cooperators but selection does not act for the good of the group. It is even possible that this defecting strategy will lead the species to extinction. When all of the population is composed of defectors this is an **evolutionarily stable strategy** (ESS): it is resistant to invasion by a mutant cooperator.

Using these figures, we can construct a decision matrix for A in the light of what B may do (see Figure 11.9). An inspection of this shows that the best course of action for A is to defect; since A and B are interchangeable it follows that this is also the best course of action for B. Here we get to the heart of the dilemma: individually rational action leads to joint calamity when both parties pursue it.

In the early 1970s, John Maynard Smith began to explore the potential of game theory – of which the prisoners' dilemma is part – to explain animal conflict (Maynard Smith, 1974). His ideas about 'hawk and dove strategies' could have been applied to the prisoners' dilemma but since they belonged to biology and the prisoners' dilemma belonged (at the time) to economics they were ignored. In fact one of Maynard Smith's strategies, 'the retaliator', is very similar to the strategy of 'tit for tat' that was to do so well in later prisoners' dilemma tournaments. A similar convergence

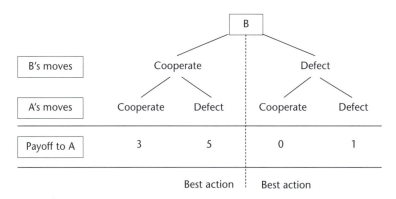

FIGURE 11.9 A decision tree for a prisoners' dilemma scenario.
The best action for A is to defect if A does not know in advance how B will play.

in thinking was taking place as Robert Trivers started working upon his ideas on reciprocal altruism in 1971.

In 1979 Robert Axelrod realised that the prisoners' dilemma only yielded defect as the rational move if the game was only played once. He argued that cooperation could evolve if pairs of individuals in a prisoners' dilemma met repeatedly. It could follow that from repeated interactions both would learn to cooperate and so reap the greater reward. Axelrod set up a tournament to test the success of strategies for playing the prisoners' dilemma many times. He invited academics from all over the world to submit a strategy that would compete with the others in many repeated rounds. Sixty-two programs were submitted and let loose on Axelrod's computer. The important features of this first tournament are as follows:

1. Each pair that meets would be very likely to meet again. In other words, future encounters are important.
2. Each strategy would be matched against each other and itself.
3. The scoring points were set as follows: R = 3, T = 5, S = 0, P = 1. It was decided that the convention R > (T + S)/2 should be adopted, which means that mutual cooperation yields greater rewards than alternately sharing the payoffs from 'treachery' and 'suckers' moves.

The strategy that won was submitted by Anatol Rapaport – a Canadian who had studied game theory and its application to the arms race. The strategy was a simple one called **tit for tat**. This strategy says:

1. Cooperate on the first move.
2. Never be the first to defect.
3. In the face of defection retaliate the next move but then cooperate if the other player returns to cooperation.

Although tit for tat won the overall contest it never won a single match. The reason is easy to see. The best a tit for tat strategist can do is draw with its opponent. In the face of a defector it loses the first encounter but then scores as many points afterwards. By definition it can never strike out to take the lead because it is never the first to defect. Tit for tat wins overall because even if it loses it is never far behind. The problem with 'nastier' strategies is that they have to play each other. If players constantly defect on each other their total reward is low. For a while, it

looked like tit for tat could serve as a model for moral behaviour: it does pay to cooperate after all. Tit for tat might win a general contest but could it ever invade an aboriginal population of selfish defectors? The answer is yes if a few tit for tat strategists appeared and met each other. For a time it looked like tit for tat could demonstrate the possibility of the evolution of human morality: niceness could prosper.

Problems with tit for tat

This strategy seemed to hold high hopes for modelling the behaviour of humans and other animals. After publication of these results in Axelrod's book, *The Evolution of Cooperation* (1984), criticism began to mount. It became clear that tit for tat was not an ESS and its success was sensitive to the precise details of the rules Axelrod had set. In other words it was possible to devise a strategy that would beat tit for tat. In fact in the original tournament the more forgiving strategy of tit for two tats ('Slow to be angered and quick to forgive', as the singer Joan Armatrading said) would have done better. Strangely this was the very strategy that Axelrod offered as an example when notifying others of the tournament but no one opted to submit it – perhaps thinking they could do better. Also if the rules were such that the competition was based on elimination, tit for tat would quickly be eliminated.

Another problem is that tit for tat is very sensitive to errors. Whether or not an opponent has defected or cooperated on the previous move requires the transfer of a message. In the real world, as opposed to the cyberworld of computer tournaments, messages become corrupted and occasional mistakes are made. At an error rate in signalling of 1 per cent tit for tat still came out on top. At 10 per cent tit for tat is no longer the champion. Under these circumstances tit for two tats or generous tit for tat (GTT) can prevail. Here the strategy is to forgive up to two defections. This prevents the effect of noise or mistakes from leading to mutual and constant recrimination. But this strategy has its problems too since GTT allows cooperators to flourish. Cooperators will do as well as GTT and it is conceivable that their numbers might drift to a position where they form the majority. It is now that defectors could strike. They can gain from cooperators who then dwindle in number. If there are not enough GTTs left, then the population will become all defecting. So playing tit for tat may drive away defectors if communication is clear but if errors are made GTT begins to flourish. GTT allows cooperators to grow in number, defectors can then thrive upon cooperators and we could be left with a population of all defectors.

The central problem with all these games is that it is difficult to design a strategy that would always win whatever the others in the pool. If only the others are known we could design a winner but this is not the case. The theoretical promise of using prisoners' dilemma-type games to model the evolution of altruism out of an initial pool of selfish behaviour has been set back by such problems (Hagen and Hammerstein, 2006).

11.3.3 Applications of game theory

The theory of games has found widespread application in modelling animal and human behaviour, the latter in such diverse fields as sexual strategies, economic decision-making and political thinking (see Barash, 2003). A consideration of mutualism provides instructive examples.

Mutualism results whenever the reward for cooperation exceeds that for free-riding defection (i.e. R > T). Given this, however, two versions follow depending on the values of S and P (suckers' pay-off and punishment) sometimes called 'by-product mutualism' and 'synergistic mutualism' (also sometimes called an 'assurance game').

As shown in Figure 11.10, in the case of by-product mutualism it is best for player A (ego) to cooperate whatever B does, and unilateral cooperation has a higher payoff than mutual defection. An example of this might be constructing an irrigation channel that will boost one's own crop production and that of others whether others help or not. The term 'by-product' is used since the benefits to others are the result of the actions of ego which ego is inclined to do anyway. Another example might be hunting: if A hunts and shares food with B this is still preferable to neither hunting at all. The other type of mutualism shown in Figure 11.10 is called synergistic since the same actions serve to increase the payoffs. Here

		Player B (the other)		
		Cooperate	Defect (by-product mutualism)	Defect (synergistic mutualism)
Player A (ego)	Cooperate	4,4	2,2	2,2
	Defect	2,2	1,1	3,3

FIGURE 11.10 Payoffs for two types of mutualism: by-product and synergistic.

R > T (cooperating with a cooperator is preferable to defecting) but P > S (defecting with a defector is preferable to cooperating with a defector) and overall R > P (mutual cooperation is preferable to mutual defection). Here, then, there are two Nash equilibrium points: mutual cooperation and mutual defection, since a move away from these two positions by any one player will make them worse off; but there is only one Pareto optimum: both cooperate.

An interesting application of these ideas to human cooperative behaviour was made by Alvard and Nolin (2002) in their study of whaling practised by the inhabitants of Lamalera, an island in Indonesia. The behaviour of the islanders can be modelled as a series of choices; men can cooperate and hunt for whales or go off individually and try to catch fish. So long as the crew is large enough (n > 8) the rewards of whaling are worthwhile. The researchers lived on the island for about a year and were able to quantify return rates as kg of fish or whale meat per hour of activity. A summary is shown in Table 11.3.

Obviously hunting whales as a solitary activity is not feasible and so S = 0. The study provides an interesting illustration of mutualism in action but also identifies the social norms and institutions (for example, rules about sharing the product) that are vital to maintain this cooperative system. The

Table 11.3 Reward structure for fishing and whaling from the island of Lamalera

	Others hunt whales	Others fish
Ego hunts whales	R > 0.39 kg/hr	S = 0
Ego fishes	T = 0.32 kg/hr	P = 0.32 kg/hr
SOURCE: Data from Alvard and Nolin (2002).		

importance of communication in these types of scenarios is also endorsed by Eric Smith (2010) who argues that one of the primary functions of language itself, with its ability to clarify problems, define norms, rules and punishments and express commitment, may be the facilitation of solutions to collective action problems.

Environmental problems and the prisoners' dilemma

It would be foolish to imagine that all human interactions resolve into prisoners' dilemmas but the diversity of situations to which the model can be applied is quite surprising. The abuse of environmental assets for individual gain is one such situation.

In 1968 the biologist Garrett Hardin coined the phrase '**tragedy of the commons**' as a metaphor for the nature of many environmental problems (Hardin, 1968). Hardin used the idea of the free grazing of common land such as found in medieval Europe to convey the result of individuals each maximising their own self-interest. Suppose three herdsmen each graze three cattle on a piece of land over which they have grazing rights. Equilibrium will be established between the rate of growth of grass and soil formation and the loss of grass and soil by grazing. At a certain level of grazing, the equilibrium is stable and cattle can be fattened without irreversible damage to the environment. But then it occurs to one of the herdsmen that he could place an extra beast out to graze. All the herdsmen would suffer to some degree, in the sense that their animals would not reach their former size, but the loss to the individual herdsman would be more than compensated by the possession of an extra animal. The tragedy arises because all the herdsmen think like this. The result is overgrazing and a collapse of that ecosystem.

Thirty years after Hardin's original conception we can recognise that this is another form of the prisoners' dilemma. By maximising his own utility each herdsman is in effect defecting. When they all defect they are all worse off. Hardin's idea is actually more useful as a metaphor than as a realistic description of how the commons were managed in the small-scale communities of the Middle Ages (where shame and social coercion would have been used to regulate against cheating), but, looking around, we

can observe the tragedy of the commons in action in relation to many environmental problems of the 20th century. For years the seas have been treated as belonging to nobody, and therefore have been regarded as a suitable dumping ground for pollutants. In some places, overfishing has devastated fish stocks. Similarly, when we burn fossil fuels we extract considerable benefit from the energy released and pass the cost on to the global commons to be shared by others. To use the terminology of economists, the costs are externalised and the polluter only pays a fraction of the true damage cost. To overcome the tragedy of the commons, and make the polluter pay the full costs of the damage caused, is the problem of problems in green economics. According to Pearce et al. (1989), we need to internalise the externalities. Interpreted in terms of game theory, we need to adjust the rewards and penalties to encourage people and institutions to cooperate rather than defect. If, for example, carbon dioxide emissions or car usage were heavily taxed, the temptation to defect would not be so great.

Differences between prisoners' dilemma and social dilemma situations

Situations where the cost of defecting (not paying taxes, overgrazing, overfishing, releasing CO_2) are borne by the wider community are sometimes called social dilemmas, and although they may appear similar there are some important differences between prisoners' dilemma and social dilemma situations. In the former the cost is borne by one player, whilst in the latter the cost is shared among many, including the player. Moreover, in prisoners' dilemma situations defection or cooperation is not always anonymous. Your actions may be recorded by the other player and filed away as part of your reputation. Prima facie, then, it looks like social dilemmas will be less conducive to cooperation than prisoners' dilemmas and social dilemmas will need to be accompanied by group exhortations to act in the group interest. We are all probably good reciprocal altruists and cooperators amongst our circle of acquaintances but people need more coercive measures to avoid defecting and dropping the costs onto a wider anonymous group – hence fines for not paying taxes, and fishing quotas. Politicians periodically have to remind people about the common good. As J. F. Kennedy said in his inaugural address of 1961: 'my fellow Americans: ask not what your country can do for you – ask what you can do for your country'. Social dilemmas are also sometimes called 'collective action patterns'; the variation in terminology in this whole area reflects the fact that cooperation has been approached numerous times independently by a number of different disciplines such as evolutionary biology, psychology and economics. But in general terms, the larger the social group that bears the cost of individual selfishness then the less effective shame becomes and the more individuals can hide behind their relative anonymity.

Ridley applies game theory to environmental problems and draws out the political lesson (by his own admission 'suddenly and rashly') that one solution to the prisoners' dilemma is to establish ownership rights over the commons and confer them on individuals or groups. According to Ridley (1996), through ownership, and effective communication within the owning group, comes the incentive to cooperate out of self-interest. The conclusion from Ridley's extrapolation from biological game theory to free-market political economy will not be to everyone's taste, but it represents one serious contribution to the problem of how to channel self-interest to the greater good.

More generally, it is interesting how commentators on the tragedy of the commons infer different solutions depending on their political orientation. Free-market conservatives such as Ridley use it as an illustration of the need for private ownership: if only the metaphorical shepherds owned their own bit of the resource then they would ensure, out of self-interest, that it was managed properly. Meanwhile, liberals argue the need for a central regulating agency to curb excessive self-interest that might damage the common good.

11.3.4 Indirect reciprocity and reputation

Ridley's analysis highlights the importance of effective communication to encourage cooperation. Recent work on reciprocal altruism confirms this point in a novel way. We have seen that reciprocal altruism can be expected to operate when animals

have a good chance of meeting again to return favours. Such situations arise where animals gather in small colonies. Early human groups were almost certainly like this. But the commonest criticism of this model of reciprocal altruism is that real-life encounters are often between individuals who have little chance of meeting again. So can the idea of reciprocal altruism be extended to cover non-repeated exchanges between individuals?

Nowak and Sigmund (1998) think that it can. They found, using computer simulations, that altruism could spread through a population of players who had little chance of meeting again so long as they were able to observe instances of altruism in others. The simulation involved creating donors who would help the recipient if the recipient was observed to have helped others in previous exchanges. The logic of this manoeuvre is that a donor will be motivated to donate help to someone with a track record of altruism since help is likely to be fed back indirectly in the future. But more importantly, each act of altruism increases the 'image score' of the donor in the minds of others and so increases the probability that she or he will be the recipient of help in the future from other observers. The problem with large groups is that any player will know the image score of only a fraction of the population. By making some simplifying assumptions Nowak and Sigmund derived a remarkable relationship. They set initial conditions such that a player who was observed to have helped in the last encounter was awarded a score of +1 and that a player who had defected was awarded 0. In the next encounter the new player defected on those with a score of 0 but cooperated with those on a score of +1. If we call the fraction of the population that any player knows the score of q, the cost to the donor c and the benefit to the recipient b, then they found that cooperative behaviour can become an ESS when $q > c/b$. That is:

$$\text{The probability of one player knowing the image score of the other} > \frac{\text{Cost to donor}}{\text{Benefit to recipient}}$$

The interesting feature of this relationship is its similarity to Hamilton's equation whereby altruism spreads if $r > c/b$ (see Chapter 3).

11.3.5 Game theory and the moral passions

Given the conditions under which we would expect reciprocal altruism to flourish – repeated encounters and sufficient cognitive ability to recognise helpers and cheaters – we could expect to find reciprocal altruism and indirect reciprocity practised frequently in human populations. It is also likely that interactions that favoured altruists, and hence strategies such as tit for tat, were common in the life of early hominids. It is possible then that such interactions have left their mark on the mental life of humans perhaps in the form of distinct modules dealing with emotions. If so, we should find that we are keen to cooperate, but equally determined to punish defectors; we should remember those who gave us favours and not forget those who cheated. Evidence for the importance of detecting cheaters in our psychological makeup was provided in Chapter 9 when the modular mind was considered. This section explores further the possibility that there is some evidence that humans are indeed 'wired' with developmental genetic algorithms that are adapted to a life of reciprocal social exchanges.

Trivers (1985) has argued that the practice of reciprocal altruism formed such an important feature of human evolution that it has left its mark on our emotional system. Humans constantly exchange goods and help, over long periods of time with numerous individuals. Calculating the costs and benefits and deciding how to act requires complex cognitive and psychological mechanisms. Trivers suggests that a number of features of our emotional response to social life can be related to the calculus of reciprocal exchanges and our emotional response serves to cement the system together for the benefit of each individual. Some typical emotional responses that are amenable to this sort of analysis are guilt, moralistic aggression, gratitude and sympathy. In addition we have a highly codified system of justice to ensure fair play.

We experience moral outrage and are motivated to seek retribution when an altruistic act is not reciprocated. Selection may thus have favoured a show of aggression when cheating is discovered in order to coerce the cheater back into line. If we are the cheater in a reciprocal exchange, by say not returning a favour or not discharging an obligation, we may, if

detected, be cut off from future exchanges that would have been to our benefit. So perhaps guilt is a reaction designed to motivate the cheater to compensate for misdeeds and behave reciprocally afterwards. In this way it serves as a counterbalance against the tempting option to defect in a prisoners' dilemma. To encourage reciprocation in the first place we have a sense of sympathy. A sense of sympathy motivates an individual to an altruistic act and gratitude for a favour received provides a sense that a debt is owed and favours must be returned. Figure 11.11 shows a conjectured matrix for the emotional reactions that are likely to be associated with various moves in prisoners' dilemma-type scenarios. This topic is explored more deeply in Chapter 21.

If detection of cheating was so important in the social environment of our evolution we might expect that other components of our mental apparatus are finely tuned to detect cheating. Mealey, Daood and Krage (1996) investigated if our memory of faces is enhanced by a knowledge that the face belongs to a cheater. They presented a sample of 124 college students with facial photographs of 36 Caucasian males. Each photograph was supplied with a brief (fictitious) description of the individual giving details of status and a past history of trustworthy or cheating behaviour. Students were allowed about 10 seconds to inspect each face. Of the 36 pictures seen 12 were described in the category of trustworthy, 12 neutral

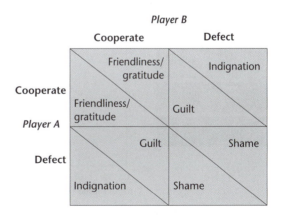

FIGURE 11.11 Some possible emotional reactions to moves in the prisoners' dilemma.
The reaction first listed is to the player on the left. It is suggested that guilt, for example, may accompany defection as a means of persuading the defectors to return to cooperation in future encounters.

and 12 as threatening or likely to cheat. One week later the subjects were shown the pictures again (together with new ones) and asked if they remembered the face. The overall finding was that both male and females were more likely to remember a cheating rather than a trustworthy face. The effect was significant for males and females but stronger for males (p = 0.0261). This work supports the general notion that our perceptual apparatus is adapted to be efficient at recognising cheaters. But puzzling features remain. In the same study the authors found that if pictures of high-status males were used then the enhanced recognition of cheaters disappeared for males and was even reversed for female subjects in that they were now able to recollect trustworthy faces more reliably.

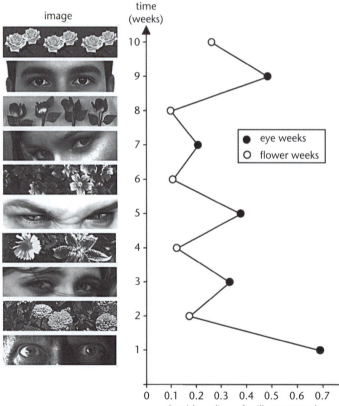

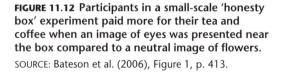

FIGURE 11.12 Participants in a small-scale 'honesty box' experiment paid more for their tea and coffee when an image of eyes was presented near the box compared to a neutral image of flowers.
SOURCE: Bateson et al. (2006), Figure 1, p. 413.

11.3.6 Reputation – someone's watching

In Figure 11.11 the emotion of shame or guilt is predicted to arise when individuals have behaved selfishly in a context where someone else has behaved generously, and, importantly, both players know this. The selfish individual may experience shame even if the other person is unable to work out who they are. But if they are known then their reputation has suffered a damaging blow. Given that we are predicted to be mindful of our reputation, could the awareness of being watched increase cooperative behaviour? The answer from a series of laboratory-based economic games seems to be yes. Subjects behave far more cooperatively and exhibit greater generosity to others in such games when they are conscious of being observed (see Hayley and Fessler, 2005). In one study, this was even found in games where the observer was a robot (named Kismet) looking very unhuman-like – apart from a pair of bulging eyes (Burnham and Hare, 2007). In a neat little study Melissa Bateson et al. (2006) put these ideas to the test in a real-world context by measuring the responses of her colleagues in a university psychology department to the option of paying for tea and coffee in an honesty box. Above the box, situated where the tea and coffee could be made and purchased, were placed instructions on how to pay accompanied by an image of either a flower scene or a pair of eyes. The basic image (flower or eyes) was alternated each week and also changed in terms of types of flowers and the identity of the eyes to avoid any specific image effects. The total amount paid each week (per litre of milk consumed as a measure of tea and coffee taken) was measured and plotted alongside the image displayed (see Figure 11.12).

The net result was that people paid 2.76 times more in weeks featuring an eye image compared to weeks displaying a flower image. It seems plausible then to suppose that humans are responding to visual cues of being watched at a deep, possibly subconscious level, and this exerts a powerful effect directing more pro-social behaviour, possibly through a concern for reputation.

The experiment is open, however, to another interpretation which is that eyes are effective, for whatever reason, at drawing attention to the written instructions, in this case written instruction on how to pay. Two later studies throw more light on this. One concerned whether the presence of eye images alongside a notice in a cafeteria to clear away one's tray and litter after use would enhance cooperative behaviour. In this case the eye images accompanied two sets of written instructions: one was to clear away litter; the other was a neutral one about not consuming one's own food. The group found that in both cases the eyes images induced greater pro-social behaviour irrespective of the written instructions (Ernest-Jones et al., 2011). The effect seems very robust. In a more recent study charity buckets, featuring stick-on eye images or neutral star images, were placed near supermarket checkouts. People donated roughly 48 per cent more when facing buckets with eye images pasted on compared to neutral images (Figure 11.13). Interestingly, the effect was stronger when the supermarket was quieter (Powell et al., 2012).

A number of local authorities in the UK and the USA have now used cardboard images of policemen placed in stored to reduce crime (Figure 11.14).

In a sense, the eye images used in these experiments are false social cues since they are not real eyes. But the fact that humans do respond should not perhaps surprise us. Pornography, for example, can cause sexual arousal even though the images are only two dimensional. Some experiments using eye images have, however, yielded neutral findings with

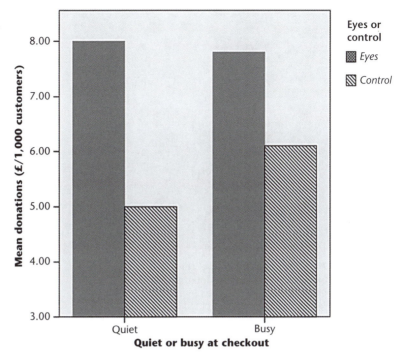

FIGURE 11.13 Propensity to donate to a charity bucket placed near a supermarket checkout.
People donated more when the bucket had two stick-on eyes compared to neutral symbols. The effect was stronger when the supermarket was quiet, consistent with the idea that eyes have a greater effect when they contribute more to the sense of being watched.
SOURCE: Data from Powell et al. (2012), Figure 3.

no eye effects recorded (for example, see Raihani and Bshary, 2012). Discrepancies in the literature may be partly explained by the duration of the response to cardboard eyes. Adam Sparks and Pat Barclay (2013) used economic games with 188 undergraduates and showed that a short exposure to eye images increased generosity in 'dictator games' (where individuals are awarded a sum of money and can choose how much to give away to a fellow player) compared to the control of no eye images present. But interestingly, a long exposure to eye images cancelled this induced generosity effect. There is possibly then some sort of habituation effect to false social cues, this may explain why cardboard policemen placed in stores seem to have some initial effect but then often end up stolen (Figure 11.14).

FIGURE 11.14 PC Bobb, a cardboard policemen placed in a store in NW England to deter shoplifters.
Several police authorities report that the presence of cardboard police officers has the effect of reducing crime in shops. Sadly, this one was stolen.

SOURCE: *The Telegraph*, 28 September 2012. Copyright Cavendish Press (Manchester) Ltd. Reprinted with kind permission.

11.3.7 Altruism and costly signalling

The phenomenon of costly signalling was originally developed by Zahavi (1975) to explain courtship behaviour (see Chapter 4). However, the concept can also be applied to altruistic behaviour. The essential idea is that if altruistic behaviour can bring about other benefits (such as attracting mates) then it may acquire adaptive advantages for a donor to display altruistic tendencies irrespective of the genetic commonality between the donor and recipient or the expectation of any future return on favours given. Put simply, altruism may be seen as a sexy trait. In an ingenious set of experiments, Wendy Iredale, Mark van Vugt and Robin Dunbar (2008) examined people's contributions to a charity in the presence or absence of an observer of the same and opposite sex. Men were found to contribute more to a charity when observed by an attractive member of the opposite sex compared to the same sex or no observer. Female contributions did not vary significantly between any of these conditions (Figure 11.15).

The interesting questions that follow are what costly signallers are actually advertising, and what benefits accrue to the displayer. There are two general classes of possibility that can be advanced: one is that costly signalling is advertising the cooperative nature of the signaller with the aim of attracting other cooperators to mutually reap the benefits of cooperative behaviour; the other is that costly signalling is advertising other qualities such as the skills and wherewithal to accumulate and 'waste' assets. The first of these Michael Price calls 'auto-signalling' and the second 'other signalling' (Price, 2011).

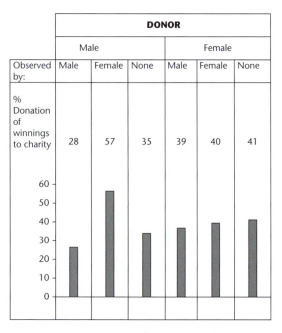

	DONOR					
	Male			Female		
Observed by:	Male	Female	None	Male	Female	None
% Donation of winnings to charity	28	57	35	39	40	41

FIGURE 11.15 Percentage of money won in a game donated to charity according to sex of donor and onlooker.
Men were seen to donate significantly more when observed by a female. For men across the three conditions F (2,42) = 4.60, p = 0.01, η^2 = 0.18; for women F (2,42) = 0.03, p = 0.97, η^2 = 0.00.

SOURCE: Data from Iredale, van Vught and Dunbar (2008).

Summary

- Altruism can be understood biologically in terms of kin selection and reciprocal altruism. In the former, genes direct individuals to help other individuals that are related and so share genes in common. In the latter, individuals help unrelated individuals in the expectation of favours returned. Kin selection and reciprocal altruism offer insights into the biological basis of altruism in humans.

- There is evidence that our closeness to relatives and our altruistic inclinations towards them are moderated according to their coefficient of genetic relatedness.

- A number of hypotheses have been designed to address the phenomenon of grandparenting. Grandparental solicitude tends to follow genetic relatedness.

- Strong evidence has emerged that humans possess kin estimation protocols that are activated early in life. Kin estimation influences both altruism and sexual attractiveness.

- Game theory shows that strategies involving cooperation such as tit for tat could offer a model for the evolution of human morality.

- The undoubted importance of altruism and cooperation in the environment of evolutionary adaptation made it imperative that cheaters could be detected. The need to detect cheaters could have shaped our emotional life in ways such as our sense of justice and our experience of gratitude and sympathy.

- The prisoners' dilemma is a useful model to explain human interaction and some current environmental problems.

- There is evidence that altruistic tendencies are enhanced when someone is watching a potential donor. Altruism may also function in such contexts as a costly signalling device to aid sexual attractiveness.

Key Words

- Evolutionarily stable strategy (ESS)
- Game theory
- Grandparental solicitude
- Hamilton's rule
- Mutualism
- Paternity certainty
- Patrilocal
- Prisoners' dilemma
- Reciprocal altruism
- Reciprocity
- Tit for tat
- Tragedy of the commons

Further reading

Barash, D. P. (2003). *The Survival Game*. New York, Henry Holt.
Readable introduction to game theory and its manifold applications.

Ridley, M. (1996). *The Origins of Virtue*. Viking, London.

Discusses game theory and human cooperation. Interesting and controversial application of game theory to politics and environmental issues.

Davies, N. B., J. R. Krebs and S. A. West (2012). *An Introduction to Behavioural Ecology*. 4th edition. Chichester, UK, Wiley-Blackwell.

Classic textbook. Not specifically concerned with humans but chapters 11, 12 and 13 deal very well with the theories of altruism and cooperation.

Conflict and Crime

I know indeed what evil I intend to do
But stronger than all my afterthoughts is my fury,
Fury that brings upon mortals the greatest evils.
(**Euripides**, *The Medea*, **c. 431** BC)

The Greeks knew a thing or two about dysfunctional families, and it is not surprising that, on several occasions, Freud turned to their myths and legends to find labels for what he supposed were problems of the human psyche. He posited, for example, an Oedipus complex, whereby males subconsciously desire the death of their father in order to sleep with their mother, and an Electra complex involving the secret desires of daughters for their fathers. Most evolutionary psychologists have not been very impressed by these ideas, seeing them as both unlikely and misleading. Much human behaviour, especially that involving conflict, is, however, maladaptive. In the *Medea* story above, Jason abandons his wife Medea for a more desirable bride. Medea, in her anger, kills the bride's father, the bride, and even her own children by Jason. Medea can be taken to signify both a wronged woman's fury and the dark and inexplicable irrational forces in human nature. *The Medea* is of course a work of fiction, but people do such things. We need not, however, invoke a Medea complex; Darwinism (as you might expect) also has something to say about murder and infanticide.

This chapter, then, is concerned with the application of evolutionary theory to understanding interactions involving conflict between offspring and their parents, between siblings, between spouses, and between individuals and the wider society. To tackle these problems, theoretical perspectives established in earlier chapters will be deployed. In Chapter 3, it was noted that the existence of altruism posed a special problem for Darwinian theory and that the breakthrough came with Hamilton's notion of inclusive fitness. The concept of inclusive fitness also helps us to understand conflict between related individuals. It seems obvious that, since offspring contain the genes of their parents, parents will be bound to look after their genetic investment. But parents have loyalties that are divided between care for their current offspring and the need to maintain their own health in order to produce future offspring. It follows that offspring may demand more care than parents are willing to give, and in this situation we should expect to observe a mixture of altruism and conflict. It is also easy to see that partners in a sexual relationship may have different interests and strategies, and it is these conflicting interests that help to explain violence and strife within marriages.

One extreme manifestation of human conflict is **homicide**. From an analytical point of view, homicide has the advantage that considerable statistical information on it is available. Two American psychologists, Margo Wilson and Martin Daly, have pioneered the use of homicide statistics to test evolutionary hypotheses, and this chapter also considers their work.

12.1 Parent–offspring interactions

12.1.1 Parental altruism

Everyone is aware that the animal world abounds with examples of parents (often mothers) making great sacrifices in their efforts to protect and nurture their young. Perhaps the most extreme form of motherly care can be seen in some spider species (of which there are about ten) where the young eat their mother at the end of the brood care period. In the case of *Stegodyphus mimosarum*, the spiderlings do not eat each other or spiders of other species, yet they devour their mother with relish; it seems that this 'gerontophagy' serves as the mother's final act of parental care. From a Darwinian perspective, parents will cherish their progeny because their inclusive fitness is thereby increased. In turn, parents will be loved by their offspring because they can provide help to increase the fitness of the offspring. Parents will also be loved, although with less fervour, because they can increase the inclusive fitness of current offspring by producing more siblings. Altruism between parents and offspring is thus covered by Hamilton's theory of inclusive fitness. Parents will donate help *b* at a cost *c* to themselves as long as:

$$\frac{b}{c} > \frac{1}{r}$$

where *r* = coefficient of relatedness between parents and offspring. For diploid outbred offspring (that is, parents who are not related), *r* = 0.5.

12.1.2 Parent–offspring conflict theory (POCT)

The widespread occurrence of parental care, and the clear biological function it served, probably diverted biologists for many years from the fact that conflict between parents and offspring may also have a biological basis. It awaited the work of Trivers to point out that **parent–offspring conflict** is also predicted from evolutionary theory (Trivers, 1974). The theory of such conflicts is worth examining for the light it throws on human behaviour, especially **maternal–foetal conflict**.

To follow Trivers' original argument, suppose one parent (the mother) gives birth and care to one infant each breeding season and that the infant

needs and benefits from the care. The problem faced by the parent is when to cease providing care. A theoretical solution requires that we examine the costs and benefits of parental care to the parent (P) and the offspring (A). Figure 12.1 shows how *r* values are distributed between parent and offspring.

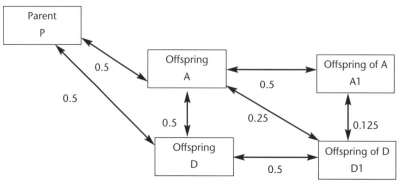

FIGURE 12.1 Coefficients of genetic relatedness (*r* values) between parents and offspring, between siblings and between nieces or nephews.

Suppose we define the benefit to the parent in terms of units of probability that A will survive and reproduce, and the cost to the parent as units of reduced probability that it can produce another offspring D that will survive and reproduce. As A matures, the law of diminishing returns dictates that further care to A will bring fewer benefits than investing in a new offspring D. Eventually, the interests of parent P (or equivalently the helper genes carried by P) are best served by withdrawing from A and investing in D. Although the parent ultimately cares equally for A and D (they are both related to the parent with *r* = 0.5), A is concerned about D only half as much as it is concerned about itself since the *r* value between A and A is 1 but that between A and D is 0.5. Similarly, A is twice as concerned about its own offspring than about the offspring of D. The upshot of all this is that A and its parent P will have a conflict about how much care is diverted to A. Offspring A will demand more care from P than it is in the interests of P to give. P will desire to give less help to A in order to focus resources on D. The analysis predicts that offspring will reach an age at which they will demand more investment than their parents are willing to give.

It has proved difficult to test Trivers' theory quantitatively because of the difficulties of quantifying costs and benefits. There is qualitative evidence, however, that is in broad agreement with theoretical expectations. One prediction from conflict theory is that offspring will prefer parents to direct resources at themselves (especially when they are young) rather than expending effort in producing more offspring. Young human parents often jokingly complain that 'kids are the best contraceptives'. Among chimpanzees, this may literally be true. Tutin (1979) noted that immature chimps often attempt to interrupt the copulations of their parents.

A similar phenomenon is observed in weaning conflicts, something Trivers (1974) used as an illustration of his ideas. When a mother is breastfeeding she is supplying resources to her infant at a possible cost of allocating resources to future children. Moreover, since breastfeeding suppresses fertility it serves to delay future investment as well to the benefit of current offspring. Parents and infants will differ in what they regard as the optimum allocation of resources, so conflicts over the timing of the cessation of lactation and the beginning of weaning are predicted. But so far the literature on weaning in primates has yielded ambiguous results with some studies offering support and others not (see Maestripieri and Pelka, 2002, for a review). One fairly established phenomenon in keeping with POCT, however, is the exacerbation of conflict at transition points. When a female primate resumes oestrus this is a signal to offspring that she is likely to reproduce again and start directing energy to future offspring. As predicted, nipple solicitation by young has been observed in a variety of primates to increase markedly when this transition occurs (Pavé et al., 2010; Zhao et al., 2008). Humans are unusual mammals in that they raise several dependent offspring of different ages at the same time, implying the potential for a mixture of inter- and intra-brood conflict. But it is probably significant that in many pre-industrial cultures a new pregnancy often determines the timing of weaning of an existing child. Among the Turkana pastoralists of Kenya, for example, the most common reason given for weaning was pregnancy. Indeed, among these people conflict between mothers and their children tends to be high when mothers resume sexual relations with their husbands – a significant transition point signalling the potential shifting of investment from parenting towards mating (Gray, 1996).

As the age of a parent increases, and it nears the end of its reproductive life, so its chances of producing any more offspring reduce and the costs to itself of giving care to existing offspring decrease. It should follow that older parents are more willing to invest in their young than are younger ones. This seems to be supported by the fact that the abortion rate in mothers declines with the mother's age, although other factors may also be at work here (see Tullberg and Lummaa, 2001).

Towards a general model

Parent–offspring conflict can be further understood by combining a variety of theoretical perspectives: Hamilton's ideas on **kin** selection, Trivers' theory of parent–offspring conflict and life history theory (LHT). According to LHT there are two fundamental (but linked) trade-offs in this context: current versus future offspring, and quantity versus quality. The outcome of both these decisions impacts on the nature and intensity of parent–offspring conflict. Essentially conflict and competition arise because of asymmetries in what parents and offspring regard as optimal levels of investment, as these optima are often different so competition and conflict arise. The way a decision is reached is influenced by such factors as age of parents, stability and density of environmental resources, mating systems, the condition of current offspring and the condition of parents (Figure 12.2). The way these factors impact on conflict is assessed below:

Environmental resources

In environments where resources are poor, hard to obtain, highly variable, or where organisms face a high level of risk, this can be expected to cause rapidly diminishing returns on parental investment in any one offspring. If a parent lavishes increasing resources on a single offspring there is a chance in unpredictable environments that all this will be wasted if that offspring faces a high risk of extrinsic mortality from violence or disease. In these circumstances quantity will be favoured over quality. As a result offspring will compete amongst themselves and with parents for the limited resource flow available. The result is likely to err towards relatively harsh, neglectful and low-investment parenting.

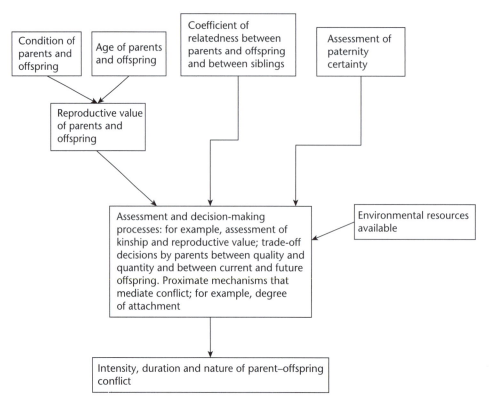

FIGURE 12.2 A generalised model for an evolutionary understanding of parent–offspring conflict.

This quality–quantity trade-off has been observed in many Western societies where children in larger families receive less parental support and childcare (Lawson and Mace, 2009). In traditional societies also, larger family size is associated with poorer growth and survival rates of offspring (Hagen et al. 2006).

Condition of offspring and parents

Parents also need to factor into their decision-making estimates of the fitness and condition of current offspring. In fundamental terms, parental investment in a child will be predicted to rise until any fitness gains (of the parent) through further investment are equal to or less than the potential fitness gains that could be achieved by investing in other current or future offspring. Harsh though it sounds, parents should be selected to eventually cease investment when offspring are unlikely to survive or reproduce

themselves. Militating against this, of course, will be a whole host of socioculturally influenced emotions, norms, ethical codes and empathetic aptitudes. Despite this, there is evidence that across all cultures disabled children experience high rates of abuse and neglect (Bugental and Happaney, 2004). In their survey of records of 60 preindustrial societies, for example, Daly and Wilson (1988a) found that the two most common reasons for infanticide were poor prospects for rearing (that is, low parental resources of various kinds) and the infant being deformed or seriously ill. The fact that both the condition of the parents and that of current offspring are predicted to impact on decisions about investment is sometimes called 'discriminating parental solicitude'. Mothers who are in good condition, have good parenting skills and access to ample resources are predicted to invest more in high-risk children than mothers without these assets. Studies on Latino children in the USA have supported this claim (see Bugental et al., 2010).

Age of parents

Parent–offspring conflict is expected to decline as the age of the parent increases. This follows from what is sometimes called the 'terminal investment hypothesis': as a parent ages so its future reproductive potential declines and so it increasingly pays to invest in current rather than future offspring. There comes a point when no future offspring are possible and so inter-brood competition at least is eliminated. There are a variety of studies that offer some support to this idea. Beaulieu and Bugental (2008) found that older mothers were more likely to invest in demanding offspring. In another study older mothers were observed to direct more praise and attention towards their children than younger mothers (Bornstein and Putnick, 2007). The problem with these types of studies of course is that as parents age they probably acquire better parenting skills more generally. But in a large scale study of 13,604 families in 10 locations in the USA, Schlomer and Belsky (2012) controlled for many of these confounding variables and found that older mothers did indeed put more effort into parenting than mating. Furthermore, high maternal effort and investment were strongly correlated with reduced levels of parent–child conflict.

Coefficients of relatedness and assessment of paternity

Low r values between parents and children are expected to negatively impact on the care a parent gives a child. One indirect line of evidence comes from studies on the reactions of men to faces of children with different degrees of resemblance to themselves. As noted in Chapter 11, men are more favourably disposed to children that they resemble. In another study Rebecca Burch and Gordon Gallup (2000) studied the behaviour of 55 men convicted of spouse abuse. They found that the severity of injuries suffered by the female partner was inversely proportional to the perceived resemblance between the father and his biological children. Significantly, in families where stepchildren were present, female injuries were worse but there was no relationship between resemblance of the father to his biologically unrelated children and spouse injury. The authors suggest that paternal resemblance may act as a cue to assess fidelity and paternity (Burch and Gallup, 2000).

There is also evidence that suggests that men's violence towards their pregnant partners is influenced by cues indicating paternal certainty. Graham-Kevan and Archer (2011), for example, found that pregnant women who were struck on their abdomen by an abusive partner reported more conflict relating to sexual jealousy than those struck on other parts of the body. The targeting of the abdomen may be an unconscious attempt to damage a child suspected not to be the man's own. Finally, Daly and Wilson's survey of 60 preindustrial societies revealed that real or suspected non-parenting was associated with infanticide in 20 out of 35 societies where infanticide was found (1988a).

One of the triumphs of the human personality is that many stepfamilies work so well against the grain of this Darwinian prediction. Nevertheless, when large statistical samples are examined abuse of parents towards children is found to be much higher amongst step-parents compared to biological parents. This important topic is considered below.

Parent–offspring conflict theory has stimulated a fresh look at behaviour that was once thought of as being unproblematic. In pregnancy, for example, it is easy to assume that the interests of the mother and foetus are virtually identical and that the considerable investment by the mother in the foetus is a clear case of kin-directed altruism. Work by Haig (1993) and others, examined in the next section, shows that the situation is more complicated and provides a useful testing ground for theories of conflict.

12.1.3 Maternal–foetal conflict

One of the most intimate relationships in the natural world must surely be that between a mother and her embryonic developing child. The mother is the life support system for the foetus: she provides it with oxygen from every breath she takes and food from every meal she eats. It is tempting to think that, in this precarious state, the interests of the mother and foetus must be identical, and that conflicts of the sort examined above can only arise after the birth of the child, when the mother will soon be in a position to produce more offspring. Even here, however, gene-level thinking brings some surprises. The Harvard biologist David Haig (1993) has applied POCT to this situation and has produced his own theory of genetic conflicts in human pregnancy. The crucial point to appreciate in Haig's analysis is that the foetus and the mother do not carry identical genes: genes that are in the foetus may not be in the mother if they were paternally derived. Even if they are maternally derived (and therefore present in the mother), they have only a 50 per cent chance of appearing in future offspring of the same mother. It is quite feasible, then, argues Haig, for foetal genes to be selected to draw more resources from the mother than is optimal for the mother's health or optimal from the mother's point of view of distributing resources among her current and future offspring. One example where Haig's theory and POCT applied to maternal–foetal conflict has met with empirical support is conflict over the supply of glucose to the foetus.

Conflicts over glucose supplied to the foetus

When a non-pregnant woman eats a meal high in carbohydrates, her blood sugar level rises rapidly, but then falls as the insulin secreted in response to the raised glucose level causes the liver to store the excess glucose as glycogen for later use. In contrast to this, a similar meal taken during late pregnancy will cause the maternal blood glucose level to rise to a higher peak than before and moreover stay high for a greater length of time, despite the fact that the insulin level is also higher. In effect, the mother is less sensitive to insulin and compensates, but not entirely, by raising her insulin level. This is puzzling. Why should a woman develop a reduced sensitivity to insulin and then bear the cost of having to produce more?

Haig's theory of genetic conflict suggests an answer to this problem. It is in the interests of the foetus to extract more blood sugar for itself than it is optimal for the mother to give. The mother is concerned for her own survival after giving birth and, in addition, is more concerned for existing and future offspring than is the foetus. The foetus can send signals to the mother via the placenta just as the mother can send signals to the foetus. Haig suggests that one result of this is that, in late pregnancy, the placenta produces allocrine hormones that decrease the sensitivity of the

mother to her own insulin, thereby allowing the glucose level to rise to benefit the growth of the foetus. The mother responds in this tug-of-war by increasing her insulin level. Further evidence in favour of this theory is that the placenta possesses insulin receptors and that, in response to a high insulin level, it produces enzymes that act to degrade insulin and thus disable the mother's counterattack. We can picture the escalation of measure and countermeasure as resulting in a pair of forces acting upon the level of some parameter such as glucose, each attempting to move it in an opposite direction towards two opposed optimum levels (Figure 12.3).

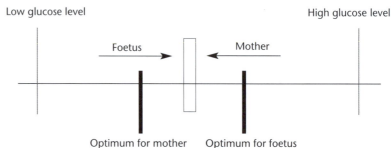

FIGURE 12.3 Schematic representation of the effort by the mother and foetus to drive the blood glucose level to different optima.
The central cursor can be thought of as moving on a sliding scale subject to pressure from the mother and foetus.

This conflict over glucose supply has potentially damaging side effects. The insulin resistance induced in the mother by the placenta increases the risk of the mother developing diabetes later in life. Indeed, about 7 per cent of pregnancies in developed countries lead to the mother contracting gestational diabetes mellitus (Feig et al., 2008). Yet mothers that develop this condition tend to have larger babies than normal mothers, supporting the idea that the placenta and foetus have engineered insulin resistance to obtain additional resources from the mother.

12.2 Sibling rivalry

The logic of genetic relatedness dictates that individuals will care for themselves and their offspring more than they will for their siblings and the offspring of their 'sibs'. At its extreme, this asymmetry could result in **siblicide**, which is in fact widespread in nature. In some species of eagle, the mother normally lays two eggs even though, in nearly all cases, only one offspring survives. The mother perhaps lays the second egg as an insurance policy against infertility in a single egg. On hatching, the elder chick kills the younger (see Mock and Parker, 1997).

The ideas introduced by Trivers can help to explain parent–offspring conflict and sibling rivalry. In the latter case, siblings will seek to divert the flow of parental resources from a sibling to themselves. Some evidence in support of this comes from the fact that tantrums are used by infants in a number of species, including baboons, chimpanzees and zebras, as an apparent attempt to increase maternal investment in themselves (Barrett and Dunbar, 1994). Frank Sulloway (1996) has developed these ideas into a remarkable theory of birth order personality differences in human families, which we have already touched on in Chapter 7. He argues that competition between siblings for parental investment leads children to occupy different personality niches to secure their needs. Firstborns, he suggests, try to maintain their initial advantage by aligning their interests with their parents and take advantage of their greater age than later children by assuming authority over them. They tend to adopt their parent's values and are more conservative, conventional and responsible than later born siblings. In contrast, later-born children develop traits that enable access to parental resources despite the presence of older siblings. To achieve this they are typically more flexible and imaginative than their elder siblings, less conservative and conventional, and more likely to be the rebels in the family.

This interesting line of research has generated a large volume of literature, some of it supportive (Davis, 1997; Saroglou and Fiasse, 2003) and some of it less so (Beer and Horn, 2000; Jefferson, Herbst and McCrae, 1998). One problem in these studies is that the comparison of children's personalities between families introduces lots of confounding effects such as differences in socioeconomic status, class and race. Recently, Healey and Ellis used a within-family methodology and found personality differences between first- and second-born children in agreement with Sulloway's model (Healey and Ellis, 2007).

All in all, the study of violence between siblings from an evolutionary perspective is methodologically taxing. On the one hand, the principle of inclusive fitness suggests that conflict between siblings (with r values of 0.25 or 0.5 for half and full siblings respectively) should be less serious than conflict between unrelated individuals where $r = 0$. But on the other hand, siblings often have something to argue about (for example, resources from a parent) and they spend a lot of time together. Consequently, as noted earlier, sibling competition is quite common and among non-human animal species can often result in the death of a sibling. One area where predictions should be easier to test is the hypothesis that conflict between pairs of step-siblings (that is, genetically unrelated children in a family) and between pairs of half siblings ($r = 0.25$) should be more intense than conflict between full siblings ($r = 0.5$). One study that offers some support for this is that of Khan and Cooke (2008) who found that living with a stepsibling was a strongly predictive factor in the incidence of inter-sibling violence. Once again there are confounding variables but this is an area ripe for research.

12.3 Human violence and homicide

The application of adaptationist thinking to human conflict within families was pioneered by Martin Daly and Margo Wilson in the 1980s. Daly and Wilson completed their PhDs in animal behaviour in the early 1970s and were inspired by E. O. Wilson's (1975) *Sociobiology* to apply evolutionary theorising to step-families.

Before we examine their work, it is important to understand the way in which adaptive reasoning can be applied in this area. In modern nations, homicide is, for the most part, damaging to the fitness of an individual. The perpetrators are likely to be found and either incarcerated or executed. Many homicides also have a negative effect on inclusive fitness since it is often one relative who kills another. To cap it all, many homicides are followed by suicide, which is hardly fitness maximising. At first sight, it would seem bleak territory on which to erect adaptationist arguments, and, unsurprisingly, homicide has usually been regarded as a result of inherent human wickedness, a failure of social upbringing or the result of some sort of pathological condition. The originality of the work of Daly and Wilson has been to realise that, amid all these causes, there may also be the effect of psychological mechanisms that can be understood in selectionist terms.

Homicide is a gruesome business, but the serious nature of the crime makes statistics on killings more reliable than those on probably any other act of violence. It is against patterns of homicidal statistics that predictions from evolutionary psychology can be tested.

12.3.1 Infanticide

Infanticide is not uncommon in the animal kingdom. The males of such animals as lions and lemurs will often kill the offspring of unrelated males to bring a female back into oestrus. Among the vertebrates, it is mostly males who kill infants, but where males are the limiting factor, the roles are reversed. Among the marsh birds called jacanas, polyandry is often found, and sex roles are often inverted. A large territorial female may have several nests containing her eggs in her territory, each nest presided over by a dutiful male. If one female displaces another and takes over her harem of males, she sets about methodically breaking the eggs of her predecessor. In these examples, infanticide can be understood as a means of increasing the fitness of the killer.

We may find such instances distasteful, but we rightly regard examples of human infanticide as being even more horrific. In most countries, infanticide is illegal, and debates continue about the legitimacy of foeticide (the destruction of a foetus in the uterus). It could, however, be that in the environment in which early humans evolved, infanticide in some circumstances represented an adaptive strategy. If we can judge the reproductive experiences of early humans by those of modern-day hunter-gatherers, we begin to appreciate the tremendous strain that raising children probably entailed. Infant mortality would be high, fertility would be low, partly as a result of the prolonged feeding of infants, and the best that most women could hope for would be two or three children after a lifetime of hard work. Under these conditions, raising a child that was defective and had little chance of reaching sexual maturity, or a child that was for some reason denied the support of a father or close family, would be an enormous burden and contrary to a woman's reproductive

interests. At a purely pragmatic level, infanticide may have sometimes been the best option to maximise the lifetime reproductive value of a woman. One might term the withdrawal of support by the father the 'Medea effect'.

If infanticide did once (as it still does in some cultures) represent a strategy for preserving future reproductive value, we might expect the frequency of infanticide to decrease as the mother's age increases. This follows from the fact that, as a mother ages, the number of future offspring she can raise falls and so each one becomes more valuable. Daly and Wilson (1988a) present evidence that this effect is at work among both the Ayoreo Indians and modern-day Canadians. Figure 12.4 shows infanticide in relation to the age of the biological mother.

The results from Figure 12.4 are in agreement with predictions, but it is difficult to rule out other effects. Women may become better mothers as they age, learning from experience; younger mothers may suffer more social stress. The effect could thus be one of socially learned skills and culturally specific stress factors rather than an adaptive response.

Reproductive value of offspring and infanticide

In the analysis above, Daly and Wilson derived testable predictions from the way in which the reproductive value of the parent varied as a function of age. The child too has reproductive value as the carrier of the parent's genes and as a potential source of grandchildren. From these considerations, predictions can also be derived.

From a gene-centred perspective, children are of value to their parents because they have the potential to breed and continue to project copies of genes into future generations. This dispassionate approach suggests that the value of a child would increase up to puberty and decrease as the child approached the end of its reproductive life. This follows from the fact that the chances of, for example, a 10-year-old girl reaching sexual maturity at 16 (a typical age for the first menstruation in a child in a hunter-gatherer society) are greater than those of a 2-year-old. The 10-year-old has benefited from 10 years of parental investment and has only 6 more years to survive the hazards of life to reach 16, whereas the 2-year-old has 14 years to go (Figure 12.5). The variation in valuation will be

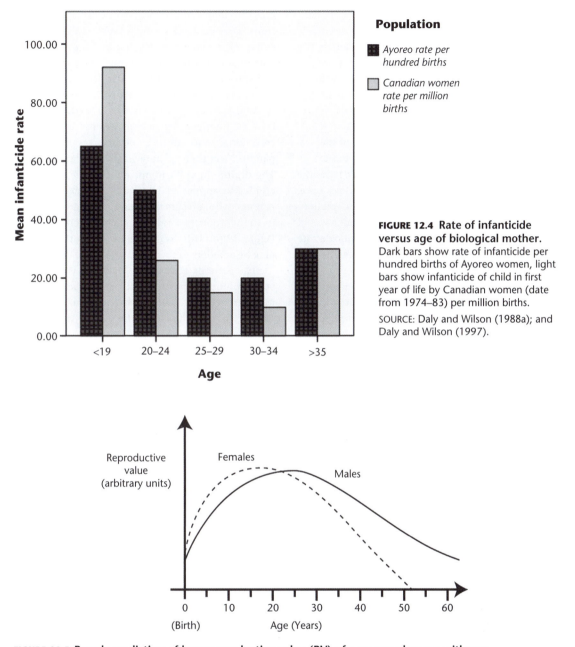

FIGURE 12.4 Rate of infanticide versus age of biological mother. Dark bars show rate of infanticide per hundred births of Ayoreo women, light bars show infanticide of child in first year of life by Canadian women (date from 1974–83) per million births.

SOURCE: Daly and Wilson (1988a); and Daly and Wilson (1997).

FIGURE 12.5 Rough prediction of how reproductive value (RV) of a person changes with age.
RV can be conceptualised as the number of offspring a child/person is likely to produce in the future at any given age. Reproductive value is quite low at birth, reflecting the risks of infant mortality, but rises quickly as the infant has an increasing probability of surviving to sexual maturity. For females, residual reproductive value peaks at, or shortly after, menarche, although peak fertility (a measure of the probability of conception) will be slightly later. RV declines with age and heads towards zero for females facing the menopause (the inclusive fitness value of grandmothers is not considered here). At birth the RV of males is slightly lower than females since they face greater childhood risks. The peak RV for males is slightly later than for females since males mature later and also tend to marry and mate with slightly younger women. The decline of male RV is not as rapid as that of females since their fertility persists beyond middle age.

more marked and steep in cultures with a high infant mortality rate, but even in industrialised countries, where infant mortality has dropped dramatically over the past 100 years, the effect will be present.

From the way in which the reproductive value of offspring varies with age, Daly and Wilson made the following predictions about parental psychology:

● In conflicts, parents will be careful in aggressive encounters with their offspring according to their reproductive value. The filicide rates should fall as the age of the child increases from zero to adolescence.
● In the environment of evolutionary adaptation, the greatest increase in reproductive value would have occurred in the first year. The filicide rate is consequently predicted to fall rapidly after the first year of life.
● Child homicide by non-relatives is not expected to vary with age in the same way as filicide since the children of other parents are neutral in respect of their reproductive value to an unrelated offender.

Figure 12.6 shows data for childhood homicide in Canada in the period 1974–83 as used by Daly and Wilson to test these predictions. The results

are consistent but do not rule out other interpretations. One could posit, for example, an 'irritation factor' that varies with the child's age. The lower risk for fathers may then represent a lower time of exposure, and the fall in risk with age may be a product of coming to terms with the difficulties of child-rearing. This theory would predict a higher level of homicide by parents against adolescents, considering the conflicts between teenagers and parents – but then again teenagers are harder to kill. The decline in child homicide with the age of the child also applies to homicides initiated by step-parents, who, presumably, have no interest anyway in the reproductive value of their stepchildren (see Figure 12.6). This too indicates that other factors may be at work.

The patterns observed by Daly and Wilson have been confirmed by numerous subsequent studies. John Archer (2013) conducted a meta-analysis of studies that examined how the relative risk of child abuse and infanticide varied with the child's age. The prediction under test was that a negative correlation should be seen between age of child risk of child abuse and infanticide. Table 12.1 shows how the studies reviewed supported this.

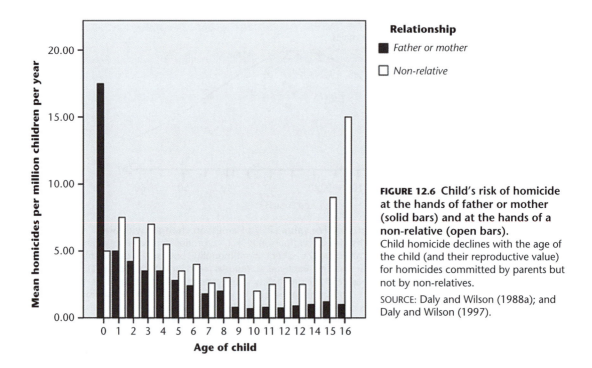

FIGURE 12.6 Child's risk of homicide at the hands of father or mother (solid bars) and at the hands of a non-relative (open bars).
Child homicide declines with the age of the child (and their reproductive value) for homicides committed by parents but not by non-relatives.

SOURCE: Daly and Wilson (1988a); and Daly and Wilson (1997).

Table 12.1 Results of a review of 13 studies linking age of child to abuse or homicide.

As predicted by the reproductive values model of Daly and Wilson, risk of violence towards children decreases with age (hence negative correlations).

Variables	Correlation coefficients	Number of studies reviewed
Child abuse and age	All negative: –0.42 to –0.97	9
Homicide and age	All negative: –0.68 to –0.76	4

12.3.2 Infanticide and step-parents: the Cinderella syndrome

Popular traditional culture abounds with tales of wicked step-parents who fail to provide proper care for their children. The story familiar to many in the West is that of Cinderella. Cinderella's biological mother has died and her father has remarried. Cinderella is raised in the house of her stepmother, who has two children of her own by a former marriage. Despite the obvious virtues of Cinderella compared with her 'ugly sisters', she is treated harshly. Fortunately, a way out of her plight is provided by the *deus ex machina* of a charming

FIGURE 12.7 Cinderella.
SOURCE: Illustrated Punch, 2/4/1892, p. 162.

prince. It would be wrong to try to construct a scientific argument on the basis of a fairy story, but it is worth noting that cultures all around the globe have their own variant of the Cinderella myth. The essence of such stories is that stepfathers and stepmothers are wicked and not to be trusted. Again, this proves little by itself since all the stories could have descended from a common archetype. Such is the bad image of step-parents in traditional folklore that, faced with a high divorce and remarriage rate in the Western countries, children's books now go out of their way to show step-parents in a more positive light.

The stories may, however, reflect a fundamental component of the human experience: the raising of children by non-genetic parents. Throughout human history, step-parents must have played a role in the raising of young when, for a variety of reasons, one parent was killed or left the family unit. A prediction that follows from selectionist thinking about parenting is that parental solicitude should be discriminating with respect to the offspring's contribution to the reproductive interests of the parent. In stepfamilies, it would follow that parents should be more protective of their biological children than their stepchildren. We are not reliant on folktales to test these predictions. Daly and Wilson (1988a, 1988b) supply a range of cross-cultural statistical evidence showing that children in stepfamilies are injured or killed in disproportionate numbers compared with those in their biological families. The effect seems to be extremely robust. As Daly and Wilson (1988a, p. 7) remark:

> Having a step-parent has turned out to be the most powerful epidemiological risk factor for severe child maltreatment yet discovered.

A child living with one or more step-parents in the USA in 1976, for example, was 100 times more likely to be fatally abused than a child of the same age living with his or her biological parents. This effect is also illustrated by the Canadian data for 1973–84 (Figure 12.8). In a recent meta-review of the literature, Archer (2013) found that 8 out of 9 child abuse studies and 9 out of 10 homicide studies showed more violence to children by step-parents compared to genetic parents.

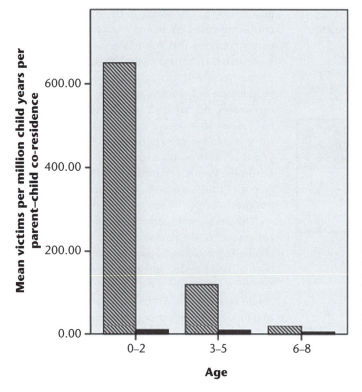

Parent type

▨ *Step-parent*

■ *Natural parent*

FIGURE 12.8 Risk of a child being killed by a step-parent compared with a natural parent in relation to the child's age.

The homicide rate by step-parents vastly exceeds that of natural parents.

SOURCE: Daly and Wilson (1988a); and Daly and Wilson (1997).

These findings have, understandably, been met with shock and incredulity, and there have been numerous attempts to show they are not sound. One obvious objection is that families containing step-parents may be unrepresentative of family life since, by definition, step-parents are the products of some breakdown of an existing family unit. The previous family unit may have failed because of violent behaviour by one of the parents, and hence the sample is not representative. Temrin et al. (2011), for example, have proposed that stepfamilies are likely to contain a higher proportion of parents that are disposed to violence and so violence to stepchildren could result from this and not any low co-efficient of genetic relatedness between child and parent. Hence although the evolutionarily derived 'discriminative solicitude hypothesis' is consistent with patterns of violence in stepfamilies, so might be the alternative 'anti-sociality hypothesis'. Hilton et al. (2015) tested these competing ideas by deriving two sets of predictions. Firstly, if the violence in a family comes solely from the antisocial tendencies of an aggressive parent,

then violence should be equally directed towards genetic and stepchildren; secondly, if the observed bias in violence towards stepchildren derives from this antisocial personality effect, then the differences in abuse rates between step and genetic children should reduce as any independent measure of aggression in the step-parent also reduces (this follows from the idea that in the antisocial personality hypothesis stepfamilies will be over-represented in parents with violent tendencies). To test these predictions the group collected data from 387 men who had at least one step- or genetically related child under 18 and had committed a violent offence against the child. They were also able to attach a measure of 'anti-sociality' to each father (independent of the abuse of the child) by examining police records and noting their criminal history. The results showed that stepchildren faced a greater risk than genetic children in the same family – contrary to the anti-sociality hypothesis which predicts roughly equal violence to any child due to the personality of the father. In addition, the greater risk to stepchildren persisted irrespective of

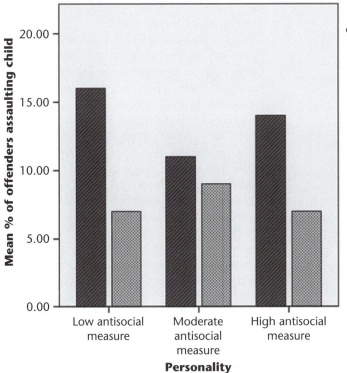

Stepchild or genetic child

☐ *Step*

☐ *Genetic*

FIGURE 12.9   **The tendency of stepfathers to assault either step- or genetic children.** Sample was 118 offenders who had both step- and genetic children. At all levels of personality stepchildren were abused more frequently tending to rule out explanations based on antisocial personality. The higher risk to stepchildren did not disappear as the level of inherently antisocial personality trait reduced, also suggesting that it is kinship that is the main driving factor in accordance with the discriminative parental solicitude model. Personality measure was based on criminal records.

SOURCE: Data from Hilton, N. Z., G. T. Harris and M. E. Rice (2015). 'The step-father effect in child abuse: comparing discriminative parental solicitude and antisociality.' *Psychology of Violence* 5(1), Figure 2, p. 8, copyright © 2015 by the American Psychological Association. Adapted with permission.

the antisocial personality tendencies of the perpetrator (see Figure 12.9).

Another prediction that follows from kinship models of violence is that the killing of stepchildren will be a result of hostile aggression, whereas when parents kill their genetic children it will more likely be due to instrumental reasons (such as low reproductive value) and hence will be by different means. There is some support for this: in studies by Daly and Wilson (1994) and Harris et al. (2007) stepchildren were far more likely to be killed by beating, whereas genetic children are more likely to have been killed by suffocation, drowning or strangulation. As a contrary note, we might observe that this is what we would also expect from the antisocial personality hypothesis: violent parents end up in stepfamilies.

In summary, then, the discriminative parental solicitude and kinship models of violence in families seem to be strongly supported by the data on violence in stepfamilies. The ultimate causation may be based on genetic relatedness but the proximate mechanism could possibly include weaker bonds of attachment

between stepchildren and their step-parents. Some anomalies remain: children who are adopted, for example, do not appear to face any greater risk than natural children in a family. Possibly the risks here are reduced by the rigorous screening of would-be adoptive parents in Western countries – a factor illustrating again the difficulty of interpreting the data on crime in the family.

It hardly needs saying that this does not cast a shadow on all stepfamilies or that violence in these circumstances is somehow excusable. Fortunately, homicide is still an extreme rarity; it is probably more of a puzzle to explain, in selectionist terms, why so many stepfamilies do work and so many children receive love and affection from non-biological parents. Crime statistics can be depressing, so it is good to realise that counter-Darwinian (or at least not obviously selected) behaviour is also part of being human.

Conflict is not always a product of genetic relatedness or the lack of it. Violence can follow from competition over resources and differences in mating strategies. The next two sections examine these causal factors.

12.4 Young men behaving badly: age-specific violence and crime

> I would there were no age between sixteen and
> three-and-twenty, or that youth would sleep out the
> rest; for there is nothing in the between but
> getting wenches with child, wronging the ancientry,
> stealing, fighting.
> (Shakespeare, *A Winter's Tale*, Act III, sc. iii)

In 1983 Travis Hirschi and Michael Gottfredson published a landmark paper on how crime rates vary with the age of the offender. They produced a series of curves which they demonstrated to be remarkably consistent across time, place and culture. Generally, crimes against property and persons rose rapidly in early adolescence, peaked around 20 years of age and then declined more slowly thereafter. Males and females had a similar shaped curve but male rates were typically many times higher than female rates at all ages. The authors were at a loss to explain how these curves could be explained by concepts and theories then available in the discipline of criminology. Since this initial publication the age–crime curve has been shown to have a familiar and similar shape in a wide range of industrialised societies

(see Ellis and Walsh, 2000). Figure 12.10 shows the age–crime curve for the USA in 2010.

Explaining the shape of the age–crime curve has proven surprisingly difficult (Walsh and Beaver, 2009). In relation to the disparity between males and females, Kanazawa (2009) argues that the male desire for resources and status originally evolved as a means of attracting mates and this explains why the vast majority of crime worldwide is committed by men. Women do commit crime, but, as Anne Campbell (2009) argues, compared to men females tend to commit different types of offences and for different reasons. Whereas male crime tends to be focused on physical dominance, display, status and resource acquisition, female crime is more risk averse (females may have dependent children after all) and often is carried out to meet immediate needs such as food or rent money.

If we accept as a working hypothesis that one of the ultimate causes of male crime is the urge to acquire status, resources and wealth in the context of mating behaviour then the shape of the curve begins to make sense: crime rates rise rapidly as men come to sexual maturity and as the benefits of crime begin to satisfy the drive towards status-enhanced reproductive success. To explain the fall after the age of about 20–25 years Kanazawa and Still (2000) have advanced a simple and rather neat potential explanation of the curve. Their argument is that the benefits and costs of competitive risky behaviour (a characteristic of crime) will both rise sharply with age

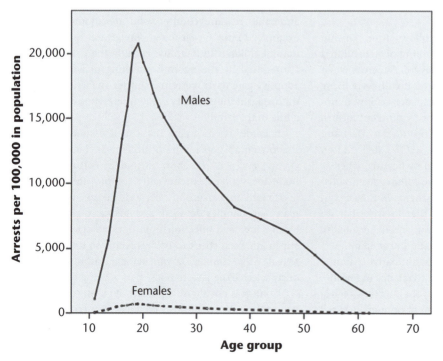

FIGURE 12.10 Arrests in the USA per 100,000 of population in specific age groups for the year 2010. Male crime greatly exceeds female crime at all age groups. Crime rates rise rapidly in adolescence and peak around 18–20 years of age. Similar patterns are found in most Westernised societies.

SOURCE: Data from Tables 4 and 5 in Snyder (2012).

in the teenage years but in slightly different ways. There are no reproductive benefits from competing before puberty for obvious reasons; but once men are capable of reproduction the benefits of competition (appearing occasionally as violence and theft) in ancestral environments would rise rapidly and remain high afterwards since males remain fertile into late adulthood. The costs (retaliation and punishments) rise in a similar way but time-shifted by a few years. According to Kanazawa and Still this is because the damage to a male's reproductive fitness rises rapidly as soon as he has produced a child since any costs then potentially impact on the welfare of that child. Hence the cost curve is time-shifted since a male typically produces a child a few years after the beginning of sexual activity. This is part of a more general pattern that men are predicted to shift some energy from mating to parenting after they become fathers (see Chapter 8). Interestingly, some studies do show that testosterone (a hormone implicated in male violence, risk taking and competitiveness) is higher in non-married compared to married men (Mazur and Michalek, 1998) – a thesis that Peter Gray (2003) calls the 'reproductive bonding hypothesis'.

The net fitness benefit curve is obtained by subtracting costs from benefits and bears a remarkable resemblance to age-specific crime rate and violence curves (compare Figure 12.10 and Figure 12.11). The analysis of Kanazawa and Still could be further tested by examining the crime patterns of young men with and without children. Although the cost curve may represent an ancestral adaptation to times when men probably were fathers by a given age, it is reasonable to expect that it is also sensitive to men's knowledge of whether they are actually fathers or not. One might expect then the overall crime curve to peak sooner and decline much more slowly for men of various ages who are not fathers.

Kanazawa (2009) extends his analysis to make a series of predictions about patterns of criminality, some of which already fit with trends known to criminologists and others which remain to be tested:

(i)   There should be a negative correlation between criminality and social class. If wealth, resources and status are all mate attractants then males of low social class lacking in these assets will on average be more motivated to seek them through crime than those with them.

(ii) Since women find taller men more attractive shorter men will be more motivated than average to compensate for this deficit through the illegitimate accumulation of wealth and status.

(iii) For similar reasons to (ii) criminality is predicted to be higher amongst unattractive men compared to attractive men.

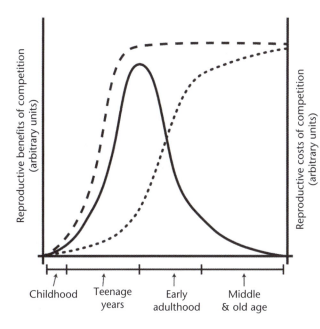

KEY:

– – – Reproductive benefits

· · · · · · · Reproductive costs

——— Net benefits =
Reproductive benefits
– Reproductive costs

FIGURE 12.11 A hypothetical costs and benefits curve for male crime as postulated by Kanazawa and Still (2000).
Reproductive benefits rise steeply at adolescence and remain high throughout adulthood (since males are fertile into middle age and beyond). The cost curve also rises steeply but with an age delay (see text). The net benefits curve (obtained by subtracting costs from benefits) rises and falls in a similar pattern to that actually observed for crime rates (see Figure 12.10).

SOURCE: Redrawn and modified with permission from Kanazawa, S. and M. C. Still (2000). 'Why men commit crimes (and why they desist).' *Sociological Theory* 18(3), 434–47.

(iv) If we take intelligence as a measure of how to deal with novelty then people of low intelligence are less likely to comprehend the novel ways (in terms of ancestral conditions) by which crimes are detected and punished in modern society (security cameras, forensic evidence and the criminal judicial system) and so people of low intelligence will tend to commit more crimes than those of higher intelligence. People of high intelligence are, of course, much less likely to be found out, so this prediction is in addition to this factor.

The model of Kanazawa and Still is persuasive but more work needs to be done in explaining why the female curve rises and falls in a broadly similar way but at a much lower peak than absolute crime rate and falls more slowly. It is plausible that the reproductive benefits to a woman from crime are lower (her reproductive output is more limited by biology) and that the costs are higher to women with children and remain higher, relatively speaking, compared to men for a longer period after birth since children are heavily dependent on female care and protection. The benefits curve for female crime would also have a different shape due to the menopause.

There is at last a growing recognition of the need to apply biological and evolutionary concepts to the discipline of criminology (see Walsh and Beaver, 2009). The work of Wilson and Daly and Kanazawa and Still has pointed the way but it will need some ingenious work to tackle such a socially structured problem as crime and also devise procedures to test evolutionary hypotheses that are not equally well explained by other accounts. One further line of enquiry is to examine if the costs and benefits underpinning the age–crime curves are also influenced by current environments as predicted by human behavioural ecology (HBE) models.

fight each other for females and females will tend to be choosy. But the precise title of the work in which Darwin expounded these ideas is 'The Descent of Man and Selection in Relation to Sex', and the phrase 'selection in relation to sex' suggests the possibility of other forces of selection related to sexual behaviour. Perhaps the best contender for a third category of sexual selection is the conflict that arises between the sexes as a result of different interests and different decisions reached about such matters as the timing of sexual reproduction, number of offspring, pair bond duration, number of other mates, and indeed whether the sex making the advances is suitable at all. In behavioural ecology this perspective has come to be known as sexual conflict theory, a prime prediction of which being the idea of sexually antagonistic selection where selective forces act out in different directions in males and females. This typically suggests, among non-human mammals at least, that we might expect males to develop adaptations to coerce females to mate against their will and to reduce their freedom to mate with other males. Recorded examples of this behaviour include forced copulation, mate guarding, copulatory plugs, and even, in the case of some insects such as bed bugs, hypodermic genitalia that inseminate a female through her abdomen. As a consequence, females are predicted to evolve counter-adaptations to resist these strategies and secure their own interests. Amongst humans concealed ovulation is a prime candidate for a sexually adaptive countermeasure.

We might begin, therefore, to see sexual selection as consisting of the four components shown in Figure 12.12. The resulting conflict of interests manifests itself in several ways. In evolutionary psychology the application of these ideas about sexual conflicts often falls under the heading of strategic interference theory – a topic considered later in this section.

12.5 Human sexual conflicts

12.5.1 Conflicts arising from sexual selection

In Darwin's original formulation there were two forces of sexual selection: intra and inter, with the usual outcome, amongst mammals at least, that males

12.5.2 Strategic interference theory

As noted earlier, conflict between men and women will arise as a result of different mating strategies. One component of this is that males and females will have different levels of sociosexuality: males will typically desire more sexual partners in any given period

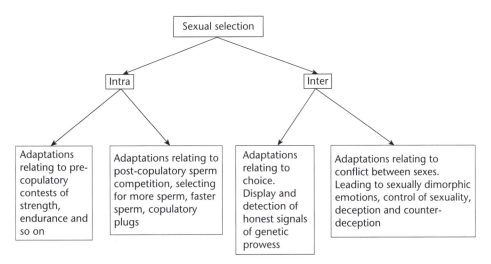

FIGURE 12.12 The existence of two sexes with different physiologies and interests leads to a variety of outcomes under the heading of sexual selection.

and will generally wish to progress from dating to sexual intercourse more rapidly than will women. These divergent strategies, themselves reflections of underlying biological realities discussed in Chapter 4, will often clash – a phenomenon explored by strategic interference theory (Figure 12.13).

Strategic interference theory (SIT) has been developed by David Buss and others as a way of understanding the outcomes of the incongruence between male and female strategies (Buss, 1989a). The theory examines the consequences of the conflict that results when one sex interferes with the successful enactment of the strategy of the other. For example, a woman may desire a mate who is kind, wealthy and ambitious. A man, realising this preference, may exaggerate his level of kindness, wealth and ambition. As a consequence a woman may end up with a mate below the level of qualities that were advertised and that she desired. Another component of SIT is the idea that men and women will be predicted to have an array of adaptations (such as the emotions of anger and jealousy) to alert each to the possibility of deception and motivate corrective action (Figure 12.13).

As well as deception, the clash of strategies, as illustrated in Figure 12.13, can result in other forms of conflict such as sexual harassment and aggression. The higher sociosexuality of men would suggest that men will harass women for sexual favours more commonly than women will men. Several studies have shown that sexual harassment in the workplace is more commonly experienced by women than by men (Juliano and Schwab, 2000) and that the largest proportion of victims tend to be women who are single, divorced or separated and under 35 years of age (Studd and Gattiker, 1991). Robust tests of predictions about the direction of sexual harassment, however, are fraught with methodological difficulties. Part of the problem is that harassment is difficult to define and is experienced differently, with probable different rates of reporting, by men and women. Interestingly, if SIT is correct women will experience more distress from harassment than men (since it is an expression of a male sexual strategy of low-commitment sexual relations and a threat to a typical female strategy of requiring a demonstration of status, resources and commitment) and so women are predicted to report harassment more frequently. The situation is further complicated by the social and organisational status of the harasser and the harassed. SIT would predict that since women seek status in a prospective partner then harassment from a high-status male would be less unwelcome than harassment from a low-status male. This would be an effect very difficult to test given the factors that impinge upon whether such complaints would be reported. Unsurprisingly, there are mixed findings about whether the realities of these complex situations accord with SIT (see Browne, 2006).

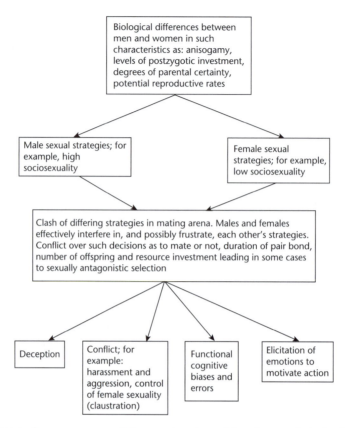

FIGURE 12.13 The origins and consequences of divergent male and female sexual strategies as envisaged by sexual conflict theories and strategic interference theory.

12.5.3 Marriage as a reproductive contract and control of female sexuality

Marriage is a cross-cultural phenomenon and, although ceremonies differ in their details and marriage law varies across cultures, marriage has a set of predictable features across virtually all societies. Marriage entails or confers: mutual obligations between husband and wife; rights of sexual access that are usually but not always exclusive of others; and the legitimisation of children.

Viewed through Darwinian eyes, marriage begins to look like a reproductive contract. This is seen at its clearest when fears arise that the contract has been breached, such as when one male is cuckolded by another. In reproductive terms, the consequences of cuckoldry are more serious for the male than the female: the male risks donating parental investment to offspring who are not his own. This male predicament is reflected in laws dealing with the response of the male to incidents where he finds his partner has

been unfaithful. In the USA, a man who kills another caught in the act of adultery with his wife is often given the lesser sentence of manslaughter rather than murder. This is a pattern found in numerous countries: sentences tend to be more lenient for acts of violence committed by a man who finds his wife in flagrante delicto. The law carries the assumption that, in these circumstances, a reasonable man cannot be held totally responsible for his actions.

Violence after the event may serve as a threat to deter would-be philanderers or to rein in the affections of a wife whose gaze may be wandering. Before the event, however, males of many species guard the sexuality of their partners. Research on animals has shown that, in numerous species where parental investment is common, males have evolved anti-cuckoldry techniques. Male swallows, for example, follow their mates closely while they are fertile, but when incubation begins, mate guarding ceases and males pursue neighbouring fertile females. When the same males perceive that the threat of cuckoldry is high, such as when they are experimentally temporarily removed, they seem to compensate by increasing the frequency of copulation with their partner. Indeed, male swallows do have something to fear: Moller (1987) estimated that about 25 per cent of the nestlings of communal swallows may be the result of extra pair copulation.

If anti-cuckoldry tactics have evolved in avian brains where members live in colonies, both birds are ostensibly monogamous and both sexes contribute to parental care, might we not expect similar concerns to have evolved in humans? The answer is probably yes. Numerous cultural practices can be interpreted as reflecting anxieties about paternity. These include: veiling; chaperoning; purdah and incarceration; foot binding; and, finally, the obscene practice of female genital mutilation (FGM).

12.5.4 Jealousy and violence

> Not poppy, nor mandragora,
> Nor all the drousy syrups of the world
> Shall ever medicine thee to that sweet sleep
> Which thou owedst yesterday.
> (Shakespeare, *Othello*, Act III, sc. iii, 333)

Iago, who speaks the lines above, is one of the most evil of Shakespeare's characters, and he understands well the power of **jealousy** acting on the human mind. Tormented by his suspicions, Othello first kills his wife Desdemona and then himself. In the light of selectionist thinking, the emotion of jealousy in males is an adaptive response to the risk that past and future parental investment may be 'wasted' on offspring that are not their biological progeny. In females, given the certainty of maternity, jealousy should be related to the fact that a male partner may be expending resources elsewhere when they could be devoted to herself and her offspring. Men also lose the maternal investment that they would otherwise gain for their offspring since this is directed at a child who is not the true offspring of the male. It is to be expected that jealousy and its consequences would be asymmetrically distributed between the sexes. We should also expect it to be a particularly strong emotion in men since humans show a higher level of paternal investment than any other of the 200 species of primates. In a near-monogamous mating system, men have more to lose than women do.

To test for sexual differences in the experience of jealousy, Buss et al. (1992) issued questionnaires to undergraduates at the University of Michigan asking them to rank the level of distress caused by either the sexual or the emotional infidelity of a partner. The result suggested that men tend to be more concerned about sexual infidelity and women more about emotional infidelity.

The same effect was observed when subjects were 'wired up' and tested for physiological responses to the suggestion that they imagine their partner behaving unfaithfully either sexually or emotionally. The difference was less marked for women, but men consistently and significantly showed heightened distress at thoughts of sexual, compared with emotional, infidelity.

Such effects are what would be predicted from an evolutionary model of the emotions. Men would tend to be more concerned about the sexual activity of their partners since it is through extra pair sex that a male's investment is threatened. Women, on the other hand, should be less concerned about the physical act of sex per se than about any emotional ties that might lead her partner and his investment away from herself.

Following Buss's suggestions, the evidence supporting sex differences in human jealousy has typically come from three types of study:

1. Forced choice studies where men and women are obliged to state which situation (sexual infidelity or emotional infidelity) they would find most distressing.
2. Physiological response studies where physiological parameters (such as heart rate, blood pressure and palm sweating) are monitored as subjects are asked to imagine various types of jealousy-inducing scenarios.
3. Continuous rating scale questionnaire studies where men and women grade their response to situations that invoke jealousy.

Reviewing such studies, Harris (2003) casts doubt on whether the evidence really did support predictions from evolutionary psychology as advanced by Buss and others. In essence she concluded that in some studies the differences were small or even reversed. More recently Robert Pietrzak et al. (2002) have suggested that such ambiguities in the literature may be a result of the fact that different tests have been applied at different times to different subject groups. Accordingly, these researchers set about applying all three types of test to the same group of subjects. This time the three tests gave consistent findings. On the forced choice experiment, males reported more distress at the thought of sexual infidelity and females more distress at the thought of emotional infidelity. The physiological measures used by the group were heart rate, skin conductance, skin temperature and frowning activity. Relative to measures of these variables when neutral imagery was used, men demonstrated a greater increase than women in heart rate, skin conductance and frowning during sexual infidelity imagery. In contrast, and as predicted, women were more responsive than men on all these measures in response to images of emotional infidelity. Figure 12.14 shows the results for four measures on the continuous rating scale measurements in relation to sexual and emotional infidelity.

Although the results were internally consistent and as predicted from evolutionary expectations,

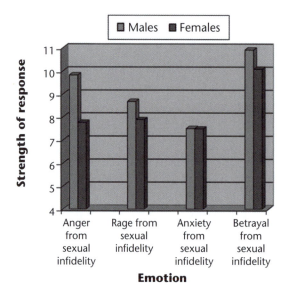

FIGURE 12.14 Differences between males and females in their emotional response to sexual and emotional infidelity scenarios.

SOURCE: Data from Pietrzak et al. (2002).

For sexual infidelity differences (M > F) were significant for anger ($p < 0.001$), rage ($p < 0.001$) and betrayal ($p < 0.01$).

the drawback remains that the investigation was carried out on a narrow subject group: US college undergraduates from a largely middle-class background. Perhaps for these reasons a Darwinian approach to jealousy and its central postulate, that it is a sexually dimorphic emotion evolved to respond to the different problems faced by males and females, has attracted numerous attempts to demonstrate its falsity and the superiority of alternative accounts. Consequently, the issue now is favoured with a large body of literature. Two reviews of this stand out: that of Christine Harris (2003) is favourable to the various studies that challenge evolutionary accounts; whilst Buss (2014) lists a total of 26 studies that offer support. Faced with seriously diverging conclusions it is hard at present to see a clear consensus emerging. One advantage of the evolutionary approach is that it is fruitful in suggesting new testing possibilities. Easton et al. (2007), for example, looked at the case histories of people diagnosed with morbid jealousy – in many ways a methodological improvement over studies asking people to state their feelings from imagined scenarios. They predicted that men would be more jealous than women in relation to rivals that offered status and resources; whilst more women than men were predicted to be jealous of rivals that offered youth and physical attractiveness. Both predictions were supported. In relation to homicides resulting from jealousy the data points in more ambiguous directions. It is clear that more men than women commit homicide as a result of jealousy (Daly and Wilson, 1998) but men are more likely to commit homicide anyway. Harris (2003) reviewed 20 studies and found no difference in jealousy-motivated homicide rates between men and women when base rates were taken into account. Similarly Felson (1997) looked at 317 cases of homicide and found little difference in the proportion where men killed women as a result of sexual jealousy compared to the proportion where females killed a male partner as a result of sexual jealousy. Perhaps what is needed here is an examination of the causes of jealousy in both cases – that is, physical or emotional infidelity – where sex differences would be predicted. More recently, Sagarin et al. (2012) conducted a meta-analysis of 40 individual studies and concluded

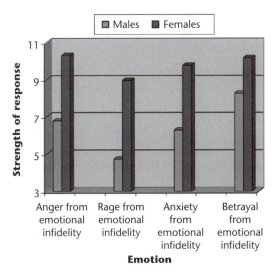

FIGURE 12.15 Male and female emotion ratings to emotional infidelity scenario.

For emotional infidelity differences were significant (F > M) for anger (p < 0.01) and anxiety (p < 0.02).

SOURCE: Data from Pietrzak et al. (2002). 'Sex differences in human jealousy: a coordinated study of forced-choice, continuous rating-scale, and physiological responses on the same subjects.' *Evolution and Human Behavior* **23**(2): 83–95.

that there was a high level of support for jealousy as a sexually dimorphic emotion driven by different sexual interests.

Most studies on this sexually dimorphic 'jealousy effect' have been conducted on populations in industrialised countries, often using students as subjects. One recent exception is work by Brooke Scelza (2013) on a small 'natural fertility' population: the Himba of Namibia. Consistent with other work, Scelza also found that men were more concerned about sexual infidelity than emotional infidelity by a wide margin. Interestingly, women were also slightly more concerned about sexual infidelity than emotional infidelity, but their concern about emotional infidelity was still greater than that of men.

Despite there being some criticism of the artificial nature of tests used to explore this effect (laboratory situations and imagined scenarios), the overall thrust of the work seems quite robust.

12.5.5 Deception and signalling

When my love swears that she is made of truth,
I do believe her though I know she lies,
That she might think me some untutored youth,
Unlearned in the world's false subtleties.
Thus vainly thinking that she thinks me young,
Although she knows my days are past the best,
Simply I credit her false-speaking tongue:
On both sides thus is simple truth suppressed...
(Shakespeare, Sonnet 138)

With typical incisiveness and compaction of thought, Shakespeare neatly captures the deceptions and counter-deceptions that figure in human sexual politics. He pretends to believe the lies of his lover in the hope that she will think him less worldly and more innocent than he really is, yet he knows she does not fall for this, but both persist in this mutual game of deception and counter-deception.

Four hundred years later we are still at it and signalling is as important as ever. Humans mate assortatively (that is, mating is not random and high-quality phenotypes tend to end up with other high-quality phenotypes) and so the quality of mate that an individual can secure will depend on signals conveying information about the individual's own qualities. Under these circumstances, we might predict that selective forces will arise to signal a quality higher than is the case (that is, a dishonest signal); and also, since everyone is both a signaller and a receiver, adaptations to distinguish reliable from unreliable signals. There is now quite a large body of literature of the evolution of honest and dishonest signals but one consensus that has emerged is that traits that are costly to produce will reliably signal quality. Hereafter, however, the dynamics could get complicated: males might begin by displaying a costly trait signalling high quality and females select on the basis of this; but then the race is on for males to evolve ways of displaying the trait without paying the cost, leading to a loss in fitness in females that fall for this. This itself serves as a selective force, causing females to shift to select on the basis of a new trait that is more honest, and so the cycle continues. Similarly selection should favour traits that enhance the ability of males to control the reproductive behaviour of females whilst simultaneously favouring traits that enable females to avoid such control.

There is now a flourishing line of research examining how the sexes deceive reach other but in different ways (Mulder and Rauch, 2009). Tooke and Camire (1991), for example, in a study on undergraduates, found that males were more likely to exaggerate their career prospects and level of commitment compared to women, whilst women were more likely to send false signals about their physical appearance. In a later study, Keenan et al. (1997) found that females tended to assume that males would deceive them by exaggerating their financial assets and future prospects. More recently, Amanda Johnson and co-workers found that the ability to deceive was related to self-awareness. They recruited novice actors to give a short videoed presentation where they had to fake their characteristics by, for example, exaggerating their wealth and physical fitness. The actors then filled in a questionnaire that enabled a measure of their self-awareness to be taken. Independent participants then rated the believability of their performances. The researchers found a significant positive correlation between private self-awareness and the skill to deceive others (Johnson et al., 2005). Such results have ramifications outside of sexual conflict theory. Could consciousness and theory of mind have arisen as adaptive solutions to the problem of deceiving sexual partners whilst simultaneously guarding against being deceived? It is a potentially fruitful area of study that could lead to further hypotheses: we might expect, for example, that females will be better at detecting deception in mating contexts than males since they have more to lose from an unsuitable match.

The feelings experienced by males and females, on realising they have been deceived, should also differ since different types of deception carry different fitness consequences for each sex. To test for this effect, Haselton et al. (2005) issued questionnaires to 217 undergraduate subjects asking them to rate their distress on a scale of 1–7 (1 = not at all distressing, 7 = very upsetting) at the prospect of being deceived in a specific way by a dating partner. Table 12.2 shows a sample of the predictions, their evolutionary foundations and the outcomes of the study. The study

Table 12.2 Sample of results from a study testing strategic interference theory
Men and women are predicted to experience different levels of distress following deception depending on the degree to which such deception thwarts their own sexual strategy.

Type of deception	Adaptive reasoning	Level of distress outcome predicted	Level of male distress (+/– SD)	Level of female distress (+/– SD)	p value (supported or not)
Resource deception. Stating a higher level of resources than is the case	Women tend to value resources in a relationship more than men (see Section 14.3) and will be more upset by this type of deception	Women > Men	3.17 (1.40)	4.38 (1.42)	< 0.01 (supported)
Pre-copulation statements about depth of feelings. Stating deeper feelings than are felt	Women desire commitment to longer-term relationships more than men	Women > Men	4.35 (1.82)	6.74 (0.75)	< 0.01 (supported)
Pre-copulation signals about sexual access	Men desire more short-term mating opportunities and have a higher sociosexuality	Men > Women	4.69 (1.68)	3.24 (1.52)	< 0.01 (supported)
SOURCE: Results from Haselton et al. (2005), Table 2, p. 7.					

offers support for SIT but given that these are antici-pated feelings it would also be useful, though more difficult, in future studies to measure actual behaviour and distress in real contexts. It would also be interesting to look for sex differences in adaptations to detect for the types of deception that males and females can expect – are there different sensitivities here, for example?

Amid this welter of honest and deceptive signals humans still have to make mating choices and this forms the subject of the next, necessarily lengthy, section of this book.

Summary

- A knowledge of genetic relatedness (r) is useful in helping to understand both altruistic and antagonistic behaviour. Conflict will arise when the reproductive interests of two individuals differ. This is predicted between sexual partners and even between parents and offspring since offspring may demand resources at the expense of current and future siblings.

- Maternal–foetal conflict provides a good example of the dynamic balance that is reached when the optima for resource allocation differ between two genomes.

- Human violence resulting in death provides a set of data that allows predictions from evolutionary psychology to be tested. There is evidence, for example, that the risk of a child being killed by its mother declines as the age of the child and the mother increase. These facts are in keeping with the concept of reproductive value, but there may be other factors at work.

- Whereas kinship is usually predicted to reduce levels of violent behaviour, there were probably circumstances in the environment of evolutionary adaptation of early hominids in which, for example, infanticide served as an adaptive strategy to maximise the lifetime reproductive value of a parent. One of the most robust findings of the work of Daly and Wilson is that stepchildren experience a much higher risk of infanticide than genetic children. Psychological mechanisms that were once adaptive in the life history of early hominids may, when activated, now give rise to maladaptive consequences in cultures in which homicide is punished. Daly and Wilson have published pioneering work on homicide, but the nature of the statistics used makes it difficult to control for all the social and cultural factors at work.

- A remarkably robust finding in Westernised countries is the age–crime curve showing that crimes committed by young men rise rapidly in number in adolescence, peak at about the age of 19 and then fall. Crime rates are also considerably higher for men than for women. One potentially useful way of explaining this is the male drive for status and resources – commodities that would once have signalled high reproductive value. Much more work is needed to test this approach.

- Much of the coercion and violence exercised by men over women can be understood in terms of paternity assurance. Marriage, from a Darwinian perspective, can be regarded as a reproductive contract between men and women to serve similar but, in important ways, different interests. Men have probably been selected to monitor and jealously guard the sexuality of their partners, whereas it is probable that women are more concerned about resources. Jealousy is an emotion experienced by both sexes but with more violent repercussions in cases where men suspect sexual infidelity on the part of their partner. Divorce statistics are consistent with the contract hypothesis and reflect age-related fertility differences between men and women.

▶ One of the consequences of intersexual selection is conflict. The sexes are predicted to differ in the sort of strategies that serve their needs. The study of this conflict has been formalised into strategic interference theory. This has proven valuable in explaining, for example, the types of deception each sex practises on the other.

Key Words

- Homicide
- Infanticide
- Jealousy
- Kin
- Maternal–foetal conflict
- Parent–offspring conflict
- Siblicide

Further reading

Daly, M. and M. Wilson (1988a). *Homicide*. New York, Aldine de Gruyter.
Daly, M. and M. Wilson (1998). *The Truth About Cinderella*. London, Orion.

Both these books contain the ground-breaking interpretations of crime statistics made famous by Daly and Wilson.

Walsh, A. and K. M. Beaver (2009). *Biosocial Criminology*. New York, Routledge.

Several chapters deal with an evolutionary approach to crime. Chapter 8 also considers other biological explanations of the age–crime curve.

Salmo, C. A. and T. K. Shackelford (eds.) (2008). *Family Relationships: An Evolutionary Perspective*. New York, NY, Oxford University Press.

Edited volume of articles looking at family relationships from a variety of angles, including conflict and sexual aversions.

Shackelford, T. K. and V. A. Weekes-Shackelford (eds.) (2012). *The Oxford Handbook of Evolutionary Perspectives on Violence, Homicide, and War*. New York, NY, Oxford University Press.

Heavyweight review of the theoretical and empirical literature by major scholars in the field.

Mating and Mate Choice

Human Sexual Behaviour: Anthropological Perspectives

We must acknowledge, as it seems to me, that man with all his noble qualities, with sympathy which feels for the most debased, with benevolence which extends not only to other men but to the humblest living creature, with his god-like intellect which has penetrated into the movements and constitution of the solar system – with all these exalted powers – Man still bears in his bodily frame the indelible stamp of his lowly origin.
(Charles Darwin, 1871, p. 405)

The first part of this chapter examines the physical and historical evidence on human sexuality by employing perspectives often found in biological or physical anthropology, perspectives that routinely focus on Darwin's 'indelible stamp'. Its goal is to establish the species-typical mating strategies employed by human males and females. It has already been established in Chapter 4 that males and females, by virtue of basic differences in biology, are likely to have different interests, so it should come as no surprise to find that these strategies are different.

13.1 Cultural distribution of mating systems in contemporary traditional or preindustrial societies

Humans living in today's industrial or more developed countries are living in conditions far removed from those prevailing in environments where the basic plan of the human genotype was forged. In addition, many such cultures are strongly influenced by relatively recent ideologies and belief systems such as Judaism, Christianity and Islam with their injunctions in favour of or against specific mating systems (Judaism, Christianity and Islam prohibit **polyandry**, for example, Christianity and (by and large) Judaism prohibit **polygyny**, but polygyny is permitted under Islam). If we want to ascertain the sexual behaviour of humans before our mode of life was transformed by the Neolithic revolution and then industrialisation, or before our heads were filled with strict religious and political ideologies, it makes sense to look at traditional hunter-gatherer cultures that still exist, or cultures that have been relatively immune to Western influence and maintain their traditional patterns of life. This is not to suggest that such cultures are primitive or have no ideas of their own, but we could be assured that such cultures have not been subject to the mass persuasion systems of, for example, the Church and State found in the West.

A broad sweep of different human societies reveals that, in many, the sexual behaviour observed departs from the monogamy advocated (at least in a legal sense) in most Western cultures (Figure 13.1). One problem with such quantitative surveys, however, is that terms like marriage, mating, love and partner are highly socially constructed and may not easily be transferable between cultures.

Another fruitful line of enquiry is to look at contemporary hunter-gatherer societies. Here we at least know that people face ecological problems similar to those of our ancestors 100,000 years ago. The often harsh existence of hunter-gatherers also places restrictions on the sort of social development that can take place. Patterns of mating behaviour cannot drift too far from that which is optimal for survival.

The human pair bond

Most mammalian species are polygynous and humans are unusual in the strength of the pair bond that they form. Indeed the human pair bond does

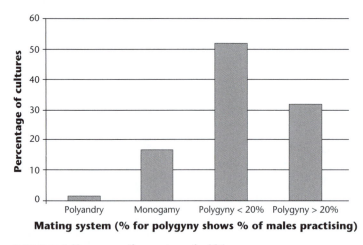

FIGURE 13.1 Human mating systems in 186 traditional cultures prior to Western influence.

SOURCE: Data from Murdock and White (1969).

seem to be a natural state. In most societies most people (about 90 per cent) end up getting married at some time (Fisher, 2012). As noted earlier, polygyny is permitted in about 80 per cent of all human societies, but in most of these ostensibly polygynous cultures only about 5–10 per cent of men have several wives simultaneously (Frayser, 1985). Moreover, polygyny tends to be associated with status, dominance hierarchies and the accumulation of wealth and it is likely in pre-Neolithic hunter-gathering societies that were relatively un-stratified and not focused on the accumulation of capital goods that **monogamy** was the norm for most men and women.

It is a speculative argument, but bipedalism may have helped push early hominins towards a more monogamous lifestyle. Once our ancestors left tree cover and moved onto more open territory, bipedal females probably carried their infants in their arms, rather than have them cling to the fur on their backs as is found in quadrupedal apes. This would have impeded their ability to protect their young and gather food for themselves and their offspring. The protection and attention of a single monogamously inclined male would have helped solve this problem.

But the pattern shown in Figure 13.1 still poses something of a puzzle. The anthropological record shows that over 80 per cent of cultures permit men to take multiple wives, yet in most of the world's developed countries monogamy is legally enforced.

Now in traditional cultures having many wives is associated with wealth and status. Even after the Neolithic revolution as inequalities in wealth intensified so polygynous marriage reached new extremes with the harems of the ruling elite. Yet in modern societies, also characterised by massive inequalities in wealth, monogamy is presented as a desirable norm. Hence the question arises that if polygyny is part of our historic evolved psychology and is favoured by inequalities in wealth, how come monogamy has been accepted as the norm in so many successful modern cultures?

Henrich et al. (2012) provide a persuasive answer by suggesting that monogamy in modern cultures is a socially imposed norm which has been favoured by cultural evolution. In their argument, polygynous cultures contain internal conflict due to the high degree of male–male intrasexual selection (see Chapter 4). Low-status males, deprived of a wife and otherwise facing genetic oblivion, are thereby incentivised to pursue high-risk strategies to secure both resources and mates. This is predicted to lead to higher rates of theft, rape, murder, sexual slavery and prostitution in polygynous cultures compared to monogamous ones – predictions that are worthy of investigation. In contrast, in monogamous cultures the higher number of males actually married tends to reduce crime and induce males to invest in children and generally contribute more to the overall wealth of the culture. In this model, then, monogamous cultures outcompete polygynous ones and monogamy spreads by former polygynous cultures copying the culturally successful practice of monogamy. The global spread of monogamy outside of the West has been relatively recent: laws prohibiting polygyny were introduced by Japan in 1880, by China in 1953, by 1955 in India and 1963 in Nepal.

Monogamy, however, is not always lifelong. In the West, as divorce is made legally easier and more socially acceptable, so the rate has risen. In the UK, for example, by 2010 about 42 per cent of all marriages ended in divorce (Office of National Statistics, 2013). Human monogamy then seems to be a system that does not last and is one carrying a significant degree of infidelity. A variety of surveys on American marriages show indices of philandering, adultery and extra-marital affairs ranging from 20–40 per cent for men and 20–33 per cent for women (Kinsey et al., 1953; Fisher, 1992). Extra pair copulations have been reported in a wide range of other societies (Fisher, 1992) and also in other species, such as birds and foxes, that were once labelled as socially monogamous (Mock and Fujioka, 1990).

13.2 Physical comparisons between humans and other primates

> The next time a new species of primate is discovered we should be able to deduce its social behaviour by examining its testes and dimorphism in body and canine size. (Reynolds and Harvey, 1994, p. 66)

This claim is indeed a remarkable one since it suggests that complex social characteristics can be inferred from simple measurable parameters. If the claim is reliable, the same technique should also, since humans have a shared ancestry with the primates, throw considerable light on ancestral human sexual behaviour. We will now examine the potential of this claim and how it can be applied to the sexual behaviour of humans and other primates.

13.2.1 Armaments or ornaments? Exploring insights from sexual selection

In Chapter 4 we noted how such factors as the operational sex ratio, the relative reproductive rates of males and females and relative parental investment all serve to determine the strength and direction of sexual selection. Among humans, a male's investment in producing and rearing offspring is less than that of female's, hence, after each successful fertilisation a male is free, in principle, to re-enter the mating pool sooner than females. The returning male will then encounter a relative shortage of females for two reasons: a) males have a longer reproductive period than females, b) in any group a proportion of females will already be pregnant or have suppressed fertility whilst they are lactating and caring for an infant. Hence, the operational sex ratio in

human groups will be male biased (that is, M/F > 1). Other factors may reduce the supply of males such as warfare or selective migration, but these issues notwithstanding, this analysis suggests that males will compete with other males, and females, if allowed, will exercise mate choice. But there are also examples in the animal kingdom of females displaying sexually selected traits. This may arise if females compete for resources required for reproduction (for example, nesting sites), when males provide significant amounts of paternal investment, or when mating with multiple partners enhances a female's reproductive success. In Olive baboons, for example, Domb and Pagel (2001) found compelling evidence that the size of a female's genital swellings at the time of **oestrus** was an honest display of her reproductive value; the larger the swelling the younger the age of first conception, the greater number of offspring and the higher the proportion of offspring that survived. Significantly, larger swellings also excited more male interest and male v. male competition.

What is really at stake, and quite difficult to predict, is the relative importance of the three primary mechanisms of sexual selection, namely: same-sex contests to gain access to and monopolise mates (that is, intrasexual selection); discriminating mating choices (intersexual selection) and post-copulatory sperm competition. The consequences are important since as a result sexual selection could bring about either: weapons, musculature and aggression for forcing away competitors; ornaments and displays to appeal to the opposite sex; features of sperm production that enable success in sperm competition; or some combination of all three (Figure 13.2).

Factors influencing the dimensions of sexual selection

A variety of ecological and corresponding physical factors will play a role in determining where males of a given species will lie on the 3D diagram shown in Figure 13.2. If females form small groups they could be collectively defensible by a single male and this will select for skill and strength in male v. male contests, driving down the importance of mate choice and sperm competition. But the mating environment is also important. Most birds, for example, are highly mobile and live in three dimensions and so

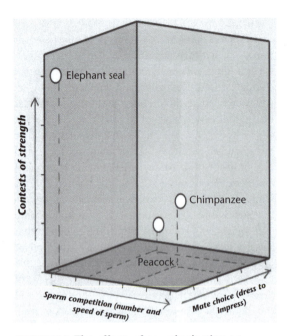

FIGURE 13.2 The effects of sexual selection on males.

Three species are shown for comparison. In the case of uni-male, multi-female elephant seal groups, males face intense male v. male physical competition whilst females exert little choice over the victor and have little choice in mating. Consequently, sperm competition and mate choice factors are of low importance compared to physical contests. In the case of peafowl (the *Pavo* genus) males display in front of females and most females choose the same few males; sperm competition and physical aggression between peacocks is of low importance compared to the development of sexually appealing ornaments to satisfy female choice. Chimpanzees, living in polygynandrous groups, are more complex and mating involves physical aggression and dominance between males, some choice exercised by females and post-copulatory sperm competition (since females will behave polyandrously), hence all dimensions of sexual selection are significantly at work (see Stumpf and Boesch, 2005).

SOURCE: Based on an idea in Puts (2010).

mates are hard to defend. Consequently, the frequency of aggressive male contest is low in this taxon and instead males are obliged to develop features to appeal to female preferences – the extreme form being lekking species (see Chapter 4). The mobility of humans is obviously mostly two dimensional and early hominins would probably have lived in small-ish (100–150) multi-male, multi-female groups with female breeding asynchrony. Such conditions point

at least to the possibility of male contests and the monopolisation of females (that is, polygyny). Now, for whatever ecological and/or sexual reasons, men are stronger than women, and so selection should favour female traits to attract men rather than to try to monopolise them. This latter point suggests that, whereas the study of female v. female physical contests to secure men is probably a research dead end, it does make good theoretical sense in the study of human mate choice to examine those qualities in females that make them attractive to men and so search for their adaptive underpinnings.

Indeed, even a cursory inspection of modern and traditional cultures shows that both males and females display physical and other attributes, and that males are extremely susceptible to physical signals of female sexual attractiveness. It would seem then that gaudy lekking peacocks and plain peahens, or highly dimorphic polygynous elephant seals, do not provide good models for human mating. One consideration that helps an understanding of display and choice in humans is that female fecundity is biologically variable between individual females and strongly related to age, predicting that in seeking long-term relationships women should display and men respond to signs of youth and fertility. This points to the need to carefully examine the balance of a whole range of factors, such as **sex ratios**, parental investment, variation in mating quality, searching costs and encounter rates, to understanding human sexual display and choice (Waynforth, 2011). Table 13.1 shows how display and choice may be related to these key features of mating. This is considered further in Chapter 14.

Body size dimorphism

As discussed in Chapter 4, intrasexual selection will often give rise to characteristics, such as a large body size, that assist in same-sex contests. Figure 13.3 shows how body size dimorphism varies in relation to breeding system for primates. This variation is consistent with the idea that the large differences between measures of dimorphism in monogamous and polygynous (single-male) contexts can be explained by the more intense competition between males for females in the latter. Even in multi-male groups, some competition is observed to secure a

Table 13.1 Factors impacting on mating displays and sex-specific choosiness

Socioecological conditions and population parameters	Displaying sex	Choosy sex
Low male investment in gametes (anisogamy)	Males	Females
Male parental investment lower than female parental investment	Males	Females
Male reproductive rate higher than female reproductive rate	Males	Females
Male-biased OSR (that is, surplus of males, M/F > 1)	Males	Females
Female-biased OSR (that is, a surplus of females)	Females	Males
Competition between females for resources that enhance reproductive success	Females	Males
High variation in male quality	Males	Females
High variation in female quality and fecundity (for example, age related)	Females	Males
High encounter rates High parental investment from both sexes High variation in mate quality in both sexes	Both sexes	Both sexes
OSR = operational sex ratio		

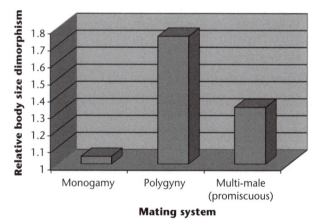

FIGURE 13.3 Body size dimorphism (adult body weight of male divided by adult body weight of female) versus mating system for various species of primate.

SOURCE: Data from Harvey and Bradbury (1991); and Smith and Cheverud (2002), Table 1, p. 1098.

place in dominance hierarchies. This points to the possibility of using body size dimorphism in humans as evidence for an ancestral mating system, a point considered later in this chapter.

13.2.2 Testis size

The significance of **testis** size is that it indicates the degree of sperm competition (see Chapter 4) in the species. In the 1970s, the biologist R. V. Short

suggested that the difference in testis size for primates could be understood in terms of the intensity of sperm competition (Short, 1979). To obtain reliable indicators, testis size has to be controlled for body weight since larger mammals will generally have larger testes in order to produce enough testosterone for the larger volume of blood in the animal and a larger volume of ejaculate to counteract the dilution effect of the larger reproductive tract of the female.

When these effects are controlled for, and relative testis size is measured, the results support the suggestion of Short that relatively larger testes are selected for in multi-male groups where sperm competition will take place in the reproductive tract of the female (Figure 13.4). A single male in a harem does not need to produce as much sperm as a male in a multi-male group since, for him, the battle has already been won through some combination of body size and canine size, and rival sperm are unlikely to be a threat. In contrast, in promiscuous multi-male chimpanzee groups, females will mate with several males each day when in oestrus. It is possible that the sexual swellings that advertise oestrus in many female primates, which males find irresistible, actually promote sperm competition since the best way for a female to produce a son who is a good sperm competitor is to encourage competition among his potential fathers.

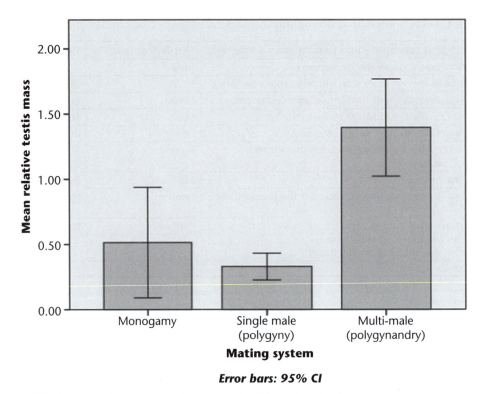

FIGURE 13.4 Relative testis size versus mating system for 29 species of primates.
The y axis refers to deviations from the allometric line of testis size against body weight. Thus, a value of 1 would indicate a testis size expected for an average primate of a given body weight. Values above 1 indicate that testes are larger than would be expected. Expected testis weight was calculated from the allometric equation $T = 0.035 B^{0.72}$, where T = testis weight and B = body weight (grammes). The difference between monogamous and single-male groups is not significant, but both differ significantly from multi-male groups.
SOURCE: Data from Harcourt et al. (1981), Table 1, p. 56.

13.2.3 Testis size and bodily dimorphism applied to humans

Diamond (1991, p. 62) has called the theory of testis size and sperm competition 'one of the triumphs of modern physical anthropology'. As we have seen, the theory has great explanatory power, and we will now apply it to humans.

Table 13.2 shows some key data on testis size and bodily dimorphism for humans and the great apes. The fact that men are slightly heavier than women could reflect a number of features of our evolutionary ancestry. It could indicate the protective role of men in open savannah environments, it could be the result of food-gathering specialisation whereby men hunted and women gathered, or it could reflect male competition for females in uni-male or multi-male

groups. The dimorphism for humans is mild, however, compared with that for gorillas; this would indicate that *Homo sapiens* did not evolve in a system of uni-male harem mating.

Also, if early humans routinely competed to control groups of females, we would not only expect a higher level of body size dimorphism but also smaller testes. The testes of gorillas are about the size of kidney beans and relative to their body size are less than half the average for humans. The erect penis of a 400-pound male gorilla is also only about one inch long. On the other hand, if early humans had behaved like chimpanzees in multi-male groups, we would expect larger testes. In fact, if human males had the same relative size of testis as chimps, their testes would be roughly as large as tennis balls. Figure 13.5 shows a comparative and

Table 13.2 Physical characteristics of humans and the great apes in relation to mating and reproduction

Species	Male body weight (kg)	Female body weight (kg)	Dimorphism M/F	Mating system	Weight of testes (g)	Weight of testes as % of body weight	Approx. number of sperm per ejaculate (x 10⁷)	Estimated global population
Humans (Homo sapiens)	70	63	1.1	Monogamy and polygyny?	25–50	0.04–0.08	25	More than 6,000,000,000
Common chimps (Pan troglodytes)	40	30	1.3	Multi-male in promiscuous groups	120	0.3	60	Fewer than 110,000
Orang-utan (Pongo pygmaeus)	84	38	2.2	Uni-male Temporary liaisons	35	0.05	7	Fewer than 25,000
Gorilla (Gorilla gorilla)	160	89	1.8	Uni-male Polygyny	30	0.02	5	Fewer than 120,000

NOTE: The last column shows the precarious state of the natural populations of our nearest relatives.

SOURCE: Data from Harcourt et al. (1981); Foley (1989); and Warner et al. (1974).

schematic representation of human, gorilla, orang-utan and chimp males. The size of the large circle relative to the female shows the degree of **sexual dimorphism** in the species. The length of the arrows and the pair of dark shapes show the relative size of the penises and testes of the males. One unexplained feature of male morphology is the large size of the human penis.

From a comparison of human testis size with other primates, Short (1994, p. 13) concludes that

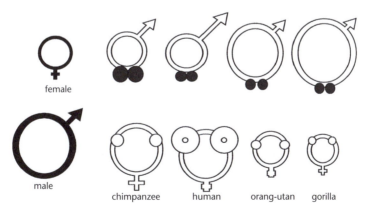

female

male

chimpanzee   human   orang-utan   gorilla

FIGURE 13.5 Sexual dimorphism: how the great apes see each other.
The size of the large circle for each species indicates body size dimorphism. The top row, for example, shows the relative size of the female (dark circle on left) to the male of each species. The size of the pair of solid circles indicates size of testes relative to other males of each species. The small circles on the females indicates the relative size of breasts; the size of the arrows indicates relative penis size for males. The large size of the human penis compared to other primates, and the large size of human female breasts are distinctive features on which there is as yet no explanatory consensus.

SOURCE: Data from Short and Balban (eds.) (1994); Foley (1989); Warner et al. (1974); and Dixson (1987).

we are not 'inherently monogamous ... neither are we adapted to a multi-male promiscuous mating system'. His view is that 'we are basically a polygynous primate in which the polygyny usually takes the form of serial monogamy'. This is to examine the problem from a male perspective of course. The fact that humans do have relatively large testicles (larger than gorillas) does imply that sperm competition has been a feature of our ancestry. Now, 'it takes two to tango' as the saying goes; so if sperm have competed in our past then women must have mated (polyandrously) with more than one man. But as we saw in Chapters 5 and 6 infant humans need bi-parental care and this is facilitated by a strong pair bond. The way to reconcile these differing indications might be to conclude that ancestral human mating was ostensibly monogamous but plagued by 'adultery' or covert extra pair copulations. Not so different, if we take a clinical view, from the situation in many societies today.

13.2.4 Sperm swimming speeds and morphology

Speed

In species where sperm competition is intense there is a selective pressure driving males to produce large quantities of sperm on the basis that the more sperm produced by an individual male the better the chance of fertilisation involving his sperm. Upping the quantity is one strategy; but, if sperm from several males are competing in the reproductive tract of a female, might there also be a selection in favour of faster sperm? After all, the first sperm to reach the egg is surely at an advantage. Nascimento et al. (2008) investigated the swimming speed of sperm for four different species of primate: chimpanzees, rhesus macaques, western lowland gorillas and humans. Figure 13.6 shows their results. Both chimps and

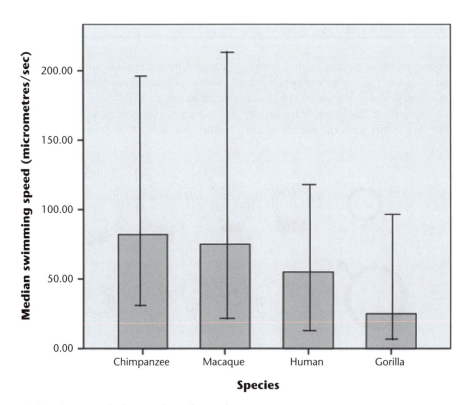

FIGURE 13.6 Swimming speed of sperm from four primate taxa.
Bar height shows median values and line shows range. High swimming speeds are plausibly associated with high sperm competition. Each group is significantly different (p < 0.05) from the other.
SOURCE: Data from Nascimento et al. (2008), Figure 1a, p. 300.

macaques live in multi-male-multi-female sexual groups which means that females will mate with several males over the period of a few days. As expected from sperm competition theory, sperm from these species do have the highest median speeds. Male gorillas in uni-male/multi-female groups produce the slowest sperm since males here experience little sperm competition. Humans lie between the two positions, suggesting once again that sperm competition was not entirely absent in our evolutionary past.

head   midpiece

FIGURE 13.7 Mammalian sperm showing midpiece.

Morphology

Sperm motility is driven by mitochondria packed into the midpiece section (Figure 13.7). Overall, sperm size is remarkably constant across mammalian species and does not seem to vary with body size or mating system. But it might be supposed that a large midpiece section should predict a faster swimming speed. This is exactly what Firman and Simmons (2010) found in their experiments on sperm from house mice. It might also be expected that since motility is related to mating system then midpiece volume might also vary in relation to the mating behaviour of the species. Anderson and Dixson (2002) measured the midpiece volume of sperm from 31 species of primate and were able to show that midpiece volume and testis size were strongly correlated – a finding in keeping with the idea that large testes indicate high levels of sperm competition. The authors also tested the hypothesis that there might be a difference in midpiece volume between primates where females mate with single partners and where females mate with multiple partners. The difference was significant: with groups where females mate with multiple partners (and therefore create the conditions for sperm competition) males have a higher midpiece volume. Their work also showed that humans lie closer to the single partner than multiple-partner grouping (Figure 13.8).

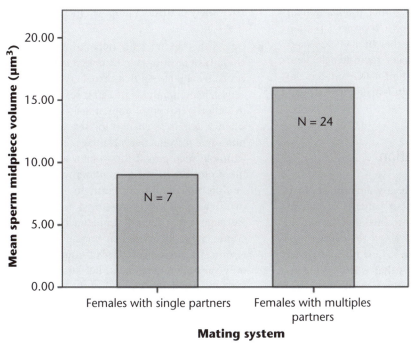

FIGURE 13.8 Sperm midpiece volume in relation to mating system. Midpiece volume is significantly different (F = 3.46, p < 0.05) between primate groups where females have single partners compared to those with multiple partners. N = number of species in each group.

SOURCE: Data from Anderson and Dixson (2002), Figure 1, p. 1701.

One mystery concerning the fate of sperm is that the typical lifespan of ejaculated sperm is about 48 hours, about twice as long as the 24-hour lifespan of an egg. This makes mating outside the peak fertile period of the **ovarian cycle** a puzzle. Not only is the process inefficient in terms of producing a zygote, but it runs the risk of ageing sperm fertilising a freshly ovulated egg, or freshly ejaculated sperm fertilising a senescent egg – both scenarios increasing the risk of genetic damage in the zygote. Possibly the risk is reduced by an effective filtering mechanism.

13.2.5 The evidence from immunology

Recent immunological research has also thrown light on the likely patterns of mating amongst early humans. Nuann et al. (2000) examined the number of white blood cells in 41 species of non-human primate and found that the greater the number of mating partners that females had the higher the white blood cell count. They reasoned that the basis for this was that 'promiscuous' species were susceptible to sexually transmitted diseases and so required a more complex immune system. The authors observed that the immune system of humans was more like the polygamous gorilla (*Gorilla gorilla*) and the monogamous gibbon (*Hylobates lar*) than the chimpanzee; suggesting that humans are marginally polygamous (uni-male, multi-female) and marginally monogamous (uni-male, uni-female) rather than multi-male, multi-female in their mating behaviour.

13.3 Concealed ovulation

The size of male testes and bodily dimorphism may yield useful information about patterns of human mating, but it could also be expected that the female body might hold some clues about the mating propensities of early *Homo sapiens*. One notable aspect of human female sexuality is that women are continually sexually receptive throughout the ovarian cycle. Another is that the precise moment of **ovulation** is concealed from both male and female. Clear testimony to this is the lucrative business of

manufacturing contraceptives on the one hand and test kits for ovulation on the other. Yet humans are part of a minority among mammals in this respect. In most mammalian species, oestrus is announced with an explosive fanfare of signals. In the case of female baboons, for example, the skin around the vagina swells and turns bright red, she emits distinctive odours and, just to reinforce the point, she presents her rear to any male she happens to fancy.

The concealment of ovulation is a feature that humans share with about 32 other species of primate, but for at least 18 other species, ovulation is advertised boldly and conspicuously. We can confidently expect concealment and advertisement to have some adaptive significance, and a number of intriguing theories have been proposed to account for both. Concealed ovulation in humans has led to a great variety of hypotheses. The problem is not a trivial one since the concealment of ovulation only serves to make sexual activity even more inefficient. As sexual activity is costly to an organism (in terms of energy and time expended, exposure to risks of transmission of disease and vulnerability to predators), there must be some sound evolutionary reasons for masking the time when sex could be more productive. It is likely therefore that concealed ovulation is part of a female's mating strategy.

The variety of theories proposed include:

● One is the 'sex for food' hypothesis. Given the ubiquity of prostitutes in history, it has been suggested by Hill (1982) that sexual crypsis allowed early female hominids to barter sex for food. If ovulation were conspicuous, a male would know when a gift in exchange for sex would bring him some reproductive advantage, and he could withhold such gifts if the woman were not fertile. Concealing ovulation allows women to be in a position almost constantly to exchange sex (with the prospect of paternity) for resources. Hill argues that there is ample evidence in the ethnographic literature that human males trade resources such as meat for sex. Women gain not only valuable nourishment, but also the opportunity to copulate with some of the best males in the group, since males good at provisioning are also presumably able in other respects. There is some evidence that, in hunter-gatherer societies, this still happens, in that women will exchange sex with the best hunter in return for meat (Hill and Kaplan, 1988).

- An alternative to this is the 'anticontraceptive hypothesis'. Burley (1979) has suggested that sexual crypsis evolved after women had developed enough cognitive ability to associate conception with copulation. If women were aware of ovulation and could thus refrain from sex to reduce the risks of childbirth to themselves, they would leave fewer descendants than those who were unaware and hence could not exercise this choice. In this view, concealed ovulation is the product of genes outwitting consciousness, the triumph of matter over mind.

- Benshoof and Thornhill (1979) suggest that, by concealing ovulation from her ostensibly monogamous partner, a female can mate with another male (who may appear to have superior traits) without alerting her first mate. They suggest that a woman may unconsciously 'know' when she is ovulating, and this could help her to assess when an extramarital liaison would be rewarding. In this scenario, sexual crypsis becomes a strategy employed by the female to enable her to extract paternal care from a male who thinks that she will bear his children while enjoying the ability to choose what she regards as the best genetic father. All males are potentially genetic fathers, but not all males are able to provide paternal care; some, for example, may already be paired off.

- Richard Alexander and Katherine Noonan at the University of Michigan have proposed a view of sexual crypsis as a tactic developed by women to divert men from a strategy of low-investing, competitive polygyny towards a more caring and high-investing monogamy. For a male to be assured of paternity, he needs to remain in close proximity to his partner for prolonged periods of time. It is no use a man wandering off to have sex with another woman, since she may not be ovulating and his wife back at home may become the subject of the attentions of like-minded philandering males. So the male stays at home, finds his partner constantly desirable and has the reward of a high degree of confidence in his paternity. This is the

so-called 'daddy at home' hypothesis (Alexander and Noonan, 1979).

- It is common in primates to find males killing offspring that are not their own in order to bring the bereaved mother into oestrus again. Support for the need for females to evolve a countermeasure against this comes from studies on the 'postconception oestrus' of grey langurs (*Presbytis entellus*) and red colobus monkeys (*Colobus badius*). In both of these species, an oestrus signal is given even when the females are pregnant. This could be interpreted as a measure to confuse males; if so, it is one of the few examples of 'dishonest signals' given by females of their reproductive status. With this in mind, Hrdy (1979) has suggested that sexual crypsis served to confuse the issue of paternity for early hominids and thus prevent infanticide by males who, if they were confident that they were not the genetic fathers of offspring, could engage in this practice. This is sometimes known as the 'nice daddy' theory.

There is clearly no shortage of ideas to account for concealed ovulation, and the literature is vast and difficult. Even a superficial glance at primate sexual behaviour shows that concealed ovulation can be found in a variety of mating systems, such as those of monogamous night monkeys, polygynous langurs and multi-male vervets. It is entirely feasible, therefore, that the various hypotheses are not entirely exclusive. One, such as the 'nice daddy' theory, could explain the origin of concealment, and another, for example the 'daddy at home', its maintenance. In disentangling these arguments, much progress has been made by two Swedish biologists, Birgitta Sillen-Tullberg and Anders Moller (1993). By looking at the probable phylogeny of the anthropoid primates, they claim to be able to evaluate the various hypotheses.

The crucial test is of course when concealed ovulation first appeared. Using procedures to disentangle primitive, derived and convergent characteristics, Sillen-Tullberg and Moller constructed a phylogenetic tree of changes in the visual signs of ovulation. They concluded that the primitive state for all primates was probably slight signs of ovulation, and that concealment had evolved independently 8–11 times (the range indicating the effect of slightly different assumptions in the modelling).

Sillen-Tullberg and Moller draw three sets of conclusions from their work:

1. Ovulatory signals disappear more often in a non-monogamous context. This tends to support Hrdy's (1979) infanticide hypothesis.
2. The fact that, once monogamy has been established, ovulatory signals do not usually disappear throws doubt on the 'daddy at home' hypothesis of Alexander and Noonan (1979) as the origin of concealed ovulation.
3. Monogamy evolves more often in lineages that lack ovulatory signals. The lack of ovulatory signals could be an important condition for the emergence of monogamy.

In short, if we accept the conclusions of Sillen-Tullberg and Moller, it seems that concealed ovulation may have changed and even reversed its function during primate evolution. It began as a female strategy to allow ancestral hominid women to mate with many males yet remain secure in the knowledge that the confusion over paternity would protect their infants from infanticide. Once concealed ovulation was established, a woman could then choose a resourceful and caring male and use concealed ovulation to entice him to stay with her to provide care. Women became continually sexually active, and men correspondingly found women continually desirable – even without signs of ovulation. It is an interesting thought that the continual sexual

receptivity of human females and the high frequency of non-reproductive sex in which humans engage began as a female strategy to outwit men. An ancestral male, anxious to secure his paternity, could no longer afford the time budget to guard a group of females and prevent sexual access by other males. Instead, he was forced into a monogamous relationship, providing care for his partner and what he thought were his own offspring, in return expecting sexual fidelity from his wife. This in turn led to a male psychology that was particularly sensitive to any evidence of unfaithfulness, with repercussions for the emotional life of men. Such is the origin of the strength of sexual jealousy that today it still accounts for a good deal of human conflict (see Chapter 12). Some recent evidence suggests that ovulation may not be totally concealed and this is explored in Chapter 14.

The evidence of this chapter suggests that human mating is complex and does not easily fit into a simple system. We may have to face the fact that asking if human sexual behaviour is naturally monogamous, polygynous or polygynandrous is a bit like asking if water is naturally a solid, liquid, or a gas – it all depends on the local conditions. In the case of water this is matter of the physics and chemistry of bond structures; in the case of human sexuality this variability is perhaps due to our ability to behaviourally adapt in directions that enhance our fitness in the circumstances we create for ourselves. The next chapter looks at current human sexual behaviour in contemporary contexts.

Summary

🔺 An evolutionary approach to human sexuality helps us to understand the mating strategies pursued by ancestral and contemporary males and females.

🔺 Data on sexual dimorphism, male testis size, sperm swimming speed, immunological evidence and sperm morphology can be used to infer ancestral mating behaviour. The fact that males are slightly larger than females tends to suggest some degree of intrasexual competition among males for mates. Human testes are too large, however, to point to a uni-male mating system, such as found among gorillas, and too small to be consistent with multi-male and multi-female 'promiscuous' mating groups.

🔺 Concepts drawn from sexual selection theory applied to mammals, such as male v. male contests, male display and female choosiness, must be applied cautiously to human mating. In the case of humans both sexes can be expected to display and exhibit choosiness depending on such parameters as the operational sex ratio, variation in mate quality, encounter rates and degree of parental investment in offspring.

There are many features of female sexuality, such as concealed ovulation, that remain enigmatic. The balance of current theories suggests that sexual crypsis (concealed ovulation) began among early hominid females in a uni-male setting, enabling them to choose desirable males for mating without the risk of infanticide from suspicious males. In effect, concealed ovulation served as a strategy whereby women resisted signalling to men the period of maximum fertility and thereby confused paternity estimations. Once established, it is suggested that women could then use crypsis to extract more care from a male.

Key Words

- Monogamy
- Oestrus
- Ovarian cycle
- Ovulation
- Polyandry
- Polygyny
- Sex ratios
- Sexual dimorphism
- Testis

Further reading

Baker, R. R. and M. A. Bellis (1995). *Human Sperm Competition*. London, Chapman & Hall.
A book that makes some controversial claims about human sexuality based on unusual and original research.

Betzig, L. (ed.) (1997). *Human Nature: A Critical Reader*. Oxford, Oxford University Press.
A useful book that contains numerous original articles on human sexuality together with a critique in retrospect by the original authors.

Gray, P. B. and J. R. Garcia (2013). *Evolution and Human Sexual Behavior*. Cambridge, MA, Harvard University Press.
Good integration of recent views and evidence on human sexuality into a broad-ranging survey.

Ridley, M. (1993). *The Red Queen*. London, Viking.
A delightful book that explores the nature of sexual selection and its application to humans.

Short, R. and M. Potts (1999). *Ever Since Adam and Eve: The Evolution of Human Sexuality*. Cambridge, Cambridge University Press.
Superbly illustrated and authoritative work. A humane account of the evolution and significance of human sexuality.

Ryan, C. and C. Jetha (2010). *Sex at Dawn*. New York, NY, Harper.
Provocative and hugely entertaining account of the evolutionary origins of human sexuality. A book that takes issue with what the authors call the standard evolutionary psychology narrative of human sexuality.

Human Mate Choice: The Evolutionary Logic of Sexual Desire

Had we but world enough and time,
This coyness, lady, were no crime.
We would sit down, and think which way
To walk, and pass our long love's day
(**Andrew Marvell,** *To His Coy Mistress,* **c. 1650**)

The onset of sexual desire exerts a powerful force over human lives and European literature is full of *carpe diem*-style poems written by men eloquently exhorting their mistresses to seize the day (i.e. come to bed). One notable feature of sexual attraction is that it is highly discriminating – we have strong preferences about who will and who will not suffice as a potential partner in both the short and longer term. Furthermore, however idealised our view of romantic love, there is abundant evidence that humans employ a variety of hard-headed criteria in assessing the desirability of a mate, including income, occupation, intelligence, age and, perhaps above all, physical appearance. The appearance of the human body is an important reservoir of information and has exerted a potent influence over the whole of our culture. Western art since the time of the Greeks and certainly since the Renaissance has celebrated the beauty, grace and symbolic significance of the human body. We are enthralled and fascinated by the appearance of members of our own species. Whether there are universal and cross-cultural standards in the aesthetics of the body is still a debatable point, but certainly in Western culture there exists a multimillion dollar industry to help us better to shape our bodies towards our ideals or disguise the effects of age. Corporate advertising worked out long ago that one of the best ways to market a product is to associate it with a handsome specimen of one or both sexes. Where nature falls short, the surgeon's knife can be recruited to the cause and the statistics on cosmetic surgery are a testament to the West's near-obsession with physical beauty. Figure 14.1 shows the top ten cosmetic surgical procedures for men and women in the UK carried out in 2013.

So where does this aesthetic sense come from? Somewhat surprisingly, Darwin himself, reviewing standards in a variety of cultures, concluded that there was no universal standard for beauty in the human body. A common modern response to the question is to repeat the mantra 'beauty is in the eye of the beholder' and while this does convey a fragment of the truth, it is also misleading. If standards did vary enormously between individuals, then anyone could become a glamour or fashion model; yet common sense tells us that attractive people are viewed to be attractive by a large consensus. Furthermore, some recent meta-analyses of the literature on attractiveness have shown that there is massive agreement in attractiveness ratings within and between cultures (Langlois et al., 2000). Beauty does not seem to reside

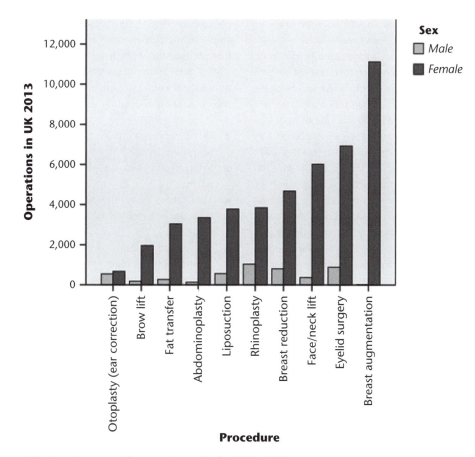

FIGURE 14.1 The top ten cosmetic procedures in the UK in 2013.
Overall operations on women outnumber those of men by about 10: 1.
SOURCE: British Association of Aesthetic Surgeons (2014).

in the eye of the beholder, nor is it simply a response to culturally induced norms.

To a Darwinian, the way to proceed on this question is clear enough: our perceptual apparatus will be designed to respond positively to features that are honest indicators of fitness – for aesthetics read reproductive potential. We are descended from ancestors who made wise choices in selecting their partners and, to the extent that we have inherited their desires and inclinations, we can expect any criteria we use in choosing a mate to be fitness enhancing. This chapter is focused around both these issues: the aesthetic judgement exercised, and the decision-making criteria used, when males and females choose a sexual partner.

14.1 Evolution and sexual desire: some expectations and approaches

Darwinians view attractiveness in terms of reproductive fitness rather than as the relationship between an object, an observer and some abstract Platonic form. Features that are positive indicators of reproductive fitness in a potential mate should be viewed as attractive by males and females. In this sense, beauty is more than skin deep – it is to be found in the 'eye' of the genes. Despite the mild degree of polygyny indicated by the evidence presented in Chapter 13 and the few cases of opportunistic extreme polygyny, it is clear that, in most relationships, men and women make an appreciable investment of time and energy.

Consequently, both sexes should be choosy about future partners, but in different ways.

Of all the features used in appraising a potential mate, two in particular have produced robust empirical findings that reveal inherent differences between male and female taste. These are physical attractiveness and the status of males. In the case of male status, the application of the principles established in Chapters 4 and 13 predicts that, since females make a heavy investment in raising young and bi-parental care is needed following birth, females will be attracted to males who show signs of being able to bring resources to the relationship. This ability could be expected to be indicated by the dominance and status of a male within the group. Dominance could be selected for by intrasexual selection if males compete with each other, and by intersexual selection if females exert a preference for dominant males. A crude indication of dominance would be size, and we have already noted that humans are mildly dimorphic. In the complex social groups of early humans, however, there was bound to be a whole set of parameters, such as strength, intelligence, alliances and resource-holding and provisioning capabilities that indicated the social status of the male. Some of these would be subtle and context dependent, and a female would be best served by a perceptual apparatus that enabled her to assess rank and status using context-specific cues and signals.

If females respond to indicators of potential provisioning and status, males should be attracted to females that appear fecund and physically capable of caring for children. Since the period of female fertility (roughly 13–45 years of age) occupies a narrower age band than that of the male (13–65 years), we would also expect the age of prospective partners to be evaluated differently by each sex. Men should be fussier about age than women and hence rate physical features that correlate with youth and fertility higher on a scale of importance than should women. Indeed, studies have shown that attractiveness ratings decline with age for both men and women, but that these effects are more pronounced for women than men (McLellan and McKelvie, 1993). This seems to be a double standard but one that may have deep evolutionary foundations: the preference for youthfulness in women is to be expected not only because youthful women are more fertile, but also because

younger women have a longer period of fertility ahead of them than older women and could therefore produce more offspring by the admiring male.

To test these expectations, we can examine human preferences using data from a number of sources and types of study:

- What people say about their desires in response to questionnaires
- What people look for when they advertise for a partner
- Medical evidence on whether features deemed to be attractive really do correlate with health and fecundity
- The use of stimulus pictures (line drawings and photographs) to elicit body shape and facial preferences.

In the sections that follow, each of these approaches will be employed.

14.2 Questionnaire approaches

14.2.1 Cross-cultural comparisons

The use of a questionnaire on sexual desire in one culture lays itself open to the objection that responses reflect cultural practices and the norms of socialisation rather than universal constants of human nature. In an effort to circumvent this problem, David Buss (1989b) conducted a questionnaire survey of men and women in 37 different cultures across Africa, Europe, North American Oceania and South America, and hence across a wide diversity of religious, ethnic, racial and economic groups. As might be expected, numerous problems were encountered with collecting such data, but Buss's work remains one of the most comprehensive attempts so far to examine the sensitivity of expressed mating preferences to cultural variation. From the general considerations noted above, Buss tested several hypotheses (Table 14.1). The results show moderate to strong support for all the hypotheses.

In a more recent examination of mate selection preferences, Schwarz and Hassebrauck (2012) claim that many previous studies suffer from several problems. Firstly, they tend to rely on a relatively small

Table 14.1 Predictions on cross-cultural mate choice preferences

Prediction	Adaptive significance in relation to reproductive success	Number of cultures (percentage of total in brackets)	
		Supporting hypothesis (p < 0.05)	Contrary (con) or not significant (NS)
Women should value earning potential in a mate more highly than men do	The likelihood of a woman's offspring surviving and their subsequent health can be increased by allocation of resources to the woman and her children	36 (97%)	1 NS (3%)
Men should value physical attractiveness more highly than women do	The fitness and reproductive potential of a female is more heavily influenced by age than for a man. Attractiveness is a strong indicator of age and fertility	34 (92%)	3 NS (8%)
Men, on the whole, are likely to prefer women younger than themselves	Men reach sexual maturity later than women do. Also as above	37 (100%)	0 (0%)
Men will value chastity in a partner more than women	'Mummy's babies, daddy's maybes'. For a male to have raised a child not his own would have been, and still is, highly damaging to his reproductive fitness	23 (62%)	14 NS (38%)
Women should rate ambition and drive in a prospective partner more highly than men do in their partners	Ambition and drive are linked to the ability to secure resources and offer protection, both of which would be fitness enhancing to a woman	29 (78%)	3 con (8%) 5 NS (13%)

SOURCE: Buss, D. M. (1989b). 'Sex differences in human mate preferences: evolutionary hypotheses tested in 37 cultures.' *Behavioural and Brain Sciences* **12**: 1–49.

number of preference choice items (typically 18); secondly, the age profile of participants tends to be quite narrow (typically college students) and so rules out an examination of how preferences might vary with age; and thirdly, the role of relationship status (that is, single or attached) is not adequately factored in. To overcome these problems their own study used a questionnaire containing 82 selection criteria to examine the preferences of 21,245 participants in Germany aged between 18 and 65 years who were not in a close relationship. In keeping with other studies they found that women are more demanding than men, with a far greater number of mate selection criteria being more important to women than to men. Typical factors that women rated as more important

than men were: wealth and generosity, intellectualism, dominance, sociability, reliability, sense of humour. The only factors more important to men than to women were physical attractiveness and 'creativity and homeliness'. Interestingly, age and level of education had very little effect on the pattern of these desirable features suggesting a remarkable stability of preference across the lifespan (see Figure 14.2).

14.2.2 Age difference preferences

Data on age difference preferences from the study by Buss (1989b) discussed above also allow a calculation of mean age preferences for mating. On average, men

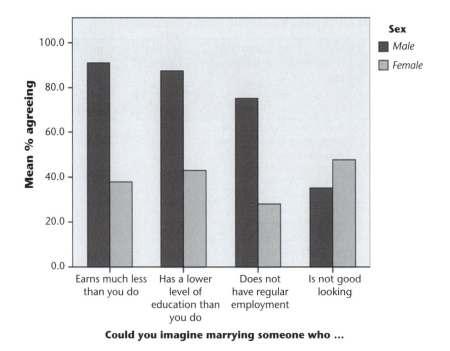

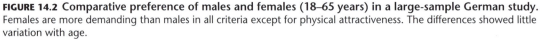

FIGURE 14.2 Comparative preference of males and females (18–65 years) in a large-sample German study. Females are more demanding than males in all criteria except for physical attractiveness. The differences showed little variation with age.

SOURCE: Data from Schwarz and Hassebrauck (2012).

prefer to marry women who are 24.83 years old when they are 27.49 years old, that is, 2.66 years younger than themselves. Women, on the other hand, prefer to marry men who are 3.42 years older. Interestingly, there are data available for 27 of the 33 countries sampled for actual ages of marrying; the actual difference is that men marry women 2.99 years younger than themselves. The fact that these figures agree so closely suggests that, at least in terms of age, preferences and practice are reassuringly similar. Other studies have supported this: women seem to prefer men a few years older than themselves (Waynforth and Dunbar, 1995).

A variety of other studies have shown that as men age they express romantic interest in women progressively younger than their own age (Kenrick and Keefe, 1992). An evolutionary explanation of these effects is quite straightforward: men of any age should be adapted to find women of high fertility and reproductive value attractive so long as they are fertile themselves, and since male fertility persists well past the age of female menopause older men should still

be attracted to women in a younger fertile age group. The age preferences of females are less obvious but are usually taken to reflect the fact that older men will be in a stronger position to provide protection and resources. In addition, there is no biological point to, say, a 50-year-old woman feeling strongly inclined to form a partnership with a 20-year-old man. The age-related attractiveness of women is complicated by the fact that two key biological factors – residual reproductive value, a measure of future possibilities of conception and child-rearing; and fertility, a measure of sexual intercourse leading to a conception – have two different optima. In Westernised societies residual reproductive value peaks at the age of menarche at about 12.5 years (see Section 8.3.3 and Figure 12.5), whilst peak fertility has been estimated to occur in the early to mid-twenties (Wood, 1989; Dunson et al., 2002). It might be expected then that a man's age-related sexual interest in a woman is a compromise between the two optima.

In order to investigate age-related sexual preferences Jan Antfolk et al. (2015) looked at the desires and behaviour of men and women aged 18–43. In an interesting variation on such studies this group looked at both ideal preferences (something fairly unconstrained) and actual sexual behaviour (something that reflects the preferences of others and hence the constraints of competition). The group argued for a number of expectations:

1. For reasons of resources, status and protection women will prefer men slightly older than themselves.
2. Men, irrespective of their age, will tend to be sexually interested in women in their early to mid-twenties.
3. Where there are differences between the age of actual partners and expressed preferences, then, since women are the selective limiting resource for men (see Chapter 4), women will more closely realise their preferences than men.

The group examined the responses of 12,656 Finnish men and women. Table 14.2 shows a selection from their data.

The results are interesting. Contrary to the simple heuristic that 'men prefer women younger than themselves', and in keeping with the fertility hypothesis, young men aged 18 actually preferred women in the 20–22 year age group. Above the age of 25, and in keeping with other studies, men prefer women increasingly younger than themselves. But in keeping with the suggested stronger influence of female choice, the discrepancy between preference and actuality was smaller for women than for men.

The fact that men over 25 prefer increasingly younger women offers some support for the influence of residual reproductive value and fertility optima on mate choice. But as found in other studies there are other factors at work since the age preference expressed by men did not remain constant somewhere in the early twenties. This may reflect the influence of female choice, male competition (from men in their twenties) and social expectations. It could be, for example, that 40-year-old men have learnt not to waste too much effort seeking (on the whole) unavailable 22-year-old women; even though for some the learning process is slow.

There are, however, alternative 'social constructivist' explanations. In expressing their preferences both sexes may be falling in with sociocultural expectations or reacting strategically to economic and political conditions. One phenomenon often cited to support this approach is the putative growth in 'toy boy' relationships whereby wealthy women (often celebrities) enter into relationships with much younger males.

To assess the merits of these two approaches, Michael Dunn et al. (2010) looked at online dating sites to gather age difference preferences among males and females across 14 different countries and two religious groupings (Christian and Muslim). Differences between males and females were similar across all the cultures and religious groups examined (Figure 14.3). The countries sampled covered a broad range of economic systems and levels of wealth and development and so it is arguable that this consistency points to a biological causation rather than a sociocultural one. They found very little evidence for a 'toy boy effect' – although it is conceivable that this strategy is confined to an elite group of wealthy females. It would be useful to sample such females in a similar range of cultures. Figure 14.3 also shows that the preferences are not simply driven by reproductive biology

Table 14.2 Comparison of expressed age preferences with age of real sexual partners for a sample of 12,656 Finnish men and women

Respondents' age	Men (range shows 95% CI)		Women (range shows 95% CI)	
	Age preference	Age of real partners	Age preference	Age of real partners
18	20–22	17.5–18.5	21–22	20–21
25	25–26	24–25	27–28	27–28
35	28–31	31–33	35–37	36–37
40	32–35	36–38	40–42	40–42
SOURCE: Data taken from Antfolk et al. (2015).				

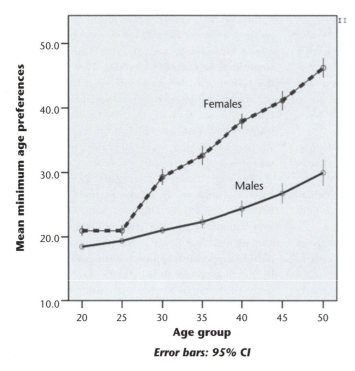

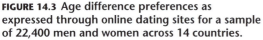

Age group

Error bars: 95% CI

FIGURE 14.3 Age difference preferences as expressed through online dating sites for a sample of 22,400 men and women across 14 countries. The x axis shows the age of men or women expressing a preference, the y axis the minimum age acceptable for a potential relationship. Men have a lower age preference than women for all ages and this difference increases with age.

SOURCE: Data taken from Dunn et al. (2010), Table 1, p. 387. Data points are means (n = 14) of the mean of responses from each of 14 countries, error bars show 95 per cent confidence intervals (CI).

since the age group preference expressed by males does actually rise with age so that a 50-year-old tends to cite as a minimum level of acceptability a woman around 30 years of age. Of course this may also contain a concession to what is socially acceptable or even likely in the mating market.

There are clearly problems with many studies based on questionnaires, particularly when unselective samples are used. One sometimes gets the impression that American undergraduates are constantly being plagued by interviewees asking about their sex lives. Nevertheless, the findings tend to be in agreement with evolutionary expectations. If, as social science critics would say, responses are conditioned by social norms, we still have the problem of explaining why so many social norms correspond with evolutionary predictions.

14.3 The use of published advertisements

An intriguing way to gather information on mating preferences is to inspect the content of 'lonely hearts' advertisements in the personal columns of newspapers and magazines. A typical advertisement reads:

> *Single prof. male, 38, graduate, nonsmoker, seeks younger slim woman for friendship and romance.*

Notice that the advertisement offers information about the advertiser as well as his preferences for a mate. Such information carries some advantages over questionnaire response surveys, in that it is less intrusive and less subject to the well-known phenomenon that interviewees will tend to comply with what they take to be the expectations of the questioner. Moreover, the data are 'serious', in that they represent the attempts of real people to secure real partners. Against this must be placed the fact that the data are selective and probably do not represent a survey across the entire population profile. Greenless and McGrew (1994) examined 1,599 such advertisements in the columns of *Private Eye* magazine. The results for physical appearance and financial security are shown in Figure 14.4.

The results are consistent with the questionnaire surveys of Buss and others, and support the following hypotheses:

1. Women more than men seek cues to financial security.
2. Men more than women offer financial security.
3. Women more than men advertise traits of physical appearance.
4. Men more than women seek indications of physical appearance.

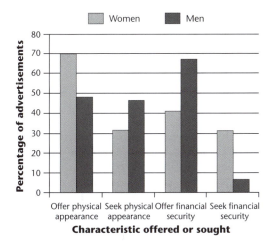

FIGURE 14.4 Percentage of advertisers seeking and offering physical appearance and financial security.
SOURCE: Data from Greenless and McGrew (1994).

It also appears that men of high status seek, and are able to attract, women of higher reproductive value. At the anecdotal level, most people know of ageing male rock stars or celebrities who marry women many years their junior. In a study of a computer dating service in Germany, Grammer (1992) found a positive correlation between the income of men using the service and the number of years separating their age from the (younger) women they sought.

Qualities advertised and qualities sought

In a variation on these types of studies, De Backer et al. (2008) examined the match between what men and women advertise and what they themselves seek in a partner. The question here is whether what men and women advertise about themselves are those same qualities actually sought by the opposite sex. In other words, do males and females know what the opposite sex is looking for? The study examined 800 advertisements in various Belgian newspapers published in the years 2000 and 2001. Figure 14.5 shows how what was advertised and sought by both sexes compares.

Overall the correlation is quite good. A few discrepancies of note are that females tended to advertise wealth and intelligence more frequently than

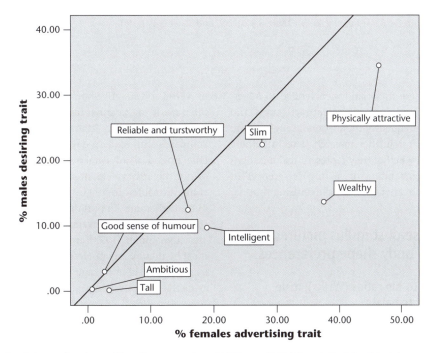

a) Females advertising and males seeking. Spearman rho = 0.881, p = 0.004 (two tailed) for this data.

FIGURE 14.5 Correlation between traits that females and males advertise and those that males and females seek (respectively) from a sample of 800 personal ads in various Belgian newspapers.
Line shows y = x , i.e. where % advertising = percentage seeking

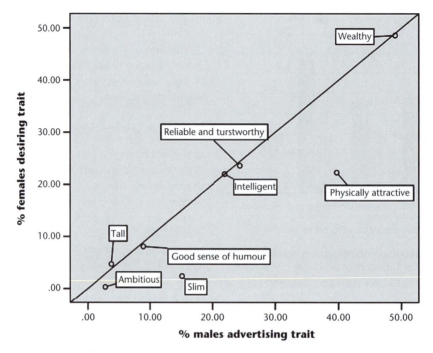

b) Males advertising and females seeking. Spearman rho = 0.913, p = 0.002 (two tailed) for this data.

FIGURE 14.5 (*Continued*)

SOURCE: Data from De Backer et al. (2008), Table 3.1, p. 89, and Table 3.2, p. 90.

men sought these traits; and men tended to advertise their attributes of physical attractiveness and slimness more frequently than women sought them. There is room for further research here into these deviations and whether they represent real misalignments in understanding of mating preferences or whether they are artefacts of the advertising process.

14.4 The use of stimulus pictures to investigate body shape preferences

14.4.1 Waist-to-hip ratios (WHRs): male assessment of females

It is clear that good looks are important to men seeking a partner, but what type of looks should be preferred? The belief that there is a vast variability in notions of beauty between cultures in time and place has tended to thwart the scientific search for

universal and adaptive norms for beauty. Reference is often made to the rather fleshy nude women appearing in paintings by Titian and Rubens, which are compared with modern-day models to emphasise the changing ideals of attractiveness (although whether or not the artists were attempting to depict an ideal is questionable). In 1993, however, Devendra Singh published some important work suggesting that there may be some universals in what the sexes find attractive. Singh (1993) argued that two conditions must be met by any universal ideal. First, there must be some plausible linkage of features designated as attractive to physiological mechanisms regulating some component of reproductive fitness, and hence a positive correlation between variation in attractiveness and variation in reproductive potential; in other words, attraction must equate with fitness. Second, males should possess mechanisms to judge such features, and these should be assigned a high degree of importance in the estimation of attractiveness.

Singh (1993) argued that the distribution of body fat on the waist and hips meets the conditions above. More specifically, he suggested that the **waist-to-hip ratio** (WHR) is an important indicator of fitness and attractiveness. The WHR (that is, waist measurement divided by hip circumference) for healthy pre-menopausal women usually lies between 0.67 and 0.80, whereas for men it usually lies between 0.85 and 0.95. It is well known that obesity is associated with a higher than average health risk, but what is more surprising is that the distribution of fat in obese women, and hence the WHR, is also a crucial factor in predicting their health status. Singh collected a body of evidence to show that women with a WHR below 0.85 tended to be in a lower risk category compared with women with a WHR above 0.85 for a range of disorders such as heart disease, diabetes, gall bladder disease and selected carcinomas. Since Singh's original papers, medical evidence does suggest that WHR is associated with the health status of females. Table 14.3 shows a sample of this research and also includes **body mass index** (BMI) – another variable thought to be responsible for attractiveness ratings.

If the WHR does have an adaptive significance, it should not be subject to the vagaries of fashion.

To test this, Singh examined the statistics of Miss America winners and *Playboy* centrefold models.

The results confirm the widespread suspicion that the weight of fashionable models has fallen over the past 60 years. What is equally significant is the fact that the WHR has remained relatively constant. This suggests that while weight as an indicator of attraction is, to some degree, subject to fashionable change, the WHR is far more resilient. It follows that the WHR is a possible candidate for a universal norm of female beauty. It is stable among women deemed to be attractive, and it correlates with physical health and fertility. The next step is to ascertain whether it is a factor in the assessment of beauty.

To test perceptions of attractiveness in relation to weight and WHR, Singh (1993) presented subjects with a series of line drawings (Figure 14.6), with the instruction that they rank the figures in order of attractiveness. The results were unambiguous: for each category of weight, the lowest WHR was found to be most attractive. In addition, an overweight woman with a low WHR was found to be more attractive than a thin woman with a high WHR. This again suggests that attractiveness is more strongly correlated with the distribution of body fat, as it affects the WHR, than with overall weight per se.

Table 14.3 Research pertaining to reproductive and health implications of high female WHRs and varying BMI values

Female WHR and related health and reproductive effects	Reference
Healthy premenopausal women have WHRs in range 0.67–0.80	Lanska et al. (1985)
WHR approaches male range after menopause	Arechiga et al. (2001)
Elevated WHR is a risk factor for cardiovascular disorders, adult onset diabetes, hypertension, cancer, ovarian cancer, breast cancer and gall bladder disease	Huang et al. (1999) Misra and Vikram (2003)
Women with higher WHRs have more irregular menstrual cycles	van Hooff et al. (2000)
Women participating in donor insemination programmes have lower probability of conception if their WHR is above 0.8 (after age and BMI are controlled for)	Zaadstra et al. (1993)
Women suffering from polycystic ovary syndrome who have impaired oestrogen production have higher WHRs than average. When such women are given hormone treatment with an oestrogen-progestogen compound, their WHRs decrease over time	Pasquali et al. (1999)
Female BMI and related health and reproductive effects	Reference
High BMI (that is, obesity) associated with complications in pregnancy, menstrual irregularities and infertility	National Heart, Lung and Blood Institute (1998)
Low BMI related to menstrual difficulties and non-ovulation	DeSouza and Metzger (1991)
Very low BMI (anorexia nervosa) associated with higher miscarriage rates, higher premature birth rate and lower birth weight	Bulik et al. (1999)

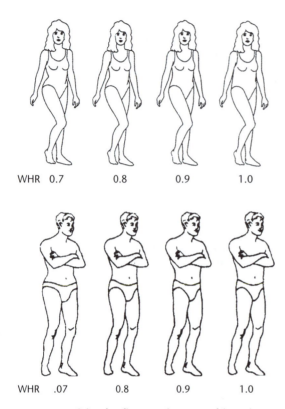

FIGURE 14.6 Stimulus figures given to subjects in Singh's study.

A series of four such figures was presented in three categories: underweight, normal and overweight. Only the normal weight category is shown here for each sex.

SOURCE: Female images: after Singh, D. (1993) 'Adaptive significance of female attractiveness.' *Journal of Personality and Social Psychology* 65: 293–307; copyright © 1993 by the American Psychological Association. Adapted with permission. Male images: Singh, D. (1995a) 'Female judgement of male attractiveness and desirability for relationships: role of waist-to-hip ratios and financial status.' *Journal of Personality and Social Psychology* **69**(6): 1089–101; copyright © 1995 by the American Psychological Association. Adapted with permission.

14.4.2 Waist-to-hip ratios (WHRs): female assessment of males

The prevalence of magazines depicting women in various states of undress would seem to imply that men are more easily aroused by visual stimuli than women. Magazines devoted to pictures of men exist but cater largely for the homosexual market rather than for serious viewing by women. This does not, however, mean that women are insensitive to male

physique. Singh (1995) also applied the concept of WHR to the female assessment of male attractiveness. It is well established that fat distribution in humans is sexually dimorphic and that, among the hominoids, this is a feature unique to humans. After puberty, women deposit fat preferentially around the buttocks and thighs, while men deposit fat in the upper body regions, such as the abdomen, shoulders and nape of the neck. Such gynoid (female) and android (male) body shapes vary surprisingly little with climate and race. Singh reviewed evidence to show that the WHR is correlated with other aspects of human physiology, such as health and hormone levels (Figure 14.7).

Singh presented women with line drawings (Figure 14.6) of men in various weight categories and with varying WHRs, asking them to rate the men's attractiveness. In all body weight categories, figures with a female-like, low WHR were rated as being least attractive. The most attractive figure was found to be in the normal weight category with a WHR of 0.9 (Figure 14.6). Interestingly, when females were asked to assign attributes such as health, ambition and intelligence, they tended to assign high values to the high WHR categories. Men with WHRs of 0.9 not only look better, but are also assumed to be brighter and healthier.

Singh's initial work was carried out on Caucasian males. To test whether different cultures shared the same WHR preference, Singh and Luis (1995) applied the same procedures to a group of Indonesian men recently arrived in the USA (94 per cent of whom were of Chinese descent) and a group of African-American men. The findings were virtually identical to those of the Caucasian study. The normal weight group was found to be the more attractive overall and again, within all groups, a WHR of 0.7 was found to be most attractive. Women reached conclusions similar to those of men when evaluating the line drawings. It appeared to Singh that neither ethnicity nor gender significantly affected the WHR dimension of attractiveness.

14.4.3 The cultural variability of attractiveness judgements

Learning from the media

Singh's work has aroused a great deal of interest and some controversy. One possible objection is that men are exposed to a barrage of images of women

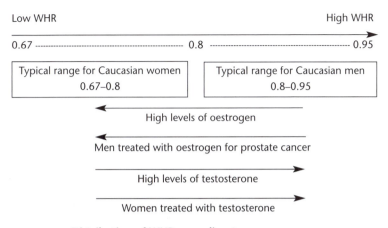

Low WHR   High WHR

0.67 ------------------------------------- 0.8 ------------------------------------- 0.95

Typical range for Caucasian women
0.67–0.8

Typical range for Caucasian men
0.8–0.95

High levels of oestrogen

Men treated with oestrogen for prostate cancer

High levels of testosterone

Women treated with testosterone

FIGURE 14.7 Distribution of WHR according to gender and hormone levels.
Low WHRs are indicative of high oestrogen levels, high WHRs indicate high levels of testosterone.

SOURCE: Data from Singh (1995).

from the media, so preferences may be culturally determined. It is possible, for example, that many men learn their cues for what counts as attractive from the behaviours of other men and respond positively to the features of media-generated stars and icons.

Even if WHR ideals do seem to remain steady in a given culture, other aspects of body shape seem variable in their estimation. As Swami and Furnham (2007) observe, ideals for female body shape seem to change through time even in the same culture. After the First World War, an ideal of female body shape emerged that was androgynous in form (de-emphasising the waist and breast). This is sometimes explained as women asserting their equality with men by emphasising the masculinity of their body shape (women over 30 were given the vote for the first time in the UK in 1918; in the USA the presidential election of 1920 was the first time all women in America could exercise the vote). This trend is illustrated by the fact that the mean of the vital statistics of Miss America winners in the 1920s was about 32–25–35. Interestingly, this gives a WHR of 0.71, but the breast measurements are much lower than those found in the 1950s. The icon of female beauty in the 1950s was Marilyn Monroe who had a curvier hourglass figure than the ideals of the 1920s. In the 1960s, the height of Miss America winners increased while their weight decreased. Such was this trend that, by

the 1980s, the majority of Miss America winners had a BMI of less than 18.5, a figure considered clinically underweight.

WHR preferences in a hunter-gatherer culture

To test if male preferences for a female WHR of around 0.7 is universal and culturally invariant, Marlowe and Wetsman (2001) presented drawings of women to men of the Hadza people. The Hadza are hunter-gatherers living in Tanzania. In this culture, labour is sexually divided: women dig for tubers and gather wild berries and fruit, while men search for honey and hunt for game. Marlowe and Wetsman were mindful of the problems that beset many of these stimulus image experiments; namely, separating WHR and BMI effects. The problem is that higher WHRs also suggest a higher BMI and so a higher overall level of fatness. In their study, they found that Hadza men preferred a higher WHR as most attractive compared to that chosen by US subjects (see Figure 14.8). They argue that several considerations could explain these findings. First, among hunter-gatherer tribes, where

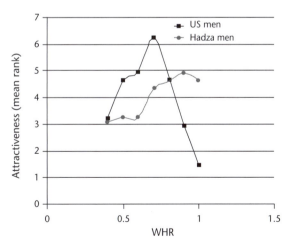

FIGURE 14.8 Attractiveness of female body shape in relation to WHR judged from line drawings by US and Hadza males.

SOURCE: Data in Marlowe and Wetsman (2001), p. 483.

women's work is energetically expensive, thinness could indicate poor health due to parasites and disease, compromising their ability to gather food and tend to children. Second, in all cultures, a high WHR is associated with pregnancy and is therefore expected to be undesirable in a prospective mate. But the effect of any indication of pregnancy will be mediated by the total fertility rate (TFR) of women in each culture. In the USA, for example the TFR is 2 (meaning that the average women raises 2 children), but among the Hadza, the TFR is 6.2. The lower the TRF, the more important it is to attend to signs of pregnancy: in the USA, a man could typically raise only 1 more child with an already pregnant woman, but a Hadza male has the potential to sire another 5.2 children with an already pregnant woman. Marlowe and Wetsman (2001) reason that this effect may help explain why a high WHR would be more damaging to the reproductive prospects of an American male and hence would be regarded as less attractive than the same WHR judged by a Hadza male.

In a more recent study, Marlowe and colleagues took into account the different body shapes of native Hadza women and typical US women. Compared to Caucasian women, Hadza women have buttocks that protrude more prominently. This may be significant since estimates of WHR from a frontal image depend on assumptions about body shape, so that, for example, for the same frontal WHR, more protruding buttocks will lower the overall WHR if measured circumferentially. Marlowe et al. (2005) presented US and Hadza men with side images as well as frontal images. A sample of the results is shown in Table 14.4.

It is notable that in this study differences in optimally attractive WHR remain between US and Hadza men (0.68 v. 0.79) but this is a smaller difference than found in previous studies (Marlowe and Wetsman, 2001). Note also from Table 14.4 that Hadza men prefer a lower side-view WHR than US men – reflecting the fact that Hadza females have a different body shape to US females (they have more protruding buttocks). Another significant finding from this study was the fact that the mean WHR of Hadza women is higher than the mean WHR for US women (see Table 14.5).

There may be genetic reasons for the differences in US and Hadza female WHRs, themselves reflecting adaptive pressure resulting from differing levels of optimality in different environments. As Marlowe et al. (2005) conjecture, Hadza women may have higher WHRs than US women as a result of the need for a larger gut to digest the bulky fibrous tubers that figure prominently in the Hadza diet. Or a more male WHR shape (signifying a relatively smaller pelvis) may reflect the increased importance of mobility over longer distances for Hadza women compared to Caucasian women. Whatever the underlying causes of these differences, it is interesting that the WHR preferences of men in the US and the Hadza culture correspond quite closely with the means of young women in their respective populations (compare Table 14.4 with Table 14.5). Marlowe et al.'s interpretation of these findings is to suggest that a fixed WHR preference is not a human universal, but rather that men adjust their evaluation of ideal WHRs according to local ecologies. This could consist of a simple mean tracking programme such that men alter their perceptions of ideal according to local averages. Such a mechanism would be consistent with other work that shows a preference of averages in faces and other objects (see Halberstadt and Rhodes, 2000).

Table 14.4 Comparison of preferred frontal and side-view WHRs for US and Hadza men

	US males	Hadza males
Frontal WHR	0.70	0.90
Side-view WHR	0.65	0.63
Weighted mean and hence overall WHR	0.68	0.79
SOURCE: Data from Marlowe et al. (2005).		

Table 14.5 A comparison of mean WHR for samples of Hadza and American women

Sample population	Age (range)	Mean WHR +/– 1 SD
Hadza women	17–82	0.83 +/– 0.06
Hadza women	17–24	0.79 +/– 0.04
Young American students	18–23	0.73 +/– 0.04
American nurses	23–50	0.74 +/– 0.08
SOURCE: Data from Marlowe et al. (2005).		

WHR and BMI: confounding effects and cultural variability

One criticism of Singh's work on WHRs is that it ignores the crucial role played by the BMI in judgements of attractiveness. The BMI is the height of a person in metres divided by the weight in kilograms squared.

BMI also has health implications and so may be expected to influence attractiveness judgements. For Caucasian women in Western Europe and the USA, the optimal BMI (from a health point of view) is estimated at 19–20 kg/m² (Tovee and Cornelissen, 2001). The 'normal' range for BMI is 18.5–30 and it serves as a rough measure of whether someone is overweight or underweight – a BMI > 30 kg/m² is usually taken to indicate obesity.

In the early days of WHR and BMI research there was some controversy about the relative effects of WHR and BMI on attractiveness ratings. It was suggested, for example, that the problem with Singh's original line drawings was that as WHR is changed, then so too is the overall body mass of the figure. In Singh's overweight group of line drawings, for example, altering the width of the waist also alters the apparent BMI, making it difficult to establish if the determination of attractiveness comes from the variation in WHR or BMI or both (Tovee and Cornelissen, 1999).

In a study on the Shiwiar people of the Ecuadorian Upper Amazon, Laurence Sugiyama (2004) employed an experimental procedure based on Singh's original drawings designed to isolate effects of BMI and WHR. He concluded that Shiwiar males preferred females with a BMI higher than local averages. Thus, if WHR and BMI were not independently assessed, this would translate to high WHR figures, since such figures give the impression of high BMI values. When differences in body weight were controlled, Shiwiar men exhibited a preference for lower than locally average female WHR figures.

In another study, Streeter and McBurney (2003) designed an experimental procedure in which they claimed the effects of weight and WHR could be isolated. They manipulated a photograph of a woman to obtain a whole range of hip, waist and chest measurements. Subjects were also asked to estimate the weight of the woman. The authors found that although weight accounted for 66 per cent of the total variance in attractiveness estimations, the effect of WHR was still clearly apparent. For both male and female judges, the attractiveness rating of the photograph peaked at a WHR of 0.7. In their review of the literature on attractiveness in Westernised cultures, Weeden and Sabini (2005) concluded that BMI and WHR probably both made independent contributions to female attractiveness, but that BMI is likely to be the stronger predictor of the two. They also concluded that for females BMI and WHR were maximally attractive over ranges at the lower end of typical female distributions (i.e. WHRs close to 0.7 and BMIs close to 20).

Other work that demonstrates the variability of BMI preferences comes from Martin Tovee and colleagues (Tovee et al., 2006). They looked at the WHR and BMI preferences of four groups of males: 100 British Caucasians, 35 ethnic Zulus, 52 Zulus who had migrated to Britain and 60 British men of African descent. Some interesting results were found for the relationship between BMI and perceived attractiveness according to the ethnic group (Figure 14.9).

As can be seen from Figure 14.9 (a), for UK Caucasians, the attractiveness of BMI-related images peaks at around 20.85 kg/m² – a figure near the optimum for health and fertility; but for native Zulu people, the attractiveness of the same images peaks at a BMI of 26.52 kg/m². Britons of African ethnic origin resemble UK Caucasians in their preferences, whilst Zulus recently migrated to the UK lie somewhere in between – suggesting perhaps a gradual adjustment of preference to UK norms.

Figure 14.9 (b) shows differences in preferences between groups for figures of a high BMI value (32 kg/m²) that would be labelled as obese in the UK. Tovee et al. (2006) explain the preference for high BMI figures among native South African Zulu men by pointing to the different health associations with BMI in the UK and rural South Africa (see Table 14.6). In essence, a high BMI carries positive connotations and implications in rural South Africa but potentially negative ones in the UK.

From such findings and those on WHR in the same study, Tovee et al. (2006) conclude that this provides evidence for flexible attractiveness preferences that shift according to local conditions and corresponding health associations. It is interesting to

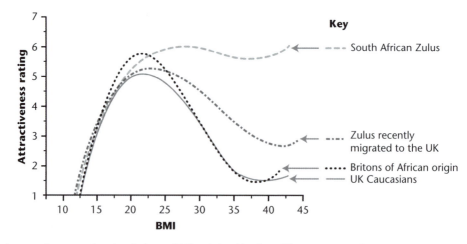

(a) Female attractiveness ratings in relation to BMI as judged by four different groups of men

Note: for most adults a BMI of 25 to 29.9 is considered overweight; a BMI of 30 to 39.9 is considered obese; a BMI of 40 or above is considered severely obese.

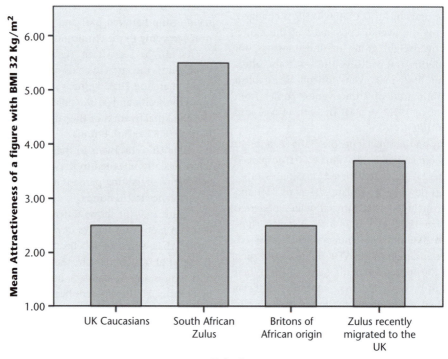

(b) Attractiveness rating (on a nine-point Likert scale) of a female figure of BMI 32 kg/m² for four male ethnic groups

FIGURE 14.9 Comparison of attractiveness ratings of female images by four groups of observers according to BMI.

SOURCE: Data from Tovee et al. (2006).

Table 14.6 Suggested associations between BMI and socioeconomic status (SES) in two different cultures

	UK	Rural South Africa
High BMI	Associated with low SES, poverty and poor diet	Associated with high SES
Low BMI	Associated with lower levels of cancer, higher SES and longer-term female health	Thought to signal weight loss due to disease and parasite infections

note that the recent Zulu immigrant to the UK has a response midway between native Zulus and resident Britons. This shows that the change in preferences is not a genetic drift of any sort but a phenotypic change in the cognitive appraisals of body shape. The human brain may therefore be receptive to cues about what constitutes optimality or normality in the local population. One way of achieving this is through imitation: recent arrivals to a new cultural context imitate the norms prevailing there. There is also evidence that frequent exposure to particular body shapes conditions judgements of attractiveness. Krzysztof Kościński (2012) explored this theme by hypothesising that people develop a template of a typical phenotype in the population based on perceptual experience; this template is then used as a standard in attractiveness judgements. To test this conjecture in a real, albeit niche, ecological context Kościński looked at the female body shape preferences of two male groups: swimmers and non-swimmers. They found unequivocally that male swimmers, but not non-swimmers, preferred the typical body shape developed by female swimmers (broad shoulders, narrow hips, above average arm circumference).

This cultural variability in ideals of male and female body shape does pose a challenge to the strict nativist view of evolutionary psychology that our preferences were adapted to conditions in past environments. We could posit mental mechanisms that respond to what is adaptive in local contexts but this assumes that there was sufficient variability in conditions and food availability in the Old Stone Age for this context-sensitive mechanism to evolve.

One way of resolving the issue of the relative contribution of WHR and BMI to attractiveness may be to see them as part of a nested hierarchy of cues or a series of filters of decreasing bandwidths. WHR could serve as a broad pass filter to initially distinguish male from female, pregnant from non-pregnant females and then young from old females. The BMI effect could then kick in to give some context-specific information about the health and nutritional status of the person. Another interpretation is that the mind uses a system of perceptual heuristics. Sugiyama (2004), for example, has argued that since most studies show a positive correlation between poor health, low fertility and increased WHR then perhaps the simplest perceptual heuristic might be 'find a WHR value lower than the local average attractive'. His study on the Shiwiar people of the Ecuadorian Upper Amazon provides some support for this.

Other have questioned whether such rules really represent adaptations. Gray et al. (2003), for example, have argued that such norm-dependent preferences may simply be a product of generic sensory biases. Women generally have WHRs lower than those of men so an even lower WHR may represent a supernormal stimulus such as observed in many animal behaviour studies (Eibl-Eibesfeldt, 1970). The idea though that a simple heuristic results from a visual inspection of the average WHR in the population or as a consequence of a lower than average WHR serving as a supernormal stimulus is challenged by the findings of Karremans et al. (2010) who found that congenitally blind men also prefer a low WHR in females. This does not entirely rule out the effects of exposure to local WHR values since blind men can get information in a variety of non-visual ways. The concept of a supernormal stimulus may lie behind the appeal of high-heeled shoes (Box 14.1).

14.5 Female sexual strategies across reproductive cycles

When males reach sexual maturity gametogenesis occurs continually at the phenomenal rate of about 3,000 sperm produced per second. Human females, in contrast, produce gametes in cycles. **Ovulation** occurs

Box 14.1 High heels, biomechanical gait and supernormal stimuli

A supernormal stimulus is one which produces a more vigorous perceptual response than the normal stimulus of its type. The phenomenon was observed by Konrad Lorenz who found that if he placed an oversized egg just outside the nest of a ground-nesting bird such as the oyster catcher then the bird will attempt to retrieve it back to the nest in preference to normal-sized eggs. Later Tinbergen found that birds who laid small blue eggs preferred to sit on ridiculously large plaster copies painted blue in a vain attempt to hatch them. One interpretation of this strange behaviour is that since large eggs have a better chance of hatching than small ones the bird is following a simple decision rule of 'prefer the largest egg'. The idea has since been applied to a range of animal behaviours (Barrett, 2010). In relation to humans, Doyle (2009) and Doyle and Pazhooli (2012) have argued that artificially enlarged breasts and high heels both act as **supernormal stimuli**.

A cursory inspection of footwear practices in Westernised cultures shows that many women wear high heels and that the wearing of high heels is often associated with displays of sexuality. There have been times in historical periods when men have worn high heels (such as the Cuban boots once worn by The Beatles in the 1960s) but high stiletto heels are primarily worn by women. Their linkage with sexuality led the feminist Germaine Greer to denounce them as symbols of male oppression and sexualisation of women. Clearly fashion trends and societal attitudes to women and sex play a part in their varying popularity, but are there evolutionary factors also at work that provide reasons for this practice?

It is easy to speculate: high heels appear to increase height; give the appearance of smaller feet and longer legs; and, due to the effects on posture, cause the chest to be thrust forward, the buttocks backwards and the calf muscles to tighten. All of these could plausibly be linked to the enhancement of fitness-related

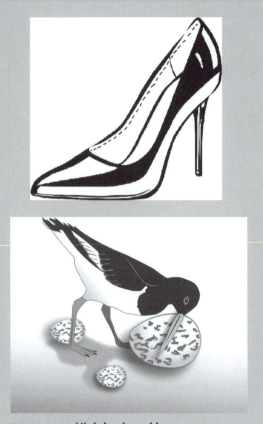

FIGURE 14.10 High heels and large eggs. Might they both be supernormal stimuli?

cues of attractiveness, but systematic empirical evidence is lacking. One line of enquiry that has received some attention, however, is the difference in gait between men and women and the role of high heels in emphasising this. Anatomical differences between the sexes translate into sex differences in gait even without the wearing of high heels. Researchers have attached point lights to male and female bodies and made films of both sexes walking, removing other cues about sexuality and simply revealing patterns of locomotion. When presented with these images subjects were able to identify the sex of the walker with a reliability approaching 70 per cent (Pollick et al., 2005). When the images are analysed they show women to have a greater sway of the hips and reduced knee extension compared to men, whilst men take longer strides and move their upper bodies more than women.

At a very rough level of analysis we might expect heterosexual men and women to find attractive those physical features, such as WHR, chin size, upper body strength, shoulder width and breasts, which are sexually dimorphic and hence crudely distinguish men and women. So, focusing on the sex differences in gait, Paul Morris et al. (2013) explored two areas: firstly, whether the effects of high heels on gait facilitated a better male–female discrimination (if other cues of the sex of the walker were concealed); secondly, whether high heels enhanced perceptions of femininity and attractiveness.

The group recorded point-light displays (only a selection of moving lights attached to the body can be seen) of 12 women walking with and without high heels and played these to participants who were asked to rate the attractiveness of the walkers (a good interactive simulation of this can be found at www.biomotionlab.ca/Demos/BMLwalker.html). They found that the wearing of heels raised the attractiveness rating for all walkers as judged by both male and female viewers (Figure 14.11). With a separate group of subjects, viewers were asked to identify the sex of the walker (not knowing that they were all female). The wearing of heels reduced the attribution error rate: with heels 17 per cent of the female walkers were classed as males, without heels the incorrect rate was 28 per cent.

One interpretation of these results is that the wearing of high heels exaggerates the distinctive features of the natural female gait and this acts as a supernormal stimulus or releaser (in the sense envisaged by Lorenz and Tinbergen), causing arousal in males. As the authors point out, there are complex cultural meanings of high heels – they often serve as a rite of passage, for example, for girls having reached puberty.

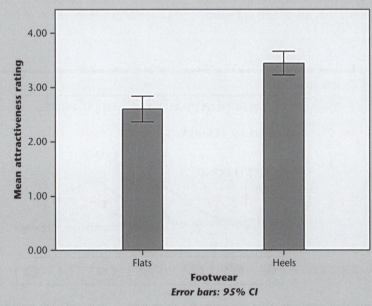

Error bars: 95% CI

FIGURE 14.11 Attractiveness ratings of 12 female walkers (concealed except for points of light) with flat shoes and then with heels.
Attractiveness significantly increased with the wearing of heels (p < 0.001).
SOURCE: Data in Morris et al. (2013), Fig. 2, p. 178.

after the lining of the uterus thickens and develops a viable blood supply. If pregnancy does not occur then the lining of the uterus is sloughed off, a process called menstruation, and another cycle begins. The adaptive function of menstruation is something of a puzzle and is considered in Chapter 8. Hence human females experience two linked cycles: the menstrual cycle and the ovarian cycle, both of which occur over a period of around 28 days. Figure 14.12 shows a standard biological description of the ovarian cycle.

14.5.1 Heightened sexual interest

In Chapter 13 it was noted that human female ovulation is largely concealed compared to that of other primates and that the evolution of this concealment may have been linked to a tendency towards monogamy. There is mounting evidence, however, that female behaviour does change across the ovarian cycle and that the period of peak fertility may not be entirely concealed from males and females.

One remarkable fact about female reproductive biology is that the fertile window in a monthly cycle is very short, lasting only about a day. If we also take account of the fact that the lifespan of sperm

in the female reproductive tract is about 3–5 days, then copulation only has a chance of resulting in conception about 20 per cent of the days during a monthly cycle. Given this low probability of conception per act of intercourse, one might expect behavioural adaptations that promote sexual reproduction with a desirable partner (that is, one of high fitness value) and somehow increase the chances of success. There is some support for this notion. Several studies report that women experience increased sexual desire near ovulation (Regan, 1996), and are more aroused by visual stimuli at mid-cycle (Slob et al., 1996). A study by Wilcox et al. (2004) looked at the sexual behaviour of 68 women who were either using an intrauterine device (IUD) or who had been sterilized by tubal ligation (that is, they were not using chemical means for birth control). The group found that sexual intercourse was significantly more frequent during the follicular phase, peaked at ovulation and declined thereafter (see Figure 14.13). The fact that these women had undergone some form of (non-chemical) contraception meant that there was no particular reason why they should try to time intercourse around their most fertile period.

As Wilcox et al. (2004) point out, there could be a variety of proximate causes of this elevated interest

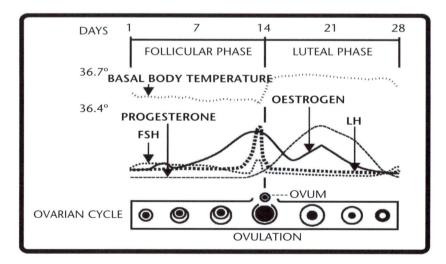

FIGURE 14.12 The ovarian cycle.
Ovulation itself occurs about a day after a surge in LH levels which causes the follicle and wall of the ovary to rupture releasing the oocyte (egg). A small rise in basal body temperature is also observed.

FSH = Follicle-stimulating hormone; LH = luteinising hormone

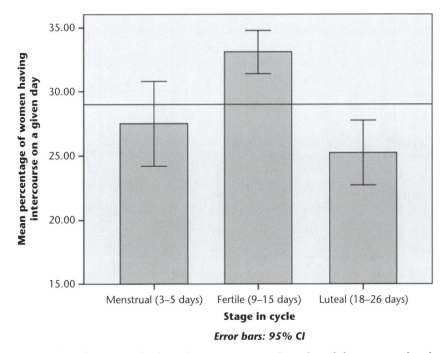

FIGURE 14.13 Proportion of women who have intercourse on a given day of the menstrual cycle relative to the day of ovulation.
Bars are shown representing average frequency of intercourse in the three stages identified. (ANOVA shows p < 0.001.) The horizontal line shows mean frequency of intercourse on non-bleeding days (29 per cent), n = 68. The results show that intercourse is more common during the fertile phase.
SOURCE: Adapted from Wilcox et al. (2004), Figure 1, p. 1540.

in sexual activity. It could be that women's libido fluctuates in a cyclical manner and in functional synchrony with ovulation; or possibly women appear more attractive to men around ovulation and men are responding to subtle behavioural or pheremonic cues; alternatively, intercourse could accelerate ovulation such that ovulation is a response to more frequent intercourse rather than a cause. Intriguingly, there is occasional support in the literature for each of these possibilities (see Gangestad et al., 2005; and Wilcox et al., 2004). Against these indicative studies must be balanced by the findings of Brewis and Meyer (2005) who examined the self-reported sexual behaviour of over 20,000 ovulating women across 13 countries. They found no significant difference in the frequency of intercourse in the pre-ovulatory, ovulatory and post-ovulatory stages of the ovarian cycle.

But supporting evidence for enhanced female desire at least during the fertile stage comes from

work by Gueguen (2009b) on the receptivity of women to courtship solicitation. In this study attractive men were chosen to act as confederates and randomly approach unsuspecting women in the age range 18–25 years in the pedestrianised area of the city of Vannes in France. The male confederates were instructed to say that they found the woman attractive and ask for a phone number to secure a later date. One minute after they parted a female assistant approached the woman, explained the scenario and asked for details concerning her menstrual cycle and if oral contraception was used. Despite what must have been the generation of disappointment for many women, the experiment did pass scrutiny by an ethics committee. The research showed that women were more likely to accept the solicitation and provide a phone number during their fertile stage of the ovarian cycle (Figure 14.14).

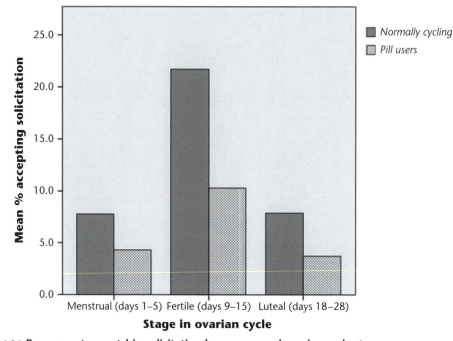

FIGURE 14.14 Responses to courtship solicitation by women and ovarian cycle stages.
Women were observed to be much more likely to respond to a courtship solicitation by an attractive male (and provide a phone number) when they were at the most fertile stage of their ovarian cycle.
SOURCE: Data from Gueguen (2009b), Table 1.

14.5.2 Extra pair copulation theory

Extra pair copulation (EPC) theory proposes that females have adaptations, such as concealed ovulation, that allow them to copulate with someone other than their primary partner, so securing, perhaps, genes superior to those of her main partner or more compatible (for example, in the major histocompatibility complex (MHC) region) to her own. In the context of the ovarian cycle, EPC theory may explain why generalised increases in libido mid-cycle are difficult to observe experimentally. In the work of Brewis and Meyer (2005), noted earlier, they were careful to point out their failure to detect associations between ovulation and frequency of intercourse applied to women and their main pair-bonded partner.

Gangestad, Thornhill and Garver-Apgar (2005) argue that predictions from EPC theory need to incorporate a consideration of the changing values of risks and benefits. Ancestrally, there would always be women who would gain benefits in terms of the genetic fitness of their offspring from EPC

with men who were displaying reliable indicators of high genetic fitness. There are, however, costs associated with EPC: the potential loss of investment from the main partner and the possibility of physical violence. But this ratio of benefits to costs changes across the ovarian cycle. During the infertile phase the costs are present and the benefits absent; during the fertile phase both costs and benefits are in operation. Hence, any inherent predilection towards EPC (which does assume that in the past the benefits outweighed the costs for at least some women) will be most pronounced during the fertile phase.

Evidence for this effect would be forthcoming if it could be shown that women do prefer high-fitness men more during their most fertile phases. Work on odour and symmetry points in this direction. Differences in symmetry between people and for a given individual across time tend to be small. But it is plausible that degrees of symmetry and **asymmetry** are indicators of fitness (see Chapter 15) and that they correlate with other indicators, such as odour, which can be more readily communicated and detected. Gangestad and

Thornhill (1998) obtained measures of the symmetry of 80 men, asked them to wear a T-shirt for two nights and then asked 82 women to rate the attractiveness of the scent from the T-shirt. They then compared the attractiveness of the scent with the degree of symmetry of the male responsible and plotted this against the probability of conception of the female making the judgement (estimated from her self-reported position in the menstrual cycle). They found a significant correlation between preference for odours associated with symmetry and conception likelihood.

14.5.3 Signs of ovulation – evolved signals or leaky cues?

Figure 14.12 shows that ovulation is accompanied by a slight rise in basal body temperature, a fact that underlies the use of basal body thermometers by women who wish to identify when they are ovulating. But might there be other signals detectable to others? There are now a variety of studies that suggest that ovulation may reveal itself in subtle ways that impact on male and female behaviour. Based on their work on the attractiveness of female odours, for example, Havlíček et al. (2006) conclude that the fertile period in humans should now be considered 'non-advertized, rather than concealed' (p. 81). Table 14.7 gives a sample of some recent work.

Despite these numerous findings there remains one overriding and important question unanswered: are signs of ovulation functional signals selected to advertise a women's fertile status, or are they 'leaky cues' – signs that women have evolved to suppress but nevertheless ones that males have evolved to

Table 14.7 Signs of ovulation as suggested in some recent published work

Study area	Findings	Reference
Lap dancing	Women lap dancers in clubs in Albuquerque who were ovulating received about $335 for a five-hour shift on high-fertility days compared to about $260 per five-hour shift on low-fertility days (see Box 14.2).	Miller, Tybur and Jordan (2007)
Flirtatious behaviour	Women in their fertile phase were significantly more likely to say 'yes' to a request for a dance from an approaching male confederate (that is, someone part of the experimental team) than women outside their fertile phase.	Guergen, N. (2009)
Walking and gait	Some studies show that women are perceived to walk in a more attractive and feminine way during the fertile phase of their ovarian cycle.	Grammer et al. (2003) Provost et al. (2008)
Attractiveness of clothing	Photographs of women taken during their period of high fertility were judged higher in terms of 'trying to look attractive' than photographs of women taken at low-fertility phases when these were presented to an independent group of judges.	Haselton et al. (2007)
Pleasantness of odours	The odour of T-shirts or underarm cotton pads worn by women was rated 'sexier', or more pleasant, or more attractive by male judges when women were at period of high, compared to low, fertility.	Singh and Bronstad (2001) Thornhill et al. (2003); Havlíček et al. (2006)
Mate guarding and jealousy	Women reported that their partners were more vigilant, monopolising and possessive when they (the women) were at a period of high, compared to low, fertility.	Gangestad et al. (2002); Haselton and Gangestad (2006)
Voice pitch elevation	Bryant and Haselton recorded 69 women saying the simple sentence 'Hi, I'm a student at UCLA' during fertile and infertile stages of their ovarian cycle. The mean basic pitch of the voice in the non-fertile period was 206Hz and it rose in the fertile phase to 211Hz ($p = 0.02$).	Bryant and Haselton (2009)
Facial attractiveness	There is evidence that skin texture and facial shape change during the menstrual cycle. Furthermore, face shapes at the peak of ovulation were rated as more attractive than those outside this phase.	S. C. Roberts et al. (2004); Oberzaucher et al. (2012)

Box 14.2 Ovulating female lap dancers receive higher earnings from tips

In a fascinating study, Miller et al. (2007) collected data from 18 professional lap dancers, working in the Albuquerque area of New Mexico, concerning their earning on any night and their stage in their ovulatory cycle. Lap dancers earn most of their money from tips given by male clients. If a man likes the look of one of the dancers he can request a more personalised dance for which he pays. It is an intimate and legal form of sex work, and crucially for this study the dancers provide a whole variety of stimuli – visual, tactile and olfactory – in the process of persuading male clients to request further performances. The results showed a peak in earnings during oestrus for normally cycling dancers but not for those using contraceptive pills. Figure 14.15 also shows how peak earnings were significantly higher for normally cycling women compared to pill users during their fertile phase.

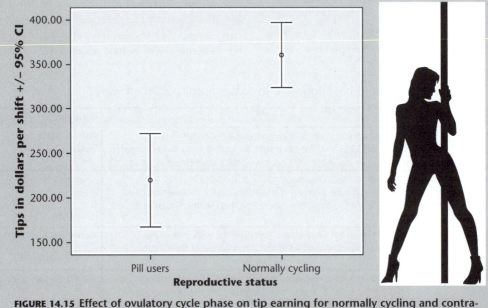

FIGURE 14.15 **Effect of ovulatory cycle phase on tip earning for normally cycling and contraceptive-pill-using lap dancers.**
The figure shows the earning at the fertile phase of the ovarian cycle. Bars show 95 per cent confidence intervals. Tips are significantly higher for the normally cycling group compared to pill users. The results also showed that earning peaks during the fertile phase, suggesting that ovulatory cues are transmitted, received and acted upon by male viewers.
SOURCE: Data from Miller et al. (2007).

detect? In the case of the former option more work needs to be done on modelling what fitness benefits a woman would gain by advertising her reproductive status, especially given that the most obvious signs of ovulation as found in other primates are largely suppressed. Moreover, even if there are fitness benefits then there are surely costs as well, in that such signals could attract the unwanted attention of males and compromise a female's ability to make her own mating decisions. The advantages of males detecting ovulation are perhaps more obvious: mating with a woman in her fertile period will clearly bring more reproductive rewards than mating outside this period.

Another possibility is that signs of ovulation are honest signals, part of a repertoire of signs that advertise the overall quality of the female and not just her position in the ovarian cycle. In their study of wild baboons (*Papio cynocephalus anubis*), Leah Domb and Mark Pagel (2001), for example, found that the size of genital swellings in females during ovulation was a good predictor of both the overall fertility of the female and the mating effort of males directed towards them. Applying this model, it may be that ovulatory signals are temporary enhancements of features that correlate with female fitness. Oestrogen levels are related to female quality, for example, and oestrogen also affects voice pitch and odour; so as oestrogen levels rise during ovulation these more basic signs are also enhanced.

The position that humans finds themselves in, where ovulation seems to affect both male and female behaviour, might be the result of a balance of complex factors and costs and benefits distributed differently between males and females, such as the pressure to display quality and an arms race between females trying to conceal ovulation and males trying to detect it. The interesting thing about work in this area is that it involves the study of physical changes not easily attributed to cultural influences and so, like physical anthropology more generally, it provides valuable insights into the evolutionary underpinnings of modern social behaviour (Haselton and Gildersleeve, 2011).

This chapter has established that humans make mating decisions based on body image. The next chapter considers facial attractiveness.

Summary

- Although males and females share many mate choice criteria, evolutionary considerations would predict a difference between the attributes sought by males and by females in their prospective partners. Females are predicted to look for high-status males who are good providers, whereas males are predicted to look for young, healthy and fertile females who are good child-bearers. Men are also much more likely to seek sex without commitment. There is considerable empirical support for these and related predictions.

- Two useful sources of information about mating preferences have come from cross-cultural questionnaire surveys and the analysis of personal advertisements.

- Physical beauty in both males and females is expected to be correlated with signs of reproductive fitness. Evidence is now accumulating that body shapes found to be attractive are those which carry fitness indicators.

- Waist-to-hip ratio (WHR) seems to be a key determinant in attractiveness judgements. Evidence suggests, in relation to female WHR, that this is not an invariant ratio sought by all human males but instead is modified according to cultural and ecological contexts. More specifically, attractive WHRs in females are strongly associated with WHRs of young fertile females in that population.

- Body mass index (BMI) also impacts upon the perception of attractiveness, but cultural assumptions about, and connotations of, BMI also influence attractiveness judgements made by males.

- The ovarian cycle in females has been shown to influence the sexual behaviour of both men and women. This suggests that ovulation is not entirely concealed.

Key Words

- Asymmetry
- Body mass index (BMI)
- Extra pair copulation
- Ovulation
- Supernormal stimuli
- Waist-to-hip ratio (WHR)

Further reading

Buss, D. M. (1994). *The Evolution of Desire*. New York, HarperCollins.
Based on research from Buss and others. Clearly explains expected sex differences in desire.

Rhodes, G. (2006). 'The evolutionary psychology of facial beauty.' *Annual Review of Psychology* **57**: 199–226.
Good review article.

Rhodes, G. and L. A. Zebrowitz (eds.) (2002). *Facial Attractiveness*, Westport, CT, Ablex Publishing.
Series of articles on the subject. Well-organised book and easy to use.

Swami, V. and A. Furnham (2007). *The Psychology of Physical Attraction*. London, Psychology Press.
Excellent coverage of recent research in this area. Shows how evolutionary psychology can be informed by social psychology and cultural considerations.

Geher, G. and G. Miller (eds.) (2012). *Mating Intelligence: Sex, Relationships, and the Mind's Reproductive System*. London, UK, Psychology Press.
Series of chapters arguing for the existence of a discrete type of mating intelligence.

Facial Attractiveness

Look not in my eyes, for fear
They mirror true the sight I see,
And there you find your face too clear
And love it and be lost like me.
(**A. E. Housman**, *A Shropshire Lad*, **XV**)

The human face is a potent source of information and the strong responses we have to the facial appearance of other people might suggest that this has been the subject of natural selection. In this chapter we will look at what features tend to correlate with attractiveness and what fitness signals attractive faces might supply.

15.1 Symmetry and fluctuating asymmetry

If you stand in front of a mirror and shield each half of your face in turn, you will notice that, unless you are very lucky, your face is not perfectly symmetrical. But does symmetry bear any relationship to attractiveness? Early research on this issue was clouded by the fact that when faces were made perfectly symmetrical (by simply presenting two mirror images of each half of the face), artefacts of this procedure, such as shading effects, adversely affected the viewers' judgements. Better techniques have reliably shown that symmetry does indeed enhance attractiveness (Grammer and Thornhill, 1994; Mealey et al., 1999; Perrett et al., 1999; Scheib et al., 1999). Furthermore, the importance of symmetry extends to the rest of the body. Gangestad and Thornhill (1994)

found a positive correlation between the symmetry of men, measured on such features as foot length, ear length, hand breadth and so on, and estimations by women of their attractiveness. In the light of this evidence, the obvious question that arises is why symmetry should be attractive. The most probable explanation is that it requires a sound metabolism and a good deal of physiological precision to grow perfectly symmetrical features. The development of bilateral characters such as feet, wings, fins and so on is open to a variety of stressful influences, such as parasite infection and poor nutrition, which result in an asymmetrical finished product. On this basis, the extent of symmetry observed in such characters may serve as an honest signal of phenotypic and genotypic quality.

Research on the importance of symmetry often uses the concept of **fluctuating asymmetry** (FA). FA refers to bilateral characters (that is, one feature on each side of the body) for which the population mean of asymmetry (right measure minus left measure) is zero, variability about the mean is nearly normal, and the degree of asymmetry in an organism is not under direct genetic control but may fluctuate from one generation to the next. So FA might be usefully measured from, say, ear measurements but not bicep measurements, since it is perfectly normal that right-handed people (the majority) will have slightly larger biceps on the right arm, but there is no reason to expect any systematic difference in ear sizes across the population. There is now abundant evidence that FA is increased by mutations, parasitic infections and environmental stress, so FA consequently becomes a negative indicator of phenotypic quality (Manning et al., 1996). The measurement of absolute and relative fluctuating asymmetry is shown in Figure 15.1.

Manning et al. (1997) have also suggested that since the maintenance of symmetry requires metabolic energy, those males who have 'energy-thrifty genotypes' are better placed to maintain a low FA. Those males with high resting metabolic rates should show greater signs of asymmetry, since less energy is available to divert to symmetry maintenance. A preliminary study of 30 males supported this prediction. There now seems to be much evidence that high levels of FA negatively influence attractiveness (see Little et al., 2000).

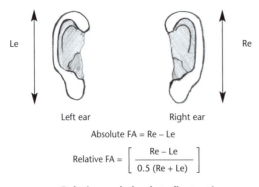

FIGURE 15.1 Relative and absolute fluctuating asymmetry.
Relative FA is usually taken as the positive value of the relative difference between right- and left-sided measurements.

Over the last 25 years a large body of literature has built up exploring the connection between FA and measures of human health and fecundity. Van Dongen and Gangestad (2011) conducted a meta-review of the literature and came to several conclusions. One was that, despite a probable publication bias where significant effects are preferentially submitted for publication, the linkage between FA and health and fecundity was real and robust. Another,

however, was that the 'effect size' of FA as a predictor of health and fecundity measures was quite small, with correlation typically between 0.1 and 0.15.

15.2 Averageness

In 1883, Galton wondered if certain groups had distinctive facial characteristics. To investigate this, he took photographs of two groups: criminals and (curiously) vegetarians. Galton overlaid the photographic plates of each subject, creating a composite face for each group. His efforts to extract distinguishing features of criminals and vegetarians proved fruitless, but he did notice that the composite or '**averaged**' faces were quite attractive. Over the past 20 years, computer-assisted technology has enabled this line of research to proceed apace. Using a computerised technique of merging faces, Langlois and Roggman (1990) found that not only were composite faces more attractive than individual ones, but also the more faces that went into making a composite, the more attractive, up to a point, the face became. A face composed of 32 faces, for example, was more attractive than one composed of two.

But why should composite faces be attractive at all? Symons (1979) suggested that the average was rated as attractive because the average of any trait would tend to be optimally adapted for any trait, since the mean of a distribution presumably represents the best solution to the adaptive problem. Thornhill and Gangestad (1993) add to this the fact that protein heterozygosity tends to be highest in individuals who exhibit the average expression of heritable traits that are continuously distributed. It could be, then, that **facial averageness** correlates with attractiveness because averageness is an indication of resistance to parasites. The reasoning here is that, in any population, the majority of pathogens will be best adapted to the most common biochemical pathways of their hosts. Sex is one defence against this: by outbreeding, parents produce offspring that are different from themselves; any parasites that were successful with parents may be less so with their offspring, since the offspring will have a whole new array of proteins and a different defence system.

It follows that the greater the genetic polymorphism at the population level – that is, the more alleles at any given locus on the genome of the species – the more likely it is that at least some individuals will stay ahead of parasites. If genetic variation confers fitness on a population, the same thing may be true at the level of an individual: the more heterozygous an individual, the more variability to be found in its genome, the more varied the proteins that are produced and thus the harder it will be for a parasite to exploit its host efficiently.

The experimental problem to overcome in such studies is the fact that there are at least two variables at work: averageness and symmetry. Composite faces also tend to be more symmetrical since asymmetries are ironed out. Symmetry, as already noted, is a reliable cue to physiological fitness that takes the form of resistance to disease and a lack of harmful mutations that compromise development stability. Yet there must also be other factors at work, for whereas symmetrical faces are regarded as being more attractive than faces with gross asymmetry, computer-generated, perfectly symmetrical faces are not perceived as being as attractive as natural faces with a slight asymmetry (see Perrett et al., 1994). There is now some consensus that averageness and symmetry both work independently to enhance attractiveness (for a review, see Rhodes, 2006).

15.3 Attractiveness and health

Work on facial attractiveness has clearly laid out the sort of features in male and female faces that are attractive to the opposite sex. There is also a good deal of theoretical speculation and reasoning linking such features as averageness, neoteny (see later), hormone markers and fluctuating asymmetry to reproductive success. But robust empirical studies linking attractiveness to health and lifetime reproductive success (measured by, say, longevity and number of children) are harder to find. This probably reflects the difficulty of collecting data on fertility and health from populations with healthcare and readily available birth control. One recent study by Lena Pflüger et al. (2012) helps fill this gap. This group looked at the facial attractiveness and reproductive output of 88 women in a rural community of mountain farmers in Austria. They collected data on the reproductive success of the women (measured by

number of pregnancies and number of children) and also obtained images of their faces in young adulthood (19–23) and later in life (post-menopause). Male students at the University of Vienna then rated the attractiveness of the young and older faces (on a scale 0–100). Reproductive success was then correlated with attractiveness. Significant positive correlations were found between attractiveness and number of children and number of pregnancies for those women who had not used hormonal contraceptives, but not for those who had. These positive correlations also remained when adjustments were made for income (both husband and wife) and for years of marriage.

15.4 Testosterone: dominance, attractiveness and the immunocompetence hypothesis

15.4.1 Testosterone as a negative handicap

In a world where intersexual selection and female choice operate it will be advantageous for males to send out honest signals to females in the form of secondary sexual characteristics that advertise their degree of fitness (see Chapter 4). This signal could take the form of a handicap that only males with a certain level of fitness can bear.

This type of reasoning lay behind the immunocompetence hypothesis suggested by Folstad and Karter (1992). They argued that testosterone could serve as just such a handicap: it suppresses the immune system such that only well-adapted males can tolerate a high level of the hormone; additionally it reveals its presence by affecting the growth of secondary sexual display characteristics. In this system, cheaters could not invade since their immunocompetence would be compromised and their fitness reduced. Figure 15.2 shows how the system is conjectured to work.

Evidence in support of this scheme for non-human animals comes from the work of Saino et al. (1997) on barn swallows. The length of tail of a barn swallow seems to be an honest advertisement of testosterone. Males with short tails were not only less attractive to females, but also, when injected with testosterone, suffered a higher mortality than their

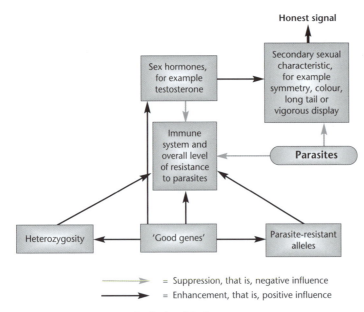

FIGURE 15.2 Conjectured relationship between resistance to parasites, secondary display characteristics and honest signals according to the immunocompetence hypothesis.
The hypothesis initially received considerable attention but recent evidence suggests the effect may be weak at best.

is a hormone playing a role in the life history-driven shift of resources between somatic (that is, maintenance) and mating effort. But the idea of testosterone as a negative handicap prized by females relies on a series of five conjectured relationships:

1. Testosterone influences facial traits regarded as masculine.
2. Testosterone imposes metabolic costs and is also an immunosuppressant.
3. Only males with a high-quality immune system can afford high levels of testosterone.
4. High-quality males produce high-quality offspring (that is, immunocompetence is heritable).
5. Women, at least under some conditions, will favour masculine traits.

long-tailed counterparts. But other conclusive studies are hard to find. In their review of the topic M. L. Roberts et al. (2004) concluded that for non-human animals there was at best weak evidence that testosterone acted to suppress the immune system.

In respect of humans the evidence is even more ambiguous. To be sure, testosterone is a powerful hormone with a variety of effects. It is responsible for the masculinisation of the male face during development (for example, protruding cheekbones, strong brow ridges and angular jaw) and hence offers a signal of its presence to other males and females. In both men and women it promotes the diversion of energy to skeletal muscle, especially the sexually dimorphic distribution of muscle mass in the chest, upper arms and shoulders (see Chapter 8). Male upper body musculature may have evolved under the force of intrasexual selection, but is also displays competitive ability to women. Testosterone also promotes behaviour related to competitiveness such as dominance-seeking and assertiveness. Generally, it

If the argument is to work then every link in the chain must be sound, yet as Scott et al. (2013) argue sometimes the evidence is not overly strong. With respect to stage 1 above, it is clear that levels of testosterone increase in men during adolescence and so do masculine features. There is also some supportive experimental evidence. Ian Penton-Voak and Jennie Chen (2004) took facial photographs of 50 Caucasian men whose testosterone levels (from samples of saliva) were also measured. The group was then divided into 25 high and 25 low testosterone males. Composite facial images were then constructed from the faces in each group. When presented with these images, volunteer observers tended to significantly judge the high testosterone faces as more masculine than those in the low testosterone group. One of the few studies to relate testosterone levels to objective measures of facial shape (and not simply a subjective impression of masculinity or attractiveness) was carried out by Nicholas Pound et al. (2009). This group found a correlation of about 0.36 between masculinity and testosterone levels. If such correlation levels are typical of each of the five stages listed above then their multiplicative combination yields rather a small overall effect

(0.36⁵ = approx. 0.006) (Bussière, 2013).
Reviewing the literature on a wide range of
species, Isabel Scott et al. (2013) concludes
that there is little evidence of a general and
straightforward linkage between testosterone, genetically mediated immunity, health
and trait size.

15.4.2 Facial dimorphism and attractiveness research

Since there is sexual dimorphism in face
shape, it is possible to enhance the femininity or masculinity of any face by manipulating features such as chin size and cheekbone
prominence using computer imaging. Perrett et al.
(1998) did just this, looking for the effects of masculinisation and feminisation on perceptions of
attractiveness. They initially found that feminis-
ing a female face increased its attractiveness, a
preference they saw as contrary to the averageness
hypothesis: female faces became more rather than
less attractive if they departed from the average in
a feminine direction. One surprising result of initial work on this phenomenon was that when an
average male shape was feminised, it too became
more attractive. Later work by this same group and
others has pointed towards more subtle changes in
female preferences according to how male faces are
manipulated and there is now a flourishing area of
research devoted to investigating changing perceptions of attractiveness in masculinised and feminised male faces.

The typical procedures involved in transforming
facial images along such dimensions as age, race and
gender are explained by Tiddeman et al. (2001). To
masculinise or feminise a face, for example, averaged
male and female faces are constructed by combining features from a number of individual male and
female faces of similar age and ethnicity. Then differences between these archetypal male and female
faces can be calculated for various facial dimensions
(e.g. brow ridge, chin size, nose width, etc.). These
differences can then be used to transform an individual facial image in a more masculine or feminine
direction (Figure 15.3).

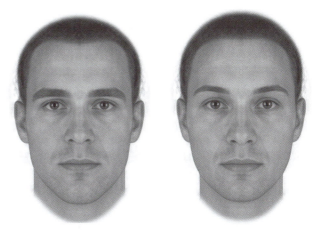

**FIGURE 15.3 Example of a male face morphed to
appear more masculine (left) or more feminine
(right).**
Reproduced by kind permission of Dr Lisa DeBruine,
Institute of Neuroscience and Psychology, University of
Glasgow.

15.4.3 Trade-off theory

What qualities do females look for in a mating partner? In theory, we might expect a female to desire
both sets of the following characteristics:

- A partner who is caring, cooperative, honest, loy-
  al and would make a good parent ('dads')
- A partner who has a good immune system (as
  indicated by secondary sexual characteristics)
  and who will therefore provide a good set of
  genes for her children. Note high testosterone
  can only be supported by males with an efficient
  immune system since testosterone is an immunosuppressant ('cads').

The problem with finding a male with both these characteristics is that there may be a conflict between the
second and the first set. If the immunocompetence
hypothesis is correct then high levels of testosterone
indicate a good immune system. But testosterone is
also implicated in some antisocial qualities such as
aggression. There is some evidence that evolution
may have solved this by providing females with what
we might call an 'onboard context-influenced variable response attractiveness detection unit'.

Several researchers have now suggested that preferences women exhibit for male facial features result from optimising this trade-off between costs and benefits (Jones et al., 2008; DeBruine et al., 2010a). One way of testing this is to appreciate that the balance of costs and benefits is likely to be influenced by contextual factors such as the probability of conception, the type of relationship envisaged (long term or short term) and the risks to general health posed by the environment. Hence the outcome of this cost–benefit analysis might vary in predictable ways.

Probability of conception

Penton-Voak and Perrett (2000) presented five computer-constructed images of male faces of varying levels of masculinity and femininity in a national UK magazine and asked female readers to choose the most attractive face. Subjects were also asked details of their menstrual cycle. The responses were then ordered into two categories: judgements made at the time of low conception risk (days 0–5 and 15–28 of the ovarian cycle) and high conception risk (the follicular phase of days 6–14). Figure 15.4 shows their findings.

Although there was some ambiguity in the instructions to subjects ('most attractive' could be interpreted as qualities for a long- or short-term sexual partner, for example), the results provide evidence for cyclical changes in female preferences. There seems to be a shift to attend to signs of high testosterone and (possibly by implication) heritable immunocompetence when conception is most likely.

Several other studies support his effect. Victor Johnston and colleagues, for example, used an ingenious technique to construct a short movie film showing faces gradually morphed from male to female. As the frame containing the facial image unrolled, subjects (females) were asked to rate males faces using criteria such as physical attractiveness, intelligence and sensitivity. As predicted, there was a shift in female preference to a more masculine male face in their high-risk phase of the menstrual cycle (Johnston et al., 2001).

If this work is robust, it seems that ancestral females desired immunocompetent 'cads' during their most fertile phase, and caring 'dads' the rest of the time. This internal unconscious mechanism, coupled with sexual crypsis, ensured that ancestral females could get the best of both worlds.

Since this work by Penton-Voak, Perrett and Johnston, the whole area of menstrual cycle shifts in women's preferences for masculine attributes has received considerable attention (DeBruine et al., 2010b). The balance of evidence has, generally, been supportive, although one study by Harris (2011) failed to replicate these effects.

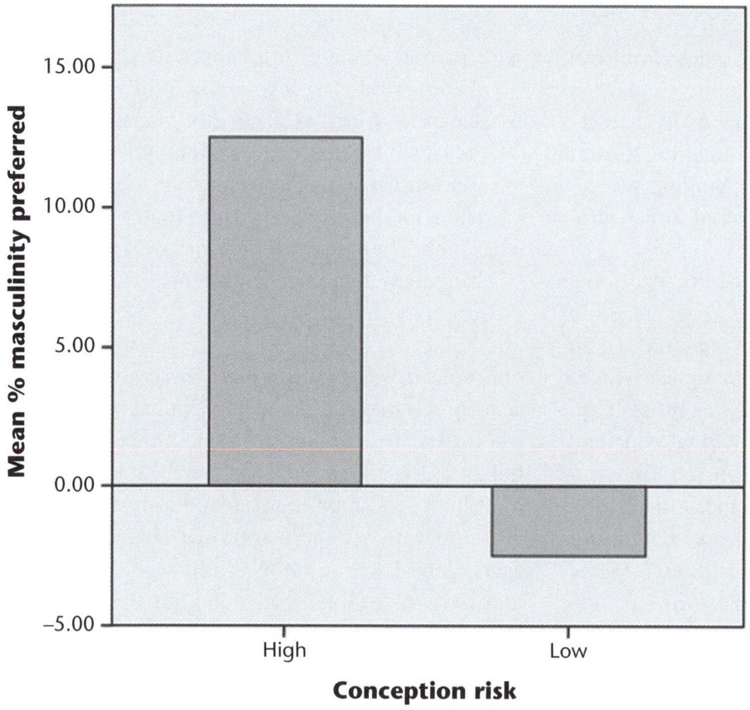

FIGURE 15.4 Average percentage of masculinity preferred by female subjects choosing a face as most attractive in relation to their conception risk. When the risk of conception was high women on average preferred a face masculinised by 12.5 per cent. When conception risk was low this figure was minus 2.5 per cent, i.e. a slightly feminised face.
SOURCE: Data from Penton-Voak and Perrett (2000).

In addition, women's preferences for other masculine features have been noted to intensify during the fertile phase of the menstrual cycle. 'In a variation on Gangestad and Thornhill's T-shirt experiment (1998), briefly discussed in Chapter 14, a recent study by Randy Thornhill et al. (2013) asked female subjects to smell T-shirts that had been worn by men and rate the attractiveness of the odour. For these same men levels of testosterone were measured. They found a positive correlation between the probability of conception of women at the time they rated the smell of the T-shirt and the testosterone level of the male that had worn it ($r = 0.32$, $p = 0.016$). Other features investigated supporting this effect are: voice pitch (Feinberg et al., 2006); dominant men's body odours (Havlíček et al., 2005); masculine body movements as indicated by point-light displays (Provost et al., 2008); and masculine body shapes (Little et al., 2007).

Type of relationship

If a woman is thinking of only a short-term relationship then the expected costs of a masculine but low-investing male are less than the costs of such a male in a longer-term relationship. Consequently, it could be expected that preference for masculine features is higher for short-term than for long-term partners. Support for this contention comes from work by Little et al. (2002) who found that women already with a partner or looking for a short-term relationship showed a preference for masculinised faces. This preference was absent amongst women taking oral contraceptives suggesting that the effect is mediated by hormones.

Environmental health risks

One prediction from trade-off theory is that women's preferences should incline towards masculine faces in environments likely to induce poor health (e.g. where pathogens are prevalent and mortality rates are high). This follows from the idea that masculine men are more likely (theoretically at least) to father healthy children with higher survival chances. This prediction was tested by Lisa DeBruine et al. (2010a) who looked at the preferences of over 4,000 women of white ethnic background from 30 countries recruited to an online questionnaire survey. True to expectation, they found that across the 30 countries

sampled masculinity preferences increased as indicators of social and environmental health decreased. But as the authors point out, this correlation allows for other interpretations. Countries with poor health indicators may also be areas where crime rates are high and gender inequality disadvantages women. In these cultures, therefore, the driver behind masculinity preference may not be related to the desire for healthy offspring but rather a strong and protective partner.

There are also other possibilities related to intersexual competition. Brooks et al. (2011) looked at the effects of national income inequality and national homicide rates on women's facial preferences. Using some of the data already collected by DeBruine et al. (2010a) they found that homicide rate and income inequality were better predictors of masculine facial preference than health indicators alone. They interpret this in terms of intersexual competition where women are hypothesised to prefer masculine faces as cues for dominance in environments where male–male competitive aggression has a strong effect on male wealth and status.

At present this area contains some uncertainties. It looks like there is weak or inconsistent evidence in relation to the immunocompetence hypothesis but stronger evidence that women's preferences do actually shift over the menstrual cycle, and in relation to type of relationship and contextual factors. The next section considers other work that has some bearing on this tension between explanations that look to masculine features as indicators of health and genetic worth, and features that indicate strength and dominance.

15.4.4 **Beauty or the beast? Female choice and male dominance**

The experimental uncertainty in the linkage between female mate choice and male testosterone has led some authorities to suggest that testosterone and associated facial masculinity may be indicators of male competitive ability and not simply (if at all) immunocompetence. As Puts (2010) notes, it is possible that the liberties in mate choice that people in modern industrialised societies enjoy may have biased research towards attractiveness judgements and hence choice by males and females, neglecting the fact that in ancestral times male v. male competition may have been a more important determinant of mating success.

Correcting this imbalance opens up several potentially very productive lines of research. The recorded changes in female facial preferences across the ovarian cycle may indicate some facultative adjustment of preferences in line with cost–benefit ratios of different types of male in competitive environments and not just an estimation of the quality of their immune system. Similarly, cross-cultural variation in facial preferences may correlate with income inequalities, male and female roles, and the level of stratification in a society. It is certainly a testable hypothesis that in societies where status follows from intense male v. male contests then indicators of male competitiveness (through, for example, testosterone-influenced facial masculinity) may be highly prized by women.

Physical evidence points to the importance of male contests, at least in the past. Men are stronger, larger, faster and more physically aggressive than women. The average man, for example, is stronger than 99.9 per cent of all women (Lassek and Gaulin, 2009). Table 13.2 shows the sexual dimorphism in body mass of humans compared to nearest primate relatives. At first sight, the M/F body mass ratio for humans of about 1.1 seems fairly modest compared to chimpanzees (1.3) and orang-utans (2.2). However, this figure underestimates human sexual dimorphism in features most relevant to physical competition. Human females are different from other female primates in having large stores of body fat – possibly functioning as an energy reserve to supply demanding offspring or as a form of sexual ornamentation (for example, fat deposited in breasts and on thighs and buttocks). When fat-free mass is considered, the M/F mass ratio rises to 1.4. The distribution of this excess mass is also significant: men have 60 per cent more total muscle mass than women, 80 per cent greater arm muscle mass and 90 per cent greater upper body strength (Lassek and Gaulin, 2009).

Such differences might suggest an evolutionary history of fighting among males – something that the relative sexual autonomy enjoyed by males and females in recent times may have led researchers to neglect. But even today, about 95 per cent of same-sex homicides from every society for which data is available are committed by men (Daly and Wilson, 1988a). Indeed, as explored in Chapter 12, male violence peaks during reproductive years. Human males may not grow fighting 'weapons' – such as the antlers possessed by deer or the large canines of male gorillas – but men actually manufacture weapons such as clubs, spears and other projectiles. Puts (2010) suggest that we should consider these weapons as part of the male phenotype. Moreover, several features possessed by males, such as beards and low voices, that have often been examined, with ambiguous results, from a female choice perspective, in reality may be signs of dominance directed towards other males.

If males are built to fight other males and this is a primary source of sexual dimorphism, the question then arises as to how, in practice, physical prowess and dominance could have secured extra mates. The lack of oestrus signalling from females would tend to rule out fighting for short-term sexual access (no point if short-term mating could not guarantee offspring). Instead we might expect male strength to be used to facilitate control over longer-term sexual partners or the achievement of a dominant position such that successful males enjoyed access to females with little fear of costly challenges from other males.

It is possible of course that features that once arose in the context of male rivalry and aggression come to be preferred by females and so have been further subject to intersexual selection, but more evidence is needed to support this idea. Furthermore, research into women's preferences for masculine traits such as beards, low voices and masculinised faces often yield conflicting results. To be sure, as noted above, there is some evidence for preferences for more masculine features when women are at their most fertile stage (near ovulation), but in surveying the literature Puts (2010) notes that the effect of perceptions of dominance is even greater (see Figure 15.5).

Other lines of evidence also point to the role of male force and dominance in mating. In numerous societies males have controlled (and in many still do) female sexuality through a variety of strategies such as familial control over marriage partners, female exogamy (even in many Westernised cultures wedding traditions involve the father 'giving away' the bride), mate guarding and strict rules about the appearance and behaviour of women.

Masculine traits compared (a versus b) for effect on attractiveness or dominance		Effect size (Cohen's d) on attractiveness	Effect size (Cohen's d) on dominance	Reference
a	b			
Full beard	Clean shaven	−0.25	1.6	Neave and Shields (2008)
Masculinised voice	Feminised voice	0.25	2.4	Puts et al. (2006)
Masculinised face	Feminised face	1.25	4.25	DeBruine et al. (2006)
Muscular build	Typical build	1.5	2.6	Frederick and Haselton (2007)

FIGURE 15.5 Effect of increasing masculinity as recorded in studies on male beard, voice pitch, facial shape and body build on perceptions of dominance and attractiveness.

In all these studies the effect on dominance is greater, supporting the contention that masculine features arose initially in the context of male competitiveness with other males rather than sexual ornamentation. Cohen's d is a standard measure of effect size and indicates the relative size of the difference between the means of values in two groups. It is defined (here) as (mean a – mean b)/pooled standard deviations.

SOURCE: Data from Puts, D. A. (2010). 'Beauty and the beast: mechanisms of sexual selection in humans.' *Evolution and Human Behavior* **31**(3), 157–75.

Further support for the importance of overall physical sexual dimorphism comes from work by Holzleitner et al. (2014) linking face shape to body dimensions. This group looked at correlations between face shape and BMI and height from measurements made on 40 Caucasian men and women. This enabled faces to be constructed that were typically associated with (and could therefore possibly signal) BMI and height variations and hence masculinity in body shape. Twenty female Caucasian students were then asked to rate the masculinity of the faces. The group found that faces were rated as more masculine if they corresponded in their construction with taller and heavier men. In essence this study suggests that facial appearance may also give clues to the masculinity of body size and shape.

FIGURE 15.6 The Abduction of the Sabine Women (copy of a painting by Poussin 1594–1665).

The forcible abduction of women (bride capture) is still a feature of many tribal societies and judging by the greater diversity of mitochondrial DNA compared to nuclear DNA (see Section 5.4.2) was probably a frequent occurrence in early human societies. The rape or abduction of the Sabine women was one of the founding myths of ancient Rome, where here the word 'rape' is a rendering of the Latin *raptio* meaning abduction rather than sexual violation in the modern sense. In the story, as recounted by Livy, early male Romans (8th century BC) needed wives and when negotiations failed they seized women from the neighbouring tribe of the Sabines. According to Livy no sexual assault took place and the Sabine women were persuaded to accept Roman husbands and take on Roman citizenship. The story illustrates what once may have been a common occurrence in human evolutionary history: male physical strength used to secure female mating partners.

SOURCE: Reproduced by kind permission of Bonhams.

15.5 Neoteny

Neoteny refers to the display or retention beyond a young age of youthful features. Often these features are found to be attractive or cute. Some early theorising on this phenomenon was done by Konrad Lorenz (1943) who interpreted the attraction that adults feel towards baby faces in terms of his theory of fixed action patterns. Lorenz suggested that the facial features of human babies, such as a large forehead, large eyes, small chin and bulging cheeks, serve as social releasers that act upon innate releasing mechanisms to activate nurturant and affectionate behaviour (Lorenz called this mechanism the Kindchenschema, or 'baby schema' in English). This idea could in part explain why humans find baby-like faces, be they on humans, young animals or even toys such as teddy bears, alluring and there is some experimental support for this notion (Eibl-Eibesfeldt, 1989; Archer, 1992). Sternglanz et al. (1977) investigated the effect of varying the features of a baby's face on its perceived attractiveness as judged by American college students. By altering various parameters of line drawings, the overall conclusion was, as Lorenz suggested, a marked preference for faces with large eyes, large foreheads and small chins (Figure 15.7).

Later, Eibl-Eibesfeldt (1989) proposed that the qualities that males find attractive in human females, such as a small, upturned nose, large eyes and a small chin, correspond to 'infantile' features. Such features are found in the young of other creatures that both sexes find attractive or 'cute'. It may be that women's faces have targeted in on the perceptual bias in male brains and developed these features to evoke the caring response that males feel towards their young. On the other hand men are also predicted to search for signs of youthfulness and fertility in female faces and possibly neotenous features serve as stimuli for youthfulness. This effect is sometime called 'overgeneralising' in that cues designed to elicit care in the context of childrearing are generalised to other contexts.

The importance of youthful features and the preference for neotenous faces is supported by numerous academic studies (for example, Cunningham, 1986) and has been confirmed cross-culturally as well. Cunningham et al. (1995), for example, found remarkably similar evaluations of attractiveness of faces in general, and specifically the impact of neoteny, from subjects from a wide variety of cultural backgrounds: white American, Asian, Japanese, Taiwanese, Korean and black Americans. More recently, Borgi et al. (2014) have found that a preference for infantile features in humans and other animals can be detected in children of 3–6 years of age.

The neoteny effect has also been used to explain the facial characteristics of cartoon characters. In what has become a famous essay, the late Stephen Jay Gould claimed to trace the increasing neoteny of the face of Mickey Mouse from the 1930s to the 1980s, resulting in the contemporary depiction of an animal with a large head in relation to body size and large eyes in relation to head height – both neotenous features (Gould, 1982). Box 15.1 explores this idea further with the suggestion that the humble teddy bear has been slowly evolving towards higher neoteny since its origins in the US in 1902. It is highly likely that the attraction humans experience towards other animals (especially companion animals) is founded in part on the response to neoteny (try the exercise in Figure 15.8). Most common pet species (especially

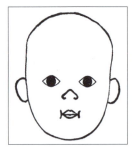

FIGURE 15.7 Drawing of an infantile face showing features found to be attractive: high forehead, large eyes, small nose, small chin.

FIGURE 15.8 Contrasting images of dog faces. Which face do you find most attractive? The answer is probably the one on the right since its features are more neotenous (large eyes, high forehead) compared to the one on the left.

Box 15.1 The evolution of the teddy bear

The teddy bear is a much-loved children's soft toy that takes its name from the former American President Theodore Roosevelt, whose nickname was Teddy. The bear connection arose from a hunting trip in 1902 attended by Roosevelt and other huntsmen. During the trip Roosevelt refused to shoot a bear which had been tied to a tree (to ensure the trip was a success for the President). A cartoon depicting the event appeared in the Washington Post, which gave a man called Morris Michtom the idea of producing a new soft toy. As a cultural artefact the teddy bear provides an interesting case study of cultural evolution in action.

Early bears did look like real American black bears with small eyes and extended snouts, but gradually their features softened. In the early 1980s there was an exhibition of historical teddy bears at the Cambridge Folk Museum in the UK. On hand at the local university were two ethologists, Robert Hinde and L. A. Barden, who took advantage of access to the specimens to make measurements of the way their features had changed over time.

Two ratios were measured: eyes to crown/eyes to base of head, and snout to back of head/crown to base of head (Figure 15.9). The higher the ratio of the former (that is, a high forehead), and the lower the ratio of the latter (that is, flatter face) the more baby-like is the face. Their results are shown in Figure 15.10. It can be seen that from its inception around 1902 the teddy bear has become progressively more neotenous in appearance. The study was based on a limited sample and made the rather large assumption that the dated examples in the exhibition were representative of teddies of that period. A more systematic survey with a larger sample has yet to be conducted.

An obvious question is what is driving this evolutionary process. It could be that the move towards more baby-like features gradually provided a better fit with what young children want and this exerted a selective force such that neotenous bears were preferentially purchased and hence produced. Alternatively, since it is adults that do the purchasing it may be the elicitation of positive and nurturant feelings

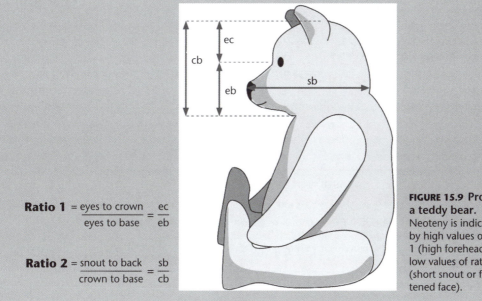

Ratio 1 $= \dfrac{\text{eyes to crown}}{\text{eyes to base}} = \dfrac{\text{ec}}{\text{eb}}$

Ratio 2 $= \dfrac{\text{snout to back}}{\text{crown to base}} = \dfrac{\text{sb}}{\text{cb}}$

FIGURE 15.9 Profile of a teddy bear. Neoteny is indicated by high values of ratio 1 (high forehead) and low values of ratio 2 (short snout or flattened face).

by neotenous bears that have driven these changes. Partly to address this question, Morris et al. (1995) collected a sample of teddy bears of varying degrees of neoteny and presented these to girls and boys in the age categories of 4, 6 and 8 years old. Interestingly, when presented with a choice only children of 6 and 8 years of age actually showed a preference for baby-faced teddies. Children aged 4 years actually showed a slight, but statistically non-significant, preference for adult features in a toy bear. These findings are explicable if we consider that 4-year-olds are not expected to provide care and nurturing and so perhaps the response to neoteny is muted or undeveloped. It seems that

the evolution of the teddy bear has not been driven by the preferences of children under 5 years old but more likely by the response of adults who actually buy the toys.

This case study is interesting in that it also enables parallels to be drawn with, and tested against, biological evolution. Clearly teddy bears do not reproduce by themselves but there is a reproductive system turning them out and they compete with each other in their market niche. One could, therefore, argue that they illustrate some key stages in Darwinian evolution: variation, inheritance, modification and adaptation. Chapter 20 considers cultural evolution in more depth. See also Whiten et al. (2011).

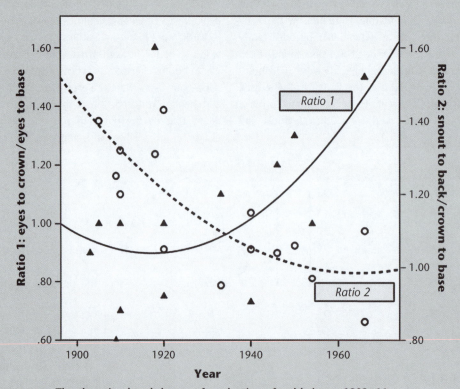

FIGURE 15.10 The changing head shapes of a selection of teddy bears 1903–66.
As time passed since the innovation of the teddy bear around 1902, ratio 1 (closed triangles, solid line) has increased, whilst ratio 2 (open circles, dashed line) has decreased; this indicates that teddy bears have become more neotenous in appearance.
SOURCE: Data from Hinde and Barden (1985), Figures 1 A and B, p. 1372. Lines of best fit added.

cats and dogs) have behavioural and morphological infantile features which are partly retained into adulthood and the resulting neotenous features may have been part of the domestication process

What is interesting about the fondness for baby-like faces is that it is found in both men and women. Brosch, Sander and Scherer (2007) looked at the ability of infant faces to capture the attention of viewers compared to adult faces. They found that infant faces were indeed superior to adult faces in arresting attention but they found no difference between men and women. Recently, however, the advent of specialised computer software has enabled facial images to be generated and altered in subtle ways; so that, for example, the geometry of faces agreed to be 'cute' by independent observers can be established and then used to manipulate images of real faces to make them more or less cute according to these measurements. Using such techniques Sprengelmeyer et al. (2009) found that in a 19–26-year age group young women were better able to detect differences in cuteness than young men. They also found that cuteness sensitivity of postmenopausal women was not significantly different to that of men (either in a 19–26- or 53–60-year age group). Premenopausal women taking oral contraceptives (which raise levels of oestrogen and progesterone) were more sensitive to cuteness than equivalent groups not taking oral contraceptives. All this points to a role for such hormones as oestrogen and progesterone in heightening the response to cute features.

In another study, Lobmaier et al. (2010) used images of real babies to generate composite faces grouped into two categories: most cute and least cute. These were then used as prototypes so that a spectrum of cuteness could be created (Figure 15.11).

Subjects were then presented with pairs of images (such as shown in Figure 15.11) that differed by varying degrees of cuteness and were asked to choose the 'cutest' image. They found that women were consistently better able to choose the cutest face.

A sex bias in the attention given to infant faces was also observed in experiments carried out by Rodrigo Cárdenas et al. (2013). About 30 men and 30 women were asked to look at images of pairs of faces from an adult (male or female) and a baby (boy or girl). Whilst the subjects were looking, eye movements were tracked enabling a measure of the time spent selectively looking at the baby face or the adult face. It was found that women looked longer at infant faces when the face was paired with either a male or female adult face. Males only looked longer at an infant face if it was paired with an adult male face; pairing with an adult female face led to the female face being inspected for longer.

There is evidence that the caring response elicited by neotenous features also extends to adult faces. Keating et al. (2003) manipulated photographs of real adult faces (white male and female faces, and black male and female faces) to make them more or less neotenous. These photographs were then attached to printed resumes or CVs. The resumes in turn were attached to stamped addressed envelopes and then deliberately 'lost' in towns in the US and in Kenya. The idea was that faces that elicited a caring or nurturing response would be more likely to have stimulated the finders to take the effort to place

FIGURE 15.11 Example of an infant face transformed in cuteness.
A collection of images was used to create composite prototypes for two extremes of baby faces: most cute and least cute. The geometric differences between these prototypes could then be used to reduce or enhance cuteness. The image on the right has been transformed to enhance cuteness by 100 per cent compared to image on the left.

SOURCE: Images kindly supplied by Professor Janez Lobmaier (University of Bern, personal communication, 2015).

the resumes in the envelopes and return them. The group found that in all cases, except mature black faces, resumes with neotenous facial images attached were more likely to be returned.

15.6 Similarity and difference

15.6.1 MHC and facial attractiveness

A noted earlier we should expect to find attractive faces correlating with signs of health and parasite-resistant alleles. Given that heterozygosity in the major histocompatibility complex (MHC) is also thought to be an indication of an efficient immune system we might also expect individuals to be on the lookout for a partner with an MHC system that is different and hence complementary to their own (Box 15.2). But how can this be achieved?

There is evidence of both mice and humans being attracted to smells that indicate differences between potential mates in the major histocompatibility complex (MHC) (see Chapter 17). If heterozygosity is a desirable goal in the genome of one's offspring, then might individuals rate the attractiveness of faces of people who have MHC genes dissimilar to their own more highly than faces from MHC-similar individuals? Given that establishing a mechanism for the MHC to affect odour has so far proven quite difficult, the idea that the MHC might influence facial appearance would seem to be a long shot. Yet some studies have pointed to this as a possibility. Roberts et al. (2005a)

Box 15.2 The MHC system

Humans are under constant attack from parasites and other infectious agents, which are kept at bay for most of the time by a highly complex immune system. It is to be expected, then, that breeding with someone who could provide our offspring with genes that help build an efficient immune system would be a desirable goal. With this in mind, over the last 20 years or so numerous studies have been carried out on the role of the major histocompatibility complex (MHC) in human mate choice. The MHC is a set of protein products that sits on the surface of cells and plays a crucial role in the immune response in vertebrates. The MHC genes responsible for these surface proteins are found on chromosome 6 and consist of about 4 million bases organised into about 180 genes. These MHC genes code for peptides (short protein-like molecules) that activate receptors on T cells (white blood cells) and prime them to recognise and destroy foreign bodies. The interesting feature of this set of genes is their high degree of polymorphism; in this case hundreds or even thousands of different alleles exist. The nucleotide diversity in this region of the genome is at least 100 times greater than the genome average. It is as if there is no single stable solution as to what MHC genes should look like.

Several ideas and lines of reasoning have been put forward to explain this enormous diversity:

- One idea is that this polymorphism is maintained by heterozygote advantage. MHC gene expression is 'co-dominant' meaning that each gene in the diploid pair is expressed and actually makes some protein. Hence if a cell is heterozygous it produces more types of MHC proteins than a homozygous cell; and the more types of proteins that are produced and used to activate T cells the wider the variety of foreign agents these cells can detect and help destroy.
- Another hypothesis suggests that the MHC polymorphism is maintained as a consequence of frequency-dependent selection resulting from a genetic arms race between the immune system and pathogens. In this model, pathogens mutate rapidly to eliminate from themselves peptides that look similar to the variants being produced by the host's MHC system. One they do this they can evade detection and spread rapidly, until, that is, they encounter a host with a formerly

rare MHC allele that confers resistance; this rare allele then becomes more common. In this way stability is never achieved but instead huge numbers of MHC variant genes will always exist. In addition, temporal and spatial fluctuations in pathogen type and intensity could select for MHC diversity as humans encounter differ pathogen species.

- Another suggestion is that MHC gene polymorphism arises from sexual selection. In this scenario, individuals in each sex will prefer partners who have an MHC genotype different from their own. This mechanism would serve to increase the heterozygosity of offspring and, as noted above, heterozygosity enhances the immune response (Penn et al., 2002). In essence, parents can produce offspring with MHC genotypes different to their own, thereby serving to offer protection from pathogens that may have become adapted to the parental genotype. An added twist here is that incest avoidance would also predict that individuals should prefer mating with partners with MHC genes different to their own (see Chapter 17).

- MHC diversity could also be maintained by foetal selection *in utero*, whereby foetuses with too little MHC heterozygosity are discarded as being potentially vulnerable and unfit. Here the selective pressure is the mother's quality filtering mechanism.

These several ideas are not mutually exclusive. MHC genetic polymorphism could originally be the result of pathogen pressure selecting and maintaining variety but then maintained by sexual selection, incest avoidance or some combination of both (see Havlíček and Roberts, 2009, for a review).

looked at female ratings of male faces in relation to whether the men where MHC similar or MHC dissimilar to the female subject doing the rating. Surprisingly, MHC-similar men were rated as more attractive. In another study, Roberts et al. (2005b) looked at the facial attractiveness ratings of men (as judged by 50 female volunteers) in relation to the heterozygosity of the men's MHC genes measured at three loci. They found that men who were heterozygous at all three loci were judged to be more attractive than those with only one or two heterozygous loci. More recently, Lie et al. (2008) measured the heterozygosity of 77 males and 77 females at markers on MHC-related sites and also markers on different chromosomes. The male and female images of these subjects were then rated for attributes by the opposite sex. They found that the attractiveness of males was positively correlated with MHC heterozygosity but not heterozygosity on other chromosomes. The effect was not significant for females.

In relation to the matching of compatible MHC genotypes between prospective partners, there arises the interesting possibility that females might benefit from the sampling of many males to find the one most immunologically compatible. It is known that during their passage to the uterus sperm induce a host reaction and the female's body perceives sperm as foreign agents and proceeds to attack them with leukocytes (Barratt et al., 2009). It is a possibility that the female's reproductive system makes judgements about the health of sperm and suitability of MHC genes based on some sort of chemical signature, and so allows healthy and compatible sperm through this immune response barrier. If this is the case then the standard narrative that human females have little to gain from multiple matings will need revision with serious implications for the understanding of human sexuality.

If, by whatever mechanism (physical appearance, odour, or sperm filtering), males and females are attracted to MHC-dissimilar individuals then one might expect married couples to share fewer MHC genes than any pair of people taken at random from the general population. However, despite numerous studies, this effect remains elusive (Havlíček and Craig Roberts, 2009). The overall impression from work on MHC-related effects is that is there a phenomenon at work here but that it is poorly understood and it is likely that conflicting evidence reflects the influence of a whole range of confounding factors that also strongly affect mate choice, as well as methodological differences between the studies.

Reconciling all these studies is going to prove tricky. If they are reliable, the impression is that MHC-correlated facial preferences tend to be assortative (that is, individuals like the faces of individuals who are MHC similar); whilst odour preferences are disassortative (individuals like odours from MHC-dissimilar individuals). Interestingly, assortative mating in humans has been reported on a number of occasions. Humans tend to end up in relationships with people who are similar to themselves in a variety of physical, social and psychological traits (see Swami and Furnham, 2008, chapter 7). This subject also falls under the heading of imprinting and is examined in the next section.

15.6.2 Imprinting

Imprinting is a genetically driven learning process. It usually consists of a relatively short sensitive phase occurring early in the development of an organism, and the learning experience acquired during this phase has long-lasting or even irreversible effects. It can be divided into filial and sexual imprinting (Figure 15.12).

Filial imprinting

One of the earliest forms of imprinting to be studied was filial imprinting – the tendency of a newly born animal to imprint on the first object it sees. It was first reported by the British amateur biologist Douglas Spalding (1841–1877), rediscovered by the ethologist Oskar Heinroth and studied in detail and brought to public awareness by Konrad Lorenz. Lorenz demonstrated that newly hatched goslings would imprint on himself (or more likely his boots) and follow him around, ignoring their natural mother. Such a mechanism has obvious advantages in natural conditions since the first object most young animals will see is their mother and it makes obvious adaptive sense to stay close to their primary caregiver. In a sense familial imprinting is a quick heuristic fix to the problem of who to trust early in life.

Whereas Lorenz's goslings followed him along the ground, others have used the phenomenon to induce birds into following a flying object. In 1993, for example, the Canadian inventor and light aircraft enthusiast Bill Lishman used imprinting to lead a flock of Canadian geese on a migration from Ontario in Canada to Northern Virginia. This feat formed the basis of the movie *Fly Away Home* and inspired the techniques used to make the French film *Winged Migration* (Lishman, 1996).

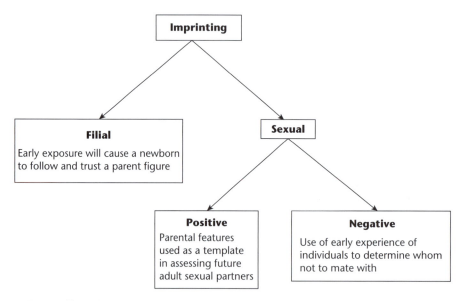

FIGURE 15.12 Types of imprinting.

Sexual imprinting

Another well-studied form of imprinting is **sexual imprinting**. This can be divided into two types: positive imprinting, whereby a young animal uses the observed phenotype of one of the parents to assess future sexual partners (ten Cate et al., 2006); and negative imprinting where young children (for example) develop an aversion to mating with any adult or sibling that they have grown up with (the Westermarck effect, see Chapter 17).

Positive sexual imprinting

For many animals one of the obvious benefits of positive imprinting is species recognition and the avoidance of hybridisation. In taxa with many similar species (types that have only recently diverged and speciated) it is imperative to mate with the correct species and avoid the negative fitness costs of producing hybrid offspring. This may be the functional cause of the many types of positive imprinting recorded in the literature. In some African lakes, such as Lake Victoria and Lake Malawi, there are large numbers of different cichlid species of fish living in close proximity and distinguished by colouration but otherwise very similar in body shape and behavioural ecology. This raises the question as to how reproductive isolation (the avoidance of cross-breeding) can be maintained. In tackling this question Verzijden and ten Cate (2007) demonstrated that positive sexual imprinting was probably at work. They set up cross-fostering experiments on two similar species – *Pundamilia pundamilia* and *Pundamilia nyererei* – and were able to show that young females developed sexual preferences for the males of the species of their foster mother, even if the foster mother was of a different species.

Similar experiments have been performed on mammals. Working with sheep and goats, Kendrick et al. (1998) showed that male goats (*Capra aegagrus hircus*) raised by a foster mother ewe (*Ovis aries*) later in life preferred to mate with sheep. Similarly, male sheep foster mothered by female goats preferred to mate with goats when they grew up. A similar cross-species preference effect was observed for female sheep and goats raised by foster goats and sheep respectively. On the other hand, such mechanisms cannot be universal since brood parasites such as the cuckoo (*Cuculus canorus*) do not imprint on their host species: baby cuckoos still grow up to mate with other adult cuckoos.

Whether species recognition ever was an important cause of positive imprinting in humans is unclear. Since the extinction of the Neanderthals about 35,000 years ago modern human populations have not lived in close proximity to similar hominin species. It is possible, however, that imprinting lives on as a vestigial trait.

Homogamy

Do people end up marrying someone whom they resemble? It has been suggested that assortative mating (a phenomenon sometimes called **homogamy**) might be a way of optimising outbreeding. Too much inbreeding can reduce fitness through the accumulation of homozygous recessive genes (hence incest avoidance and, possibly, MHC-dissimilar odour preferences), but excessive outbreeding brings its own costs (think of the ultimate outbreeding with another species that brings about hybrid sterility) and reducing outbreeding (through phenotypic matching) can stabilise successful collections of genes (see Jaffe, 2001). The preference for similar faces may also come from sexual imprinting. If the faces of parents are imprinted on children, and then children use these templates as standards of beauty, then as adults they may prefer faces like their own since such faces also resemble those of their parents. Indeed, several studies have reported that women's faces resemble those of their long-term male partners above the level expected by chance alone (for example, Penton-Voak and Perrett, 2000a; Bereczkei et al., 2002; Little et al., 2003). It is tempting to attribute these findings as evidence of positive sexual imprinting, since if men and women resemble their parents and use their experience of their parents' faces as a guide to what to find attractive, then this type of assortative mating (being attracted to someone like yourself) is just what one would expect. But there are alternative explanations. If, for example, due to sexual competition, attractive people tend to end up with attractive people and less attractive ones with others similarly less attractive, then since some features are found to be universally attractive (such as facial symmetry or prominent cheekbones) attractive people will tend to resemble each other. This process is a type of phenotypic matching as opposed to imprinting. Assortative mating like this would also result if preferences for facial traits and the traits themselves are heritable. Imagine a women who inherits a preference for certain shape of nose: she chooses a man with this type of nose and produces a daughter who inherits both the nose shape and the nose preference. She in turn will look for a partner with this nose shape, who will, by implication, look like herself. Alvarez and Jaffe (2004), however, found that the phenomenon of homogamy persisted even when the expected similar attractiveness of partners' faces was taken into account.

There are various methodological and theoretical problems that need to be resolved before positive sexual imprinting can be accepted as part of the behavioural ecology of humans. A basic premise of Darwinian approaches to human mating is that physical attractiveness is some sort of signal of genetic and phenotypic quality. But if imprinting did occur then a child born of parents of poor genetic and phenotypic quality (who would conceivably show physical traits associated with this) would in future be seeking mating partners with these signs of lower quality. As such, she or he would be at a disadvantage compared to same-sex rivals who did not have genes for positive sexual imprinting but simply rated the attractiveness of potential mates using more universal criteria. One factor in its favour, however, would be that positive imprinting of parental characteristics at least focuses on something that must have been successful since the parents have produced an offspring. This might be favoured in harsh environments where successful parents were indeed above average in health-related qualities.

One further problem is that by itself positive imprinting would increase the risk of inbreeding since it would enhance the attractiveness of opposite-sex

siblings and other related individuals. In the final analysis, the various mechanisms at work shaping what people find attractive need not be mutually exclusive. Positive sexual imprinting could serve as a first broad pass filter – something that ensures an individual mates with the opposite sex and the same species. Negative imprinting could then function to inhibit mating between co-socialised individuals. Finally there could be a whole series of other layers of fine filters or algorithms conferring sensitivity to markers of genetic fitness. In a review of recent studies, Rantala and Marcinkowska (2011) found the evidence for positive sexual imprinting in humans inconclusive and in sore need of more studies (like that of Bereczkei et al. (2004)) designed to disentangle imprinting and genetic effects. It is also important to realise that positive and negative sexual imprinting do not negate each other; both could work in tandem to promote optimal outbreeding. Positive sexual imprinting ensures that an individual will mate with the same species and helps to keep together local adaptations, whilst negative sexual imprinting ensures that specific individuals who are close kin are avoided. How a balance is achieved and whether the two mechanisms can adjust to determine the level of outbreeding that is optimal in a given environment is not understood (see Rantala and Marcinkowska, 2011).

Although the evidence on positive sexual imprinting is tentative and suggestive, studies on negative sexual imprinting have revealed far more robust findings, and this is the subject of Chapter 17. A far more difficult problem for evolutionary theory is to explain non-reproductive mate choices made by men and women when they fall in love with a member of their own sex. This is the subject of the next chapter.

Summary

▶ Physical beauty in both males and females is expected to be correlated with signs of reproductive fitness. Evidence is now accumulating that body shapes found to be attractive are those which carry fitness indicators. Facial beauty in both sexes is likely to be related to overall fitness, health and immunocompetence. Faces may convey honest signals about fitness and reproductive value.

- Symmetrical faces tend to be regarded as attractive – this may be because symmetry is a sign of a healthy immune system and a body relatively free from stress.

- Composite or averaged faces tend to be perceived as attractive – this may be because the average is also the optimum for continuously distributed traits, or it may reflect a bias in perception where the average of frequently seen objects is registered as attractive.

- The role of testosterone in enhancing facial attractiveness is a problematic area. There is some evidence that women during their fertile phase are attracted to faces signifying high testosterone. One argument is that testosterone is an honest signal and handicap indicating a good immune system in those males that can afford to carry high testosterone levels. Another view is that testosterone is a sign of dominance.

- Both males and females are sensitive to neotenous faces. Increasing neoteny also tends to increase attractiveness or perceptions of cuteness. The fundamental reason for this is still not established. It could be an elicitation of nurturant feeling towards vulnerable younger members of the species, or the action of supernormal stimuli signalling youthfulness.

- The role of the major histocompatibility (MHC) complex in mate choice is difficult to unravel. The evidence so far suggests people like the faces of other people with a similar MHC complex, but prefer odours from people with a different MHC complex.

- Sexual imprinting is likely to be present in a negative sense that we have an aversion to sexual relations with those we grew up with (see Chapter 17). There is mixed evidence concerning positive sexual imprinting.

Key Words

- **Averaged**
- **Facial averageness**
- **Fluctuating asymmetry (FA)**
- **Homogamy**
- **Neoteny**
- **Sexual imprinting**

Further reading

Rhodes, G. and L. A. Zebrowitz (2002). *Facial Attractiveness: Evolutionary, Cognitive, and Social Perspectives* (Vol. 1). Ablex Publishing Corporation.

Edited volume of articles. Successfully conveys the varied nature of research findings in this field.

Perrett, D. (2010). *In Your Face: The New Science of Human Attraction*. Palgrave Macmillan.

Accessible, readable and yet scholarly. Well-illustrated account of research into facial attractiveness. Tackles the subject from a number of directions.

The Paradox of Homosexuality

Oh who is that young sinner with the handcuffs on his wrists?
And what has he been after that they groan and shake their fists?
And wherefore is he wearing such a conscience-stricken air?
Oh they're taking him to prison for the colour of his hair.
A. E. Housman (Additional Poems XVIII)

When Housman wrote those lines around 1895 he had in mind the trial and imprisonment of Oscar Wilde. Like Wilde, Housman was gay though he never admitted it publicly; even the poem from which the lines were taken was not published until after his death. Interestingly, Housman was one of six children and had a brother and sister who were also homosexual. Perhaps the metaphor of the 'colour of his hair' for sexual preference revealed his suspicion that his sexual orientation had a biological origin.

Sexual orientation is a controversial and hotly debated topic in biology, psychology, philosophy and anthropology. Terms such as heterosexual, homosexual, bisexual and asexual, which might seem innocuous enough at first glance, have been known to cause offence when used in a manner insensitive to context and in disregard of other measures of self-identity. The American novelist Gore Vidal, for example, famously insisted that homosexual and heterosexual were adjectives and not nouns. Such sensitivities are understandable given the history of bigotry and prejudice that has surrounded sexual behaviour. It is not the intention here to engage in any philosophical analysis of these issues but it is appropriate to touch on some features of the contested concepts associated with them. For example, sexual preference and sexual orientation are not interchangeable terms. It

is usually agreed that the term 'sexual orientation' implies little choice whilst sexual preference does suggest some component of choice. Sexual orientation, which describes a relationship between the self and others, also needs to be distinguished from sexual identity, which is a measure or indication of self-identity. It is possible that sexual orientation, sexual behaviour and sexual identity are discordant. For example, a man who is strongly attracted to other men but who only has sexual relations with women exhibits a discordance between orientation (homosexual) and behaviour (heterosexual). People experiencing these misalignments are sometimes referred to as 'closeted'.

One way to avoid the imputation of sexual identity when actually referring to orientation is to use the terms heterosexual, homosexual and bisexual as adjectives describing behaviours rather than nouns referring to types (Diamond, 2010). This highlights the important point that a person's identity is not solely defined by sexual orientation, be that homosexual or heterosexual. Another way is to use the terms **androphilia** (man-loving) and **gynephilia** (woman-loving); these terms then define the behaviour by the object desired rather than the person's own identity. They are often used when describing non-Western cultures to circumvent the problem that the meaning of terms such as homosexual and heterosexual vary cross-culturally. They also make no reference to the sex of the actor, which has some biological advantages in that if androphilia, for example, is a gene-influenced trait then it could be present in both men and women.

There is also a debate about the integrity of the traditional types of orientation: homosexual, heterosexual, bisexual, transgender and the less commonly used asexual. As long ago as 1948 Alfred Kinsey thought it preferable to classify people along a 7-point scale from homosexual to heterosexual. But even this makes assumptions that one is a trade-off for the other, in that decreasing masculinity means increasing femininity. It has been argued that possibly one can be both very masculine and very feminine (Shively and DeCecco, 1977). Finally, there is even the suggestion made by philosophers such as Michael Foucault that the categories of sexual orientation are not 'natural kinds' but rather social constructs (see Stein, 1998; and Kirby, 2003).

Lots of problems remain in this highly contested field, but the scientific consensus now seems to be that sexual orientation is due to a combination of genetic, hormonal and environmental influences. It is also generally accepted that homosexuality (to take one orientation) is not a product of faulty child-rearing or a troubled family background or adverse life experiences as was once misguidedly thought. It also seems that sexual orientation is not a matter of choice – although behaviour may be. A heterosexual woman may, for example, choose to become a nun and so exhibit asexual behaviour. In the analysis here the terms homosexual and heterosexual should be taken implicitly as adjectives describing behaviour even if for convenience the terms sometimes appear as nouns. Most of the discussion in this section will be on male homosexuality since most research within a Darwinian framework on same-sex relationships has focused on this topic.

The incidence or prevalence of homosexuality is difficult to ascertain precisely, partly because responses to surveys will depend upon the precise nature of the questions asked and also because people may not self-report in a candid way (a phenomenon strongly influenced by culture and the age group of the respondents). In addition, there is some fluidity in the terms homosexual and heterosexual.

In the 2011 UK census, for example, 1.1 per cent of the population surveyed identified themselves as gay or lesbian, and 0.4 per cent as bisexual. In the USA, a Gallup poll published in 2012 found that 3.4 per cent of US adults identified themselves as lesbian, gay, bisexual or transgender (LGBT); but young people (18–29) were three times more likely to identify as LGBT than people over 65 (6.4 per cent and 1.9 per cent respectively) (Gates and Newport, 2012).

If homosexuality derives, at least in part, from genetic influences then it poses something of a paradox for Darwinism. Homosexual men and women who completely exclude opposite-sex partners obviously do not reproduce at all; but various studies have shown that homosexual men and women in general (who sometimes do become fathers and mothers through temporary heterosexual relationships) have fewer offspring than heterosexuals (see Bell and Weinberg, 1978; Iemmola and Camperio-Ciani, 2009). Hence the problem for Darwinism: natural selection should, over time, eliminate those genes that reduce fecundity and reproductive fitness. If, however, homosexuality is a result of environmental effects or life experiences then the problem largely disappears (apart from the problem of explaining how a phenotype arises that responds to influences in this manner). So two important and related questions arise: 'Does homosexuality have a genetic basis?' and 'If it does, how can this Darwinian paradox be resolved?'

16.1 A genetic basis for homosexuality

A standard way of investigating the influence of genetic factors on a trait is through twin studies, especially the comparison between monozygotic (MZ) and dizygotic (DZ) twins. Table 16.1 provides a summary of a few studies on this.

Table 16.1 shows considerable variation between studies, reflecting no doubt different methodologies and sample sizes. Of the studies cited in Table 16.1, that by Långström et al. (2010) is by far the largest to date, involving the study of the sexual orientation of 2,320 MZ twins and 1,506 DZ twins. Such studies tend to show a higher concordance for MZ twins than for DZ twins, a result to be expected if genetic factors were at work. But if there is a genetic basis it

Table 16.1 Concordance rates for a variety of paired siblings
Concordance measures the likelihood of one of a pair exhibiting a condition if the other does.
Hence Långström et al. (2010), for example, found that if one male monozygotic twin was
homosexual then the chance of the other twin being the same was 18 per cent.

Relationship	Study and concordance rates for homosexuality			
	Bailey and Pillard (1991)	King and McDonald (1992)	Whitam et al. (1993)	Långström et al. (2010)
Male (MZ) twins	52%	10%	65%	18%
Male (DZ) twins	22%	8%	28%	11%
Male siblings	9.2%			
Adoptive brothers	11%			
Female (MZ) twins			75%	22%
Female (DZ) twins				17%

is unlikely to follow a simple Mendelian pattern of inheritance (see Mustanski et al. (2003) for a review).

Efforts to locate an area of the genome responsible for any genetic influence on homosexuality have met with mixed results. Because of the evidence pointing to maternal transmission (see later in this section), early efforts focused on the X chromosome since males can only inherit the X chromosome from their mother. One study by Dean Harmer et al. (1993), working at the National Cancer Institute in the USA, found that of 40 homosexual brothers 33 shared a set of five markers on part of the X chromosome known as Xq28 (a region at the tip of the long arm of the X chromosome carrying about 8Mb of information). The study was repeated a few years later with different samples and reached similar conclusions (Hu et al., 1995). Additionally, in this second study a linkage was observed between markers on Xq28 and male homosexuality but not female homosexuality. But another study on Canadian gay men failed to find any correlation between homosexuality and these markers (Rice et al., 1999). The reaction of the press in the USA and the UK to these findings makes an interesting study in its own right (see Conrad and Markens, 2001).

Later Mustanski et al. (2005) carried out a full genome scan of 456 individuals from 146 families with two or more gay brothers. They found only faint evidence for the previously reported linkage with Xq28 but suggested instead the involvement of genes on chromosome 7 in an area known as 7q36.

In summary, twin studies point to genetic factors at work but the identification of regions on chromosomes associated with homosexuality is still in its infancy. We now turn to the next question, which is how the fitness costs of homosexuality can be overcome.

16.2 Kin selection models

We noted in Chapter 3 how altruistic behaviour directed towards kin may be favoured at the genetic level if the conditions of Hamilton's equation ($rb > c$) are satisfied. This has led to the hypothesis (first proposed by Wilson (1975)) that androphilic males may increase their inclusive fitness by directing altruistic behaviour towards their kin, thereby increasing their kin's direct fitness. The prediction is, in essence, that androphilic males will be more generous and altruistically inclined towards their kin than gynephilic males.

Evidence relating to this prediction has not so far been very supportive. Bobrow and Bailey (2001), for example, used questionnaires to compare the altruistic tendencies of 57 heterosexual men with 66 homosexual men in Illinois, USA. They found no difference between the two groups on scales measuring generosity, affinity and avuncular tendencies. In fact, on another measure they found than homosexual men were less inclined to give monetary gifts to their siblings than heterosexual men. In response to other questions gay men reported feeling more emotionally distant from their fathers and older siblings than straight men. In a similar study in the UK, Rahman and Hull (2005) recruited 60 heterosexual and 60 homosexual men living in London and asked them to complete a questionnaire designed to measure their generosity, affinity and benevolent feelings. Once again they found no difference on these measures between gay and straight men.

Taken together the results suggest an absence of elevated levels of kin-directed altruism as predicted by kin selection theory. However, the cultures of the participants (USA and UK) may not resemble in crucial respects the environments in which male androphilia, and putative associated higher levels of kin altruism, evolved. More pointedly, despite the progressive reforms of recent decades, homophobic attitudes still persist and may act to alienate gay men from their families and so suppress any altruistic tendencies. Mindful of these problems with Western cultures, Vasey et al. (2007) examined the altruistic tendencies of a sample of androphilic and gynephilic males in the Polynesian Nation of Independent Samoa. Here androphilic men are known as fa'afafine (literally 'in the manner of women'). They found that in terms of overall generosity towards, and resources given to, close kin (siblings and parents) androphilic males did not differ from gynephilic males. They did detect, however, stronger avuncular tendencies (measured by such things as money given to nieces and nephews and frequency of babysitting) among androphilic men compared to gynephilic men. The authors quote one fa'afafine participant as saying 'My sister has a daughter and a newborn son now. Soon she will go back to work and I will be the mother' (quoted in Vasey et al. (2007), p.164). One confounding variable here (noted by the authors) is that, understandably, none of the fa'afafine men had children of their own whereas most of the gynephilic men did. So it is possible that the androphilic men simply had more time to demonstrate their generosity towards their nephews and nieces compared to gynephilic men.

Kin selection theory, then, has met with mixed results. A more general problem with the theory is that it is difficult to see why altruism directed towards children of siblings should be associated with homosexuality and not simply asexuality (as is found in the social insects, for example).

16.3 Sexually antagonistic selection and the 'fertile female' hypothesis

In this context, the term 'sexually antagonistic' suggests that a genetic feature (an allele or group of alleles) may reduce fitness when present in one sex but enhance fitness when present in the opposite sex. We noted above the mixed success of locating genes on the X chromosome that may promote homosexuality. Suppose for a moment, however, that genetic influences promoting homosexual behaviour were somewhere on the X chromosome. Since this reduces the fitness of males we might expect, if **sexually antagonistic selection** were at work, to find increased fitness in those female relatives of the homosexual male that shared that same X chromosome (or relevant regions of it). Andrea Camperio-Ciani of northern Italy has now led several studies investigating this effect. In one study he found that females on the maternal side of the family of homosexual males were significantly more fecund than females on the paternal side (Camperio-Ciani et al., 2004). Mothers of gay men, for example, produced 2.7 babies compared with 2.3 babies born to mothers of straight men (p = 0.02): a result in keeping with the notion that homosexuality may be linked to the X chromosome since females on the father's side of a male's family do not share X-chromosome genes with that male (see Figure 16.2). Moreover, the team found that homosexual males reported that there were more homosexual men on their mother's side of the family compared to their father's side (see Figure 16.1).

In a second study conducted a few years later with a larger sample, similar results were found (Iemmola and Camperio-Ciani, 2009). The sample included 152 homosexual and 98 heterosexual males. As expected, male homosexuals had fewer children than heterosexual males (a mean of 0.012 compared to 0.91 respectively). Also, as predicted, they found that female relatives on the mother's side of male homosexuals had a higher fecundity than relatives on the father's side (Figure 16.3).

This is strong evidence then for alleles on the X chromosome serving to increase fertility when present in women (possibly through enhanced androphilia) whilst serving to incline men who also possess these alleles towards homosexuality. The situation here is not one of simple determinism, however. The team

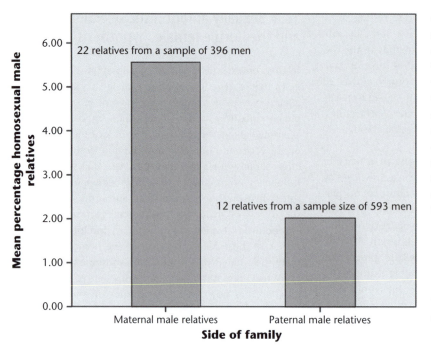

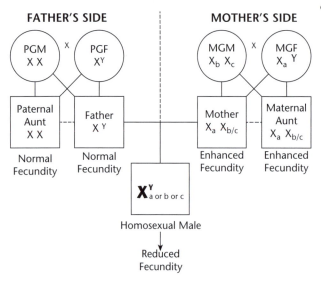

FIGURE 16.1 Frequency of homosexuality on mother's and father's side of subject's family.
The numbers in the sample for male relatives on the mother's side are those males who could potentially share an X chromosome with the homosexual male. Note on the mother's side sons of maternal uncles were excluded since they do not share an X chromosome with the homosexual subjects; this explains why there are more male relatives on father's side than mother's side. $X^2 = 8.92$, $df = 1$, $p < 0.005$.

SOURCE: Plotted with data from Camperio-Ciani et al. (2004), Table 1, p. 2218.

FIGURE 16.2 How a male inherits his X chromosome.
A homosexual male with reduced fertility inherits his X chromosome from his mother. There may be a region on the X chromosome that in females (here a mother and a maternal aunt) enhances fecundity, serving to maintain this region of the X chromosome in the gene pool. This is a slightly simplified diagram showing no crossing over.

estimate that the X chromosome effect might explain about 14 per cent of the total variance in male sexual orientation with the 'older brother effect' (see later) accounting for about 7 per cent. This leaves about 79 per cent of the variance unaccounted for and presumably due to other factors such as individual life experiences or other as-yet-unknown biological effects.

16.4 Balanced polymorphism and heterozygotic advantage

As we have noted elsewhere in this volume (see Chapter 18), one way of explaining the presence of alleles that can sometimes act to reduce fitness is through the notion of heterozygotic advantage. There may be alleles which in the homozygous state lead to homosexuality that in the heterozygous state somehow confer increased fitness, an idea first proposed by Hutchinson in the late 1950s (Hutchinson, 1959). More recently, McKnight (1997) developed the idea to suggest that genes inclining their bearers towards homosexuality in the homozygous state could be pleiotropic (having more than one effect), and when present in the heterozygous state carry with them the development of characteristics such as sensitivity, empathy and communication skills that women find attractive. These ideas were further developed by Edward Miller. He argued that since homosexual men seek out

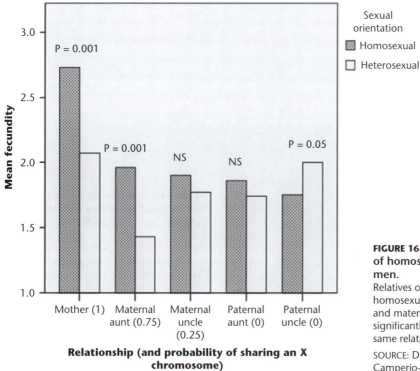

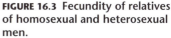

FIGURE 16.3 Fecundity of relatives of homosexual and heterosexual men.

Relatives on the female side of a homosexual male (that is, mother and maternal aunts) tend to have a significantly higher fecundity than the same relatives of heterosexual men.

SOURCE: Data from Iemmola and Camperio-Ciani (2009).

other males then this points to a fairly sophisticated system of gender recognition and discrimination already in place. So rather than imagining that this system evolved *de novo* with homosexuality it seems more likely that it is a system of androphilia (attraction to men) usually found in females but here switched on, as it were, in males. Consistent with this is the well-recognised fact that *in utero* humans begin development as females (hence both sexes end up with nipples) but that testosterone from the genitals of male foetuses causes the masculinisation of the developing male child. With this context in mind, Miller speculated that there may exist a number of pleiotropic alleles which, if inherited altogether, partially prevent androgenisation and hence incline the male towards homosexuality. If, however, only a few of these alleles are inherited then the adult male may show elevated levels of kindness, empathy and sensitivity which, whilst he remains a heterosexual, make him attractive to females and so increase his reproductive fitness (Miller, 2000).

From speculation we move to empirical tests. These ideas were tested by Zietsch et al. (2008) using a large sample (N = 4,904) of Australian twins (1,824 male and 3,080 female). They asked people to rate themselves on a 7-point Kinsey scale (0 = 'I am attracted to women only' to 6 = 'I am attracted to men only'). This gave a sample of 11 per cent men and 13 per cent women who were primarily attracted to their own sex and were therefore categorised as non-heterosexual. They also asked questions about their psychological self-identity and mating success as measured by number of lifetime sexual partners. They found that sex-atypical gender identity was significantly associated with a greater number of sexual partners for both heterosexual men and women. In other words, conventionally 'straight' men and women who were heterosexual in their sexual orientation but who identified psychologically with the traits of the opposite sex were sexually more successful. They also reasoned that if Miller was correct then heterosexual men with homosexual brothers (HeHo) would have a higher mating success than heterosexual men with

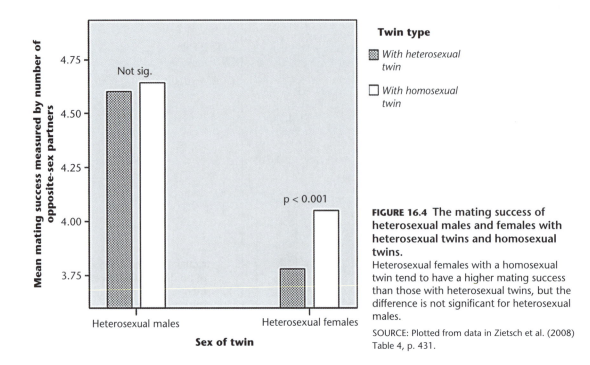

FIGURE 16.4 The mating success of heterosexual males and females with heterosexual twins and homosexual twins.

Heterosexual females with a homosexual twin tend to have a higher mating success than those with heterosexual twins, but the difference is not significant for heterosexual males.

SOURCE: Plotted from data in Zietsch et al. (2008) Table 4, p. 431.

heterosexual brothers (HeHe). This follows from the idea that if homosexuality is an abundance of pleiotropic genes preventing androgenisation then a brother who is heterosexual but has a homosexual brother has a good chance of sharing some of them. Here they found that heterosexual males and females with a non-heterosexual twin did indeed have a higher mating success than heterosexual males and females with a heterosexual twin, but the effect was only significant for females (Figure 16.4).

Miller's ideas were also explored by Santtila et al. (2009) on a sample of 1,827 males of which 160 had androphilic tendencies. In this study mating success was measured using a variety of indices such as number of children, age at first intercourse and estimated number of sex partners in the last five years. The results, however, were unsupportive of Miller's hypothesis: the mating success of heterosexual men in the HeHo group was not significantly different to heterosexual men in the HeHe group (Table 16.2).

Table 16.2 Mating success of heterosexual males (He) with a heterosexual (HeHe) and homosexual (HeHo) twin

	HeHe Means (SE)	HeHo Means (SE)	Sig
No. of children	0.79 (0.03)	0.83 (0.11)	NS
Age at first intercourse	18.20 (0.10)	18.17 (0.31)	NS
No. of sex partners in last 5 years	2.95 (0.15)	2.66 (0.60)	NS
SOURCE: Data extracted from Santtila et al. (2009), Table 2, p. 62.			

16.5 Fraternal birth order effects

One robust finding concerning male homosexuality is that male homosexuals tend to have more older brothers than heterosexuals. It seems that being born male after your mother has produced several older brothers increases a son's chances of adopting a homosexual orientation (see Blanchard, 1997, for a review of this research). The data needs careful attention since a man with older brothers will also tend to have older sisters and establishing if it is brothers or sisters that drive the effect requires some effort. In a meta-review

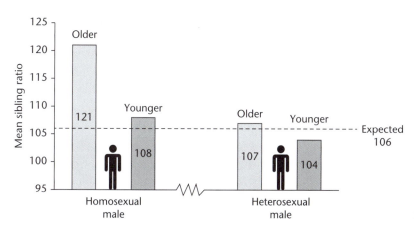

FIGURE 16.5 Sex ratios of older and younger siblings for homosexual and heterosexual subjects. The horizontal reference line (expected = 106) represents the expected sex ratio for Caucasian births of 106 male live births per 100 female live births. Older siblings of homosexual males tend to be biased towards boys (significant at p = 0.015).

SOURCE: Data from Blanchard (2004).

of the literature, Blanchard (2004) looked at the sex ratios (male/females) of the older and younger brothers and sisters of homosexual men. The sex ratio in the countries from which this data was taken averages at 106 (that is, 106 boys born for every 100 girls), so if homosexual men have an excess of older brothers (as opposed to sisters) then the sex ratio of older siblings will be higher than 106. But this should not be the case for younger siblings or siblings of heterosexual men. Figure 16.5 shows the findings.

One hypothesis to explain this trend is called the 'maternal immunity hypothesis' (Blanchard and Klassen, 1997) and is based on the idea that foetal cells or cell fragments may enter the circulation of the mother during pregnancy and promote an immune reaction. Some of these molecules from the male foetus may be involved in the androgenisation of the male brain. If the mother becomes immune to these molecules and then these maternal 'anti-male' antibodies cross the placenta they may interrupt the masculinisation of the brain of the foetus and hence disrupt the development of attraction-to-women circuitry. If present, this effect would increase with each male child born (the mother's immune system 'remembers' previous antigens) and so each older brother increases the chance that the next male child

will have androphilic tendencies. The older brother effect does appear to be robust (it was observed, for example, in the data set of the study by Iemmola and Camperio-Ciani (2009) on fertile females discussed earlier) but the precise biochemistry is far from clear. Another problem is that this seems to be a side effect of the immune response and it is difficult to see how it provides an answer to the Darwinian paradox. We might also ask why, if this effect has been a stable feature of the hominin lineage, natural selection has not devised a mechanism to subvert the effect to reduce the impact of maternal anti-masculinising antibodies. One possible solution may be to co-opt an idea proposed by Michael Ruse in the late 1980s that homosexuality may result from parental manipulation of children to ensure that they help their siblings raise their own children and increase the total fertility of the grandparents – essentially an idea based on kin selection but deriving from parental interests in manipulation rather than from the genetic perspective of the homosexual child (Ruse, 1988). Applying this idea we might hypothesise that natural selection had not tackled this anti-masculinising effect precisely because it (through kin-directed altruism) increases the reproductive fitness of the mother by promoting the production of more grandchildren than would have otherwise been the case if she produced all-heterosexual offspring. Another idea, suggested by Miller (2000) is that birth order effects may be a way of promoting different personality types that occupy different niches and so avoid sibling–sibling conflict.

16.6 Does genetic causation matter?

On the whole, knowledge of the way the world works and the processes shaping our lives is an unalloyed good and searching for the underlying causes of behaviour is part of this quest. But what bearing, it may be pertinent to ask, does even a partial attribution of genetic causes to homosexuality have on the important issue of gay civil rights? There are philosophical and empirical dimensions to this question. Philosophical approaches look at the complicated cluster of beliefs about morality, freedom of choice and expression, and the whole notion of human rights; the aim being to derive an ethically sound framework to inform policy. This approach, important as it is, is outside the scope of this book (but see Chapter 21).

There is, however, an empirical approach that can inform our ethical judgement: we could ask how people's belief about the causes of sexual orientation affect their attitudes towards gay civil rights. Just such a study was carried out by Donald Haider-Markel and Mark Joslyn (2008) who used data from a survey carried out on 1,515 adults by The Pew Research Center for the People and the Press (Pew Research Center, 2003). They found that people's views about the controllability of sexual behaviour strongly affected their attitudes. Those groups who thought that homosexuality was controllable, and hence a matter of personal choice (or influenced by upbringing), tended to have negative attitudes towards gays and gay rights. Those who thought that homosexuality was less controllable and largely a matter of biological causation had more favourable feelings towards gays. They also found that political orientation and ideology strongly influenced the

perspective adopted, with liberals tending to accept biological explanations whilst conservatives and people with strong religious beliefs were more inclined to stress the influence of environment and personal choice (after all, a sin is hardly a sin if there is no choice in the matter). Table 16.3 shows logistical regression coefficients for the strength of the associations they found.

It is tempting to think that an emphasis on genetic causes will help shape more tolerant attitudes as homosexuality is perceived as 'natural' and not a product of a malign environment or morally questionable lifestyle choice. This relies of course on the science proving to be reliable. But we should not overlook the fact that environmental causes are still causes. It is also fundamental to note that the identification of a genetic component does not solve the ethical issues about treatment of sexual minorities (although it may inform them). It is quite feasible to imagine a response from some quarters decades in the future that advocates (erroneously in the author's view) that if homosexuality is rooted in genetics then a genetic 'cure' using gene technologies can be applied. This is not an appealing line of reasoning. The laudable aim of equality of interests for lesbian, gay, bisexual and transgender (LGBT) people should not stand or fall upon a particular set of scientific findings.

Table 16.3 Logistic partial regression coefficients for correlations between characteristics of respondents and their belief in a biological causation for homosexuality and attitudes to gay rights.
A positive value implies support for both positions, a negative value suggests antithesis. So for example, the higher the level of education the more likely the respondent is to agree with a biological causation of homosexuality and accept the importance of gay civil rights; the more religious a person is, in contrast, the less inclined they are to accept biological causation and the more they are opposed to gay civil rights.

Characteristic (or strength of characteristic) of respondent	Genetic/biological attribution regression coefficient (significance of correlation)	Support for gay civil rights regression coefficient (significance of correlation)
Education	+0.280 (p < 0.01)	+0.285 (p < 0.01)
Liberal	+0.312 (p < 0.1)	+0.802 (p < 0.01)
Has gay friend	+0.754 (p < 0.01)	+0.64 (p < 0.01)
Conservative	−0.452 (p < 0.01)	−0.581 (p < 0.01)
Religiosity	−0.144 (p < 0.01)	−0.399 (p < 0.01)
SOURCE: Data from Haider-Markel and Joslyn (2008).		

FIGURE 16.6 A gay wedding. The Netherlands was the first country to legalise same-sex marriages in 2001. Since then, and at the time of writing, some 14 countries world-wide allow same-sex couples to marry. In Africa the only country that recognises same-sex marriage is South Africa; as yet no Asian countries give recognition. Traditionally, evolutionary approaches to sexual relations have concentrated on heterosexual couples, but there is now evidence for an adaptive biological basis to homosexual relations as well.

SOURCE: www.fotosearch.com. Reproduced with permission.

Summary

➤ Homosexuality presents something of an enigma for evolutionary biology since there is strong evidence that it is heritable (that is, has a genetic basis) and yet also reduces fecundity. The challenge then is to explain why such alleles persist.

▶ An early idea that homosexuality is maintained by kin selection and, for example, homosexual males assist in the child-rearing of relatives and so gain inclusive fitness, is not well supported by the evidence.

▶ There is some support for the idea that female relatives of homosexual men have enhanced fertility and so alleles tending to incline males towards homosexuality may reside on the X chromosome and be maintained by sexually antagonistic selection: they reduce the fertility of males but increase the fertility of females.

▶ Another idea is that homosexuality may be maintained by heterozygotic advantage. There is some limited evidence in favour of this approach.

▶ Homosexuality in men is associated with a high number of older brothers giving rise to the fraternal birth order effect. The cause of this effect and its adaptive function are not understood.

Key Words

- Androphilia
- Gynephilia
- Sexually antagonistic selection

Further reading

McKnight, J. (1997). *Straight Science?: Homosexuality, Evolution and Adaptation*. Psychology Press.

Slightly dated but one of the few books to deal with the subject in a comprehensive fashion.

Peters, N. J. (2006). *Conundrum: The Evolution of Homosexuality*. Author House.

Accessible and readable overview. Both a scientific and a sensitive approach to the subject.

Incest Avoidance and the Westermarck Effect

Oh, on these eyes
Shed light no more, ye everlasting skies
That know my sin! I have sinned in birth and breath
I have sinned with Woman. I have sinned with Death.
(**Sophocles**, *Oedipus Rex*, **428** BC)

The lines above are spoken by Oedipus when he discovers that he has unwittingly married his mother, Jocasta, and has fathered several children with her. Jocasta hangs herself and Oedipus gouges out his own eyes. For the ancient Greeks, as for most human cultures, incest was crime against the natural order and there were strong social rules to prohibit it.

Yet as a problem in biology and cultural anthropology, the formation of rules governing incest and its avoidance seems guaranteed to generate controversy (see, for example, Leavitt, 1990). The reason is probably that so many disciplines and paradigms can claim to have some purchase on the phenomenon and have, accordingly, produced so many competing theories. Moreover, these explanations often invoke core or foundational principles in these disciplines, increasing the odds of what is at stake. Evolutionary biologists see incest avoidance as a biological phenomenon designed to avoid inbreeding reductions in fitness; anthropologists can see it as a cultural phenomenon designed to stabilise the family or form family alliances and so elevate man from a state of nature to culture; Freudians can supply interpretations in terms of the Oedipal stage of infantile sexuality; meanwhile, sociologists often view marriage rules and their variability across cultures as a means of preserving wealth and power. In this chapter, we will consider some of these competing approaches, mainly focusing, of course, on the evidence bearing upon evolutionary explanations.

In discussing **incest** and **inbreeding** it is wise at the onset to distinguish between these two terms. An incestuous relationship is a sexual union between human biological relatives who are more closely related than legislation in that culture permits. What may be incest in one culture (or even a US state) may not be in another. Inbreeding is a more value-free biological term to describe breeding between related individuals. In this discussion, the term 'incest' will imply that humans are the focus; inbreeding will be used to refer to human and non-human animals.

17.1 Early views about inbreeding and the incest taboo

In the late 19th century, it was widely believed that inbreeding, either among humans or within other animal species, did not produce any undesirable biological effects. In 1949, the cultural anthropologist Leslie White could still insist that inbreeding does not cause degeneration and that the 'testimony of biologists' was 'conclusive' on this score. The prevailing view from about 1910 to 1960 was that incest avoidance was entirely a cultural restriction with no basis in biology.

But if the taboo against incest was not biologically based, then what caused it? Claude Lévi-Strauss (1956) suggested that the incest taboo was

FIGURE 17.1 Sigmund Freud (1856–1939) taken in 1922.

Freud is regarded as the father of psychoanalysis. His view was that infants pass through a series of stages that are essential to normal adult functioning. One of these, for boys, was the Oedipal stage. Freud saw the incest taboo as a necessary way of curbing natural sexual desires. Understandably, therefore, he vehemently opposed Westermarck's more Darwinian view.

SOURCE: Photograph taken by Max Halberstadt (https://commons.wikimedia.org/wiki/File:Sigmund_Freud_LIFE.jpg). With thanks to Wikimedia Commons.

17.2 Westermarck's alternative Darwinian explanation

Although the views of White, Lévi-Strauss and Freud were influential, they were not the only ones in this period. The main rival to these ideas, one which approached the problem from an entirely different Darwinian perspective, was the theory of Edvard Westermarck (1862–1939). Westermarck was born in Helsinki and at the age of 25 taught himself English in order to study the works of Darwin, Morgan and Lubbock in the original language. In 1887, he came to the British Museum in London and studied to produce his doctoral thesis on human marriage. His work first saw publication in 1891 as *The History of Human Marriage*, a work later expanded in 1922 to three volumes.

Westermarck suggested that men do not mate with their mothers and sisters because they are

FIGURE 17.2 Edvard Westermarck (1862–1939) (c.1930).

SOURCE: From LSE Library's collections, IMAGELIBRARY/265. Courtesy of the library of the London School of Economics and Political Science.

a means to encourage outbreeding, not for biological gain but to help cement social solidarity among different families. In *Totem and Taboo* (1913/1950), Freud suggested an alternative function for the taboo. He argued that since daughters naturally desired their fathers (the Electra complex) and sons their mothers (the Oedipal complex), then the conflicts resulting from giving free rein to these instincts would make family life impossible. Hence, humans had no alternative but to renounce and repress these desires.

disposed not to find them sexually attractive. He proposed that humans (unconsciously) use a simple rule for deciding whether or not another individual is related. If humans avoid mating with individuals they have been reared with during childhood, there is a good chance that they will also avoid mating with close relatives. Children who are raised together will, therefore, experience a sort of negative imprinting against finding each other sexually attractive. At the cultural level, this instinctive disposition transformed itself, according to Westermarck, into a conscious taboo against incest.

Freud firmly rejected Westermarck's ideas since they threatened to undermine his Oedipal complex, one of the cornerstones of his whole psychoanalytical theory. But other anthropologists and psychologists were also dismissive. James Frazer, for example, in his *Totemism and Exogamy* (1910), argued that if the inbreeding aversion was a deep human instinct, it would not be necessary to erect a taboo against it.

17.3 Testing Westermarck's hypothesis

In many ways, the fate of Westermarck's ideas in the 20th century run parallel to the broader fortunes of the whole of evolutionary thinking in psychology: an initial upsurge generated by Darwin and his immediate followers, followed by a long eclipse until the late 1960s, and then a resurgence over the last 40 years. By about 1970, it was realised that the **Westermarck effect** may be real after all and deserved careful attention. His primary thesis on incest avoidance can be broken down into a series of linked hypotheses:

1. Inbreeding is injurious to offspring (**inbreeding depression**)
2. Early association inhibits inbreeding by generating an aversion
3. This aversion is an evolutionary adaptation whose function is to reduce the risk of inbreeding depression
4. This aversion manifests itself at the cultural level as an incest taboo (sometimes called the 'representational problem').

We will examine each of these suggestions in turn.

17.3.1 Inbreeding as injurious to offspring (inbreeding depression)

By the early 1960s, evidence was mounting that the human gene pool did contain many recessive and dangerous alleles; consequently inbreeding did carry biological disadvantages. Current thinking suggests that each human carries around two or three lethal recessive alleles. The risks of inbreeding are indicated by a measure called the coefficient of inbreeding (F).

The coefficient of inbreeding

The **coefficient of inbreeding** is the percentage of gene loci at which progeny of inbreeding will be homozygous (that is, the proportion of gene sites on the chromosome where they will have inherited two identical copies of any gene) over and above the baseline level of homozygosity of the general population. Another way of saying this is that F is the probability that an individual has both alleles of a gene (that is, a specific locus) identical by descent from the same allele in a common ancestor. As an illustration, we may consider two cousins. The probability that each cousin has an identical specific gene acquired from a common ancestor is 1/8 (see Chapter 3). If the two cousins mate and produce an offspring then the chance that the offspring will acquire both copies of this gene is 1/2; so the coefficient of inbreeding is $1/8 \times 1/2 = 1/16$ or 0.0625. A simple rule of thumb is that the coefficient of inbreeding in any progeny is one half of the coefficient of genetic relatedness (r – see Chapter 3) of the parents. We should realise, however, that the actual F values may be higher than this if the whole population has been subject to inbreeding. So for example, the F value for offspring of two siblings might be higher than 0.25 if the parents of the siblings were themselves related in some way. In such cases, a correcting formula can be applied (see Bittles, 2004). This may explain why disorders among children of Pakistani first cousin marriages in the UK are higher than for European first cousins – in Pakistan there is a long tradition of first cousin marriages (see Table 17.1).

The assessment of the biological consequences of human inbreeding is made difficult by a lack of data, itself a probable reflection of the strong social antipathy to incest as well as the difficulties of establishing

Table 17.1 Scientific studies on effects of inbreeding in humans

Study and reference	Finding
Couples in India and Pakistan (Bittles, 1995)	Higher fertility rates for consanguineous, second cousin marriages
Review (Bittles and Makov, 1988)	Progeny of first cousins have morbidity levels 1–4 per cent higher than unrelated couples
Review of combined data from Europe, Asia, Africa, Middle East and South America (Bittles and Neel, 1994)	Mean excess mortality at first cousin level 4.4 per cent
Five-year study of Pakistani community in Birmingham, UK (Bundey and Alam, 1993)	Rate of lethal or chronic disorders among offspring of first cousin marriages at 10 per cent, rate among non-first cousin offspring 3 per cent. Rate among European first cousin marriage offspring 5 per cent
Study of the highly inbred, isolated Amish community in Lancaster County, Pennsylvania (McKusick et al., 1964; McKusick, 2000)	Rate of Ellis-van Creveld syndrome per live birth 300 in 60,000 compared to US average of 1 in 60,000
Postma et al. (2010) looked at the breeding success of couples living in two isolated villages (Cavergno and Bignasco) in the Swiss Alps. They used genealogical records to look at relatedness of parents, family size and reproductive success of offspring	They found that whilst parents who were related to each other produced a normal-sized family, the daughters of the related parents did have significantly fewer children

true paternity. Finding controls is also difficult, since incestuous relationships often involve confounding variables such as low maternal age, mental abnormalities and low socioeconomic status. Despite these difficulties, it is now quite clear that inbreeding carries health implications for offspring even at the level of first cousin marriages. Table 17.1 shows the results of some recent studies into the effects of inbreeding.

The apparent anomaly in Table 17.1, that second cousin and closer marriages have a higher fertility, is resolved when we note that such marriages tend to occur earlier in the woman's life and so her reproductive period is longer. Overall though, we can see that health effects become manifest even at the level of first cousin marriages.

Across the globe, cousin marriage is quite common and accounts for about 10 per cent of all marriages, although its frequency locally varies enormously between cultures. Different religions often have different views about the acceptability of cousin marriage. The Hindu Marriage Act (enacted by the Indian parliament in 1955), for example, generally bans first cousin marriages but allows them if local customs approve. In contrast, the Quran permits first cousin marriage and the practice is very common in Muslim cultures. In the UK, for example,

about 55 per cent of marriages amongst people of Pakistani ethnic origin are between first cousins. Both secular and protestant authorities in the UK permit cousin marriages, although from time to time anxieties have been expressed about the health implications for offspring of cousin marriages (Dyer, 2005). In the Western world, the USA is the only country that bans first cousin marriage in that 31 states currently prohibit such relationships. Before the American Civil War cousin marriage was legal in all states, but shortly afterwards increasing notice was given to a series of reports (such as that delivered before the war by S. M. Bemiss to the American Medical Association in 1858) suggesting that the offspring of such consanguineous unions experienced physical and mental deterioration. So gradually, despite studies that indicated such effects would be small (for example, those of George Darwin (1875) and George Louis Arner (1908)) the number of states banning cousin marriage increased rapidly up to the 1920s (see Ottenheimer, 1996).

Interestingly, Darwin chose inbreeding as his marriage option since he married Emma Wedgwood, his first cousin. As a result, he constantly fretted about the effect this might have on his children, some of whom were rather frail. He even tried, unsuccessfully, through the influence of his local MP and friend John Lubbock, to have first cousin marriages recorded in the 1871 UK census for the benefit of scientific research – reasoning that the number of surviving offspring from first cousin marriages could be used to infer the health of children (Box 17.1).

Royal families abound with instances of inbreeding. The current monarch of Britain, Elizabeth II, is a third cousin to her husband Prince Philip, since they are both great-great-grandchildren of Queen Victoria. Prince Philip's mother was also inbred (F = 0.0703), since her grandmothers were sisters (daughters of Victoria and Albert).

The Hapsburg dynasty was one of the most important royal families of Europe and also illustrates the perils of inbreeding. The Spanish branch was founded by Philip I (1478–1506) and five kings later finally died out with the childless and infertile Charles II (1661–1700). The six kings of this dynasty contracted a total of 11 marriages, 9 of which were consanguineous to a degree of third cousin or higher, including two uncle–niece marriages and one double first cousin marriage. As the royal parents intermarried so the next in line became increasingly inbred culminating in Charles II who according to calculations by Alvarez et al. (2009) had an inbreeding coefficient (F) value of 0.254; that is, 25.4 per cent of his autosomal genome was in all likelihood homozygous. This is an F value that could be expected from brother–sister marriage. Charles II was by all accounts feeble and malformed, suffering from a variety of mental and physical problems. He died aged 39 leaving no offspring despite two marriages. It seems very likely that Charles II suffered from recessive genetic disorders expressed due to his high genomic homozygosity.

Box 17.1 Darwin and cousin marriage

On 29 January 1839 Charles Darwin married his cousin Emma Wedgwood and thereafter periodically fretted that an even more weakened version of his own feeble constitution would be passed on to his children. The facts suggest that this anxiety may have had some foundation: of Darwin's ten children, three died in childhood, and although six of the surviving seven children enjoyed a long married life, three of those marriages were childless. Darwin was also convinced of the beneficial effects for plants and animals of outcrossing. In one of his monographs – *The Variation of Animals and Plants under Domestication* (1868) – he wrote that 'the crossing of animals and plants which are not closely related to each other is highly beneficial or even necessary, and that interbreeding prolonged during many generations is highly injurious' (quoted in Kuper (2009), p. 1439). In respect of human mating, Darwin's concern was such that he lobbied his scientific friend and parliamentarian, John Lubbock, to suggest that the new Census Bill planned for 1870 should ascertain if a married couple were cousins. Darwin thought that if this information were available then naturalists could look at the effect of cousin marriage on family size (Desmond and Moore, 1991). Lubbock's proposal to the House of Commons was, according to Charles's son George, rejected 'amidst the scornful laughter of the House, on the ground that the idle curiosity of philosophers was not to be satisfied' (quoted in Kuper (2010), p. 95).

Faced with this rebuttal, Darwin then encouraged his son George to pursue the matter through private research. George Darwin (1845–1912) was a mathematician, an amateur geologist and an admirer of the eugenics theories of his own cousin Francis Galton. Charles suggested a design for the research: if George collected the rates of cousin marriages amongst parents of people in asylums with the rates in the general (more healthy) population then a measured surplus in the former would be clear evidence of the injurious effects of consanguineous marriages. By ingenious means (such as using

FIGURE 17.3 Emma Darwin and her son Leonard.
This picture was taken in 1853 when Emma was 45 years of age and Leonard 3 years. Leonard was her eighth child of ten children and by this date two of her children had already died. Leonard went on to become the Chairman of the British Eugenics Society 1911–1928.

SOURCE: Reproduced with permission from John van Wyhe (ed.) 2002–. The Complete Work of Charles Darwin Online (http://darwin-online.org.uk/).

surname matching to identify cousin marriages) George carried out the research, one of the earliest examples of statistics applied to social problems. He found initially that cousin marriage varied according to social class: rates were about 4.5 per cent for the aristocracy, 3.5 per cent for the landed gentry and upper middle class (Darwin's own social group) and about 2 per cent for rural populations and artisans. The data from the asylums gave interesting results: only about 4 per cent of patients were the offspring of first cousin marriages, a rate not much higher than the national average rate of cousin marriage. Charles and George concluded that the effects of consanguineous cousin marriage, on insanity rates at least, was slight (Kuper, 2010). Later work by George Darwin and Karl Pearson also only showed slight effects. George, mindful of the fact that the upper classes went in for cousin marriage more frequently than the lower orders, even concluded that such baneful effects that might exist were mitigated and largely eliminated by the more comfortable conditions enjoyed by the best families. Reassured, the English aristocracy could, therefore, sleep easy in their beds.

Were children of the Darwin/Wedgwood dynasty affected by consanguinity?

The children of Charles and Emma had an inbreeding coefficient of 0.0630, slightly higher than the expected first cousin coefficient of 0.0625 since the parents of Darwin's mother, Josiah and Sarah Wedgwood, were themselves third cousins. This means that about 6.3 per cent of the autosomal genome of Darwin's children could be expected to have been homozygous. In fact cousin marriage was quite common in the Wedgwood–Darwin pedigree. Three of Emma Darwin's brothers married relatives: Josiah III married his first cousin Caroline Darwin, sister of Charles; Hensleigh married his first cousin Frances MacKintosh; and then Henry married his double first cousin Jessie Wedgwood. Tim Berra and co-workers (Berra et al., 2010) have looked at the children of the various Darwin and Wedgwood marriages that took place in the nineteenth century and conclude that there is statistical evidence that inbreeding did adversely affect the health of the children of these marriages (Figure 17.4).

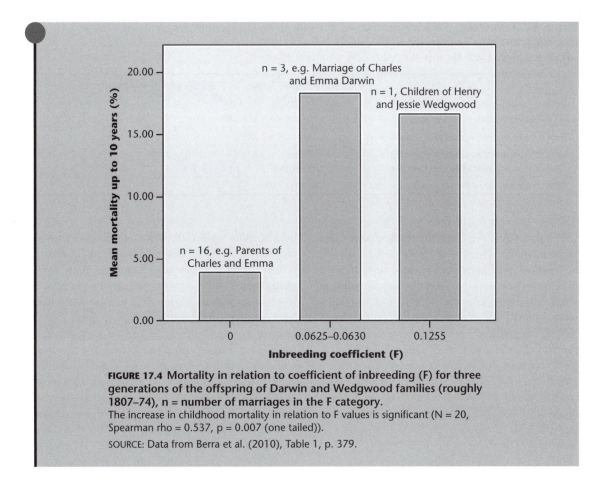

FIGURE 17.4 Mortality in relation to coefficient of inbreeding (F) for three generations of the offspring of Darwin and Wedgwood families (roughly 1807–74), n = number of marriages in the F category.
The increase in childhood mortality in relation to F values is significant (N = 20, Spearman rho = 0.537, p = 0.007 (one tailed)).
SOURCE: Data from Berra et al. (2010), Table 1, p. 379.

17.3.2 Early association inhibiting inbreeding by generating an aversion

Information on the negative impact on successful pair bonding among humans raised together comes from several sources. One that has a particularly large database (now amounting to the study of over 14,000 marriages) is Arthur Wolf's (1993) study of marriage customs in mainland China and Taiwan. Up until the early 1930s in Taiwan and the mid-1940s in mainland China, cultural practices allowed families to arrange marriages for their sons in one of two ways:

● *Major marriage*: the son would reach puberty and then marry a young female who would be brought into the house to live with the groom's parents.

● *Minor marriage*: a family would adopt a very young girl and raise her as a future daughter-in-law until both son and adopted girl reached marriageable age. In Taiwan, these girls are known as *shim-pua* or 'little daughters-in-law'.

These two types of marriage provide an invaluable data set for examining any effect of mating inhibition. The children are destined to marry one another in both cases but to measure the success of the marriage, Wolf devised a number of indices that looked at the divorce rate among such marriages and the frequency of adultery as well as the number of children produced. Generally Wolf found that the minor marriages were far less successful than the major marriages (see Table 17.2).

Table 17.2 Success of minor marriages compared to major marriages from Wolf's investigation into Taiwanese marriage (1900–25)
Minor marriages are far less successful by a variety of metrics.

	Minor marriage (N = number sampled)	Major marriage
% ending in divorce	24.2 (N = 132)	1.2 (N = 171)
% of women in marriage involved in adultery	33.1 (N = 127)	11.3 (N = 159)
Average number of children produced per woman over a 5-year period in marriage	0.89 (N = 132)	1.49 (N = 171)
SOURCE: Wolf (1970), Tables 3, 4 and 10; and Wolf (2004), Figure 4.1, p. 80.		

Wolf's study is very suggestive but there are other explanations consistent with the data. It is possible, for example, that the parents treated their adopted daughter-in-law less favourably than their biological son, leading to resentment on behalf of the future bride. Alternatively, or additionally, it is also conceivable that the adoption process was traumatic for the daughter-in-law (more so at a younger age) leading to fertility problems later in life. A more general criticism of the findings is that such *shim-pua* marriages did typically produce between four and eight children, a figure admittedly below that of major marriages but one that leads to questions about the strength of an aversion that produces so many children.

One study supportive of the thrust of Wolf's findings is that carried out by Alex Walter and Steven Buyske (2003) on Moroccan families. In the particular culture they studied, patrilateral parallel female cousins (that is, daughters of the brother of the father in a family) are often preferred as brides for the sons of the father. Housing arrangements are such that these cousins (son and cousin bride-to-be) can spend variable lengths of time together in childhood, ranging from low co-socialisation where they live in different houses, to high and early co-socialisation where they share a house and sleep in the same room for a good part of their childhood. From interviews with these cousins they found that high levels of co-socialisation reduced sexual interest in cousins for female respondents but not for male participants. They argued that this was consistent with the higher costs to females of inbreeding depression (since males can potentially produce more offspring than females).

Another source of information about the effects on marriage of co-socialisation at a young age comes

from Shepher's (1971) study of mate selection among second-generation kibbutz adolescents. The kibbutz system studied by Shepher was such that unrelated children from a variety of parents were raised together by trained nurses in heterosexual similar-age peer groups from birth to 4 years. From 6 or 7 years of age until the age of 18, the groups sometimes spanned two years of maximum age difference. The children were exposed to each other constantly and tended to live apart from their parents for most of the day, although they saw their parents in the afternoons. Shepher (1971) looked at the incidence of premarital affairs and then number of marriages among adolescents reared in the same kibbutz. Of the 2,769 marriages analysed in the whole kibbutz movement, not a single case of true inter-peer group marriage was to be found. Only five sexual couples were found from this number who were in the same peer group before the age of 6, but these were never together for more than two years. Shepher concludes that his findings are in accord with the Westermarck effect and provide evidence for negative imprinting.

We should note, however, that Shepher's work received some sharp criticism when it appeared in book form in 1983. John Hartung (1985, p. 171) was particularly severe, claiming the work to be 'one of the most sustained and pernicious affronts to empiricism since *Coming of Age in Samoa*'. The essence of Hartung's complaint is his belief that Shepher failed to properly calculate how many marriages to expect between individuals reared in a kibbutz. Much of Shepher's case rests on the fact of no marriages between couples in the same peer group; but, as Hartung notes, girls mature earlier than boys and are unlikely to find partners in the same age group anyway.

The work of Shepher has been subject to other critical appraisals. Shor and Simchai (2009), for example, conducted 60 interviews with people who grew up in a kibbutz to ascertain if they developed sexual attraction or sexual aversion to those they grew up with. They found that although few reported sexual intercourse, many reported feeling sexually attracted to their peers and few reported feelings of sexual aversion. This contrast between the two groups of findings points to the need to reconcile data on marriages with the responses of interviewees.

Cues for kinship and altruism

In Chapter 3 we explored the idea of Lieberman et al. (2007) that humans might possess 'kinship estimator' modules or neural circuits to help regulate altruism (that should preferentially be directed towards kin) and sexual interest (that should be directed away from kin), altruistic tendencies towards each other, their personal aversion to the thought of sex with co-reared peers, and their third-party moral beliefs. The essential idea was that being raised in a kibbutz in close proximity to other children would have stimulated cue for kinship (thereby affecting altruism and sexual aversion) since in ancestral environments such a long exposure to other children would have probably been in a family group and as such a fairly reliable indicator of relatedness. In terms of both altruistic tendencies and sexuality the findings were interesting:

- The longer the duration of co-residence with a peer the higher the level of altruism reported towards that peer ($r = 0.16$, $p < 0.001$)
- The longer the co-residence with opposite sex peers the greater the degree of moral condemnation expressed at the thought of sex between Kibbutz peers ($r = -0.34$, $p = 0.036$; note in estimating wrongness 1 = most wrong and 10 = least wrong, hence the negative correlation).

17.3.3 Aversion through co-socialisation as an evolutionary adaptation

The evidence so far would tend to offer some support for the idea that the Westermarck effect is an adaptation: it seems to develop in all normal children, whether related or not, and clearly there are costs associated with not having this mechanism in place. But here we can also tackle one of the early objections to Westermarck's ideas; namely, that no such inbreeding avoidance mechanism was to be found in primates. If we were to find one, this would surely count in favour of Westermarck and against the cultural explanations of Freud and Lévi-Strauss, since, presumably, animals do not have the same interests in renouncing their instincts for the good of civilization, nor is it likely that they are able to culturally transmit a Freudian-style taboo.

Inbreeding avoidance in primates

There is now plentiful evidence that primates do, to some degree, avoid inbreeding. One obvious mechanism that achieves this in the wild is sex-biased dispersal. This can be male exogamy (males departing the troop), as in the case of olive baboons, many species of macaque monkey, capuchin monkeys and some lemurs; or female exogamy, as in the case of chimpanzees, spider monkeys and muriquis (South American monkeys). It could be argued that this mechanism is a way of regulating the group size to avoid competition for resources, but this would not explain the sex bias in the individuals that leave. Reviewing the literature, Pusey (2004, p. 71) concludes: 'Nonhuman primates provide abundant evidence for an inhibition of sexual behaviour among closely related adults.'

Mechanisms of incest avoidance

Despite the empirical evidence favourable to the Westermarck effect, the mechanism by which it operates is still unclear. Westermarck himself did not speculate on the details of the proximal developmental mechanisms. Wolf (2004) explained the effect of the wife's age on fertility and marital success by using Bowlby's idea of attachment (see Chapter 8). A young female will attach herself to an older member of the family as a source of care and support. In return, this will elicit nurturing feelings from the older member. In this way, attachment brings members of a family closer together emotionally and physically. It is then the job of the Westermarck effect to ensure that this closeness does not become sexual. Wolf (2004) therefore proposes that attachment and the Westermarck effect have evolved together. The younger the wife, the stronger the attachment and the stronger the Westermarck effect in both sexes.

Freudians often feel that evolutionary psychologists misrepresent their ideas and underestimate Freud. Robert Paul (1991) and David Spain (1991), for example, both point out that Freud did not argue that adults actually desire incestuous mating (a biologically implausible supposition); rather that incestuous erotic love was the first type of sexual love to appear in children, usually began between the ages of 4 and 6, and was something that by itself could obviously not lead to mating. In Freudian theory, it is the subsequent formation of the superego at about the age of 7 or 8 that served to repress these desires. If this happens normally to the maturing child, then the outcome is a voluntary aversion to incest that is indistinguishable (in terms of what is actually manifest) from the Westermarck effect. In this sense, Freudian theory is actually positing an ontogenetic (or developmental in a Tinbergian sense) mechanism consistent with the proximate mechanism that is assumed to lie at the heart of the Westermarck effect. Here is not the place to enter into an evaluation of psychoanalytical theory, but it is worth noting that more empirically verifiable mechanisms may be at hand such as olfactory cues.

MHC and odour

Recent experiments on mating preferences in mice and humans have pointed to the possible role of olfaction in kin recognition and mate choice. Wedekind and colleagues found that females preferred the odour of men who differed from them in their major histocompatibility (MHC) haplotypes (Wedekind et al., 1995). In later experiments (involving subjects smelling T-shirts worn by volunteers), the pleasantness of the odour as experienced by both males and females was correlated with the degree of MHC dissimilarity between smellers and the wearer of the T-shirt from which the odour was sampled (Wedekind and Furi, 1997).

Since Wedekind's pioneering work there have been numerous studies attempting to replicate and extend these findings. The results, however, have been mixed. One study by Santos et al. (2005), for example, did find a female preference for male odours emanating from MHC-dissimilar males. But two other studies by Thornhill et al. (2003) and Roberts et al. (2008) found no significant differences in female ratings of odours from MHC-similar and MHC-dissimilar males. Havlíček and Craig Roberts (2009) suggest that the overall MHC odour effect may be weak, possibly as a result of humans losing much olfactory sensitivity in their evolutionary past, and so detection of any effect is strongly influenced by the methodology of the study.

In addition to the problem of replicating what seems to be a small effect, there is no agreed picture of how MHC genes influence body odour at all. There are several possibilities: MHC molecules may occur directly in body fluids such as sweat, saliva and urine; MHC molecules may bind to volatile compounds and transport them to apocrine glands (for example, sweat glands that also produce odour); or MHC molecules might influence what type of symbiotic organisms live on the surface of the skin and it is their metabolites that exude odour.

The work on MHC and odour might in time point to a subconscious role for smell in judging attractiveness, but it does not by itself support a Westermarck-style inhibitory effect from close association, since the mechanism here might be genetically based and part of a kin recognition system. Experiments by Bateson (1980, 1982) on Japanese quail, however, show that early familiarity inhibits mating, irrespective of relatedness. Experiments on mice involving cross-fostering also show avoidance of mating partners based on familiarity and not just relatedness. Strong evidence in favour of the importance of using familiar smells as a proxy for relatedness comes from experiments whereby male mice can be reared to prefer females even of their own MHC type by raising them with females with different MHC types (Yamazaki et al., 1988). So, at least in mice, it looks like there is a mechanism via olfactory cues consistent with Westermarck-style negative imprinting that disposes them not to mate with others with whom they were raised, and that such cues are obtained to some degree by learning rather than a genetic-based recognition of MHC similarity. In a review of the literature, Schneider and Hendrix (2000) think that an olfactory component to the Westermarck effect in humans is highly likely. They also make the interesting prediction that follows from this that anosmics (people who have lost their sense of smell) should be less inhibited in pursuing sexual contacts with co-socialised individuals. One might expect, for example, a higher percentage of anosmics in people who admit to incest than among the general population.

17.3.4 Inbreeding depression and cultural norms: the representational problem

Variation in laws relating to incest

We now turn to the final hypothesis of the Westermarck effect. Even if we accept that inbreeding in human populations does cause damaging effects, and that individuals develop an inhibition to breeding with those they have been socialised with in their childhood, it is not easy to see how an instinctive, unarticulated aversion can translate itself into a cogently expressed series of norms, rules and laws. The conceptual gap that opens up here is thrown into focus by the consideration that the biological aversion to inbreeding is felt between children who have been co-socialised (whether related or not), whereas the cultural prohibition is directed specifically against kin, with rules beginning to vary as we approach first cousin levels of relatedness. The incest aversion seems to be universally directed at relationships where F is 0.25 (mother–son, father–daughter, brother–sister). At the level of first cousin marriages, practices and customs vary, with some cultures prohibiting such unions and others actively encouraging them (see Table 17.3). Indeed, the whole framework of laws relating to incest seems to be based on a mixture of religious precepts and genetic ideas (see Ottenheimer, 1996).

Table 17.3 A selection of rules governing consanguineous marriages from various authorities

Religious authorities			
Ruling	**Authority**	**Sexual relationship**	**F value of progeny**
Permission required from diocese	Roman Catholic Church	First cousins	0.0625
Permitted	Protestant Church	First cousins	0.0625
Strongly encouraged	Dravidian Hindus of south India	First cousins of type 'a son with mother's brother's daughter'	0.0625
Prohibited	The Bible, Leviticus	Man and his mother, or his daughter	0.25
		Man and his father's brother's wife	0
		Man and his wife's sister	0
		Man and his brother's wife	0
Prohibited	The Koran	Uncle–niece	0.125
Allowed	The Koran	Double first cousins	0.125
Civic authorities			
Ruling	**Authority**	**Sexual relationship**	**F value**
	UK		
Criminal offence	Sexual Offences Act 1956	Sexual intercourse with a woman whom a man knows to be his granddaughter, daughter, sister, half-sister or mother	0.125–0.25
Allowed		Relatedness up to and including first cousins	0.0625
	USA		
Marriage illegal	31 states	First cousins	0.0625
Permitted	19 states	First cousins	0.0625
	China		
Prohibited	1981 Marriage Law	First cousins and closer	0.0625

From aversion to rule

Obviously, from the perspective of the target presented to natural selection, the aversion usually worked to keep close kin away from breeding with each other, despite the occasional 'misfire' as the nascent sexual desires of unrelated children were suppressed. But when and how did humans realise that the aversion could be more effectively channelled against kin marriages only? The psychologist Roger Burton may have provided a way through this impasse. Burton (1973) argued that there probably was a Westermarck-type aversion that reduced the frequency of inbreeding in early human groups. But it was precisely this mechanism that would then have allowed deleterious alleles to accumulate in the heterozygous (that is, unexpressed) condition. When incest did occur, albeit infrequently, the effects would be visible, so that gradually cultures would come to realise that inbreeding among kin produced undesirable consequences. These cultures would then see incest as transgression of the natural and moral order in some way and eventually elaborate a cultural condemnation of it. In early cultures, this prohibition would probably have taken on a religious form, as in the Bible (Leviticus 18) where God is supposed to have dictated to Moses the rules of consanguineous mating. The problem here though is that we now have to explain how cognition (the dawning realisation of the perils of incest) becomes translated into an emotion (revulsion at the idea). In some ways, the original formulation of Westermarck is more consistent. His view was that the moral norms that we articulate are a reflection of our emotional responses; we have, as it were, 'moral emotions'. We explore this below.

Moral condemnation and childhood exposure to brothers and sisters

Such a linkage, between the avoidance of incest at the personal level and the cultural prohibition against it, would tend to suggest that the moral condemnation against incest in third parties is generated by a mechanism the same as, or related to, the onset of personal aversion. From this, it would follow that children who have been raised with siblings of the opposite sex should express a stronger moral disapproval of incest in third parties than those without opposite-sex siblings. If, however, morality is a matter of absorbing cultural norms through socialisation, as cultural constructivists would suggest, then no such relationship should be observed. This theory was put to the test by Lieberman et al. (2003) in their study of the family backgrounds and moral opinions of 186 Californian undergraduates. They looked for a positive correlation between the strength of disapproval of incest and the childhood experience of being raised with a member of the opposite sex. Table 17.4 provides a summary of their findings.

As shown in Table 17.4, they found that the length of co-residence between siblings was positively associated with their moral condemnation of

Table 17.4 Moral aversion to third-party incest in relation to childhood experience

Relationship of cohabitees	Years of co-residence	Correlation between years of co-residence and degree of 'moral wrongness' expressed in relation to third-party incest	Significance of correlation p ≤
Males with sisters	0–10	0.29	0.05
	0–18	0.40	0.01
Females with brothers	0–10	0.23	0.05
	0–18	0.23	0.05
Individual with opposite sex who is not a genetic sibling (that is, not related)	0–18	0.61	0.05
SOURCE: Data from Lieberman et al. (2003).			

incest more generally. The study took great pains to control for the effects of other variables that might impact on moral sentiments such as family composition and size, sexual orientation and parental attitudes towards sexual behaviour. With all these controls in place, the correlations were positive and significant even when the opposite sex 'sibling' was known not to be a genetic relative.

These results complement the study of Wolf but differ in one important aspect. Wolf suspected that the years from birth to 6 were those in which the Westermarck effect took place. This study, however, shows that, for males, the degree of moral repugnance continues to rise up to age 18 (see Table 17.4). Lieberman et al. explain this by suggesting that, since inbreeding is more costly for females, the avoidance mechanism swings into play with a low level of socialisation and hence cues concerning relatedness. An inbreeding error is less costly for males, however, and so they delay judgement, not wishing to ignore a potential sexual partner.

The unequal costs of incest

The higher costs to females associated with inbreeding would predict, as noted, a more sensitive avoidance mechanism. Support for this notion comes from a similar study to that of Lieberman et al., but carried out independently by Daniel Fessler and David Navarrete (2004) using a data set of 250 undergraduates aged 18–39. They asked their subjects to express their attitude to incest using an 11-point Likert-type scale. In summary, they found that:

- Females raised with male siblings reported a higher aversion to incest than females raised without male siblings
- Males raised with female siblings reported a higher aversion to incest than males raised without female siblings
- Females reported a greater sibling incest aversion overall (a score of 5.33 +/– 1.98 for females compared to 4.63 +/– 2.14 for males, t = 1.44, p < 0.05)
- The effect did not depend on the relatedness of co-socialised individuals and was the same with full and stepsiblings.

As Fessler and Navarrete (2004) note, their study does not rule out the possibility that families with opposite sex siblings might be exposed to more intense parental instruction about proper sexual behaviour than those with siblings of the same sex, precisely to rule out the possibility of incest.

Disgust seems to be a powerful and possibly functional emotion in this context. Jan Antfolk et al. (2012) looked at the expression of disgust elicited by the prospect of third-party incest in relation to a number of scenarios. The group used a sample of 435 Finnish university undergraduates and disgust was measured on a Likert scale (1 = not at all disgusted, 5 = very disgusted). In keeping with previous work on this topic they found that women reported higher levels of disgust than men (medians 4.06 and 3.65 respectively, chi squared 19.89, p < 0.001).

17.4 Incest and morality

What is so significant about the Westermarck effect is that it provides an intriguing case study into how natural dispositions that have evolved for functional reasons end up as a culturally articulated, transmitted and enforced set of moral codes, or at least rules that some people in a society expect others to adhere to. Although the studies by Lieberman et al. (2003) and Fessler and Navarrete (2004) provide support for the relationship between experience and expressions of moral disgust, there remains the ticklish question of why prohibitions should exist for behaviours that no one is supposed to be motivated to perform. We noted Burton's account earlier, but Westermarck himself suggested that such taboos arise as a consequence of our ability to experience the actions of others as if they were our own. Fessler and Navarrete (2004) term this propensity 'egocentric empathy'. Such responses are indeed found when we experience disgust at the thought or sight of somebody eating a foodstuff that we find strongly unpalatable.

More generally, the proposed connection between morality and biology places Westermarck in a long line of thinkers who have argued for a naturalistic, as opposed to transcendental, approach to ethics (see Chapter 21). In the naturalists' camp, we have Aristotle, Aquinas, Hume, Adam Smith, Darwin and, more recently, E. O. Wilson and Arnhart. In the transcendentalist tradition – those who think that ethic norms lie outside nature in some way and are either intuited by our ethical sense or derived by an exercise of reason – are to be found Hobbes, Kant, Freud, J. S. Mill and Rawls. If the Westermarck effect continues to gain empirical support, it provides a signpost for how a natural science of ethics can be conceived.

Summary

- Early 19th-century views about the prohibition against incest tended to see it as a non-biological, cultural phenomenon. The exception was Westermarck, who postulated that the early co-socialisation of children inhibited sexual desire as a mechanism to avoid inbreeding.

- The Westermarck effect has several components that can be tested empirically. So far, work has tended to suggest: inbreeding does cause a depression in fitness; early co-socialisation tends to inhibit sexual desire; other primates have mechanisms to avoid inbreeding; the proximate mechanism that leads co-socialised humans to avoid mating may be olfactory or even explicable in Freudian terms; and the moral condemnation of incest varies according to the experience of the individual expressing such views.

- Charles Darwin was worried about the effects of cousin marriage on offspring, especially since he married his own cousin. There is suggestive evidence that the Darwin/Wedgewood dynasty did show reduced fitness effects from inbreeding.

- The Westermarck effect provides an interesting way in which biological fitness and moral sentiments may be related.

Key Words

- Coefficient of inbreeding (F)
- Inbreeding
- Inbreeding depression
- Incest
- Westermarck effect

Further reading

Berra, T. M., G. Alvarez and F. C. Ceballos (2010). 'Was the Darwin/Wedgwood dynasty adversely affected by consanguinity?' *BioScience* **60**(5), 376–83.

Interesting article about inbreeding in Darwin's own family tree.

Fessler, D. M. T. and C. D. Navarrete (2004). 'Third-party attitudes towards sibling incest: evidence for Westermarck's hypotheses.' *Evolution and Human Behaviour* 25: 277–94.

Research paper that examines how people react to the thought of incest in relation to their own childhood experiences.

Freud, S. (1950). *Totem and Taboo*. New York, W. W. Norton.
Famous work where Freud outlines his view of the incest taboo.

Lieberman, D., J. Tooby and L. Cosmides (2003). 'Does morality have a biological basis? An empirical test of the factors governing moral sentiments relating to incest.' *Proceedings of the Royal Society of London*. **Series B 270**: 819–26.
A research paper looking at our moral sentiments in relation to family structures.

Wolf, A. P. (1993). 'Westermarck redivivus.' *Annual Review of Anthropology* **22**: 157–75.
A paper where Wolf argues that there is strong evidence supporting Westermarck's ideas.

Wolf, A. P. and W. Durham (eds.) (2005). *Inbreeding, Incest and the Incest Taboo*. Stanford, CA, Stanford University Press.
A collection of useful article on the Westermarck effect in the light of recent research.

PART VIII

Health and Disease

Darwinian Medicine: Evolutionary Perspectives on Health and Disease

What a book a Devil's chaplain might write on the clumsy, wasteful, blundering low & horridly cruel works of nature!

(Letter of Charles Darwin to J. D. Hooker, 1858)

This chapter attempts to construct a framework for understanding how evolutionary theory can be applied to the problem of ill health and disease. The commonly used term for this type of endeavour is 'Darwinian medicine'. The term is slightly misleading since the word medicine implies medical practice and hence the treatment of ill health. Yet, as it is presently defined by the activities of its protagonists, Darwinian medicine is largely concerned with exploring, using evolutionary ideas, why humans are vulnerable to disease; it primarily focuses on questions of ultimate causation, and only then tentatively drawing preventative and therapeutic implications. Indeed, it is probably fair to say that so far it has been more successful in the area of explanation than in prescribing remedies. Perhaps this is inevitable given its relatively recent history. This notwithstanding, we should note that there have been a number of impressive calls for the integration of evolutionary thinking into medical training and public health, and convincing arguments as to how this approach should inform therapy (Nesse and Stearns, 2008; Omenn, 2010; Gluckman et al., 2011).

Williams and Nesse's (1991) article 'The Dawn of Darwinian Medicine' provides an obvious starting point for the modern application of evolutionary ideas to understanding health and disease. The reasons for the appearance of this landmark paper at this time stem probably from the more general factors underlying the rise of sociobiology and its transformation into evolutionary psychology. The success of 'gene selectionism' as both a philosophical and experimental heuristic, as outlined in works by Hamilton and Dawkins (see Chapter 3), drew attention to the fact that evolution is not concerned with the survival of species, or even the health and happiness of individuals, but rather is a matter of shifting gene frequencies. Health, therefore, is subservient to inclusive fitness. Around this period we also observe the waning influence of 'essentialism' in both biology and psychiatry. This is the realisation that a disease is not necessarily an objective ontological state but is to some degree socially and ecologically constructed. Lactose intolerance, for example, is only a 'disorder' (for some people) in cultures that drink milk, and most people in the world are lactose intolerant. This fitted quite well with evolutionary ideas that stress that adaptations are imperfect compromises, and that traits have advantages and disadvantages. George Williams's notion of **antagonistic pleiotropy** and its application to explaining senescence was also a seminal work in this area. Finally, outbreaks of diseases such as HIV, swine flu and bird flu directed attention to the evolutionary ecology of viruses and other pathogens.

With hindsight, there were many forerunners of this modern movement (Williams's paper on senescence, for example, was published in 1957). In 1975, the English physician John Harper published his book *The Evolutionary Origins of Disease* and in the 1980s and 1990s Paul Ewald, then at the University of Washington, published several papers on the evolution of infectious diseases (Ewald, 1991, 1994). Also, around the same time, Margie Profet began to publish articles using Darwinian perspectives to explore the adaptive function of menstruation, allergies and morning sickness (Profet, 1988, 1993), but there was no real sense of a field as such until after 1991. Even now, evolutionary ideas have not really penetrated medical schools to any great extent. In a survey of UK medical schools, for example, Downie

(2004) found that a minority had any evolutionary courses on the curriculum (about 37 per cent), and then only as options. More worryingly, about 10 per cent of students rejected evolution by natural selection – usually for religious reasons. Nesse and Stearns (2008) give a similar unpromising analysis of medical schools in the US. This neglect stands in sharp contrast to some of the very real findings and insights of this emerging discipline.

On a more positive note, in recent years Darwinian medicine as an academic field has expanded significantly, and there are now several journals, such as *Evolution, Medicine and Public Health,* the *Journal of Evolutionary Medicine* and *The Evolution and Medicine Review,* that serve its practitioners.

18.1 Compiling a taxonomy

In assessing the value of evolutionary approaches to the study of health and disease it will be helpful to adopt a taxonomic approach. This has been done by a number of authorities such as Nesse and Williams (1995), Gluckman et al. (2011) and Stearns and Ebert (2001) with broadly similar outcomes. The taxonomy adopted here follows their lead and is based on the idea of several key principles from which we can draw applications and illustrative examples (Table 18.1).

Table 18.1 Evolutionary perspectives on health and disease

Key principle	Applications	Examples
Uneven rates of change in overlapping domains: human evolution, cultural evolution and pathogen evolution	Mismatch between current environment and genotype. Sometimes called 'Out of Eden', genome lag or evolutionary discordance hypotheses	Lifestyle diseases due to post-Neolithic diets; especially those with high glycaemic index carbohydrates, saturated fats, added sodium chloride
	Preparedness theory – responses to phylogenetically relevant stimuli	Phobia of spiders and snakes
Natural selection is slow compared with the pace of cultural and environmental change, and the rate of change of pathogens	Bacteria, viruses and parasitic organisms have short generation times allowing for rapid mutation and selection in the face of the host's defensive systems. This is a never-ending genetic arms race	Evolution of antibiotic resistance and emergence of superbugs such as MRSA (methicillin-resistant *Staphylococcus aureus*)
	Adaptive developmental plasticity and unanticipated environments	Epigenetics and the processes of foetal development suggests that a neonate is primed to expect a certain type of environment. If the environment is not as predicted, illnesses, such as obesity and diabetes, can result

Table 18.1 cont'd

Key principle	Applications	Examples
Natural selection is an iterative process: it faces constraints and makes compromises	Life history theory and design compromises. What might seem maladaptive in the short term might be adaptive over the whole life. Some features must also be trade-offs between negative and positive effects	It pays an organism to invest in early reproduction even if this brings later senescence
	Structural compromises	Trade-off between bipedalism and large brain; and low larynx and risk of choking
	Pathway constraints. Natural selection can only act on what is already there. Adaptive landscape concept suggests it is difficult to reach another higher adaptive peak if this means travelling down an adaptive gradient	Human eye – inside out compared to more elegantly designed mollusc eye. Bipedalism and the human spine
Subjective definitions of health and optimum reproductive fitness are not the same thing Natural selection and sexual selection do not operate to optimise health or longevity. Instead they can only act according to the currency of inclusive fitness	Balancing selection. Antagonistic pleiotropy: genes may code for traits that increase reproductive success even at the cost of disease vulnerability. Diploid organisms are also subject to heterozygote advantage and so deleterious recessive genes may be maintained in the gene pool	Cystic fibrosis, sickle-cell anaemia, Huntington's disease (see Table 18.4)
	Sexual selection. The reproductive optimum in a trait (for example, level of testosterone) may not be same as optimum for health and wellbeing	High testosterone in young males may be increased by sexual selection but causes risk taking and reduced immune function
	Genetic conflicts	Imprinting may reflect fact that interests of mother, child and father are different. Haig's theory of mother–foetus conflict
	Selection arena phenomena	Loss of oocytes or spontaneous abortion may reflect a selective filtering mechanism
	Defence mechanisms mistaken as signs of ill health	Morning sickness (avoidance of toxins) Coughing and vomiting (expulsion of pathogens or toxins). Fever (elevated temperature to kill bacteria). Anxiety and smoke detector principle

The rest of this chapter is structured into three parts corresponding to the taxonomy outlined in Table 18.1:

Part I – Uneven rates of change in overlapping domains: human evolution, cultural evolution and pathogen evolution

Part II – Natural selection is an iterative process: it faces constraints and makes compromises

Part III – Subjective definitions of health and optimum reproductive fitness are not the same thing

Part I Uneven rates of change in overlapping domains: human evolution, cultural evolution and pathogen evolution

As noted in Chapter 7, natural selection acting on humans is a fairly slow process taking thousands of years to bring about significant genetic change. Slow is, of course, a relative term, but here it means that natural selection is slow relative to the cultural and environmental changes that humans have experienced and largely driven; and slow relative to the life cycles of pathogens and parasites forever poised in their microscopic multitudes to take advantage of our physiological resources. This would not be such a problem if these domains did not overlap, but they do: cultural evolution places ever-changing demands on our bodies and lifestyles; and pathogens, by their nature, thrive by exploiting higher organisms such as ourselves. We can illustrate this principle by looking at the role of microbes in human disease and the evolution of antibiotic resistance. This section concludes with a more general examination of the developmental origins of health and disease paradigm. Diet provides another important illustrative example but this is considered in the next chapter of case studies.

18.2 Microorganisms

18.2.1 Commensalists, probiotics and pathogens

Humans have evolved to cope with an environment rich in microflora. Bacteria live on our skin, in our hair and inside our bodies. Some bacteria are commensalists, meaning that they live on or inside our bodies where they gain some benefit but do us no harm. Your skin, for example, is home to about 150 species of bacteria that feed on waste skin cells and skin secretions but do you no damage. Other bacteria are mutualistic in that they gain benefits from us and we from them, these are also sometimes called probiotics. Your intestines, for example, contain about 100 trillion individual microorganisms (about ten times the number of cells in your body); most of these are useful in that they aid the digestion of food, produce vitamins and restrict the growth of other harmful bacteria (see Box 18). Finally, there are bacterial pathogens or parasites responsible for millions of human deaths through infectious diseases each year.

18.2.2 The Neolithic revolution and pathogen transmission

Humans have, with their long mammalian and primate heritage, evolved defences to deal with parasites, but the practice of animal husbandry in the Neolithic revolution made one problem more acute: namely, the movement of a damaging microorganism from an animal to a human host – so-called zoonotic disease

---

Box 18.1 Probiotic bacteria and the wisdom of breast milk

The importance of friendly intestinal bacteria is illustrated by the once-puzzling composition of human breast milk. As Figure 18.1 shows about 11 per cent by mass of the nutrients in human breast milk consists of complex **oligosaccharides** (that is, sugar-like molecules containing long chains of repeating monosaccharide units). This was once an evolutionary mystery since they are indigestible by any enzymes that the human body can produce. Compelling recent work, however, shows that whilst these sugars are indigestible to humans and many strains of pathogenic bacteria they can be used by **bifidobacteria,** which are a major constituent of healthy gut flora (Zivkovic et al., 2011).

Moreover these bifidobacteria act to discourage the growth of other harmful bacteria. In other words, human milk is selectively biased towards encouraging probiotic gastrointestinal microbiota.

An understanding of the composition of breast milk also has the potential to influence health policy on breastfeeding and infant bed sharing. The benefits of breast milk are now widely recognised and breastfeeding is encouraged. But an issue that still divides the medical community is whether infants should be encouraged to sleep with their mothers. One worry here is the belief that bed sharing may increase the risk of SIDS (sudden infant death

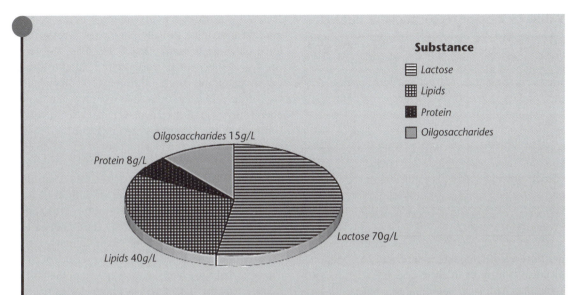

Substance

▤ Lactose

▦ Lipids

■ Protein

▨ Oilgosaccharides

Oilgosaccharides 15g/L

Protein 8g/L

Lipids 40g/L

Lactose 70g/L

FIGURE 18.1 Human breast milk composition.
The fairly high proportion of oligosaccharides was once a mystery since humans do not produce enzymes to digest them. They are, however, useful to promote the growth of friendly gut bacteria.
SOURCE: Plotted from data in Zivkovic et al. (2011); and Morrow et al. (2004).

syndrome) – although the epidemiological evidence is not conclusive. There is also the residual belief that babies should learn to be independent of their caregivers – a relic perhaps of the influential but authoritarian and behaviourist approach to child-rearing that John B. Watson advocated in the 1930s.

Approaching this topic from an evolutionary angle, Helen Ball and Kristin Klingaman (2008) have noted that the composition of human milk has a relatively low fat and high sugar content compared to that of other mammals. This suggests, they argue, that humans are a frequent-suckling species: babies digest their mothers' milk quickly and mothers, in turn, require the stimulation of frequent suckling to promote milk production. This would indicate that humans are adapted for close infant–mother contact (without long periods of separation) to maintain milk production and hence provide the child with nutrients and immunological benefits. In addition, breastfeeding causes the release of oxytocin, which helps to induce sleep in both mother and child. The researchers predicted that bed sharing would promote breastfeeding. This was supported by a study on 61 mothers and their infants delivered in a hospital in the UK: infants who shared a bed with their mother, as opposed to sleeping in a side crib or separate cot, were breastfed more frequently and, importantly, breastfeeding continued for a longer postnatal period.

transfer. Taylor et al. (2001) estimated that of the 1,415 species of organisms known to be pathogenic to humans (that is, viruses, bacteria, prions, fungi, worms and flukes) about 61 per cent were the result of zoonotic transfer. Some well-known examples of **zoonosis** include measles and smallpox (both probably originally contracted from cattle), influenza (from birds and pigs) and HIV (from primates). Two of the most recent devastating cases are the Spanish flu virus of 1918, which killed about 50 million people worldwide, and the outbreak of AIDS in the 1980s, which has since killed about 35 million humans. It begins to look as if

Table 18.2 Some recent zoonotic disease

Human pathogen or disease	Original host	Year observed in humans
Ebola	Bats, primates and antelope	1977
HIV1	Chimps	1983
H5N1 influenza	Chickens	1997
E Coli 0157: H7	Goats and/or sheep	2000
SARS coronavirus	Bats/palm civets	2003

SOURCE: Woodhouse and Antia (2008), Table 16.1, p. 219; and Warshawsky et al. (2002).

most newly emerging infectious diseases are zoonotic in origin, and most of these caused by viruses (see Table 18.2). This latter point probably reflects the fact that the RNA of viruses has a high nucleotide substitution rate and hence a high potential to evolve rapidly.

In relation to disease caused by microbes the consequences of the Neolithic revolution can be summarised as:

● High population densities facilitate microbe transfer.

● Accumulation of human waste poses higher risk of faecal–oral disease transmission (especially diarrhoeal diseases).

● Time spent inside buildings by a significant proportion of population (for example, those now not directly involved in food production) gives higher exposure to pathogens otherwise killed by sunlight, such as the influenza virus.

● Infestation of dwellings by vermin such as rats and fleas.

● A sedentary lifestyle favours disease transmission by vectors and pathogens that remain in the local environment such as the schistosomiasis fluke in fresh water.

● Zoonotic disease transfer caused by close proximity to farmed animals.

18.2.3 Pathogenicity, virulence and pathogen evolution

Evolutionary ideas can also help in the understanding of the pathogenicity and virulence of microorganisms. Rabies, for example, is a deadly disease

caused by a virus. Once symptoms begin in an infected individual (usually after a few weeks) the usual outcome is death. Thankfully, there are vaccines available which work well if administered before symptoms start. In contrast, we have the virus *Herpes simplex*: there is no known cure and once a person is infected they remain so for life. Fortunately, the symptoms are mild and the infection is rarely life threatening, but does result in uncomfortable periodic outbreaks of cold sores. So one virus kills and the other irritates. The scientific terms to describe these differences are pathogenicity and virulence. Pathogenicity is a measure of the ability of a microorganism to actually cause a disease. So commensalist bacteria, for example, have zero pathogenicity. **Virulence** is a relative measure of the degree of damage (that is, illness) caused to the host. Hence we mighty say rabies has a high virulence and high pathogenicity and *Herpes simplex* a low virulence and low pathogenicity. Significantly, rabies is carried from person to person by an animal vector (usually an infected dog that goes around biting people) whilst the herpes virus is transmitted by direct contact. This difference may be crucial in explaining virulence as we discuss below.

For many decades, up to the early 1990s, it was the common view that over time hosts and pathogens would coevolve to a state of peaceful coexistence or commensalism where virulence levels were low. Myxomatosis in Australian rabbits provides an illustration of this. When the original virus was released in the 1950s it was so successful that it killed most rabbits in about four days and reduced the Australian rabbit population by about 80 per cent after two years. But this very fact limited its spread and soon pockets of resistant rabbits appeared. Now the original strain only kills about 50 per cent of those affected. It was thought that this applied to humans too: pathogens that killed their hosts would suffer fitness costs since if the host is dead transmission rates would fall since the host is immobile. Better, from the point of view of the pathogen, to keep the host alive and moving about to ensure other humans also become infected. This become known as the theory of balanced pathogenicity: the pathogen evolves towards lower virulence and the host evolves towards immunity.

This view was convincingly challenged by Paul Ewald (1991, 1994) who used more sophisticated evolutionary reasoning to make a number of predictions about the virulence and transmission of pathogens. His ideas are discussed in the next few sections.

The relative virulence of direct contact and vector-borne pathogens

Malaria is a deadly disease caused by a protozoan parasite entering the bloodstream. It causes about 630,000 deaths per year worldwide and the World Health Organization (WHO) estimates that a child dies every minute from malaria in Africa. In contrast, the common cold is the result of infection by a virus, it is extremely prevalent but most people recover within a week and it is rarely fatal. One difference between these two diseases is that malaria is carried by a vector (a mosquito that flies from person to person) whilst the common cold virus relies on more direct contact with an infected individual or objects touched by an infected individual. The old

idea that diseases might evolve towards commensalism (see above) might work for those pathogens that require direct contact between humans for transmission, since in these cases a pathogen cannot afford to totally debilitate the host since movement is then restricted. But vector-borne diseases could easily evolve a high degree of virulence since it does not matter if the host is immobilised. In fact, extreme weakening of the host might even favour transmission since a vector such as a mosquito can feast upon the host's blood with a reduced risk of being killed. This was one of the many prediction made by Ewald's application of evolutionary reasoning to infection: vector-borne diseases will generally be more virulent than those requiring direct contact. To test this idea he collected data on mortality rates associated with vector-borne and directly transmitted diseases. As predicted, virulence was found to be higher for pathogens carried by vectors (see Figure 18.2).

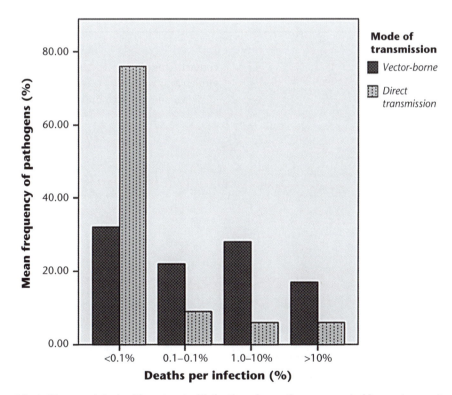

FIGURE 18.2 Mortality associated with untreated infections by pathogens carried by vectors or transmitted by direct contact.
Mortality rates are higher for vector-borne pathogens compared to those directly transmitted ($p < 0.01$, chi squared test). The high frequency of low mortality rate (0.1 per cent) of directly transmitted pathogens illustrates the idea that such pathogens need active hosts to propagate them. Death rate is estimated for every 100 untreated infections.
SOURCE: Data taken from Ewald (1994), Fig. 3.1, p. 38, and re-plotted.

The 'sit and wait strategy' of infection: virulence and survivability.

As we have seen, Ewald's arguments explain a good deal of the variance in virulence amongst microorganisms. Yet some anomalies remain showing that other factors are at work. Both the smallpox virus and the tuberculosis bacterium, for example, are transmitted person to person and not by animal vectors or water supplies, and yet both are often lethal. In 1979 the WHO declared smallpox eradicated from the human population, but tuberculosis remains. To address why such pathogens are lethal despite requiring human contact transmission, Walther and Ewald (2004) examined the operation of a 'sit and wait strategy'. If pathogens have a high virulence and yet limited modes of transmission then this might select for high durability in the external environment. Hence, although the host might be incapacitated or dead, the pathogens persist in the external environment and sit and wait for a new susceptible host to come along. This reasoning predicts, therefore, a positive correlation between virulence and durability. To test this Walther and Ewald examined a range of human respiratory pathogens. Figure 18.3 shows a plot of their data and supports this conjecture.

The work of Paul Ewald has been very persuasive but we should note that there are other hypotheses proposed to account for virulence. One is that it is an accidental by-product of the fact that the parasite may be in the wrong host or that it evolved under very different conditions to the current host. An example might be *Toxocara canis*, a helminthic worm which infects dogs but which causes blindness in humans. The effect on humans, however, is almost certainly not the result of its evolution in a human host. Another hypothesis is that high virulence is the result of 'short-sighted' evolution by colonies of parasites within an individual host – different strains

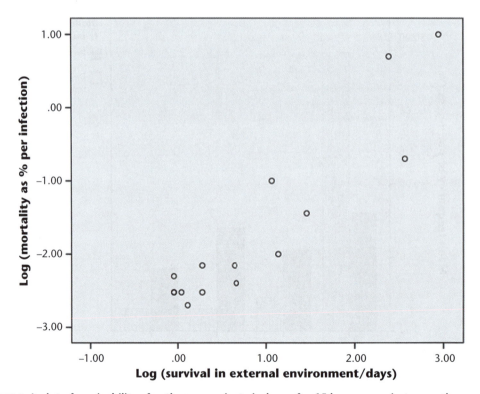

FIGURE 18.3 A plot of survivability of pathogen against virulence for 15 human respiratory pathogens.
Each measure is log (base 10) transformed due to the extreme ranges of results. The plot supports the hypothesis that enhanced survivability is correlated with high virulence, supporting the 'sit and wait' hypothesis of virulence.

SOURCE: Data from Walther and Ewald (2004), Table 3, p. 858.

compete amongst themselves and those reproducing the fastest come to dominate and cause damage even at the expense of low transmission.

Microorganisms evolve quickly but there is plenty of evidence that the human immune system has also altered in response to the impact of diseases spread after the Neolithic revolution. This explains why after contact with European colonists the native peoples of both North and South America experienced high death rates from strains of infectious diseases such as smallpox to which many settlers were immune. More recently we have developed chemical technologies to tackle the problem of microorganisms.

Antibiotics

Since the 1930s, humans have been able to supplement their own immune system with antibiotic substances. The procedure however is the victim of its own success in that now antibiotic resistance poses a real problem. The evolution of antibiotic resistance is a prime example of natural selection at work. Bacteria can quickly acquire resistance to harmful chemicals since they multiply quickly and have short generation times. Microorganisms can also make use of another process: since they not only attack humans but each other as well, they have evolved their own antibiotics. Indeed, most antibiotics that we take

are derivatives of these naturally occurring antibiotics. Penicillin, the first antibiotic to be extracted and produced in large quantities, is a good example of this. It is produced naturally by Penicillium, a genus of fungi, as part of the fungus's own defence against bacterial attack. Given that microbes have been competing among themselves for hundreds of millions of years, long before the evolution of mammals, it is likely that there are hundreds of examples of natural antibiotics and, importantly, defences against natural antibiotics in naturally occurring populations of microbes. Significantly, bacteria can also exchange DNA between species (through plasmid transfer) and so resistance to what we think of as new antibiotics can evolve quickly. A few years after the widespread use of penicillin to treat infections in the late 1940s, resistance began to be observed, including the discovery that penicillin-resistant bacteria existed even before the use of antibiotics on humans. It is estimated that the average bacterium found in soil will show resistance to about seven antibiotics (D'Costa et al., 2006). One worry is that soil-dwelling bacteria could serve as a reservoir of bacterial resistance waiting to be passed to other bacteria that humans more directly encounter. Today, clinically significant levels of resistance usually evolve within four years of the introduction of a new antibiotic (see Table 18.3 and Box 18.2).

Box 18.2 The evolution of antibiotic resistance

Table 18.3 shows how quickly antibiotic resistance can evolve. Antibiotic resistance is now one of the most pressing public health concerns in the 21st century; and although there are naturally occurring resistant strains of bacteria, it is the use of antibiotics to treat human and animal diseases that has greatly enhanced their prevalence.

To investigate the link between antibiotic use and the emergence of resistance, Goossens et al. (2005) conducted a study of outpatient antibiotic use in 26 European countries. The study found that the prescription of antibiotics varied greatly, ranging from 10 defined daily

doses (DDD) per 1,000 individuals per day in the Netherlands to 32.2 DDD per 1,000 patients per day in France. The general pattern, however, was that southern and Eastern European countries tended to use antibiotics more prolifically. Alongside this data, the group measured the frequency of antibiotic resistant *Streptococcus pneumoniae*. They found a clear correlation between reports of the incidence of resistant bacterial strains and the level of antibiotic use in that country. Similar findings have been reported by other authorities (e.g. see Cars et al., 2001). Figure 18.4 shows a plot of some of these data.

Table 18.3 The evolution of resistance to antibiotics.
Resistance is often observed a few years after discovery and introduction. Note clinical introduction is often several years after discovery. The observation of clinical resistance does not mean that the drugs are not deployed and most still have useful functions.

Antibiotic	Year of discovery	Year of introduction	Year resistance observed
Prontosil (a sulphonamide)	1932	1935	1942
Penicillin	1928	1941	1945
Streptomycin	1943	1946	1946
Chloramphenicol	1946	1949	1950
Erythromycin	1948	1952	1955
Methycillin	1959	1960	1961
Tetracycline	1944	1948	1968
Rifamycin	1957	c.1960	1962
Vancomycin	1953	1956	2002
Linezolid	1955	2000	2001
Daptomycin	1986	2003	2005
Tigecycline	1999	2005	???

SOURCE: Adapted from various sources including: Lubelcheck and Weinstein (2008); Lewis (2013).

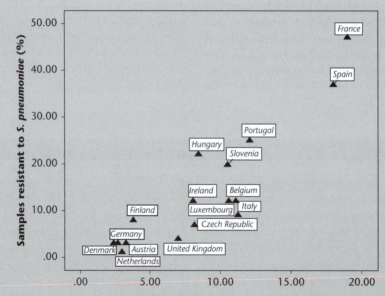

FIGURE 18.4 Correlation between antibiotic use and frequency of resistant samples for 16 European countries.
DDD – defined daily dose per 1,000 inhabitants per day. Percentage penicillin non-susceptibility of *S. pneumoniae* was obtained from the European Antimicrobial Resistance Surveillance System (EARSS) project, which surveyed the antimicrobial susceptibility among invasive (from blood and cerebrospinal fluid) pneumococcal isolates over the years 1997–2000. Doses of penicillin were for the year 1997.

SOURCE: Data on *S. pneumonia* resistance from Goossens et al. (2005). Data on penicillin doses from Cars et al. (2001). (N = 16, Spearman's rho = 0.868, p < 0.001.)

The worrying news is that bacteria may have edged into the lead. With one or two exceptions, there have been no new classes of clinically useful antibiotics discovered in the last 30 years, and there are signs of diminishing interest by large pharmaceutical companies. Yet the Infectious Disease Society of America recently reported that 70 per cent of hospital infections in the USA were resistant to one or more antibiotics (Clatworthy et al., 2007). One suggestion is that a lower selection pressure would result by targeting the virulence of bacteria (that is, their potency in causing mortality and morbidity) instead of trying to kill them outright (Clatworthy et al., 2007).

18.2.4 The hygiene hypothesis

Before the advent of antibiotics, humans staged a very effective campaign against microorganisms through a series of measures to improve public health and sanitation, especially the effective treatment of drinking water and waste water (sewage). But are modern environments perhaps too clean and sterile? In 1999 Bengt Björkstén advanced the idea that children raised in clean environments with frequent use of antibiotics and vaccinations had more allergies than children raised in less clean environments and with a lower rate of use of antibiotics and vaccinations. This became known as the 'hygiene hypothesis'. Essentially, the reasoning is that in less hygienic times the human immune system would be constantly exercised dealing with our microscopic adversaries, and would be well tuned to this task whilst at the same time avoiding any attack on the body's own tissues. In modern sterile environments the immune system may not be properly activated in childhood, leaving it vulnerable to misfiring and attacking the body's own tissues, such as: the lungs (asthma); joints (various forms of arthritis); intestines (Crohn's disease); and the pancreas (type I childhood diabetes). Significantly, these diseases do tend to occur in affluent urban populations with high standards of hygiene. The evidence that followed Björkstén's advancement of this theory was somewhat mixed, but reviewing the subject ten years after he proposed the idea Björkstén concluded that early exposure to commensal microbes did act as a stimulus towards an effective immune response. He did suggest, however,

that 'microbial deprivation hypothesis' might be a better name (Björkstén, 2009).

More recently, Molly Fox et al. (2013) has noted that the type of inflammation associated with Alzheimer's disease (AD) is similar to that of autoimmunity. The group then tested the idea that early exposure to pathogens might improve an individual's immunoregulation and so protect against later onset of AD. They used WHO data to estimate parasite stress in childhood and illness due to AD in 192 countries. They found a negative correlation between parasite prevalence and illness due to AD, supporting the hygiene hypothesis.

18.3 Developmental plasticity and unanticipated environments

We have already noted in Chapter 8 how developmental plasticity enables an organism to adapt to a future predicted environment. If, however, the prediction turns out to be wrong then the phenotype can appear to be ill matched to its surroundings with consequent indications of ill health. In this case we see not the failure of genes to keep up with rapid environmental changes but epigenetic mechanisms falling out of sequence with the even more rapid changes in modern cultures.

In relation to these ideas, Peter Gluckman, Mark Hanson and Alan Beedle (2007) have been particularly instrumental in constructing what they call 'the developmental origins of health and disease paradigm' (DOHaD). The essential idea is that through epigenetic mechanisms the developing foetus picks up intrauterine cues that guide its development in ways that are adaptive. But a mismatch between the expected environment and the actual mature environment can leave the phenotype in a mismatched state that increases the risk of various diseases (Figure 18.5).

The DOHaD paradigm helps to explain ill health effects in populations experiencing rapid environmental change. Such populations are often indigenous peoples living in colonised and later developed countries. The Pima Indians of Arizona provide an instructive example. In the latter years of the 19th century their native way of life was devastated by non-native farmers and the diversion of water supplies. Increasingly, the Pima Indians were forced to

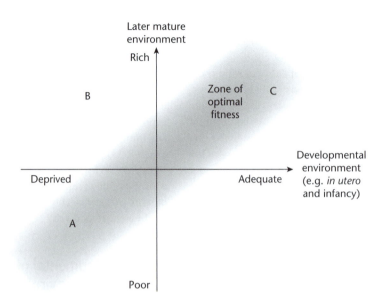

FIGURE 18.5 A tale of three phenotypes.
Individual A received signals from the uterine environment to expect poor later life conditions. Adjustments were made at birth to ensure survival and later reproduction such as low birth rate and early puberty. The predictions were correct and the phenotype was well matched to the later rather poor environment. Individual C experienced an intrauterine environment that was well resourced and hence predicted a rich later environment. Adjustments were made to ensure maximal fitness such as a high birth weight, strong bone and muscle growth, later puberty and investment in repair and growth of tissues. The predictions were correct and C is also well matched to its environment. In the case of B, however, the early environment predicted harsh adult conditions. Adjustments were made in this expectation such as low birth weight, high insulin resistance and propensity to store fat. The predictions were incorrect and in the new unexpected richer adult environment B might suffer from obesity, cardiovascular disorders and type II diabetes.
SOURCE: Based on ideas in Gluckman, P. D., M. A. Hanson, and A. S. Beedle (2007). 'Feature article. Early life events and their consequences for later disease: a life history and evolutionary perspective.' *American Journal of Human Biology* **19**, 1–19.

adopt a non-native diet including lard and white flour issued by the US government. Today over 60 per cent of adults over the age of 35 have type II diabetes and obesity is also a serious problem (Knowler et al., 1990). There is a small group of Pima Indians living a more traditional lifestyle in Mexico. Significantly, rates of obesity and diabetes in this population are far lower (Ravussin et al., 1994). One way of interpreting these results is that a rapid change in environmental conditions between generations has created a mismatch between what the growing embryo came to expect by way of future nutritional resources and the actuality of a more sedentary lifestyle with a high-fat and -sugar diet. The result being that many adults display an incidence of Western lifestyle diseases to an even greater level than white Caucasian Americans.

This idea, that for optimum health there should be a match between conditions experienced during foetal development and infancy and those in later life, is also sometimes called the 'predictive adaptive response hypothesis'. Despite the supporting evidence from studies on the Pima Indians noted above, it has been quite difficult to test the various permutations and possibilities of the potential match and mismatch future scenarios (that is, poor early/poor adult; poor early/rich adult; rich early/poor adult; and rich early/rich adult). Recently, however, Ada Hayward and Virpi Lummaa (2013) examined 127-year-old longitudinal data sets on survival, fertility and reproduction statistics from four Finnish communities from 1750 onwards. Thereby, they were able to examine the effect of varying early and later life environmental conditions on mortality and reproductive health. They found that higher mortality was associated with both early life and later life poor environments. But no support was found for the idea that survival and reproductive success would be higher if early life and later life conditions were matched.

Part II Natural selection is an iterative process: it faces constraints and makes compromises

Evolutionary constraints often arise because change by natural selection is an iterative process of continual improvement on what already exists. A particular 'design' or adaptive solution cannot be taken apart in order to start again from scratch to find a better solution since that would involve intermediary stages losing fitness and that would be selected against. In this respect, natural selection is less like a factory that can recall models or even discontinue them and bring out a completely new design, and more like a group of engineers on board a ship, navigating perilous currents and coping with changeable winds, constantly trying to improve the ship whilst always on the move and never able to return to dry dock. A useful metaphor to illustrate this problem of 'design on the move' is that of the **adaptive landscape**, first introduced by the American biologist Sewall Wright at the International Congress of Genetics in 1932. As illustrated in Figure 18.6, organisms or traits can move up a hill they find themselves on but not down to move to another one.

One example of this is the human eye where, due to a peculiarity of its evolution, light has to first pass through nerve fibres before it reaches the light sensitive photoreceptor cells of the retina. These light sensitive cells than have to send signals back through the nerves and on to the brain. The eyes of cephalopods such as the octopus evolved independently and represent a more elegant solution to the problem of forming an image and collecting information (see Box 18.3).

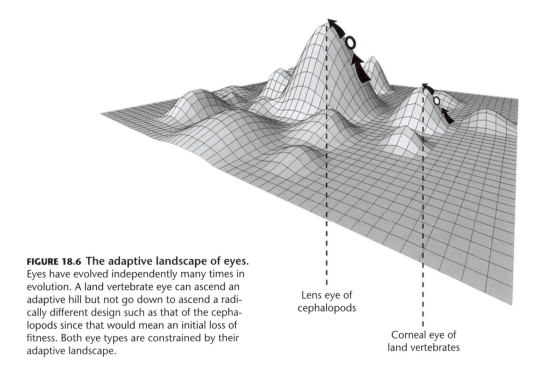

FIGURE 18.6 The adaptive landscape of eyes. Eyes have evolved independently many times in evolution. A land vertebrate eye can ascend an adaptive hill but not go down to ascend a radically different design such as that of the cephalopods since that would mean an initial loss of fitness. Both eye types are constrained by their adaptive landscape.

Lens eye of cephalopods

Corneal eye of land vertebrates

Box 18.3 The human eye – a case of unintelligent design?

Figure 18.7 shows how for the human (land vertebrate) eye the nerves enter from the rear and then lie in front of the light-sensitive cells of the retina. Hence the nerves impede the flow of light to the retina and also introduce a blind spot where they enter. In the eye of an octopus the nerve cells are more sensibly placed behind the retina. The suboptimal design of the human eye has led to a variety of problems. One of these is the blind spot – although we hardly notice this since our brains compensate for this defect. More serious is the fact that blood vessels that feed retinal cells also sit on top of the cells instead of the more logical position of behind. This again means that light has to pass through this fine layer of capillaries to reach the light sensitive cells of the retina, limiting, albeit slightly,

the intensity of light that reaches the rods and cones. The real problem here, however, is that damage to blood vessels can severely impair vision. This occurs in the case of diabetics who develop diabetic retinopathy, whereby the blood vessels proliferate and obscure vision.

Another problem related to this arrangement of nerves and cells is retinal detachment. Trauma and ageing can cause the photoreceptor layer to become detached from the pigment epithelium beneath. Today the retina can be re-fused using laser treatment, but rapid loss of vision will occur if the condition is not treated. Cephalopod eyes do not suffer from this problem since the retina is anchored to layers beneath by the axons (nerve fibres) that carry signals from the eye to the brain and pass through the tissue below the retina.

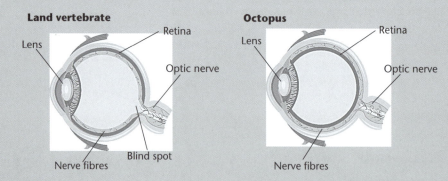

FIGURE 18.7 Comparison of the eye of a land vertebrate (for example, human) with that of an octopus.
In the case of the vertebrate the optic nerve and blood vessels pass through the retinal layer to collect an impulse from the retina and to supply it with blood. This leads to a blind spot, the need for light to pass through these fibres before reaching the retina and a relatively weak anchorage of the retina.

For other problems understandable in terms of the design weaknesses of the human eye, such as macular degeneration, see Novella (2008).

The cleric William Paley in his book *Natural Theology* once thought that 'the examination of the eye was a cure for atheism'. Now we understand better its imperfections perhaps a comment by Frank Zindler is nearer the mark:

As an organ developed via the opportunistic twists and turns of evolutionary processes, the human eye is explainable. As an organ designed and created by an infinitely wise deity, the human eye is inexcusable. (Zindler, 1986, quoted in Novella, 2008, p. 497)

Tempting though it may be, we should be wary, however, of revelling in the image of natural selection as some sort of opportunistic blunderer churning out jerry-built designs simply to secure a source of ammunition against creationists. For a start, in other contexts we may want to assume that natural selection has produced optimal solutions. Also, if what appears to be a botched job is later shown to be a cunningly clever solution then non-creationists employing this tactic risk some embarrassment. There is indeed some mounting evidence that the vertebrate eye may have some advantages not immediately obvious. Kröger and Biehlmaier (2009), for example, have shown that having all the retinal wiring inside the eye allows for a considerable saving of space – something that would have been important for early small vertebrates like fish. Moreover, Gollisch and Meister (2010) offer persuasive arguments that the vertebrate retina is not simply an image-collecting device like the plate of a camera. Instead, they suggest that the 50 or so different cell types inside the retina (hitherto something of a puzzle) serve a computational function, processing signals before they are passed down the optic nerve. Such information processing is crucial for colour vision and so it is perhaps not surprising that octopuses, whilst possessing better eyes in terms of their camera function, can only see in black and white.

18.4 Compromises of posture

Primates have existed for about 60 million years and so the transition from a quadrupedal to a bipedal gait in hominins (about 4 million years ago) is a relatively recent occurrence. The human spine was basically originally designed for a quadruped. One key adaptation to facilitate this transition was lumbar lordosis, which is the inward curve of the spine (viewing sideways) in the lower back regions. This in itself, whilst helping vertical balance, exerts considerable pressure on the intervertebral discs (the cartilaginous joints between two adjacent bony vertebrae) and the sacroiliac joints (the two joints with slight movement that link the sacrum, which supports the weight of the spine, with each ilium of the pelvis). Prolapse or herniation of these intervertebral discs (sometimes called a slipped disc) due to pressure on the spine from the weight above is a very common cause of back pain. Pregnant women face this problem acutely since in carrying a child at the front of the body (instead of underneath as in quadrupeds) the woman has to lean backwards to support this weight. Vertebral arthritis has been found in pre-historic hunter-gatherers suggesting that it is not a modern lifestyle disease but perhaps a problem relating to the transition to bipedalism and the fact that the lower spine is subject to a much higher load than would be the case for a quadruped of the same body weight as humans. Today back pain represents a major cause of pain and disability resulting in millions of days lost at work each year worldwide. Knee joints, too, often give humans problems in later life and this may be a malfunction relating to the fact that the whole weight of the human body now passes through two joints instead of the four of our quadrupedal ancestors.

There is a potential pitfall in this general line of reasoning, however. If we argue that, for example, bipedalism (a trait at least 4 million years old) has left us not fully adapted to its demands then this brings into question arguments that ancestral behaviour can be examined from an adaptationist standpoint. Perhaps we should look for things that we now do to compromise skeletal health rather than assuming the transition to bipedalism was flawed per se or that we have not yet caught up in an adaptive sense with the problems it causes.

Quantitative support for this notion, however, comes from the work of Kimberley Plomp et al. (2015) on the shape of human vertebrae. This group argued that since the nearest living relatives of humans are chimpanzees, which are quadrupedal, and that humans evolved from quadrupedal ancestors, then we might expect the shape of the vertebrae of chimpanzees and humans to reflect adaptations to quadrupedal and bipedal gaits respectively. Moreover, if there is a natural variation in the shape of human vertebrae then humans with vertebrae of a shape more closely resembling that of chimpanzees than the average could be expected to be more vulnerable to disc herniation. Now, usefully, whether or not someone has experienced a herniated disc is revealed by the presence of depressions on the surface of the discs called Schmorl's nodes. The group compared the shapes of vertebrae from skeletons of 71 humans (from medieval remains), 36 chimpanzees and 15 orang-utans. As predicted by the ancestral shape hypothesis, they found that the lumbar vertebrae of healthy humans differed significantly in shape to those of chimpanzees and orang-utans (reflecting no doubt adaptations to different styles of locomotion). But, more interestingly, they found that vertebrae from humans with evidence of pathological disc herniation were very similar in shape to those of chimps, with no statistically significant difference in shape between the first lumbar vertebrae of chimps and pathological humans. This study is one of the few that provide direct evidence that lumbar back problems may be related to the rapid evolution of bipedalism.

18.5 Evolutionary obstetrics

Compared to other primates human childbirth is a long, painful and risky process. Part of the problem is that the head of a human neonate is such a tight fit with the pelvic canal (Figure 18.8). Today it is estimated that about 12 per cent of deaths of a mother

during or following childbirth are attributable to obstructed labour. For the offspring about one million stillbirths occur each year that are attributed to the birth process (Wells et al., 2012). For many years the conventional explanation has been that this tight fit was an evolutionary compromise between the advantages of large-brained children and the need to have a slender pelvis to facilitate bipedal locomotion. This was called the **obstetrical dilemma** hypothesis, but as we shall see there are now doubts about the origins of this problem.

18.5.1 The obstetrical dilemma hypothesis

One distinctive feature of human infants is that brain growth continues after birth in a far more rapid curve than in any other primate. It was this unusual feature that led Portmann (1969) to suggest that in terms of brain development the human gestation period is more akin to 21 months than 9 months. It was possibly this remark that led to the common assumption that the human gestation period is truncated to allow the passage of a large brain; whereas, in fact, in terms of neonate body size the gestation period is not far from what is expected for a human-sized primate.

This idea of essentially premature human birth formed part of the basis of what became called the obstetrical dilemma hypothesis. This ideas suggests an evolutionary tug-of-war between two competing pressures: large brain size and bipedal locomotion. A large brain necessitates a wide pelvic canal but this compromises bipedal locomotion, which relies upon a narrow pelvis. The hypothesis suggests that the uneasy compromise between these two competing demands provides the ultimate explanation of the high levels of maternal and neonatal mortality in contemporary populations and the reason why human neonates are so physically altricial.

The obstetrical dilemma (OD) hypothesis rests on the assumption that human gestation time is shorter than expected and hence the altricial neonate is born prematurely to allow the large head to pass through the birth canal. So what is the 'expected' gestation time? If gestation

FIGURE 18.8 Average neonatal head and pelvic inlet sizes in some primates.
The drawings of *Pongo*, *Pan* and *Gorilla* are scaled so that the transverse diameters of the maternal inlets are equal.
SOURCE: Dimensions obtained from various sources including Rosenberg and Trevathan (2002); and Schultz (1969).

time is plotted against adult human brain weight then it does look as if humans give birth prematurely. Portmann estimated that a gestation period of 18–21 months would be needed for human babies to be born at about the same stage of neurological development as chimpanzees (Portmann, 1969). It is also true that in humans the head size of the neonate relative to the birth canal is larger than for any other primate, giving rise to the rotation of many babies as they pass through the birth canal during parturition (see Figure 18.8).

It is questionable, however, whether adult brain weight should be used in this comparison since most brain growth occurs postnatally. There are various other comparisons that could be made. One is to compare maternal body mass and gestation time for a range of primates. This shows that human gestation time is longer than expected for a primate of our body size by about 37 days, a fact that challenges the idea that the diversion of metabolic resources to the foetus is terminated prematurely in humans to facilitate birth. But since humans are physically altricial we still have the question of what limits further foetal growth.

18.5.2 The energetics of gestation and growth (EGG) hypothesis

Recently Holly Dunsworth and colleagues (2012) have challenged the OD hypothesis in two ways: firstly, by examining whether increases in pelvic breadth would actually mechanically and energetically impede running and walking; and secondly, by looking at the constraints on the metabolic investment in the foetus.

If further foetal growth were to be restricted by the energetic impediments to locomotion caused by a wide pelvis then we might expect female walking and running to be less efficient than that of males since females do indeed have wider pelvises. In their review of the literature on this, however, Dunsworth et al. (2012) conclude that this is not the case. Instead they propose an 'energetics of gestation and growth' (EGG) hypothesis; something quite similar in fact to the 'metabolic crossover hypothesis' proposed by Ellison (2001). In this view it is the ability

of the mother's metabolism to fuel the increasing metabolic demand of the growing foetus that is pivotal. The core of the argument is that the timing of labour is set by the point when the foetus is demanding more energy than can be supplied by the mother through the placenta.

Generally, humans are capable of sustaining a metabolic rate of between two and two-and-a-half times basal metabolic rate (BMR) over a period of a few weeks. During pregnancy the mother's metabolic rate rises to about twice BMR up to birth and then stays at this level for several months after birth even though the energy demands of the newly born child continue to rise. It is possible, then, that a gestation period of nine months has evolved in humans as the point where it pays to give birth and supply the energetic demands of the child through external rather than internal means.

Further support for the EGG hypothesis comes from the work of Karin Isler and Carl van Schaick (2012a) on caregiving and brain size in mammals. They found that humans differed from other primates in the large amounts of high-quality foodstuffs given to infants and to mothers by other group members. This provisioning of mothers during gestation and lactation and the delivery of high-quality food to infants was a major factor enabling brain growth to be supported outside of the womb. In this sense we could say that humans are cooperative breeders with group members of both sexes (for example, post-reproductive kin, older siblings) helping to provide food and protection for mother and child.

The OD and EGG hypotheses make differing claims. In the OD hypothesis it is locomotion which limits the breadth of the pelvis which in turn determines the timing of parturition according to when the baby's head can slide through the pelvic canal; in the EGG hypothesis the pelvis itself has adapted to the energetically determined size of the brain at the point when external energy sources are needed to sustain the child.

Promising though it sounds, the EGG hypothesis still has to explain why, if locomotion is not an issue, the newborn's head and the mother's pelvis are such a tight fit. Given the selective pressures to increase human brain size why did not the pelvis widen to accommodate this more comfortably and so reduce the risks of childbirth, which are considerable? Various ideas have been advanced: one suggests that a wider pelvis would entail a larger body size and that female body size is constrained by ecological factors (Gaulin and Sailer, 1985); another is that recent increases in maternal energy consumption have led to large babies and that natural selection has not yet had time to catch up (Roy, 2003); or that the timing of human birth, whilst carrying heavy risks, is optimum from the point of view of later cognitive and motor-neuronal development (Neubauer and Hublin, 2012). Hence, whilst the EGG hypothesis is plausible its case would be greatly assisted by understanding why childbirth is so difficult for humans. In this respect it would be very useful to know what the negative consequences of a broader pelvis are if locomotion is not the crucial issue.

18.5.3 An ecologically sensitive dilemma?

Some of the problems with the OD and EGG hypotheses are addressed by recent ideas from Wells et al. (2012) that revolve around the notion that maternal and foetal developmental plasticity might be misaligned.

This group argues that both stature and pelvic dimensions have fluctuated considerably over the last 30,000 years. In one classic study by Angel (1975), which is taken to be probably typical of many human populations, mean bodily stature fell by about 12 cm during the transition from the late Palaeolithic to the Neolithic period, probably as a result of the advent of agriculture. It then recovered slowly until the early 19th century and has escalated rapidly to former Palaeolithic levels in the last 200 years. The crucial part of their argument is that neonatal body proportions, pelvic size and maternal stature may not change in a directly proportional way. If changes are very slow due to adaptations to changing climates and patterns of diet then it is conceivable that changes to these three parameters might be in step and whatever obstetrical dilemma exists is neither eased nor worsened. But in the case of short-term changes (for example, taking 10–50 years) to nutritional status then intergenerational

trends in pelvis size and birth weight may be misaligned. It is conceivable, for example, that short-term changes in nutritional status might cause birth weight to increase in size faster than increases in the bodily dimensions of the mother. This cephalo-pelvic clash could also be exacerbated by disease and nutritional deficiencies. Northern populations exposed to the risk of vitamin D deficiency (due to low UV levels) have increased risk of rickets and hence face constraints on the growth of the pelvis.

One change in the Neolithic period that could have intensified the OD is the shift from a hunter-gatherer diet high in protein, fibre and low **glycaemic index** carbohydrates to a diet much richer in carbohydrates. Protein promotes growth in childhood and so adult stature may have fallen in agricultural communities as dietary protein content was reduced. On the other hand, a cereal-rich diet high in high glycaemic index carbohydrates releases glucose into the bloodstream quickly and causes rapid growth in the foetus. The scenario is supported by preliminary evidence that suggests that early agricultural communities experienced a higher rate of perinatal mortality than hunter-gathering groups, possibly as a consequence of obstructed labour (Wells et al., 2012).

If this analysis is correct the implications are worrying since increasing rates of maternal obesity and diets high in carbohydrates will only serve to exacerbate the problem in modern populations with a Western dietary lifestyle.

18.6 Vestigial structures: wisdom teeth and the appendix

Vestigial structures are hangovers from an early stage of evolution, relics of the iterative design process that appear in contemporary environments to serve no function. They illustrate the point that natural selection can only work from what is present and not effect a completely new design plan. In *The Descent of Man* Darwin listed several such structures in humans including the coccyx (tail bones), muscles of the ear, wisdom teeth, the appendix, body hair, and the arrector pili response (goose bumps). Such appendages may now seem to have little adaptive value, but can vestigial organs cause health problems? Two traits that were once routinely suggested to be vestigial, and because of this a source of problems, are wisdom teeth and the appendix. It turns out that both are perhaps slightly more complicated issues than was first thought.

In respect of wisdom teeth, the argument ran that as we learnt to use fire to cook our food, and generally source a high-quality diet, the need for a large jaw for tearing flesh and grinding food reduced, leaving modern humans with jaws too small to house 32 teeth. But genes for 32 teeth remained and their cramped condition is a source of problems. Wisdom teeth are the last to erupt in the mouth, in some people they never appear and in others they become impacted and have to be surgically removed (Biswas et al., 2010). Now whereas it is certainly true that the fossil record shows a gradual diminution in jaw size in the hominin lineage, there are also variations in contemporary populations. The anthropologist Noreen von Cramon-Taubadel looked at jaw size variations among contemporary hunter-gatherers and agriculturalists. She examined over 300 skulls from 11 different populations and found that agriculturalists leading a relatively sedentary way of life with easy-to-chew food had shorter and wider jaws (mandibles) than hunter-gatherers. This, however, was interpreted as a developmental change rather than one reflecting underlying genetic adaptations. As a child grows the sort of food it eats and the stress on jaw muscles and bone tissue during mastication will affect jaw shape. Soft foods leave the young adult with a relatively short jaw, compared to those chewing rougher and tougher foodstuffs, and hence increase the risk of encountering problems later in life as the third molars (wisdom teeth) try to erupt (von Cramon-Taubadel, 2011).

The appendix is found in several taxa of mammals including primates and rodents. In humans it is about 11 cm long and 0.8 cm in diameter and is located at the junction of the small and large intestines. Darwin suggested in 1871 that the appendix was an evolutionary remnant of a primate past where it was larger and served the useful function of helping to digest leaves. He thought it to be confined to the hominoids (a speculation later shown to be incorrect) and was the vestige of a much larger cecum

(a pouch-like structure at the start of the large intestine found in many mammals). His idea was that the appearance of the appendix was due to a reduction in cecum size as our ancestors shifted from a frugivorous diet (that needed a large cecum to aid digestion) to a more omnivorous diet with meat as an important component.

For many years following Darwin's suggestion, the appendix has been regarded as a vestigial organ – an atavistic relic of our evolutionary past. Recent research, however, suggests that the situation may be more complex. Heather Smith et al. (2013) looked at the appearance and disappearance of the appendix in the phylogeny of 361 mammalian species. They found that the appendix size correlated positively with cecum size, but that there was no correlation between the evolutionary appearance of the appendix and changes in diet or fermentation strategy. Moreover, they concluded that the appendix had evolved separately at least 32 times but was lost only seven times. This suggests that the appendix had, and continues to have, some enduring fitness value.

One idea that has gained much support recently is that the appendix is a 'safe house' for beneficial bacteria in times of intestinal infections (Figure 18.9). When the gut is flushed out following a diarrhoeal disease, the position and shape of the appendix allows it to assist in the recolonisation of the gut (Bollinger et al., 2007). This function is likely to still be of importance in developing countries where diarrhoeal diseases are common. The fact that in more developed countries the appendix can be removed without risk of later complications is probably due to good nutrition, modern medicine and the ready availability of uninfected water.

Inflammation of the appendix (appendicitis) is a serious condition which can be fatal, and despite a recent slight decline in its incidence it remains one of the commonest surgical emergencies in Europe and the USA. Its peak incidence occurs in early adulthood and so one would expect there to be a strong selective pressure against retaining such a risky tissue. It is possible that this risk is counterbalanced by the useful function of the appendix, or that appendicitis is a feature of modern lifestyle and diet and was rare in pre-Neolithic ancestral populations.

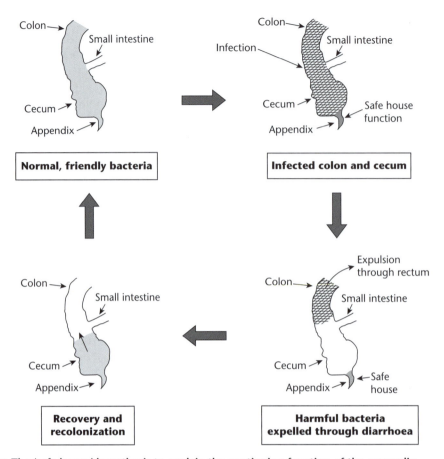

FIGURE 18.9 The 'safe house' hypothesis to explain the continuing function of the appendix.
During an infection the gut may be completely flushed, removing useful bacteria. Residual bacteria remaining in the appendix can be used as a source of friendly bacteria to repopulate the gut.
SOURCE: Based on ideas in Laurin et al. (2011).

Part III Subjective definitions of health and optimum reproductive fitness are not the same thing

Many traits shaped by natural selection may seem to us maladaptive simply because we have, understandably, a human-centred vision of what adaptations should be like (Nesse, 2005). Taking a more objective view, we know that the only currency recognised by natural selection is that of inclusive fitness. The subjective experiences of health and happiness may strongly correlate with inclusive fitness (pleasure and the satisfaction of our desires are, after all, ways natural selection gets us to do inclusive fitness-promoting

activities) but not perfectly so. Genes that promote fitness in the gene frequency sense will spread even if they reduce health, happiness and longevity. We should not expect nature to be kind to us. As Dawkins memorably noted:

> The universe that we observe has precisely the properties we should expect if there is, at bottom, no design, no purpose, no evil, no good, nothing but pitiless indifference. (Dawkins, 1996, p. 85)

This section focuses on phenomena that may appear biologically wasteful, and experiences that are felt as aversive, yet make perfect sense through the lenses of natural and sexual selection.

18.7 Antagonistic pleiotropy and heterozygotic advantage

The fact that the human genome contains only about 23,000 genes suggests that many genes must have multiple functions. Hence a mutation to a gene may enhance the fitness-promoting aspects of one function, whilst decreasing that of another. This is a commonly encountered phenomenon in many human artefacts: making a car accelerate faster means higher fuel consumption, making it safer often means adding weight and raising the cost. It is rare that one thing can be changed without some other aspect also being affected. In terms of a mutation to a gene, if it increases the overall fitness it will become fixed and widespread even if it also introduces some puzzling negative effects. The term for genes having multiple functions is pleiotropy. A key application of this concept was made by George Williams (1957) when he advanced the antagonistic pleiotropy hypothesis as a way of explaining the genetic basis of senescence. Put simply, his argument was that it is quite plausible for genes to exist that enhance reproductive fitness early in life but cause senescence later. The term antagonistic refers to the fact that the properties of these genes seem to act in opposite directions. This idea was briefly explored in Chapter 8 in the context of life history theory.

A concept closely related to antagonistic pleiotropy is heterozygote advantage – sometimes called 'over dominance' since Aa heterozygotes have a higher fitness than either aa or AA homozygotes. Both concepts have the potential to explain why some genes which have deleterious effects can persist in the gene pool. Table 18.4 shows some of the genetically based diseases to which this line of argument has variously been applied. A few of these will be discussed in more detail below.

18.7.1 Sickle-cell anaemia

A simple change to the base sequence on our DNA is known to cause the distressing condition of sickle-cell anaemia. The substitution of one amino acid (glutamine becomes valine) in haemoglobin causes an alteration in the shape of red blood cells – they appear sickle shaped – and a reduction in the oxygen-carrying capacity of the blood. The sickle-shaped cells produced when the defective gene is present on both **chromosomes** (that is, the chromosomes are homozygous) are quickly broken down by the body; blood does not flow smoothly and parts of the body are deprived of oxygen. The physical symptoms range from anaemia and physical weakness to damage to major organs, brain damage and heart failure. There is no cure for the condition, which causes the death of about 100,000 people worldwide each year. Sickle-cell anaemia is by far the most common inherited disorder amongst African-Americans and affects about 1 in 700 of all African-American children born in the USA. Its high frequency in the population, and the fact that natural selection has not eliminated it (many of those suffering die before they can reproduce), is probably due to the fact that in Africa possession of one copy of the sickle-cell gene confers some resistance to malaria. If Hb is taken to be the normal haemoglobin gene and Hbs the sickle-cell gene then people who inherit both sickle-cell genes, that is, one from their father and one from their mother, and who are therefore Hbs Hbs, suffer from sickle-cell anaemia and die young. People who inherit only one copy of the sickle-cell gene and are Hb Hbs are said to have sickle-cell trait and only some of their red blood cells are oddly shaped. It is in this latter condition that the gene gives an advantage in protecting against malaria since the malarial parasite (*Plasmodium*) cannot complete its life cycle in the mutant cells. It is the prevalence of malaria in African countries that explains why this apparently maladaptive set of genes survived in the gene pool and is now found amongst African-Americans.

18.7.2 Glucose-6-phosphate dehydrogenase deficiency

Another example related to the risk of malaria is the most common of all human genetically based enzyme defects: glucose-6-phosphate dehydrogenase deficiency (G6PD deficiency). There are several different mutations – all on the long arm of the X chromosome – which give rise to the deficiency, suggesting that it originated on multiple occasions. One form or another is found in about 6 per cent of the human population and it is responsible for

Table 18.4 Candidates for antagonistic pleiotropy.

The putative benefits of the genes responsible for the diseases listed may be evident in carriers of the disease that show no symptoms (for example, when the disease is autosomal and recessive); or, in cases where the genes are expressed, in terms of the fertility of the bearers before the disease causes serious illness (for example, Huntington's disease).

Disease	Gene symbol	Mode of inheritance	Variant implicated in disease	Incidence	Phenotypic symptoms	Suggested benefit to carrier (if heterozygote recessive) or sufferer
Huntington's disease	HTT	Autosomal dominant	Repeated CAG sequence	1/20,000 Western Europeans	Cognitive impairment, movement abnormalities	Increased fertility; possible reduced risk of cancer
Cystic fibrosis	CFTR	Autosomal recessive	Missense mutation (single nucleotide change causes different amino acid)	1/2,400 in UK	Lung infections, poor growth	Increased fertility; resistance to typhoid, cholera or tuberculosis
Sickle-cell anaemia	Hbb	Autosomal recessive	Various single nucleotide mutations	1/700 African-Americans; 1/160,000 European-Americans	Anaemia, susceptibility to infection, loss of vision	Protection against malaria in heterozygotes
Glucose-6-phosphate dehydrogenase deficiency	G6PD	X-linked recessive	Various missense mutations	1/15 world average but very variable. High in Africa and Middle East	Neonatal jaundice; damage to red blood cells	Protection against malaria
Beta-thalassemia	HB	Autosomal recessive	Various single nucleotide mutations	1/100,000 globally but variable. 1/250 in Cyprus	Anaemia	Protection against malaria in heterozygotes

SOURCE: Adapted from Carter and Nguyen (2011), Table 1; and from Jobling et al. (2014), Table 16.1, p. 519.

about 4,000 deaths globally each year (Lozano et al., 2013). The effects of the defect are complicated but in essence important biochemical pathways are disrupted and red blood cells can be destroyed. Certain drugs and, strangely, broad beans can exacerbate the condition. Populations most commonly affected are of African, Asian and Mediterranean origin. The common consensus is that the reason why these variant genes with potentially debilitating effects have not been eliminated is that the 'disease', which can often exist in a covert form, offers protection against infestation by *Plasmodium falciparum* – the deadliest of all malarial parasites spread by mosquitoes. Since the genes responsible are recessive and on the

X chromosome men are much more likely to suffer from the condition than women.

18.7.3 Huntington's disease

Huntington's disease tends to occur after peak reproductive age (onset is typically at 30–45 years of age), which tends to weaken selective pressure against it. Early symptoms are an unsteady gait and mood changes. This progresses to uncoordinated body movements (the condition was originally known as Huntington's chorea – 'chorea' comes from the Greek meaning dance), cognitive decline and then dementia. There is no known cure and life expectancy is typically about 20 years after symptoms are first noticed. The problem is found in variant copies of the Huntingtin gene (HTT) which codes for the protein Huntingtin. The DNA sequence contains a sequence of three bases (CAG) repeated multiple times (for example, CAGCAGCAGCAGCAG). Problems begin when this sequence is repeated more than about 36 times: then a protein is produced that influences the decay rate of neurons.

There is some suggestive evidence that the disease carries reproductive advantages for people early in their life. In an early Canadian study on 157 Huntington's patients compared to 170 'wild type' individuals, those with Huntington's disease had about 39 per cent more offspring than their unaffected relatives and 18 per cent more than an unrelated control sample matched for age and sex (Shokeir, 1975). A slightly later study by Walker et al. (1983) on affected individuals in South Wales (UK) also found an enhanced fertility effect.

There is also some evidence to suggest that a high number of CAG repeats in the HTT allele is associated with high levels of a tumour-suppressing protein called P53. This tallies with the finding that Huntington patients have a lower incidence of cancer than the general population (Sørensen et al., 1999).

18.7.4 Cystic fibrosis

Just how small an advantage is needed to ensure that genes that are maladaptive when an individual carrier copies from both parents can also be seen in the condition of **cystic fibrosis**. A child with cystic fibrosis (CF) is born when the relevant genes are homozygous in the recessive state; that is, it has two copies of the defective gene, one from each parent. If it has only one copy then it is said to be a carrier. People who are carriers live perfectly normal and healthy lives never realising they are carriers until they mate with another carrier. About 1 in 25 Caucasians are thought to be 'carriers' of the recessive allele for CF. The chance of two carriers meeting is thus about $(1/25)^2 = 1$ in 625 or 0.0016. The chance of a child from a union of these parents having both recessive alleles and hence displaying the condition is one quarter of 0.0016 = 0.0004, or 1 in 2,500. Hence about 1 in 2,500 of Caucasian children are born with CF. The early death of people affected by CF (that is, those who are homozygous) will not remove the allele itself. In fact the heterozygous condition only needs to carry a 2.3 per cent advantage compared to non-carriers for the recessive allele to persist indefinitely (see Strachan and Read, 1996).

Various candidates have been proposed for the selective advantage experienced by heterozygote carriers of this defective gene. These include resistance to cholera, typhoid and tuberculosis (see Poolman and Galvani, 2007). There is also some evidence of fertility effects. In one study Knudson et al. (1967) found that grandparents of CF patients were found to have an average of 4.34 offspring compared to 3.43 for the control group (p < 0.01).

Opinion on the general significance of antagonistic pleiotropy as a factor in human disease is divided. Carter and Nguyen (2011), for example, argue that the phenomenon may be very common and should be borne in mind when contemplating gene therapy lest the advantages of a gene be lost with the disadvantages. Others (for example, Valles, 2010) warn of an over-hasty extrapolation from the success of the sickle-cell anaemia model, which does seem robust, into less convincing areas. One of the problems with searching for counterbalancing advantages is that the epidemiology of the candidate disease (with perhaps the exception of sickle-cell anaemia and glucose-6-phosphate dehydrogenase deficiency) does not neatly map onto the distribution of the alleles that supposedly offer protection. In the case of CF, for example, typhoid fever and cholera have long been prevalent in tropical regions such as East Asia and India yet CF-inducing alleles are quite rare in these areas.

18.8 The perils of sexual selection

As discussed in Chapters 4 and 15 sexual selection is a powerful force and can drive traits beyond their ecological optima. In other words the magnitude of a trait at its optimum for reproductive fitness may be different to its optimum magnitude for individual survival and wellbeing. If such traits do exist in humans then the fact that they have negative health consequences becomes less of a puzzle in this light.

At a first level of approximation, humans are like many mammals in that females invest more in parenting effort, whilst males invest more in mating effort. Women also have a lower potential reproductive rate than men and so men will compete amongst themselves to gain access to desirable females (leading to intrasexual selection), and women will be careful to choose mates displaying signs of genetic and phenotypic quality (leading to intersexual selection).

These biologically based sexual strategies are now of course buffered and regulated by social institutions and customs, but their underlying force may still be apparent in the area of susceptibility to illness and risk taking.

In most Western cultures, from the moment they are born until the day they die, males face a higher risk of dying at any age than females do (Figure 18.10). Some of this excess risk may stem

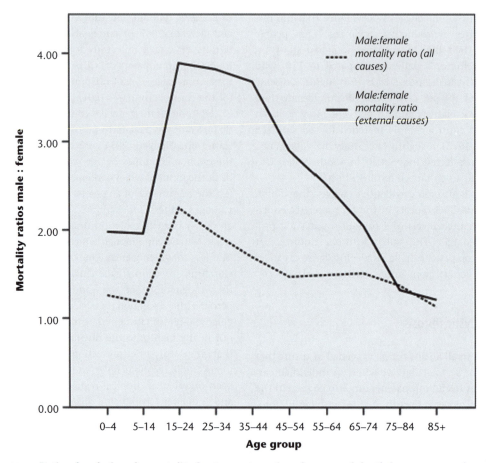

FIGURE 18.10 Ratio of male:female mortality for two categories of causes of death by age group for England and Wales, 2012.

The male:female mortality ratio is calculated as deaths of males per 1,000 in age group/deaths of females per 1,000 in same age group. Note that for all age groups there are more male deaths per 1,000 than female (that is, all ratios > 1). The ratios for external causes (accidents, falls and violence) are higher still, and peak for the age group 15–35. Proximate explanations for this effect are that men tend to be more likely to suffer from cardiovascular disorders and cancers; they are also more likely to be involved in transport accidents and deaths due to violence and crime. From an ultimate evolutionary perspective it appears that men put less investment in somatic repair and maintenance. Risk-taking behaviour may stem from competition between males or activities to impress females. The net effect is that men place more effort into mating at the expense of higher morbidity and mortality.

SOURCE: Adapted from data from the Office for National Statistics licensed under the Open Government Licence v.3.0. For a similar treatment of US data see Kruger and Nesse (2006).

from underlying genetic inequalities, such as the effect of deleterious genes on X chromosomes (since men have one copy and women two, men are denied the advantages of compensating for a deleterious allele). But the majority of disorders (for example, cardiovascular complaints) for which there is a male incidence bias could be a more general reflection of the fact that, from a life history perspective, investment in growth, mating, repair and parenting differs between men and women, with men making a lower overall investment in repair and maintenance functions compared to women (Figure 18.11).

These two differences possibly underlie (as ultimate causes) the differences in mortality statistics between men and women (Kruger and Nesse, 2006). Additionally, men are at far greater risk than women from mortality caused by external causes, such as traffic accidents and violence, which are associated with risk taking and risky lifestyles. Significantly this category peaks for men in their adolescent and young adult years (19–30). Testosterone may be the

hormonal mediation between ultimate causation (sexual selection) and risky and aggressive behaviour. If there has been intersexual selection by females preferring these traits then testosterone may be equivalent to a costly negative handicap: it is attractive to females but at the expense of male longevity.

18.9 Defence mechanisms mistaken for signs of ill health

Organisms face a constant barrage of threats throughout their lives and natural selection has put in place a series of adaptive responses to deal with these. Nesse and Williams (1995) criticise modern medical practice for putting the cart before the horse, for 'trying to find the flaws that cause the disease without understanding normal functions of the mechanisms' (p.230). The essential point is that conditions we think of as symptoms of illness, such as vomiting, might be the body's natural defence mechanism at work.

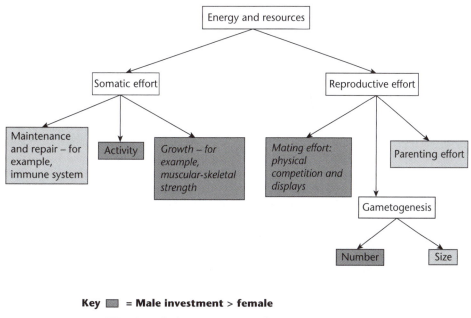

Key ▨ = **Male investment > female**

▢ = **Female investment > male**

FIGURE 18.11 Comparative male and female life history investments.
Sexual selection theory and a life history perspective predict that males will direct more effort into growth and mating than females. Activity levels for males in most societies are also usually larger than females (an energy expenditure of about 3,015 kcals/day for men compared to 2,294 kcals/day for women). For males this leaves less energy for investment in other areas related to health and maintenance, and mating effort brings its own risks.

But the body's defence mechanisms do not always function as they should. In this context we can identify two possible types of disorder: disorders that are the normal operation of defence mechanisms that have conventionally been labelled as disorders since they are unpleasant; and disorders that result from malfunctioning or dysregulated defence mechanisms. Examples of the former might include fever, anxiety, jealousy and pain in general; and examples of the latter, autoimmune diseases and anxiety disorders. If the former category is suspected then we can predict that the form of the defence response should match its function. For example, anxiety should bring about changes that are appropriate to deal with the threat. In this case the evidence is positive: increased heart rate, glucose metabolism, clotting, sweating and breathing all change in line with a flight or fight response. We can also predict that fears should be related to actual fears in ancestral environments. Fear of high places should not be expressed in creatures for which high places pose no threat.

Experiences such as coughing, nausea, vomiting, diarrhoea and pain in general are both unpleasant and usually functional, and it is their very averseness that motivates us to avoid situations that cause such responses (Williams and Nesse, 1991). More generally, of course, this is precisely why natural selection designed the circuitry behind classical and operant conditioning. In operant conditioning behaviours associated with a reward become more frequent, whilst those followed by punishment or painful consequences become less so. Many other defensive responses are products of classical conditioning. We can easily develop an aversion to cues that once preceded a series of painful experiences, such as the sight or smell or food or drink that we may have consumed to excess or that once caused illness.

It is not always easy to understand how the defensive response actually works to preserve health. An infection, for example, often results in malaise and a fever. Malaise serves to reduce activity and energy expenditure, perhaps allowing a diversion of metabolic resources to support the immune system. Fever (sometimes called pyrexia or the febrile response) is not so straightforward. One argument is that fever raises the temperature of the body and so restricts the growth of bacteria. The problem with this is that given the short generation time of bacteria, repeated fevers in different hosts over time would soon select for bacteria capable of withstanding the 1–3°C temperature rise that a fever can bring about. A more recent argument is that when cells are exposed to above normal temperatures they produce heat shock proteins. Some of these are thought to enter the circulation and enhance the immune response. It is clear, though, that high fevers are often maladaptive and the whole area remains controversial (see Kluger et al. 1998; and Blatteis, 2003).

18.10 Morning sickness as a defensive response

Observations on the health of women in all cultures reveal a common pattern: during the first trimester of pregnancy (0–3 months) many women experience the symptoms of 'pregnancy sickness': vomiting, nausea and food aversions. Since this is so common might it have a functional basis? With this perspective in mind, Profet (1992) hypothesised that pregnancy sickness was an adaptation to reduce the intake of toxin-containing foods. There are several facts that are consistent with this hypothesis. Pregnancy sickness occurs during the first trimester of development when the major organs of the foetus are under development. Moreover, it is known that the effects of drugs (for example, thalidomide) and stress from disease (for example, rubella) are at their most severe early in pregnancy.

The argument of Profet was extended and modified by Paul Sherman and Samuel Flaxman who argued that nausea and vomiting were more likely to be mechanisms protecting against biological pathogens (bacteria, fungi, viruses and other parasites) rather than mutagenic chemical substances. This follows from the fact that during pregnancy a mother's immune response is temporarily depressed; this itself being an adaptive response as the mother is carrying a foetus which is essentially akin to foreign tissue since half its genome comes from the father. The benefits of this temporary immunosuppression in ensuring the embryo is not rejected comes at the cost of increased susceptibility to infections.

Sherman and Flaxman argue that various predictions follow from this modified hypothesis. One is that nausea and vomiting symptoms should peak at the times when the embryo is most sensitive to infectious agents and when the mother is most vulnerable. This period is thought to be during the first trimester (especially weeks 5–18) and there is evidence that this is just the time when nausea and vomiting are most frequently reported (Flaxman and Sherman 2000).

Pregnancy is also associated with food cravings and aversions. Now if food aversion is a mechanism to avoid harmful substances and biological agents then such aversions should be more prevalent during the first trimester. Rodin and Radke-Sharpe (1991) quantified both cravings and aversions across all three trimesters of pregnancy and found that aversions were significantly higher in the first trimester.

Another prediction, and one that is counterintuitive without the Darwinian perspective that

unpleasant symptoms might be part of natural defences, is that nausea and vomiting should be associated with more successful pregnancy outcomes compared to pregnancies where these symptoms were absent and so unable to exert any protective effect. Ronald and Margaret Weigel (1989) conducted a meta-analytical review of data from 11 other studies on the association between aversive nausea and vomiting symptoms during pregnancy and the risk of miscarriage. They found a significant negative association (Figure 18.12).

Finally, if morning sickness and food cravings have an adaptive function then food that women typically find unpleasant should potentially contain pathogens or toxins, whilst those that are the subject of cravings should not. Based on an analysis of responses of over 5,000 women who experienced aversions, and over 6,000 women who experienced cravings, Flaxman and Sherman found strong support for this contention. They showed that more women experience an aversion to meat than a craving, and

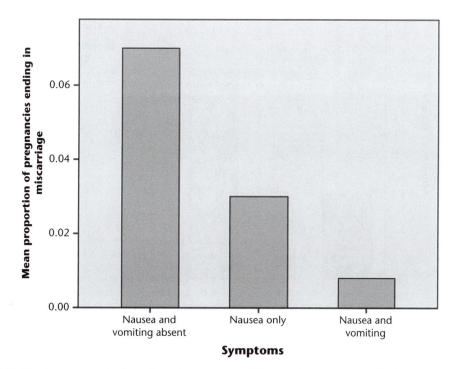

FIGURE 18.12 The frequency of miscarriages in pregnancy is inversely correlated with the presence of nausea and vomiting during pregnancy (p < 0.001).
Such evidence provides support for the adaptive function of morning sickness.
SOURCE: Plotted with data taken from Weigel and Weigel (1989).

more for a craving for fruit than an aversion. Such findings are consistent with the idea that food aversions are designed to avoid pathogens.

This same research group also argued that nausea and vomiting in pregnancy should occur least often in cultures where women are seldom exposed to food containing pathogens, and more frequently in culture where meat and other potentially dangerous foods are regularly consumed. They used dietary data available on 27 traditional societies in the Human Relations Area Files (a database of over 400 cultures based at Yale University). Figure 18.13 shows support for this prediction.

If these ideas are further sustained we face the interesting conclusion that morning sickness is not a disease but a sign of a normal pregnancy. If this is the case then therapeutic implications follow: it may not be always wise to try to medicinally eliminate vomiting and nausea since such procedures may damage what is a natural prophylactic mechanism.

18.11 Responses that are overzealous

In Chapter 9 we used Nesse's smoke detector principle, and error management theory, to explain why many defensive reactions may be false alarms. The principle relies upon the fact that the cost of an alarm or defensive mechanism (CD) must be less than the cost of damage from a real threat (CH). But in pregnancy sickness we also meet a paradox; sometimes the vomiting that results can lead, if untreated, to death by dehydration, so here CD > CH. Possibly this is explained by the fact that such extremes are unusual and for most people most of the time the system works well. One could model this by factoring in an additional probability that from time to time CD will exceed CH. Paul Ewald (1994) suggests an alternative hypothesis: that infectious agents can deregulate otherwise functional defence systems causing serious damage to the host. Diarrhoea, for example, is an effective means of expelling

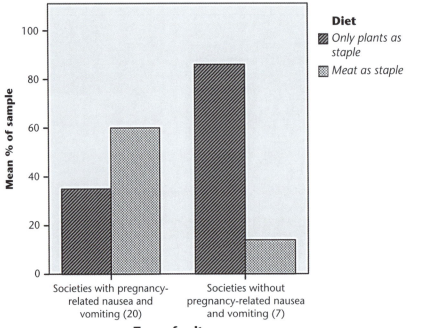

FIGURE 18.13 Diet in cultures with and without reports of pregnancy-related nausea and vomiting.
A diet high in plant foods seems to be associated with lower reports of morning sickness. Conversely, in societies where morning sickness is frequently reported meat is often a staple food (p < 0.05).

SOURCE: Plotted from data in Sherman, P. W. and S. M. Flaxman (2002). 'Nausea and vomiting of pregnancy in an evolutionary perspective.' *American Journal of Obstetrics and Gynecology* **186**(5), S190–S197.

pathogens and probably aids recovery in many cases. But when humans are infected with *Vibrio cholerae* (classic cholera) the diarrhoea is so acute that the resulting dehydration is the frequent cause of death. This on the face of it would seem a very odd defensive reaction: the cholera bacteria were successfully expelled but the patient died. Now *Vibrio cholerae* can attach to the wall of the intestine and also has a flagellum which enables it to swim quite strongly, characteristics that enable it to remain in the intestine and avoid being flushed out until it is ready. But diarrhoeal flushing does remove other bacteria and

hence reduces competition, thereby enabling *Vibrio cholerae* to multiply and ultimately disperse. In this perspective then, the pathogen is deliberately making the defensive reaction hyper severe, risking the life of the host but ensuring that bacteria reach water supplies to infect other hosts. If this interpretation is correct it supplies a valuable evolutionary perspective on the meaning of disease symptoms with a possible wider application. Ewald suggests, for example, that extreme pregnancy sickness may be the effect of an infection by *Heliocater pylori* which is then manipulating the natural defence of vomiting.

Summary

- Darwinian medicine is a relatively new field that uses evolutionary theory to explain human susceptibility to disease and illness. So far it has had more success in explaining illness than devising therapeutic regimes, although this is likely to change.

- Evolutionary approaches to the understanding of illness can be thought of as resting on a number of key principles; these include: the realisation that human genetic evolution is slow relative to the evolution of culture, the (potential) pace of environmental change, and the rate of evolution of pathogens; the fact that natural selection is an iterative process forced to face constraints and make compromises; and the appreciation that there is not a perfect overlap between optimum reproductive fitness and subjective dimensions of health and wellbeing.

- The growth in antibiotic resistance in many bacterial strains is a worrying phenomenon and almost certainly due to an unwise use of antibiotics in prescription medicine and farming.

- A relatively new field of enquiry is the idea of 'developmental origins of health and disease' (DOHaD paradigm) which suggests that humans have epigenetic mechanisms that have evolved to prepare for anticipated environments. A change in the environment from that predicted can cause ill-health effects. Research in this area is ongoing and key questions still remain concerning the adaptive nature of epigenetic plasticity, and whether ill health inevitably follows a mismatch between predicted and actual environments.

- Heterozygotic advantage, antagonistic pleiotropy and balancing selection are three related concepts that can explain why genes that ostensibly have deleterious effects on an organism can persist in the genome. It is likely that genes underlying some medical conditions (for example, sickle-cell anaemia and cystic fibrosis) persist in the gene pool for these reasons.

- Some signs of ill health may be the result of the body's natural defence mechanisms in action. Pregnancy sickness is a plausible candidate for this phenomenon; in such cases medics need to think carefully before correcting any symptoms.

Key Words

- Adaptive landscape
- Antagonistic pleiotropy
- Bifidobacteria
- Chromosomes
- Cystic fibrosis

- Glycaemic index
- Huntington's disease
- Obstetrical dilemma
- Oligosaccharides
- Pregnancy sickness

- Vestigial structures
- Virulence
- Zoonosis

Further reading

Stearns, S. C. and J. C. Koella (eds.) (2008). *Evolution in Health and Disease*, 2nd edn. Oxford, UK, Oxford University Press.

An excellent well-organised collection of chapters. Chapter 1 is a useful introduction to the whole field.

Trevathan, W. R., E. O. Smith and J. J. McKenna (eds.) (2008). *Evolutionary Medicine and Health*. Oxford, UK, Oxford University Press.

An edited book with many useful chapters.

Taylor, J. (2015). *Body by Darwin*. London, UK, University of Chicago Press.

A popular but informed account of how Darwinism informs modern medicine. Adopts an interesting case study approach.

Three Case Studies in Evolution and Health: Diet, Cancer and Mental Disorders

CHAPTER

19

Physical ills are the taxes laid upon this wretched life; some are taxed higher, and some lower, but all pay something.
(Letter of Lord Chesterfield to Dr R. C., 22 November 1757)

In the previous chapter we established a framework for thinking about how evolutionary ideas can help inform the study of health and disease. In this chapter we look in more detail at three areas where this has been successfully applied. But to put these issues in perspective it is worth looking briefly at the burden of illness suffered by humans on a global scale

19.1 The global burden of disease (DALYs)

There are a variety of ways of describing statistically the impact of disease and causes of morbidity and mortality on world populations. One of the standard ways that gives an assessment of burden of disease is the measure 'disability-adjusted life years' (DALYs). DALYs are calculated from the loss of life due to the condition and the years spent suffering adjusted by the severity of the disease. Figure 19.1 shows the percentage of total DALYs accounted for by a selection from the top 20 conditions responsible for morbidity and mortality according to whether the country is classed as high or low income.

High-income countries tend to experience DALYs in relation to degenerative ailments and 'diseases of civilisation': cardiovascular disorders, arthritis,

diabetes and cancer. Many of these may be lifestyle and diet related. Low-income countries still suffer hugely from infectious disease (especially HIV), complications of childbirth and nutritional deficiencies – conditions reflecting poverty and poor medical care. What is quite surprising perhaps is that mental and behavioural disorders (unipolar and bipolar depression, schizophrenia, anxiety and so on) figure prominently in both camps. In terms of the leading top ten causes of DALYs, mental and behavioural disorders come fifth for high-income countries and sixth for low-income countries.

In this chapter we will examine three case studies in the light of evolutionary theory: diet, cancers and mental disorders.

19.2 Case study I: diet and health in evolutionary perspective

One of the most forceful and successful applications of the concept of genome lag has been to changes in human diet. In this section we will explore evidence and theories that link the change from a Palaeolithic to a post-Neolithic diet with the incidence of modern 'lifestyle' diseases. In part we will assess the revisionist view that the Neolithic revolution, far from freeing humans from the misery and toil of subsistence

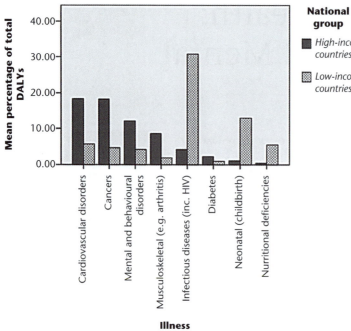

National group

■ *High-income countries*

▨ *Low-income countries*

FIGURE 19.1 A selection from the top 20 causes of DALYs in high- and low-income countries (2011).
SOURCE: Plotted with WHO data from: http://www.who.int/healthinfo/global_burden_disease/estimates/en/, accessed 9 February 2016.

living, actually worsened the human condition (Figure 19.2). As Jared Diamond said:

> In particular, recent discoveries suggest that the adoption of agriculture, supposedly our most decisive step toward a better life, was in many ways a catastrophe from which we have never recovered. With agriculture came the gross social and sexual inequality, the disease and despotism that curse our existence. (Diamond, 1987, p. 64)

19.2.1 The major dietary transitions

If we begin with the emergence of the hominins around 7 million years ago we can observe several major types of diet and dietary transitions in the path to contemporary society. In very broad terms we can detect three major periods that neatly coincide with geological and archaeological timescales:

the Pliocene, the Pleistocene and post-Neolithic.

The Pliocene and the early hominins: 7 million–2.6 million years ago

The diet of early hominins, following the split with the lineage of the chimpanzee about 5–7 million years ago up to emergence of the genus Australopithecus, was most probably omnivorous. Chimpanzees themselves, although primarily frugivorous, still consume about 65g of meat per day on average in the wild (Stanford, 1996). Further evidence comes from an analysis of carbon isotope ratios. For a variety of complex biochemical reasons the ratio of organic C^{13}/C^{12} varies according to the type of food (plant or animal) in which the carbon is found. When organisms consume these foodstuffs these carbon ratios are reflected in the isotopic composition of bones. An analysis of fossils of *Australopithecus africanus*, *A. robustus* and early *Homo ergaster* shows carbon isotope signals typical of omnivores – although the meat content was probably lower than contemporary human diets (Lee-Thorp, 2008).

The Pleistocene and the emergence of the Homo genus and *Homo sapiens*: 2.6 million–11,000 years ago

The geological period known as the Pleistocene neatly coincides with another hominin dietary pattern. Around 2.6 million years ago there is evidence that stone tools were used to butcher animal carcasses and extract marrow from bones. From this we can assume an increased dietary reliance on meat. This in itself may have facilitated (or even promoted) the move out of Africa by *Homo erectus* to more northerly latitudes where the climate would have restricted the availability of plant foods at certain times of the year. During this period genetic adaptations to a meat-rich diet probably began, leading to a reduction in gut size and a concordant increase

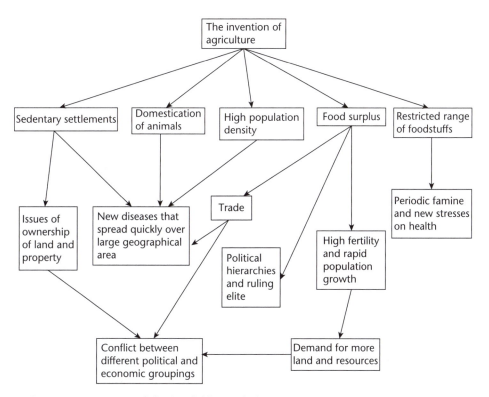

FIGURE 19.2 Some consequences of the Neolithic revolution.

in brain capacity (see Chapter 6). Significantly, from a biochemical point of view, humans are similar to obligate carnivores (such as cats) in that we are highly inefficient in de-saturating and elongating fatty acids with 18 carbon atoms in their chains (found widely in plant oils) to synthesise those containing 20 and 22 carbon atoms, which are long chain fatty acids essential to life (for example, arachidonic acid, $C20H32O2$; and docosahexaenoic acid, $C22H32O2$), and so obtain much of these nutrients from animal sources.

The actual percentage of calories obtained from meat during this period is difficult to assess and obviously would have depended on local ecologies. In relation to contemporary hunter-gatherers, Cordain et al. (2000) examined the ethnographic atlas and found that whereas only 14 per cent of cultures obtained > 50 per cent of their calories from plant foods, 73 per cent of societies obtained > 50 per cent of their calories from hunted and fished animal sources.

Neolithic and post-Neolithic cultures: 11,000 years ago–present

The Neolithic revolution began about 11,000 years ago and involved initially changes in food production technology: essentially the domestication of a range of plant and animal species. It occurred in several areas of the world independently: first the Fertile Crescent in the Near East, then Meso America and then, by about 7,000 years ago, South East Asia (Figure 19.3). The phenomenon is a remarkable example of the convergent evolution of culture. The consequences of what was in effect the invention of agriculture were enormous: a move from a nomadic lifestyle to a sedentary one based in villages, towns and cities; radical modification of local environments given over to agriculture; massive population growth; cultural developments (for example, writing and architecture); and the growth of political and bureaucratic hierarchies. Once people farmed the same area of land year after year

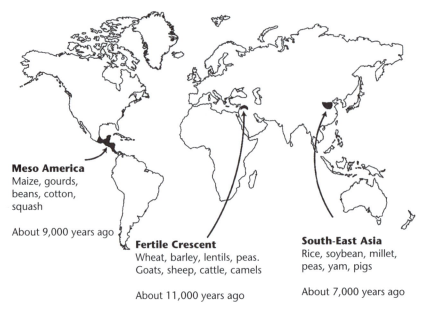

Meso America
Maize, gourds,
beans, cotton,
squash

About 9,000 years ago

Fertile Crescent
Wheat, barley, lentils, peas.
Goats, sheep, cattle, camels

About 11,000 years ago

South-East Asia
Rice, soybean, millet,
peas, yam, pigs

About 7,000 years ago

FIGURE 19.3 The domestication of plants and animals happened independently in several areas of the world.

there occurred a shift from communal to private ownership of land and resources. Land could be bought, sold and passed on, and it became important to know where your patch of land ended and your neighbour's began. Population growth, fuelled by periods of food abundance and high productivity, would have caused a demand for more land and resources, ultimately leading to conflict with neighbouring groups. Figure 19.2 illustrates some of these interactions.

This crucial period also saw a massive transformation in the type of food that we eat. Wild animals were tamed and domesticated to serve human needs (Table 19.1). Much later developments in the

Table 19.1 Animal domestication

Domesticate	Wild progenitor	Probable date and site of domestication	Reference
Dog (*Canis familiaris*)	Grey Wolf (*Canis lupus*)	11.5 KYA Levant (Modern Israel and Palestine)	Larson et al. (2010)
Cattle (*Bos Taurus* and *Bos indicus*)	Aurochs (*Bos primigenius*). Extinct	10.3–10.8 KYA Fertile crescent	Bovine Hapmap Consortium (2009)
Sheep (*Ovis aries*)	Mouflon (*Ovis orientalis*)	10–11 KYA Fertile crescent	Kijas et al. (2012)
Pig (*Sus scrofa*)	Wild boar (*Sus scrofa*)	9 KYA Near East	Larson et al. (2005) Larson et al. 2010
Chicken (*Gallus gallus domesticus*)	Red jungle fowl (*Gallus gallus*)	8 KYA China; 10 KYA Northern China	Sawai et al 2010 Xiang, H. et al. (2014)
Goat (*Capra hircus*)	Bezoar (*Capra aegagrus*)	11 KYA SE Anatolia	Driscoll et al. (2009)
KYA = thousands of years ago			
SOURCE: Jobling et al., Table 12.5, p. 400; Larson et al. (2010); Xiang et al. (2014) and Driscoll et al. (2009)			

selective breeding of plants and animals and the development of food processing technologies in the Agricultural and Industrial Revolutions of the 18th century served to accelerate what was begun by our Neolithic forebears. So much so, that most of the calories we now consume in developed Western countries come from food sources (dairy, cereals, refined sugars and processed vegetable oils) that would only have been a minor part (if any) of the pre-Neolithic diet of our ancestors (Cordain, 2007). Table 19.2 shows some post-Neolithic foodstuffs and the date of their introduction.

Table 19.2 Foodstuffs since the Neolithic revolution

Foodstuff	Date of introduction
NEOLITHIC FOODS	
Domesticated meat from mammals (sheep, goats, cows, pigs)	11,000–9,000 BP
Chickens	8,000 BP
Dairy products: milk, cheese, yoghurt	6,100–5,500 BP
Cereal grains: wheat, barley, rice, maize	10,000–9,000 BP
Wine	7,100–7,400 BP
Added salt	5,600–6,200 BP
Beer	6,000 BP
HISTORICAL AND INDUSTRIAL-ERA FOODS	
Distilled alcohol	800–1300
Potato introduced to Europe from S. America	1570
Refined sugar becomes widely available	c.1800
Invention of the hot dog	1867
Refined grains	1880
Coca-Cola	1886
Hydrogenated vegetable fats	1897
Kellogg's Corn Flakes	1906
Vegetable oils widely available	1910
Kentucky Fried Chicken	1952
High-fructose corn syrup	1970

BP = before present

SOURCE: Adapted from information in Cordain (2007); Civitello, L. (2011); and Kiple (2000).

19.2.2 Palaeolithic nutrition

A key question is whether the change from a Palaeolithic to a post-Neolithic diet has been for the better or worse. Up until the early 1980s, the conventional wisdom was that the advent of agriculture would have improved human nutrition. In 1977, for example, the American anthropologist Mark Cohen argued in his influential book *The Food Crisis in Prehistory* that restrictions on food supply at the end of the Pleistocene forced people to turn to agriculture and so accelerate the Neolithic revolution. The archaeological evidence, however, suggested otherwise: skeletal evidence of nutritional deficiencies appeared not before the Neolithic revolution but after it and so presumably as a consequence and not a cause. Today a whole raft of evidence points to the fact that the invention of farming brought about some deleterious consequences, including the spread of new infectious diseases (partly due to close contact with animals and partly due to increased human population density) and periodic bouts of famine due to a reliance on a restricted range of staple crops.

A landmark paper in the comparison of ancient and modern diets was that of Eaton and Konner published in 1985 entitled 'Palaeolithic nutrition'. These authors compared various parameters of ancestral and modern diets and found a mismatch which they suggested may underlie patterns of ill health in developed industrial societies (Eaton and Konner, 1985). In 2010 they returned to this theme to examine how closely their original ideas had been supported or otherwise by subsequent research. They reported that they felt vindicated by later findings, with recommended intake levels of various nutrients having moved closer to ancestral norms. Table 19.3 shows a selection of the sort of comparisons they made between Palaeolithic and modern diets, which has been updated using other sources.

It broad terms the modern Western diet consists of less protein, less fibre and more carbohydrate than the diet of Palaeolithic people. Energy from fat varies widely, but, as we will see later, the sort of fat consumed in modern diets is different from ancient diets. As well as the differences in these macronutrients (protein, carbohydrate and fats) it is also clear that we now obtain over half of our calories from foodstuffs (cereals, milk products, refined sugars

Table 19.3 Comparison of estimated Palaeolithic diet with the 'standard American diet' and recommended levels of intake.
Ranges are given for recommended levels for items where authorities differ.

Dietary component	Late Palaeolithic estimates	Modern US diet	Recommended levels (2010)
Energy from protein (% of diet)	20–35	15	10–35
Energy from carbohydrate (% of diet)	35–40	50	45–65
Energy from fat (% of diet)	20–35	33	25–35
Max. energy from saturated fat (% of diet)	7.5–12	11	< 10
Fibre (g/day)	> 70	15.2	25 female; 38 male
Vitamin C (mg/day)	500	87	75 female; 90 male
Refined sugar (g/day)	0	120	< 32
Sodium (g/day)	0.67	3.27	< 2.0
Potassium (g/day)	11.10	2.62	> 3.51
SOURCE: Compiled from various sources: Konner and Eaton (2010); Grotto and Zied (2010).			

and alcohol) that were simply unavailable to our Pleistocene ancestors. In many ways the modern diet is also restricted in its range of sources, relying on a few types of grain and sources of protein and fat. In contrast, the diet of the Aché (a group of hunter-gatherer people living in eastern Paraguay) includes 21 species of reptiles and amphibians, 78 species of mammal, over 150 species of bird, and a range of plant types (Kaplan et al., 2000).

19.2.3 Consequences of modern diets

Insulin resistance and glycaemic load

Glucose is the main fuel for most of our tissues and it is important that blood glucose levels are maintained at a steady level (homeostasis). After a meal, glucose is absorbed into the bloodstream and signals the release of insulin. Insulin drives the uptake of glucose by cells and especially by the liver where glucose is converted into glycogen. Insulin also stimulates cells to use glucose for energy use and to synthesise fats and proteins. Type II diabetes results when cells fail to respond to insulin adequately (that is, they show reduced insulin sensitivity) and so glucose levels start to rise leading to **hyperglycaemia**. The condition accounts for about 90 per cent of cases of diabetes in the USA, where it becomes manifest in people over about 40 years of age and is nearly always associated

with obesity. As tissues show reduced sensitivity to insulin so insulin levels have to rise to compensate; this condition is known as **hyperinsulinemia** and is one of the symptoms of type II diabetes.

The ability of a foodstuff to cause a rise in blood sugar levels is indicated by such concepts as the glycaemic index and glycaemic load. The **glycaemic index** (GI) is a concept that was introduced in 1981 to indicate how quickly carbohydrate breaks down during digestion and releases glucose into the bloodstream. By definition, glucose itself is set at 100 and the index measures the rise in blood glucose following the ingestion of 1 g of carbohydrate relative to the rise following the ingestion of 1 g of glucose. Low GI foods (< 55) include things like whole grains, beans and vegetables; high GI foods (> 70) include white bread and processed breakfast cereals. A more useful concept to assess the impact of diet on health is glycaemic load (GL = glycaemic index × carbohydrate content per serving), which takes into account the amount of food consumed (Table 19.4).

The unrefined foods that would have been available to our pre-Neolithic ancestors would have had low glycaemic indices and loads (Thorburn et al., 1987). In the typical Western diet about 30–40 per cent of the energy content now comes from high GL sugars and carbohydrates (Cordain, 2007). Most of these foods were hardly available to our pre-Neolithic ancestors and rarely consumed even as recently as 250 years ago.

Table 19.4 Glycaemic load of various modern foodstuffs.
Refined sugars and cereals would have been unavailable to our Pleistocene ancestors yet about 40 per cent of the energy in a typical Western diet now comes from these sources. Such food items have a large glycaemic load compared to fruit and vegetables and this may be implicated in insulin resistance.

Food type	Glycaemic load (= glycaemic index × carbohydrate content in 100 g)
Rice Krispies	72.0
Corn Flakes	70.1
Boiled white rice	50–61
Mars Bar	40.4
White bread	34.7
Baked potato	21.4
Banana	10-11
Carrots	4.7
Fruit (grapes, kiwi, pineapple, apple, pear, melon, orange)	5.0–11.9
SOURCE: Data taken from Cordain et al. (2005), Table 2, p. 346; and Holt et al. (1997).	

A diet with a high GL may induce hyperglycaemia and hyperinsulinemia (that is, high blood glucose and high insulin levels respectively) and in turn lead to the development of insulin resistance. Insulin resistance is in itself a serious condition and is implicated in a number of diseases – sometimes called the 'diseases of civilization' (Reaven, 1995). These include: obesity, heart disease, type II diabetes, **hypertension** and cholesterol problems such as elevated low-density lipoprotein cholesterol and reduced high-density lipoprotein cholesterol. It is significant that diseases of insulin resistance are rare or absent in contemporary hunter-gathering societies.

Dietary fat

Most fats obtained through diet and stored in the body are triacylglycerides, meaning that three fatty acid molecules (hence triacyl-) are joined to a single glycerol molecule by an ester bond. Dieticians distinguish between saturated fatty acids (SFA), monounsaturated fatty acids (MUFA), polyunsaturated fatty

acids (PUFA) and trans and cis fatty acids. Trans fats are only found in very low levels in some natural foods such as animal fats. But the hydrogenation (that is, saturation) of liquid vegetable oils to produce fats which are harder at room temperatures (to produce butter-like spreads) produces trans fats as a by-product. PUFA can also be divided into n-6PUFA (or omega 6) and n-3 PUFA (or omega 3) types according to the position of the carbon double bond.

The main source of dietary fat for a hunter-gatherer is meat. Among contemporary hunter-gatherers the meat content of the diet is very variable ranging from 20–30 per cent in the Gwi and !Kung San to 90–95 per cent in Eskimo and Nunamiut (Kaplan et al., 2000). But the meat consumed tends to be lower in **saturated fat**. The Western diet is now high in SFAs from fatty meats, cheese, milk, butter, margarine and hydrogenated fats in baked products (for example, pies and biscuits). Wild meat tends to contain higher levels of MUFAs and PUFAs than farmed meat (Figure 19.4). The ratios of n-6 and n-3 polyunsaturated fatty acids are also different. In a typical hunter-gatherer diet the n-6/n-3 ratio n has been estimated at between 2:1 and 3:1. In the contemporary Western diet, however, the n-6/n-3 ratio is approaching 10:1 (Cordain et al., 2002). This higher ratio comes about since meat from grain-fed cattle tends to be higher in n-6 PUFA and the fact that hydrogenated vegetable oils, which contribute considerably to dietary fat, are also high in n-6 PUFAs.

High intakes of SFAs and **trans fatty acids** increase the risk of cardiovascular diseases by elevating total and low-density lipoprotein cholesterol. There is also considerable evidence suggesting that the balance of n-6 and n-3 PUFAs is important in maintaining health, with a suggestion of a ratio of 6:1 as optimum – a figure lower than many contemporary diets (Wijendran and Hayes, 2004). In many countries various authorities have called for a strict limitation or ban on the use of trans fats in food (Brownell and Pomeranz, 2014).

Micronutrients, salts and fibre

The introduction of dairy food and cereals in the Neolithic revolution would have lowered the overall micronutrient density of the human diet. This is because such foods have lower concentrations of

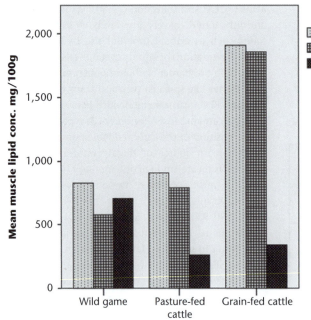

Fat type

- ▦ Saturated
- ▦ Monounsaturated
- ■ Polyunsaturated

FIGURE 19.4 Types of fat in three different kinds of mammalian muscle. Wild game is calculated as an average of elk, deer and antelope and represents the sort of meat eaten by our Pleistocene ancestors. Modern grain-fed cattle are high in saturated fat and low in polyunsaturated fat. Dietary saturated fat is a factor in the elevation of LDL serum cholesterol levels.

SOURCE: Data from Cordain et al. (2002), Table 7, p. 188.

essential vitamins and minerals compared to wild food staples such as game, seafood, fruit and vegetables. The situation worsened in the Industrial era with the processing of cereals to remove the germ and bran and then the introduction of calorie-rich but micronutrient-poor refined sugars and vegetable oils.

Sodium chloride is added to many foodstuffs and now accounts for about 90 per cent of the sodium found in the typical Western diet. In addition, the post-Neolithic diet effectively reduced the calories obtained from fruit and vegetables, substances high in potassium, and increased the calories from grains and milk products, substances low in potassium. As a consequence, the ratio of sodium ($Na+$) and

Potassium ($K+$) ions in the human diet has undergone a dramatic inversion since Palaeolithic times. (Table 19.5).

Both potassium and sodium are essential elements for human life but there is now abundant evidence that the inversion of Na/K ratios in the diet is likely to be implicated in a number of disorders such as hypertension, strokes, kidney stones and osteoporosis (see Cordain, 2007; and Frassetto et al., 2001).

In the UK people consume about 13 g of fibre per day and in the USA the figure is about 15.1 g/day (Cordain, 2007). Both levels are below the current recommended values of 20–30g per day (Lichtenstein, 2006). The modern diet is low in fibre since a good

Table 19.5 Sodium and potassium in the human diet.
Compared to our Palaeolithic ancestors we consume far more sodium and far less potassium in our diet.

Culture	Na+ (g/day)	K+ (g/day)	Na/K ratio
USA*	3.27	2.62	1.25
UK~	2.43	2.84	0.86
Palaeolithic diet	0.67	11.10	0.06
Guidelines for Americans	< 2.3	4.7	< 0.49
UK guidelines	1.6	3.5	0.46

*SOURCE: Cordain (2007); Guidelines for Americans from US Department of Health and Human Services (2005). *Dietary Guidelines for Americans, 2005.*

SOURCE: Frassetto et al. (2001); UK guidelines from the Office for National Statistics (www.gov.uk/government/uploads/system/uploads/attachment_data/file/384775/familyfood-method-rni-11dec14.pdf, accessed September 2015).

portion of our energy now comes from refined sugars, carbohydrates and dairy products which are themselves low in fibre, or from foods such as cereals which have had the fibre removed by mechanical processing. It has been estimated that hunter-gatherer diets, rich in tubers, vegetables and seeds, would have given a dietary intake of about 100g/day.

Diets low in fibre have been implicated in the worsening or causation of constipation, appendicitis, haemorrhoids and varicose veins. Moreover, studies

consistently show that diets high in fibre reduce blood pressure and serum cholesterol levels and are also associated with lower risk factors for heart disease, stroke, diabetes and obesity (see Anderson et al., 2009, for a review). Fibre also increases satiety and hence militates against over-eating. A variety of studies have also shown that high dietary fibre helps maintain intestinal health and that diets that are low in fibre are associated with colon cancer (Bingham et al., 2003).

Box 19.1 Vitamin D deficiency – a case study in genotype-environment mismatch

Rises in skin cancer have often been attributed to the mismatch between protection from UV radiation as determined by genetic skin tone and actual UV intensity – a mismatch resulting from the holiday travel of fair-skinned Caucasians to regions further south. But population migration can also work in the other direction with some people getting too little UV exposure resulting in the vitamin D deficiencies that are increasingly reported in the literature (Davies and Shaw, 2010). In the UK, for example, one study by Ford et al. (2006) based on an inner-city population in Birmingham found that one in eight white people, one in four African-Caribbean and one in three Asian adults showed signs of severe vitamin D deficiency at the end of a summer period. In another study, Das et al. (2006) looked at vitamin D deficiency in a sample of 14 white and 37 non-white healthy adolescent girls

living in or near the city of Manchester. They found that signs of vitamin D deficiency were significantly greater in the non-white sample. The data also showed that this was not likely due to differences in diet, but instead related to limited exposure to sunlight in the non-white population – a phenomenon itself probably related to cultural and religious beliefs that encourage concealing clothing and a restriction on outdoor activities (Table 19.6). The more highly pigmented skin in the non-white sample is also likely to have reduced vitamin D synthesis.

In a similar German study on vitamin D concentrations in indigenous and immigrant children and adolescents, Hintzpeter et al. (2008) found that the influence of diet on vitamin D levels was small compared to UV exposure. The researchers found that the groups most at risk were Turkish and Arab participants of both sexes

Table 19.6 Vitamin D intake and synthesis in two groups of adolescent girls. The reduced levels of vitamin D in the non-white group are likely to be due to limited sunlight exposure.

Cohort	Dietary vitamin D (µg/day)	Daily sun exposure (mins)	% body surface exposed	250HD value (nmol/l)*
White girls (n=14)	1.2	60	19	37.3
Non-white girls (n=37)	1.5	34	9	14.8
P value from Mann–Whitney U test	0.6	0.003	0.001	< 0.001

SOURCE: Data from Das, G., S. Crocombe, M. McGrath, J. L. Berry and M. Z. Mughal (2006). 'Hypovitaminosis D among healthy adolescent girls attending an inner city school.' *Archives of Disease in Childhood* 91(7), 569–72.

*25-hydroxychole califerol: levels indicative of vitamin D concentrations; a value below 30nmol/l is thought to indicate deficiencies in children.

and especially girls of Asian and African origin. The degree of veiling was not directly assessed but the researchers found a positive association between vitamin D levels and the degree of cultural integration, suggesting traditional costume may be a factor in maintaining low levels of vitamin D. Both studies point to the need for culturally sensitive public health measures to address this problem such as the use of vitamin D dietary supplements.

19.2.4 Diet and health among contemporary hunter-gatherers

If modern humans are not genetically adapted to a modern diet and this evolutionary discordance or genome lag is a significant factor in the causation of some modern diseases and patterns of ill health, then we might expect contemporary hunter-gatherers, eating a diet similar to that of our Pleistocene ancestors, to be less beset by these purportedly diet-related diseases. There is now quite a large body of evidence that suggests hunter-gatherers are indeed healthier, using some measures, than their Western counterparts (see Carrera-Bastos et al. (2011) for a review). Some of these dimensions are considered below.

Blood pressure (hypertension)

For cultures so far measured, male and female hunter-gatherers tend to have lower blood pressure means than typical Western populations. The Yanomamo people of the Amazon rainforest, who practise hunting, fishing and horticulture, have been extensively studied by anthropologists and provide an interesting comparison group (see Table 19.7). Blood pressure is indicated by two readings: systolic and diastolic, corresponding to the maximum and minimum pressure achieved during each heart beat cycle. It is usually measured in the brachial artery of the upper arm.

Moreover, in Western countries blood pressure tends to increase with age but this is hardly observed in the Yanomamo (Figure 19.5).

In the West, blood pressure readings between 120/80 and 139/89 mmHg are now considered prehypertensive and 140/90 or above as an indication of hypertension. In the USA, hypertension is alarmingly common and in 2006 affected 34 per cent of the adult population (Goldstein et al., 2011). In England the situation is probably worse: a recent survey found about 66 per cent of the adult population had blood pressure that was either prehypertensive or hypertensive (Joffres et al., 2013). Interestingly, the guidelines commonly issued by various health authorities to reduce hypertension seem to be a move towards a Palaeolithic lifestyle. For example:

- Maintain a normal body mass index (BMI) (notably Western BMI tends to be higher than those of hunter-gatherers – see above)
- Increase potassium intake and reduce sodium intake (see Section 19.2.3 above)
- Take more exercise
- Reduce alcohol intake
- Increase consumption of fruit and vegetables. The modern Western diet tends to contain fewer calories from fruit and vegetables than most hunter-gatherer diets.

Table 19.7 Comparison of blood pressure of UK adults with a hunter-gathering culture

Group	Systolic pressure (mmHg)		Diastolic pressure (mmHg)	
	Males	Females	Males	Females
Yanomamo	104	102	65	63
UK adults	130	122	73	68
Recommended	< 120	< 80		
SOURCE: UK data from Ruston et al. (2004); for Yanomamo: Carrera-Bastos et al. (2011).				

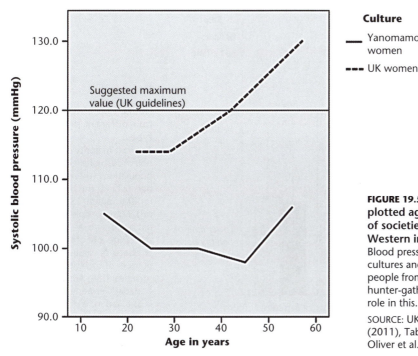

FIGURE 19.5 Systolic blood pressure plotted against age for two types of societies: hunter-gathering and Western industrialised.
Blood pressure rises with age in Western cultures and is generally higher in people from Western cultures than in hunter-gatherers. Diet may play a prime role in this.

SOURCE: UK data from Ruston et al. (2011), Table 3.1; Yanomamo data from Oliver et al. (1975).

Insulin sensitivity

There is evidence that people living in more traditional cultures maintain better insulin sensitivity as they age compared to Western populations. In several studies on the people of the island of Kitava, Papua New Guinea, Lindeberg et al. (1997, 1999) compared various health parameters with those of adult Swedes. He found that BMI values were generally lower for Kitavans than Swedes and, moreover, tended to decrease with age over 30 years for Kitavans whereas in Sweden BMI tended to increase during and beyond middle age. They also noted lower fasting plasma insulin concentrations among Kitavans than Swedes – a measure conventionally interpreted as greater sensitivity to insulin and hence a reduced risk of diabetes. Significantly, there is a virtual absence of stroke and heart disease in Kitava. The Kitavan diet is typical of the pattern noted earlier for hunter-gatherers (with the exception of fairly high intake of saturated fat due to the presence of coconut in the diet); that is: low salt and total fat, high ratio of n-3/n-6 fatty acids, high nutrient density and low glycaemic load.

Wolfgang Kopp (2006) reviewed a number of medical studies and suggested that the modern high-carbohydrate diet is mismatched with our evolved metabolism and causes reduced insulin sensitivity, raised insulin levels and in turn atherosclerosis (thickening of the walls of the arteries). More generally, the so-called diseases of Western lifestyle – type II diabetes, obesity, cardiovascular disease and cancers – are rare in hunter-gathering cultures compared to Western populations (see Carrera-Bastos et al., 2011). Hunter-gatherers also generally have lower BMI levels than Westerners (Lindeberg, 2010).

Energy expenditure, diet and BMI

The lifestyle of people living in modern industrial cultures involves much less physical labour and energy expenditure than that found in surviving subsistence-level cultures. This is probably reflected in BMI values – a widely used method of indicating nutritional status defined as: BMI = weight (kg)/ height (m)^2.

A brief survey of a number of different types of population reveals that in industrial societies both

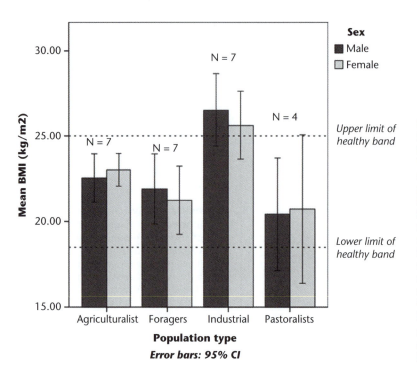

FIGURE 19.6 BMI figures for four different population types.
The vast majority of traditional, non-industrial societies have populations where most members fall into the healthy range for BMI values (18.5–25 kg/m²). Many people in industrial nations, however, lie above this range and are overweight or obese. N values refer to number of societies. Industrial societies: USA, Australia, Canada, Germany, Sweden, Russia, Japan.

SOURCE: Data from Katzmarzyk and Leonard (1998); and Leonard (2007).

males and females tend to have higher BMI values than subsistence-level populations. As shown in Figure 19.6 most subsistence-level populations have BMI values in the healthy range, whilst many people in industrial cultures have BMI values that place them in the overweight (BMI > 25 kg/m²) category. Another data set now suggests that about 30 per cent of all males and females in the USA are now obese (that is, BMI > 30 kg/m²) (Wang et al., 2011).

19.2.5 Some doubts and qualifications

An obvious objection to these claims about the health of people consuming a hunter-gathering diet is that the life expectancy of hunter-gatherers is lower than that of Westerners. In the case of the Kitawans mentioned above, for example, it is about 45 years of age, whereas in Europe and the USA it is typically between 70 and 80 years. But the problem here is that life expectancy is actually a rather misleading figure since it refers to the expected number of added years of life for a newborn child. As Ryan and Jethá (2010) point out, imagining our Paleolithic

ancestors with a life expectancy of 45 years is akin to saying they had a height expectancy of three feet. It does not mean that a typical adult died aged 45 or that a typical adult was only three feet tall. Today's average current worldwide life expectancy stands at about 66 years of age, with extremes by country from 39 in Zambia to 82 in Japan. In these terms, hunter-gatherer life expectancy is much shorter and usually between 21–37 years. But this does not mean that most hunter-gatherers will die before they reach the age of 37. This figure is the average expected added years of life after birth and as such it is heavily influenced by high infant and childhood mortality – neither of which are strongly linked, if at all, to diet. Modern medicine has drastically reduced infant and childhood mortality (and hence elevated life expectancy) to such an extent that infant mortality among hunter-gatherers is over 30 times that of nations like the UK and the USA. A better comparison might be, therefore, the mode: the age at which most people die. For hunter-gatherers this is usually between 68 and 78 (Gurven and Kaplan, 2007), and for developed Western nations in 2008 it stood at an average of 77 years for men and 82 years for women (Canudas-Romo, 2008). Essentially, Western economic development and medicine has extended lifespan through better sanitation, social stability, vaccination, antibiotics and health care. Life expectancy in hunter-gathering cultures is reduced by high childhood mortality, infections, and trauma due to accidents and violence.

But the idea that we are mismatched to post-Neolithic diets and well matched to Palaeolithic ones still has its sceptics. Is 11,000 years since the advent of agriculture insufficient time for humans to adapt to new diets? This of course depends on the intensity of the selection pressures and the degree of developmental plasticity and phenotypic variability that we have in relation to food intake. We have noted elsewhere (see Chapter 20) how adaptations to digest lactose appeared a few thousand years after the origins of animal husbandry. Similarly, relatively recent increases in the number of copies of alleles for salivary amylase (see Chapter 20) also seem to have occurred in some populations as a result of selection pressures arising from increased starch consumption in local diets. But it could be argued that such adaptations would have been relatively easy to acquire: the enzyme lactase needed to digest lactose is already produced by infant humans and it needed only a minor change (one base pair in fact) to switch off the lactase inhibitor. Similarly, the salivary amylase allele was already present and ready to be multiplied if needed.

On the other hand, if substantial genetic variation had been brought about by the rise of agriculture then one would expect to find taxonomic genetic differences between people whose ancestors have been farmers for thousands of years, such as the Chinese, and people whose ancestors have been hunter-gatherers for millennia, such as Australian Aborigines and the Ache tribes of Paraguay. Yet apart from a few genes related to infectious diseases, and the variations in amylase gene number and lactose tolerance noted above, there are no real major differences: humans are a remarkably homogenous species with relatively little genetic variance.

In support of the dietary discordance hypothesis it is often noted that humans have spent about 140,000 years, or 6,000–7,000 generations, as hunter-gatherers (and more before that if we include earlier hominin species such as *Homo erectus*), but only

about 11,000 years, or 400–500 generations, as post-Neolithic agriculturalists, and 11,000 years is insufficient to alter the adaptations that took place over the previous 140,000 years. But this comparison assumes that pre-Neolithic humans did become tightly adapted to a particular diet. The variability in conditions during the Palaeolithic period (especially as influenced by known climate fluctuations) and the wide geographical range occupied by humans (notably of course after some moved out of Africa) might suggest that we have evolved to cope with a whole range of diets, so long as they are energy-rich enough to support expensive brain tissue. In addition, if early Neolithic agriculturalists did hit upon a diet and a lifestyle so at odds with their Palaeolithic genes then this could be expected to have exerted a selection pressure to change and adapt (as it did with lactose tolerance).

This reservation is further illustrated by the fact that, that in population terms, the Neolithic revolution was a great success. After its origins in the Middle East about 10,000 years ago, its radial advance was about 1 km per year (Menozzi et al., 1978; Cavalli-Sforza, 1993). Moreover, the evidence suggests that this advance was through 'demic diffusion': agriculturalists outbred the hunter-gatherers (Sokal et al., 1991). In their survey of ill health since the Neolithic revolution Strassmann and Dunbar (1999) concluded that:

> Our analysis of the transition to agriculture uncovered no empirical evidence that it was a watershed between adaptation and maladaptation. (p. 101)

It could be countered that this suggests higher fertility and not necessarily health in the immediate post-Neolithic period. Nevertheless, proponents of the genome lag hypothesis do need to address how maladapted humans reproduced so well.

Stacked against this point, however, it is notable that many of the diseases that have been attributed to modern Western diets, such as diabetes type II, cardiovascular disorders (CVDs) and obesity, are typically late-onset conditions. Consequently, elevated mortality from, say, CVD after reproduction will exert relatively little selection pressure.

The Palaeolithic diet hypothesis seems to attract most critics when it is proposed that contemporary humans should adopt such a diet to improve their health. Advice on eating is one area where there is no shortage of fads, fashions and bandwagons. Predictably then, there are numerous popular books advocating the benefits of a Stone Age diet such as *Neander Thin: Eat Like a Caveman to Achieve a Lean, Strong, Healthy Body* (Audette, 2000) or *Metabolic Man: Ten Thousand Years from Eden* (Wharton, 2001). All this might be a step too far when uncertainties remain about what exactly our ancestors did eat and the rigidity of our adaptations to specific diets. The problem is, of course, that the current global population of over 6 billion people is only sustainable through the production and consumption of cereals. So even if the modern Western diet is suboptimal compared to that of our distant ancestors in the Old Stone Age, most of us had better get used to it.

Nevertheless, even the critics tend to agree that the modern Western diet, for whatever reason, is a substantial cause of ill health. Katherine Milton (2002), for example, argues that it is not so much the macronutrient proportions of the Palaeolithic diet that we should try to emulate but the fact that modern farming techniques and food processing technology have given us energy-dense foods too rich and with too many calories for modern sedentary lifestyles. Even this is a sort of concession to the evolutionary approach: energy inputs and physical activity are out of alignment and our bodies respond in a dysfunctional way.

19.3 Case study II: cancer in evolutionary perspective

On the origins of cancer

The origins of cancer can be traced back over a billion years to emergence of the first multicellular organisms. When organisms contain many cells two problems arise. The first is the problem of cell differentiation: how from a single fertile cell many cells can be formed with numerous different functions. The second is that cells now have to coordinate their reproductive tendencies – each cell can no longer be reckless and selfishly divide as in the case of single-celled organisms. These problems were solved by restricting cell division to specific types of cell called

stem cells and progenitor cells. Progenitor cells have only a limited number of cell divisions before they die. Stem cells, however, are essentially immortal and are a necessary source of spare cells that can aid the maintenance and repair of various tissues. This very flexibility, however, can also lead them to pose a risk of forming tumours later in life (Werbowetski-Ogilvie et al., 2012).

As a species humans are especially prone to develop cancers, with a lifetime risk of about one in three. This is considerably higher than other animals – even long-lived ones such as elephants and the great apes (although, intriguingly, domesticated species show an elevated cancer risk compared to similar wild counterparts). In recent decades, epidemiologists, oncologists and geneticists have been successful at identifying some of the proximate causes of cancer such as chemical carcinogens, ionising radiation and susceptible genes. But questions about ultimate causation, and why we should be so vulnerable, are probably best tackled using evolutionary biology.

The rapidity of evolution at the cellular level

In Chapter 18 we described how rapidly dividing and mutating microorganisms pose a special and never-ending health problem for humans. One relatively recent and important view of cancer is that it bears some similarities to infection by microorganisms, but in the case of cancer we see clonal evolution among the host's own cells. A seminal paper in this field (one of the acknowledged milestones in cancer research) came from Peter Nowell in 1976 in which he described cancer as an evolutionary process and successfully explained or predicted the sequences of clonal expansion, individual variation in response to medical variation and the evolution of therapeutic resistance (Nowell, 1976). Several decades of research have now broadly supported this idea.

One ubiquitous characteristic of neoplasms, that now makes sense in this light, is their genetic instability and variability. It is this very feature that allows the emergence of somatic cancer cells that are fitter than their neighbours (that is, normal cells in the tissue concerned) in that they rapidly proliferate and leave progeny. This insight is also consistent with the well-known clinical finding that early detection and intervention is important – since at this early stage the degree of heterogeneity, and hence the possibility of clones resistant to treatment, are both lower. Paralleling the process of the emergence of resistant microorganisms, resistance to cancer therapy has been observed among mutant clones (Merlo et al., 2006). One clinical consequence of this is that each patient's cancer might require individual therapy. The merits of the evolutionary approach to understanding cancer are gradually being recognised. Merlo et al. (2006), for example, suggest that 'evolutionary biology should be a required part of the training of cancer biologists and oncologists' (p. 933).

Constraints and compromises

It is estimated that about 1 per cent of our coding genes (that is, > 350) if suitably mutated could contribute to the emergence of cancerous cells. Now every round of cell proliferation runs the risk of an error in DNA copying, either through inherent errors in DNA-processing enzymes or as a result of the influence of mutagens. The mutation rate for single base changes has been estimated between 1.8×10^{-8} and 2.5×10^{-8} per base per generation for somatic cells (Nachman and Crowell, 2000; Kondrashov, 2003).

Natural selection has provided cells with proofreading and corrective mechanisms to prevent such errors but the process is clearly not perfect. One way of conceptualising this phenomenon is that natural selection has 'set' the mutation rate as a compromise between costs and benefits. The benefits of having a non-zero mutation rate (that is, tolerating mutations) are twofold: firstly, mutations are the raw material for evolutionary change and hence potentially useful adaptations in offspring downstream; secondly, allowing some mutations through reduces the metabolic costs needed for high-fidelity copying. Sniegowski et al. (2000) argue that the latter benefit is likely to be more important since sexual recombination reduces the effectiveness of mutation-rate modifiers. This is shown diagrammatically in Figure 19.7.

Recent work on stem cells illustrates both the problem of compromise and the rapidity of cellular evolution. In general terms, the human body has a difficult problem to solve: if a cell line continues to divide to produce new cells lost by damage then this increases the risk of cancer-inducing mutations from copying errors. One solution might be cellular

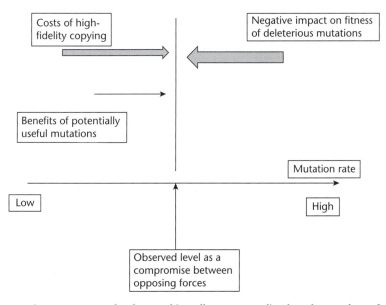

FIGURE 19.7 The mutation rate currently observed in cells conceptualised as the product of opposing forces.
Avoiding mutations imposes metabolic costs and hence the pressure to allow them through (large arrow pointing right). In addition there may be some benefit from mutations supplying variation for future adaptations (small arrow pointing right). Both of these are counterbalanced by the negative fitness impact of many mutations (large arrow to left).
SOURCE: Based on an idea in Sniegowski et al. (2000), Figure 1, p. 1056.

senescence – the programmed death of cells after so many divisions. But this of course could (and does) lead to organismal senescence and losses in fitness if ageing is premature. A complementary solution might be cell differentiation: to allow so many cell divisions to take place before cells become fully differentiated (that is, performing specific functions in the body) and non-dividing. This latter solution requires a reservoir of stem cells: these are undifferentiated cells found especially in bone marrow that act as a source of cells for repair and replenishment purposes. But the very existence of this reservoir of self-replicating cells is a source of vulnerability to mutation. It has been argued that most malignant cancers originate from normal stem cell populations (Beachy et al., 2004).

If rapidly dividing stem cells are vulnerable to mutations then we might expect that the risk of contracting a specific cancer should correlate strongly with the number of divisions that stem cells make in the specific types of tissues in which cancer can occur. This very relationship was investigated by Tomasetti and Vogelstein (2015). They plotted the lifetime risk of specific cancer types against the total lifetime number of stem cell divisions (obtained from number of stem cells × division rate per year × 80 years) in associated tissues (Figure 19.8). The correlation coefficient (r) was 0.804 which gives an r^2 value of 0.65. One tentative suggestion then is that about 65 per cent of these types of cancers can be attributed to probabilistic mutations occurring during

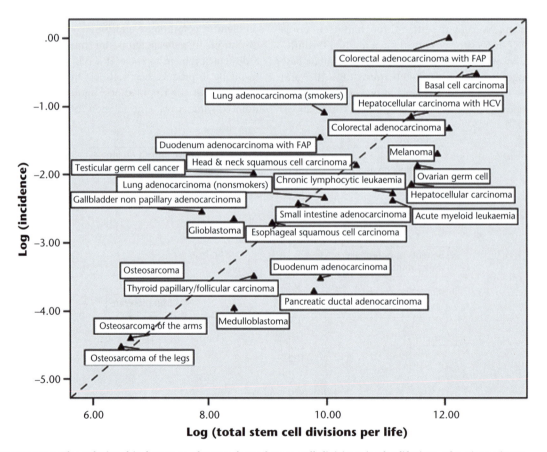

FIGURE 19.8 The relationship between the number of stem cell divisions in the lifetime of a given tissue and the lifetime risk of cancer in that tissue.
The line of best fit for this plot is log (y) = 0.54 log (x) -7.747. This suggests a relationship of R = 1.8 × 10^{-8}(N)$^{0.54}$, where R = lifetime risk, N = number of cell divisions.
SOURCE: Plotted from data in Tomasetti and Vogelstein (2015), Table S1.

cell division. The other 35 per cent are presumably due to genetic predispositions and lifestyle factors. Interestingly, the incidence for lung cancer among smokers is higher than expected from the general trend (Figure 19.8), confirming that the habit of smoking adds considerably to the baseline probability of lung cancer. Cancers such as those of the islets of the pancreas and small intestine lie towards the lower part of the cluster and it is thought these are relatively immune from environmental influences and their incidence is probabilistic. Cancers of the lung (smokers), colorectal regions, and basal (that is, skin) cells lie in the upper half of the diagram, suggesting that their incidence is higher than would be expected from random mutations occurring during normal stem cell division and hence pointing to some external influences such as smoking, diet and UV radiation respectively.

19.4 Case study III: mental disorders

Like other diseases, many mental disorders, such as autism, schizophrenia and various depressive disorders, seem to run in families. One way of conceptualising this phenomenon is through the concept of relative risk. A commonly used version of this is $Lambda_S$ which is defined as:

$$Lambda_S = \frac{\text{Risk to an individual if his/her brother or sister is affected}}{\text{Risk or incidence in general population}}$$

Table 19.8 shows some $Lambda_S$ values for a variety of disorders.

Diabetes, although obviously not a mental disorder, is included to illustrate the fact that such studies can help to establish if the condition has a heritable component. In this case, we note that type I diabetes is more strongly inherited than type II.

We should treat these statistics with caution. The fact that you are ten times more likely than average to develop schizophrenia if you have a brother or sister with the condition does not unequivocally point to a genetic foundation: it could be that you are more likely because you have been exposed to similar conditions to your brother or sister. Another

Table 19.8 Approximate risk ratios for various disorders

Disorder	Approximate risk ratio (Lambda$_S$)
Autism	>75
Type I diabetes	20
Schizophrenia	10
Bipolar disorder	10
Panic disorder	5–10
Type II diabetes	3.5
Phobic disorders	3
Major depressive disorder	2–3

SOURCE: Data from Smoller, J. W. and M. T. Tsuang (1998). 'Panic and phobic anxiety: defining phenotypes for genetic studies.' *American Journal of Psychiatry* **155**(9): 1152–62; and Mäki, P., J. Veijola, P. B. Jones, G. K. Murray, H. Koponen, P. Tienari, P … and M. Isohanni (2005). 'Predictors of schizophrenia – a review.' *British Medical Bulletin* **73**(1), 1–15.

alternative is that you carry genes that dispose you to react in certain ways to environmental stimuli. The undoubted importance of genetic factors, however, is further supported by adoption studies. One approach is to look at the children of schizophrenic parents who were given up for adoption. Tienari (1991) examined the children of 155 schizophrenic mothers who were given up for adoption compared with 155 adopted children of parents who were not schizophrenic. He found that the rate of schizophrenia was 10.3 per cent for those with schizophrenic mothers compared to 1.1 per cent of those with non-schizophrenic mothers.

19.4.1 The heritability of mental illness

Studies on the incidence of mental illness in families, such as concordance rates between monozygotic and dizygotic twins, and between other siblings, enable heritability (see Chapter 3) estimates of different mental disorders to be made. High heritability values (that is, approaching 1) suggest that the difference between an individual with the disease and one without is primarily due to genetic differences between them.

It is here then that psychiatric genetics faces its real puzzle: the heritability of many well-known psychiatric disorders is high, yet the age of their onset is often such as to considerably reduce fertility. Hence we face the problem identified earlier in this chapter of how to explain how apparently harmful alleles persist in the gene pool. Table 19.9 shows some estimates from a recent review of various studies on some common mental disorders.

The data in Table 19.9 shows that mental illness is associated with higher mortality and lower fertility than average yet in some cases with an apparent genetic component. The idea that disorders such as **schizophrenia** and depression are adaptations faces some difficulty with the data. Schizophrenics tend to have fewer children than average (Avila et al., 2001). It now seems that searching for enhanced fertility among relatives with such conditions as bipolar disorder and schizophrenia has proved fruitless (see Haukka et al., 2003). Moreover, the offspring of people with mental illness tend themselves to have higher morbidities and lower fertility (Webb et al., 2006). There is also evidence that depression does not bring about help from others, as might be proposed in an adaptive-style explanation, but can damage relationships (Reich, 2003).

The central problem for evolutionary genetic psychiatry, therefore, is explaining the high frequency, high heritability and damaging effects on fitness of many mental disorders. The problem is that over time natural selection gradually increases the frequency of high fitness alleles. It is to be expected, therefore, that alleles at most genetic loci will have gone to fixation – that is, nearly 100 per cent

prevalence of a single type. This was one of the central canons of Fisher's work on natural selection in the 1930s: genetic variation in fitness-related traits should be low, and the stronger the selection the less the heritable variation that should be observed. This collection of fixed alleles comprises the species-typical human genome and its development in different environments is what we mean by human nature. In the light of this reasoning it is difficult to argue that genes underlying mental disorders carry some counterbalancing advantage since over time they should appear in all people. If we argue, for example, that anorexia nervosa is a genetically influenced condition maintained in the gene pool because it helped our ancestors survive periods of famine, then we also need to explain why this benefit has not driven it to fixation in the population and why the heritability of the condition shows that there is still much genetic variation in its incidence. If the argument were correct then we would all be prone to anorexia nervosa and the heritability would be near zero.

The relatively small number of genes (about 25,000) possessed by humans suggests that different traits rely upon overlapping suites of genes. Hence pleiotropy, where one gene affects more than one trait, must be exceedingly common or even the norm. It is also likely that antagonistic pleiotropy, where single alleles have a positive effect on the fitness of one trait but a negative effect on the fitness of another, is also a common phenomenon. But over

Table 19.9 **Some epidemiological statistics relating to several types of mental illness.** Mortality ratios and fertility data show the impact of the illness on direct reproductive fitness; heritability indicates the degree to which genetic factors explain the appearance of the disease; paternal age effect probably reflects the effect of new mutations in sperm influencing the occurrence of the disease.

Illness	Prevalence in population (%)	Median age of onset	Mortality ratio (compared to general population)	Fertility (1 = general population)	Heritability	Paternal age effect (risk ratio for 10-year increase in father's age over 30)
Autism	0.3	1	2.0	0.05	0.90	1.4
Anorexia nervosa	0.6	15	6.2	0.33	0.56	-
Schizophrenia	0.70	22	2.6	0.40	0.81	1.4
Bipolar affective disorder	1.25	25	2.0	0.65	0.85	1.2
Unipolar depression	10.22	32	1.8	0.90	0.37	1
SOURCE: Data extracted from a review by Uher (2009); and Cross-Disorder Group of the Psychiatric Genomics Consortium (2013).						

time it is also to be expected that alleles with the greatest overall fitness effects (positive minus negative) will become fixed as the norm. Again this argument from pleiotropy implies a low heritability for disorders and so faces the similar problem of explaining their high heritability.

Against this search for compensating advantages of mental disorders, we must consider the implications of the evidence from identifiable genetic disorders and traumas. Major genetic disorders such as trisomy (thee copies of a chromosome) increase the incidence of symptoms typical of autism, schizophrenia and bipolar disorders. Similarly, brain injuries increase the risk of such conditions as schizophrenia, anxiety and depression. If these disorders were adaptations in their own right, supported by natural selection, then this would not be expected. The reason is that adaptations tend to be products of complex finely tuned systems and damage and injury would be expected to reduce their expression (imagine randomly poking the inside of a computer with a screwdriver – it is unlikely to make it work any better).

It also remains to be seen if heterozygotic advantage can ever explain more than a handful of diseases. If the phenomenon were to be widespread and of considerable antiquity then selection would favour events that reduce the costs of the homozygote. This could be achieved, for example, by chromosome crossover events that place allele A and its recessive a on the same chromosome arm so that they can be passed on together. It is possible then that what we now observe as heterozygote advantage is a transient phenomenon and current examples have only evolved recently. The difficulty of finding satisfactory adaptive-style explanations has led to an increased focus on the role of mutations.

19.4.2 Mutations and polygenic mutation selection balance

Mutations that arise *de novo* in parental germ line cells in functional genes will nearly always be harmful and subject to negative selection. The maths of the process is reasonably simple.

Let m = mutation rate (frequency per gene per person per generation)

Let p = the population frequency of mutant alleles

Let s = selection coefficient (if $s = 0$ then the mutation has no fitness consequences; $s = 1$ implies that the mutation is fatal and no offspring left. An s value of say 0.3 means that there is a 30 per cent reduction in fitness of the modified genotype).

So mutations constantly arise spontaneously and are either selected for or against, but even if they are selected against, a pool of mutants will exist at any time awaiting elimination. The result is an equilibrium frequency p of mutant alleles that have arisen but not yet been removed. It can be shown that (Keller and Miller, 2006):

$p = m/s$ if the mutation is dominant and
$p = \sqrt{m/s}$ if the mutation is recessive.

The cumulative frequency of all Mendelian disorders is quite high (relative to inherited disorders) at about 2 per cent but this is because a high number of genes (25,000 plus) are subject to mutations. If an individual and heritable Mendelian disorder is rare (for example, < 1/5,000) and explained by **mutation selection balance** (that is, $p = m/s$) then there is no paradox and we need not look for any adaptive explanation or compensating benefit: the illness is due to a balance between mutation rate and its removal by natural selection. This process explains why many Mendelian disorders (that is, disorders due to a single gene) are very rare: the mutation rate is low anyway and new damaging mutations are quickly eliminated from the gene pool. Hence, the crucial question for Darwinian approaches, as Keller and Miller (2006) argue, is whether the disorder is more common than would be expected from single-gene mutation–selection balance. Compared to simple Mendelian disorders, some mental disorders are hundreds of times more common (Table 19.10). The common nature of these diseases (compared, that is, to many Mendelian disorders), their damaging effects on fitness and their high heritability are the central puzzle of psychiatric genetics.

All in all balancing selection seems to be an unlikely cause of common mental disorders such as schizophrenia and bipolar disorder: the cost to the individual is so large that one would expect any

Table 19.10 Comparison of some Mendelian disorders with some common mental disorders.
The prevalence of two well-known Mendelian disorders is contrasted with two well-known psychiatric disorders. The low prevalence of the Mendelian disorder is explicable by a simple mutation–balance model. But the prevalence of some mental disorders is many times greater and so is a puzzle for evolutionary genetics.

Disorder type	Genetic basis	% prevalence in USA
Mendelian disorder		
Achondroplastic dwarfism	Dominant mutation on chromosome 4	0.002–0.003
Apert's syndrome	Dominant mutation on chromosome 10	< 0.001
Common mental disorder		
Schizophrenia	Unknown but heritability approx. 0.8	0.7–1.0
Bipolar disorder	Unknown but heritability approx. 0.6–0.85	0.8–1.25

counterbalancing advantages to be noticeable, yet the empirical evidence for compensating effects even in relatives is weak. An alternative theory, however, has been proposed by Keller and Miller (2006) called 'polygenic mutation–selection balance'. This model proposes that mental disorders are caused by clusters of many rare and recent mutations. They estimate that the average person carries about 500 mutations and many of these will affect brain function. Each malfunctioning allele may, by itself, not carry any counterbalancing advantage but be only weakly selected against. In other words, if mental disorders are maladaptive they always have been and there is no plus side of the equation. The net result will be a continuum in the appearance of a disorder and not a typical Mendelian pattern of discrete types. As the authors argue, mental disorders may not be a unitary category: they reflect a mixture of diagnostic convenience, historical and societal traditions, biases in perception by others (the desire to add labels) and finally real mixtures of disorder susceptibility alleles. In this view, mental disorders are not discrete 'natural kinds' but fuzzy umbrella concepts capturing a range of phenotypes that share some similarities.

Currently, there is a technical debate amongst geneticists as to whether the underlying genetic factors responsible for schizophrenia are rare or common variants, or even some combination. In a massive study of 8,008 cases of schizophrenia and 19,177 controls the International Schizophrenia Consortium concluded that schizophrenia and bipolar disorder are polygenic in nature involving thousands of common alleles, each of very small effect

(Purcell et al., 2009). Mostly these alleles will be SNPs, but there may also be a role for rare variants with high effects as part of the whole polygenic mix.

Another piece of evidence in favour of the polygenic model is the high level of co-morbidity observed between pairs of conditions such as bipolar disorder and schizophrenia, and autism and unipolar depression. Just such associations would be expected if each disorder is the result of many overlapping allele defects, but are harder to explain by other models.

19.4.3 New mutations and the effect of paternal age

Table 19.9 shows the effect of paternal age on the incidence of a genetically influenced disease and this figure provides a valuable perspective on whether the genetic variants (or mutations) involved in the disease are of old or recent origin. This in itself will then have some bearing on the evolutionary significance of those genes. If the genes are of ancient origin then they must have some resistance against the de-selective effects of natural selection and therefore might be expected to carry some counterbalancing advantage, as described by antagonistic pleiotropy and heterozygote advantage models discussed earlier. If the variants turn out to be of recent origin then it is possible that they are just part of the constant flow of *de novo* mutations that slip into the human gene pool. Eventually the rate of elimination of these undesirable alleles will be the same as their rate of formation, leaving a steady burden of moderately harmful mutations in the human genome. So how might paternal age effects help decide if mutations are ancient or recent? The answer lies in the rate of mutation that accompanies cell division. After puberty, female germ cells no longer divide and a women is left with her store of about 500 eggs. Male germ cells, in contrast, continue to divide and undergo about 23 divisions each year producing somewhere between 1,000 and 3,000 sperm per second. Each division carries a risk of a mutation appearing and it has been shown that the risk of mutations also increases with the age of the father (Crow, 2003, 2006). It follows that we might expect the frequency of incidence of genetic disorders that are due to new mutations to rise with the age of the father – hence the paternal age ratios shown in Table 19.9.

This effect was, in fact, first noticed as long ago as 1912 when the German physician and obstetrician Wilhelm Weinberg (of Hardy–Weinberg fame) noted that children born with achondroplasia (short-limbed dwarfism) tended to be born to older fathers and mothers. He even suggested that a new mutation might be responsible – remarkable foresight given the immature state of genetic theory at the time. In 1955 Lionel Penrose showed that it was the age of the father that was correlated with this condition (see Harris, 1974). The phenomenon of the accumulation of mutations in sperm is now well understood and accepted. It substantiates, for example, the requirement in many countries that male sperm donors should be under 41 years of age.

The susceptibility of sperm to mutations also explains why mitochondria are only passed down the female line (see Figure 5.8). A female's eggs are formed once at birth and so are not subject to the risks associated with constant reproduction. Mitochondria carry 13 genes but these are not subject to the cleaning effects of sexual reproduction such that mutations can be masked or eliminated. It follows that mitochondria are especially susceptible to the risks of mutations, which is probably why they only keep 13 'rump' genes – the rest having been incorporated into the nucleus. The constant replication of these genes in the germ cells forming sperm would pose an unacceptable risk of passing on mutations in the mitochondrial genome and so male mitochondrial genes are deleted in sperm and not passed on to the fertilised egg (Lane, 2015).

So, returning to Table 19.9 we can see strong paternal age effects for autism and schizophrenia pointing to newly arising mutations as a possible factor. Unipolar depression, however, is likely to have a different aetiology since no paternal age effects have been recorded. The number of alleles involved is also important. If, for example, psychiatric disorders are caused by many newly arising defective alleles each with a small effect, then each allele on its own will face only weak selection and so their persistence need not be explained by any positive effect. Sipos et al. (2004) estimated that probably between 15 and 25 per cent of all cases of schizophrenia are the result of this paternal age effect.

The effect of paternal age on the risk of various mental disorders reported in the review by Uher (2009) and summarised in Table 19.9 is also supported by more recent work of Kong et al. (2012) on mutations observed in 78 Icelandic families. This research group focused on trios of father, mother and child and the rate of arrival of new point (SNP) mutations in the child. This mammoth effort involved determining the complete genome sequences of 219 individuals. The group found a total of 4,933 new mutations (i.e. a base in a specific position found in the child but not in either parent) among 78 children, giving an average of about 63 new mutations per child. Many of the families were chosen since they had a child suffering from autism or schizophrenia without any family history of the condition. The average mutation rate for a 30-year-old father was estimated at 1.2×10^{-8} per nucleotide per generation – a figure in keeping with other estimates (see Box 3.1). Furthermore, the number of mutations increased linearly with the age of the father (Figure 19.9). This pattern needs careful interpretation since the age of the mother will strongly correlate with the age of the father (in this study $r = 0.89$) and it may be supposed that ageing mothers might also make a contribution to the *de novo* mutation load. However, when these authors performed a multiple regression analysis the correlation between a father's age and mutation rate remained extremely significant ($p < 0.001$), whilst the correlation between mutation frequency and a mother's age was non-significant ($p = 0.49$).

The group also looked at five families where the genome of father, mother and child were sequenced, allowing identification of whether the mutation occurred in the mother or father. Where the mutations could be assigned to a parent there were on average 55 from the father and 14 from the mother. It is noteworthy that the mutations from the father rose with paternal age whilst those from the mother were relatively flat (Figure 19.9).

Figure 19.10 also shows some results of work carried out by Dolores Malaspina et al. (2001) on a large cohort (87,907 individuals) of people born in Jerusalem from 1964 to 1976. They found that paternal age but not maternal age was a significant predictor of the risk of schizophrenia developing in an offspring. Later work by Malaspina et al. (2002) on a different group of subjects showed that this age effect was not due to a family history of schizophrenia (such that knowledge of this disease in the family caused fathers to delay parenting, for example) but was linked with sporadic cases (i.e. cases where an offspring suddenly developed the disease without a family background for this condition) – pointing again to the likely role of *de novo* mutations.

One important aspect of the strong correlation between mutation rate and paternal age is the implication for understanding rises in the incidence of diseases with a putative genetic basis. Changes in disease incidence is a complex phenomenon affected by many variables but this work on mutation rate points to the possibility that rises in recent decades in the incidence of some diseases with a genetic component may be related to the phenomenon, found in many Western cultures, of a rise in the mean age of couples having their first child.

The polygenic model does mean a move away from Panglossian approaches to behaviour – the idea that all commonly observed forms of behaviour, even behaviour associated with disorders, must somehow be products of positive selection. Indeed it sees all of us as affected to a greater or lesser extent by mutations. The constellation of these mutations gives rise to a spectrum of disorder severity. Above or below a clinically decided cut-off point we speak of people 'having' the disorder.

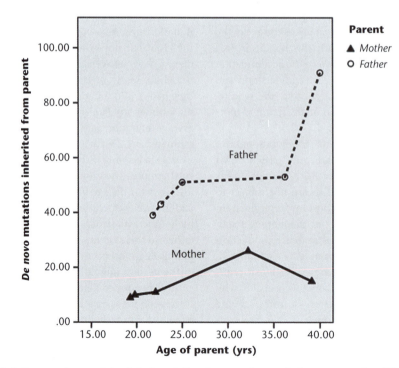

FIGURE 19.9 Mutations and parental origin in relation to age of parents from a sample of five Icelandic families.

SOURCE: Plotted from data in Kong et al. (2012).

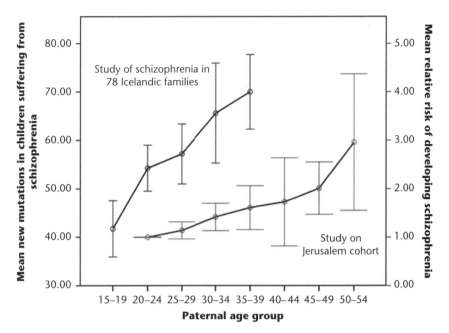

FIGURE 19.10 Two studies strongly suggestive of paternal age effects on the incidence of schizophrenia. The left-hand y axis shows number of new point mutations in a child against age of father. This is a sample of 21 children diagnosed as suffering from schizophrenia from a larger sample of 78 Icelandic families. The right-hand axis shows how the relative risk of an offspring (age group 20–24 set at 1) increases with the age group of the father from a study on a group of families from Jerusalem. Error bars show 95 per confidence intervals in both cases.

SOURCE: Left-hand axis: Data extracted and re-plotted from Kong et al. (2012), Figure 2. Right-hand axis: Data from Malaspina et al. (2001), Table 2.

Summary

▲ A strong case can be made that modern Western lifestyle diseases result from a discordance between the sort of foodstuffs humans evolved to consume and modern dietary habits.

▲ The Neolithic revolution brought about a major transition in the types of food humans eat. In broad terms, following the invention of agriculture diet became less varied, more reliant on cereals, lower in fibre content, less micro-nutrient rich and higher in saturated fats. Pre-Neolithic cultures also probably obtained more calories from meat and less from carbohydrate compared to post-Neolithic cultures.

▲ The industrialised processing of carbohydrates has led to an abundance of foods with high glycaemic indices. It is likely that the consumption of such foodstuffs is involved in the increase of diabetes type II and insulin resistance in modern populations.

▲ The diet of pre-Neolithic people was high in potassium and low in sodium. For modern diets this pattern is reversed: high sodium and low potassium. The inversion of this important ratio may be implicated in hypertension, strokes, kidney stones and osteoporosis in modern populations.

▶ Compared to hunter-gatherers, modern populations tend to have higher BMI values and higher blood pressure. This may be a consequence of modern diets high in refined carbohydrates and saturated fat as well as a consequence of lower levels of physical activity.

▶ Cancers due to mutations might have an ultimate origin in the compromise that natural selection made between the advantages of a reservoir of undifferentiated self-replicating stem cells, the advantages brought about by potentially useful mutations in the lineage of an organism and the costs of deleterious mutations.

▶ One central problem at the heart of Darwinian psychiatry is to explain why mental disorders are fairly common compared to many Mendelian disorders, have a high heritability, and yet are associated with high fitness costs. One answer to this serious puzzle might lie in the role of mutations.

Key Words

- Glycaemic index
- Hyperglycaemia
- Hyperinsulinemia
- Hypertension
- Mutation selection balance
- Saturated fat
- Schizophrenia
- Stem cells
- Trans fatty acids

Further reading

Brock, K. G. and G. M. Diggs (2013). *The Hunter-Gatherer Within: Health and the Natural Human Diet*. Fort Worth, Texas, BRIT Press.

A book that examines the Palaeolithic diet and draws some conclusions and advice about modern diets.

Jamison, K. R. (1994). *Touched With Fire: Manic Depressive Illness and the Artistic Temperament*. New York, Free Press.

An authoritative and readable work. Examines the lives of artists and writers over three centuries and the link between creativity and bipolar disorder.

McGuire, M. and A. Troisi (1998). *Darwinian Psychiatry*. Oxford, Oxford University Press.

An excellent overview of various Darwinian approaches to this subject and especially the 'out of Eden' hypothesis.

Keller, M. C. and G. Miller (2006). 'Resolving the paradox of common, harmful, heritable mental disorders: which evolutionary genetic models work best?' *Behavioral and Brain Sciences* **29**(04), 385–404.

An important paper that suggests mutations play a crucial role in mental illness.

Murphy, D. and S. Stich (2000). Darwin in the madhouse. In P. Caruthers and A. Chamberlain (eds.), *Evolution and the Human Mind*. Cambridge, Cambridge University Press.

A chapter that examines several evolutionary approaches but emphasises the application of the idea of modularity. It offers a severe criticism of DSM-IV methodology.

Nesse, M. and C. Williams (1995). *Evolution and Healing: The New Science of Darwinian Medicine*. London, Weidenfield & Nicolson.

A general work that examines a whole range of disease states (mental and physical) from an evolutionary standpoint. A useful book to start exploring this whole area.

Wakefield, J. C. (1992). 'The concept of mental disorder: on the boundary between biological facts and social values.' *American Psychologist* **47**: 3733–88.

An article that advances a useful definition of mental disorders.

PART IX

Wider Contexts

The Evolution of Culture: Genes and Memes

To every man upon this earth
Death cometh soon or late.
And how can man die better
Than facing fearful odds,
For the ashes of his fathers,
And the temples of his Gods.
(Lord Macaulay, 'Lays of Ancient Rome: Horatius', 1842)

Macaulay's famous lines remind us that humans attach remarkable importance to their beliefs and traditions, in short their culture, and that culture is something that binds individuals to groups, separates one group from another, and motivates humans to both heroic and diabolical deeds. It is easy to see that the major human achievements of the last 10,000 years, which distinguish us so markedly from the rest of the animal kingdom, are the results of cultural and not biological evolution: our genes are roughly the same as our Palaeolithic ancestors, whereas our culture has changed by an astonishing degree. Such is the disparity in these changes, and such is the complexity of cultural change, that it has proved difficult to establish the relationship between cultural evolution and genetic evolution. The subject is complicated and has stimulated a variety of models designed to address the problem. In this chapter, we will sketch out some principal lines of approach and examine some instructive case studies.

20.1 Modelling culture

If evolutionary psychology is able to say anything about human culture, we need a working definition of what culture is. The problem is that the word

culture has a range of connotations. To the historian of art and literature, it may signify something that is morally improving or uplifting. To an anthropologist, it may mean the fabric of beliefs and values common to a society. A biologist would identify it as something that is passed down by social learning. The definition of culture adopted here will be that **culture** is information (knowledge, ideas, beliefs and values) that is distinct from that stored in the human genome and which is socially transmitted. Note that this does not a priori rule out any evolutionary purchase on culture. The fine details of culture are obviously not represented in the genome, but the type of culture we generate and the functions that it serves could, in principle, be directed by our genes. This is of course an extravagant claim and one not to be swallowed lightly. The definition here, however, is more tolerant of a range of evolutionary models that have been developed to explain culture. In any case, behaviour can also be transmitted through social learning.

A number of key questions are raised when we examine the relationship of evolution to culture. We need to know what the units of cultural inheritance are, and what processes are involved in cultural transmission and evolution. In addition, we

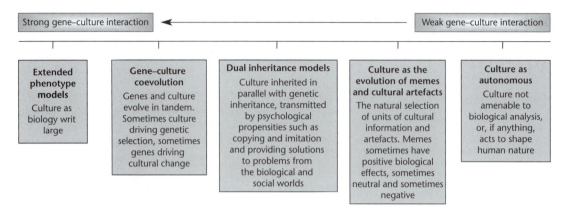

FIGURE 20.1 A spectrum of models of gene–culture interactions.

need to understand the relationship between cultural evolution and biological evolution.

A number of models have been proposed to tackle these problems and, at the risk of oversimplifying some complex work, we will reduce these to five categories and discuss each in turn. The five categories can be represented as different positions on a spectrum as shown in Figure 20.1, with models that stress the autonomy of culture to the right and models proposing a strong linkage between cultural evolution and genetic evolution on the left. Apart from the idea that culture is entirely autonomous, the other four models have significant areas of overlap and the differences between them often amount to stressing one process or feature above another.

20.2 Culture as autonomous

This model starts from the observation that, although very little genetic change has occurred over the last 10,000 years, culture has changed enormously. It focuses on the observation that cultural practices and behaviour vary considerably across the globe, despite the fact that humans are genetically very similar (modern genetics shows that the human gene pool is far less diverse than, say, that of chimpanzees). These considerations might suggest that there is no linkage between biological and cultural evolution, and that culture has its own laws of development, which are at present only dimly understood and probably best studied by

the humanities and the social sciences. In this case, the evolutionary biologist retires from the field gracefully or perhaps makes the passing observation that cultural evolution seems to be Lamarckian, in that the achievements of one generation can be passed on to the next.

Advocates of this view (characterised by Tooby and Cosmides (1992) as the standard social science model; see Chapters 1 and 2) often suggest that culture is also a primary determinant of human nature and behaviour: we are shaped by nurture and so our behaviour reflects the cultural norms around us. This approach was typified by the French sociologist Emile Durkheim (1858–1917), when he tried to dissociate social facts such as crime and suicide from individual psychology. According to his famous dictum, social facts can only be explained by other social facts.

But this model is probably too dismissive of the power of evolutionary thinking. Culture is after all the product of human minds that have been shaped by selection. Humans seem to be thriving as never before (in terms of population numbers at least) within complex cultures and so cultures must at least in part be fitness enhancing. Furthermore, many have pointed to the troubling circularity in Durkheim's view that social facts can only be explained by other social facts. On this basis, it is difficult to see how social facts ever arise. It is obvious that experience shapes our behaviour but it does so in non-random ways. Experience has to act upon a biological substrate (the brain or body) and the way the brain reacts is primarily a product of biological hardware rather than sociological software. As an analogy consider muscular strength. If we exercise our muscles, they grow larger, not smaller; this again makes perfect adaptive sense: if the body senses that muscles are in constant use, it stimulates their growth to facilitate whatever activity the phenotype has become involved in. The crucial point is that the effects of experience are facultative responses. This is seen most clearly in conditioned responses. The way we learn from experience through conditioning and association is inherently designed to help us to cope with our local (phylogenetically relevant) ecologies and make appropriate behavioural adjustments in the light of experience. Hence, as we noted in Chapter 1 on the waning influence of behaviourism, organisms have inherent learning biases and tendencies existing before the onset of experiences.

In further support of these objections, we may recall how humans can make quite rapid and subtle adjustments in their behaviour in relation to environmental and cultural cues (see Chapters 7 and 8).

20.3 Cultural evolution as the natural selection of ideas and artefacts

20.3.1 Memes

Of all the models, this is the most revolutionary. It suggests that the world of ideas (the characteristic products of culture) evolve in a Darwinian fashion but not necessarily through the medium of physical objects such as genes. To appreciate the logic of the model, consider the minimum set of conditions for some sort of Darwinian selection, and so evolution, to operate:

1. There exist in the world entities (or an entity) capable of self-replication.
2. The process of replication is not perfect, errors or changes are made and the next copy may not perfectly resemble its template.
3. The number of copies of entities that can be made depends on the structure of the entities and hence their manner of interaction with the world outside.

4. As a result of the finite nature of resources, operating spaces and so on, these entities experience differential reproductive success.

From these four minimal conditions, we should be able to witness Darwinian evolution. It is easy to appreciate that the entities above may not be strands of DNA. It is conceivable that, on other planets, the molecular basis of replication may be completely different. It is more of a shock to realise that the entities may not need to be physical at all; they may, in short, be ideas existing in and moving between brains.

Dawkins (1989) was not the first to note this insight but he articulated it forcefully in selectionist terms and coined the term **meme** to describe elements of thought or culture that replicate in human brains. As an analogy, the meme idea works surprisingly well. Memes move from brain to brain like parasites from host to host. We can catch them vertically from our parents, as in the case of rules of behaviour inculcated in childhood, or horizontally from each other, as in the case of peer pressure or conformity to fashion (Box 20.1). Some memes are truly parasitic, in the sense of damaging the survival chances of the host or the host's genes. Chastity, celibacy and self-sacrifice for noble ideals are all memes that damage their host's biological success. But this need not concern memes if memes survive: if self-sacrifice is held up as a laudable act, then others will fall under the sway of the meme, and the meme will survive. Suicide bombers may not leave many children but if the act is praised as martyrdom with the promise of rewards in another life then the behaviour may persist through cultural and not biological selection. Many memes, however, are mutualistic in that they assist their replication by ensuring the wellbeing of the host. Examples here include elementary rules of hygiene, methods of fashioning tools, avoiding disease and so on. The incest taboo may be a case of genes and memes directed to the same end if, as it seems, the Westermarck effect is based on a genetic developmental programme. The taboo becomes the meme that reinforces the genetically based mechanism of avoiding incest.

Box 20.1 The spread of memes

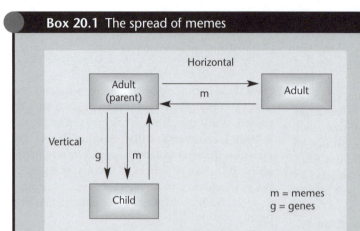

Horizontal

Adult
(parent)

m

Adult

Vertical

g   m

Child

m = memes
g = genes

Memes can be passed vertically in the case of parent to offspring or horizontally in the case of one organism to another:

- *Vertical*: When passed vertically down the generations, memes can accompany genes. In early traditional cultures, there was probably a large degree of synergy between genes and memes and this may persist into modern times. The linked memes in Catholicism that restrict birth control and also insist that offspring are raised in the faith, for example, have the dual effect of increasing the spread of memes and the genes of

those professing the memes. In these cases, sociobiological and memetic explanations yield the same results.

- *Horizontal*: In horizontal transmission, genes do not accompany memes and memes may, from a biological point of view, be fitness reducing. The meme that suggests a career is more important than children reduces biological fitness but it may nevertheless spread through imitation. A fanatical devotion to chastity may be highly successful, in that biological energies are diverted into meme replication (chastity) rather than gene replication.

We can identify vertical transmission between parents and offspring and horizontal transmission between biologically unrelated individuals. A whole host of studies have demonstrated how people tend to inherit their cultural values from parents, as well as traits such as language and social skills from other members of the community (Mesoudi et al., 2004).

Memes are unlikely to exist alone. The meme for altruism, for example, is linked with other memes that confer memetic advantage. A combination of memes Susan Blackmore (1999) calls memeplexes. Religious systems, as suggested by Dawkins and others, are complex memeplexes. They have strict rules of adherence, and employ enhancers and threats. An enhancer encourages the spread of the meme by an appeal to its virtues, such as the Biblical promise 'blessed are the merciful: for they shall obtain mercy' (Matthew, 5: 7). Advertising slogans are full of memes connected with enhancers: 'A Mars a day helps you work, rest and play' is successful in that it promises benefits from the activity (eating a Mars chocolate bar) as well as using rhythm and rhyme that appeal to our neural circuits.

20.3.2 Natural selection and artefacts

Competition between biological variants occurs in relation to finite resources: food, space, territory, mates and so on. By analogy, a single human brain (or home in the case of cultural physical objects) cannot possibly contain all cultural variants; in this sense intangible things such as ideas, music, words, values and knowledge as well as physical objects such as tools and household appliances compete for a place inside our heads and our homes respectively. Those that thrive we could argue are well adapted. Some cultural artefacts and practices, such as smoking, may from an organismal point of view seem maladaptive (smoking causes premature death) but the practice persists and is transmitted vertically and horizontally. From the point of view of a cigarette it has a successful strategy to ensure it is purchased, used and replaced; like a kind of virus, it manipulates its host to ensure its own propagation.

But, as noted earlier, many cultural variants clearly have an adaptive value that is both cultural and biological. Lumsden and Wilson (1981) used the term '**culturgen**' to describe a similar concept to the meme. As these authors note, some culturgens could have positive biological fitness benefits such as those associated with food taboos and food preparation techniques. Viewing certain animals as unclean may have a sound basis in hygiene if such animals carry parasites that can infect humans. Alternatively, certain techniques of preparing food may help to protect against parasites. It is probably no accident, for example, that food in hot countries, where ambient temperatures are such to allow bacteria to multiply rapidly on foodstuffs, is often highly spiced. Many spices (for example garlic and chilli) are known to have antimicrobial properties. In support of this thinking, Sherman and Billing (1999) conducted an analysis of 4,578 meat-based recipes from 36 countries and found that as average annual temperatures increased, so did the proportions of indigenous recipes using antibacterial spices. As a follow-up, Sherman and Hash (2001) predicted that the rise in spice use frequency with local temperature should be less marked for vegetable dishes, since the cells of dead animals are less protected against, and more conductive to, the growth of bacteria and fungi than the cells of dead plant matter. In a survey of 2,129 vegetable dishes in 36 countries, they found that although spice use did increase with ambient temperature, this was far less rapid than for meat dishes, and, furthermore, by every measure, vegetable-based recipes were less spicy than equivalent meat-based ones.

One general important question to raise is whether the meme is an ingenious and amusing analogy or whether it provides a serious set of testable hypotheses that may really help us to understand the evolution of culture. It is obvious that the analogy between the evolution of ideas and artefacts and DNA is not perfect: natural selection works because inheritance is discrete and is not blended. If any new change in the genome simply blended with other genes, novelty would be quickly lost and evolution

would grind to a halt. But it is not at all clear that memes, ideas and artefacts act like discrete units. They seem to blend and merge in ways quite unlike the information on DNA. Whereas genes have a universal language, there seems little prospect of a uniform meme language in brains.

Another serious problem is understanding the conditions by which some memes are copied well and others are ignored. In the case of biological features such as camouflage, it is easy to see why such genes have been selected. With memes, it is not at all obvious in purely memetic terms why, say, the meme for fashion styles comes and goes. Such phenomena may become clearer when we understand imitation biases, a subject that forms part of the model considered in the section below.

20.4 Dual inheritance theories

According to **dual inheritance theory**, there are two types of process that can generate design: natural (and sexual) selection acting upon genes, and a broader variety of selective processes acting upon culturally transmitted variants. One of the most influential theories of dual inheritance comes from Richerson and Boyd (2005). According to their ideas, both the psychological basis of culture (naturally selected and genetically based learning mechanisms and biases) and the pool of transmitted ideas are adaptations to solve the problems of a rapidly changing environment that early hominins faced. They argue that the way humans absorb, select and modify culture at the individual level leads to adaptive cultural evolution at the population level. In this view, culture becomes a changing reservoir of useful (but sometimes maladaptive) knowledge and skills passed between humans in tandem with genetic evolution. This model is similar to that of memetics, but more precise in the way it envisages how culture is shaped and is inherited.

20.4.1 Imitation and bias

Culture, as defined earlier, is transmitted by imitation and social learning. Indeed, human infants (as any parent will testify) are excellent mimics – far exceeding the capacity of other infant primates. But, as Rogers (1989) pointed out, there is something of a paradox at the heart of cultural transmission by imitation. Imitation is useful because it provides a rapid and cheap way of acquiring valuable cultural knowledge and skills – a process certainly faster than trying to continually figure out new solutions to life's little problems. But as the number of imitators rises (favoured by their fitness gains from copying others), so the fewer there are who are genuinely creative. There will come a point when imitators will start imitating each other and culture will stagnate and fitness levels fall as new problems are not solved. It may look as if we have reached an impasse. Mathematical models of this process show that the system reaches equilibrium when the cost of knowledge production to innovators equals the cost of being wrong to imitators (who are imitating out-of-date solutions). In such cases, the population ends up with a mixture of copiers and innovators in which (and this is the surprising point) both types have fitness levels the same as a population of innovators (Boyd and Richerson, 1995). The problem now is obvious: how can culture, which is supposed to deliver benefits by cheap and fast transmission of ecologically and socially useful knowledge, ever evolve if this process is no better than learning afresh?

Richerson and Boyd (2005) suggest that culture could escalate if innovators are given special privileges by imitators to compensate them for the costs they endure in devising original solutions. In addition, imitators would be best served by being selective: specifically, they should imitate the most successful in local populations, since the most successful must be doing something right (either imitating wisely themselves or discovering new tricks) to acquire the status they have. A moment's reflection on contemporary culture shows that this is exactly what we do: genuinely creative innovators become cultural icons that we reward handsomely and imitate. Hence also the considerable sums of money paid by advertising executives to celebrities to endorse a product: they know that this is an effective way to change our behaviour. The simple rule 'imitate the successful' is one of several learning heuristics that can assist the assimilation, transmission and evolution of culture (Box 20.2).

Box 20.2 Fast and frugal heuristics for acquiring a culture

In the tradition of Gigerenzer's work on cognitive heuristics (see Chapter 9), Richerson and Boyd (2005) identify several heuristics that individuals may use to reduce the cost and improve the effectiveness of social learning. The use of heuristics means that individuals do not have to be exceptionally clever at creating cultural solutions: simple heuristics can, like the force of natural selection, produce clever adaptations when exercised by thousands of individuals over long periods of time. Two of these adaptations are the **conformist bias** and the **prestige bias**:

- *Conformist bias*: This is a frequency-dependent bias since it is the frequency of the behaviour and not the content or the person performing it that excites attention and emulation. In the conformist bias, the strategy is to imitate the most frequent behaviour. This bias may often yield the best solution since the most advantageous form of behaviour is often likely to be the commonest. There is a mass of evidence from the social psychology literature to show that people do indeed have a strong tendency to adopt the views and beliefs of the majority (see Myers, 1993).

- *Prestige bias*: This is a model-based bias, since it is the characteristics of the person performing the behaviour (the model) rather than its content or frequency which is of interest. The logic behind this heuristic is that locally successful and prestigious people must be behaving in ways that have contributed to their success; therefore, it will pay to do what they do.

20.4.2 **Maladaptive cultural variants**

Once a system of cultural evolution (and specifically the learning mechanisms that drive it) is in place, there are costs. Many problems that humans face, such as how many wives (or husbands) to take or how family life should be organised, have low 'trial-ability'; that is, we do not have time to try out alternatives. Rather, we have to rely upon tradition and custom in the hope that they will provide tested and time-honoured solutions. But the downside of this is a tendency towards credulity: we may accept some beliefs and values that are not at all fitness enhancing. Indeed, there is no reason to expect genes and memes (to borrow a term from a related model) to be in total accord. Just as sexual and natural selection can pull in opposite directions, so too might cultural and genetic evolution. The interesting question is how far cultural or memetic fitness can evolve against the grain of biological fitness. The answer is probably a significant distance. Cultural entities have a distinct advantage over genes in that they can multiply and spread very rapidly both vertically and horizontally (that is, within a peer group). If their transmission rate is high enough, they could, in principle, considerably reduce the fitness of the biological bodies that are their hosts. The situation is analogous to the outbreak of an epidemic carried by a virulent microbe: populations can be reduced drastically until those that are left are the immune ones, or until so many are killed that the microbe can no longer be carried to new hosts.

Maladaptive culture may be generated by precisely those heuristics that serve us well over the long term. In the case of prestige bias, for example, people may emulate role models who are famous but are exemplary in the wrong ways, such as being too thin or prone to drug use. Another natural bias that is a candidate for the generation of cultural errors is the confirmation bias: a tendency whereby we tend to look for evidence that confirms our beliefs, rather than, as good Popperian falsificationists should, look for instances whereby our beliefs would be refuted. Pascal Boyer (1994) argues that the widespread acceptance of supernatural beliefs is fostered by 'abductive reasoning': a process (sometimes known by philosophers as the 'fallacy of affirming the consequent') whereby we accept a premise as correct if the implications of the premise are observed. Imagine you have a belief in the healing powers of a local deity. You pray and get better. Was your premise (that is, the existence and benevolence of your god) correct? Not necessarily: people get better anyway.

20.4.3 **Changing environments and social learning**

In the model of Boyd and Richerson (1995), cumulative cultural adaptations will be favoured when a number of conditions are found. One of these is a large spatial variability in environmental conditions, such as when we move around the Earth's surface and climatic zones. In such situations, organisms with similar genes (that is, members of the same species) can exploit spatial variation in environmental conditions without relying upon genetic variation. This is precisely what we observe in the case of humans who have colonised most of the globe. There are vast differences, for example, between the environments inhabited by the Inuit people living in the Arctic regions of Alaska and Greenland and the Maasai in the tropics of Kenya. Admittedly, there are biological differences between the two groups that partially facilitate this. The Maasai are darker skinned, taller and slimmer than the Inuit. The shorter, more rounded bodies of the Inuit help to conserve body heat, while the slim bodies of the Maasai help them to lose heat. But these differences notwithstanding, the main reason the two populations are able to survive is due to the knowledge embodied in their cultures, which is fine-tuned to local ecologies and provides cultural solutions to living in these difficult conditions. This cultural knowledge would be passed on by parents and other group members. There is an old African saying: 'It takes a village to raise a child', meaning that child-rearing is something shared by the broader community (a process known as 'alloparenting') and not just the biological parents. It is this spatial variation of environmental conditions that humans encountered and were able to master (eventually) as they spread out from Africa about 75,000 years ago.

Another condition favouring cultural adaptation is that environments must change over time at about the right rate. If environments change very slowly, then organic evolution (a slow process in humans since we have long generation times) can keep track and ensure that we remain biologically adapted. If environments change too rapidly, then cultural knowledge quickly becomes out of date and redundant. In such cases, individual learning will be favoured over sociocultural learning.

Boyd and Richerson (1995) think that the last half of the Pleistocene provided patterns of climate variation that were just right for the evolution of culture. They extend their idea to explain why complex culture as a solution to survival only appeared once: the late Pleistocene was the first time in Earth's history that a large-brained organism (the hominins) faced climate change at the crucial level of variability. In Chapter 6, we noted how several cooling episodes over the last 5 million years may have facilitated the evolution of bipedal hominins. On top of these cooling periods were fluctuations caused by Milankovitch cycles of roughly 21,000, 41,000 and 100,000 years. These cycles by themselves are unlikely to lead to adaptations for cultural learning, since they are long enough for species to cope with by changing their range or by organic evolution. However, the recent availability of high-resolution data for climatic changes over the last 80,000 years from Greenland ice cores has revealed a pattern of rapid short-scale fluctuations. It now looks likely that between about 100,000 to 10,000 years ago, the Earth's climate oscillated from glacial to nearly interglacial conditions every 1,000 years or so, with some fluctuations happening over only 100 years. Such rapid changes would have had major effects on local vegetable and animal populations as well as the type of clothing and shelter needed every few generations. The hypothesis of Boyd and Richerson (1995) is that to cope with such changes, cultural transmission became advantageous. Cultural evolution could give rise to complex adaptations much faster than genetic evolution and was therefore ideally suited to variable Pleistocene environments.

Once climatic variability (over space or time) places a premium on culturally assisted behavioural flexibility, it becomes important for juvenile humans to have sufficient time to acquire cultural skills and knowledge to function effectively in their groups. This would select for an extended juvenile period and so a delay in reaching reproductive age; humans

would become increasingly K selected (see Chapter 8). If the extended juvenile period in humans is related to the need to acquire cultural adaptations, then this is an interesting case of how cultural change (in this case, the wholesale value of culture) can drive genetic change (delayed sexual maturity). If this turns out to be correct, it provides an example of gene–culture coevolution, a model considered in the next section.

20.5 Gene–culture coevolution

In this view, culture has some autonomy from genes but is constrained by an 'elastic' leash. As the metaphor suggests, cultural change can apply a force that pulls along genetic change, but cultural change cannot drift too far away from the interests of genes before it is pulled back. This approach has spawned some highly complex mathematical models. In early models by Lumsden and Wilson (1981), culture is depicted as the outcome of the behaviour of individuals but in situations where the behaviour is itself influenced by the existing culture and 'epigenetic' rules of development. We can imagine a growing individual as bombarded by bits of culture or 'culturgens' (Lumsden and Wilson's early word for memes). Genetically based rules of development will influence a child to accept some culturgens and reject others. Genes and culture have therefore shaped the final behaviour of the adult, and, moreover, he or she then contributes in the re-creation of culture. Cultural and genetic change can take place together because culture sets the environment that may alter gene frequencies for genes that describe the epigenetic rule of development. In this manner, culture and genes roll on together (Feldman and Laland, 1996).

There are numerous variations in approach to this subject but here we will consider two examples that illustrate the general idea: lactose tolerance and adaptation to dietary starch.

20.5.1 Lactose tolerance

Lactose is a naturally occurring sugar found in milk and it can be digested so long as the body is able to produce the enzyme lactase. The enzyme itself is produced by intestinal epithelial cells that line the walls of the small intestine. It catalyses the breakdown of lactose into the smaller sugar groups glucose and galactose that can then be absorbed into the bloodstream.

Most mammals stop producing lactase at the time of weaning when offspring stop taking their mother's milk. Thereafter the young remain intolerant of milk sugar. This makes good biological sense: there is no point going to the energetic expense of synthesising lactase beyond the time when it will be useful. In addition it serves the useful purpose of forcing one child away from weaning to allow others to be produced. Following the Neolithic revolution, the milk of cattle and other animals has probably been an important component of the diet of some human populations for about 6,000 years. The prevalence of dairy farming among some human groups may have led to a selection pressure for genes allowing the absorption of lactose beyond childhood weaning. It is unlikely that early hunter-gatherers were able to continue to synthesise into adulthood the enzyme lactase that allows the digestion of milk. Although in the West we tend to assume that milk is a natural part of the diet, the majority of adults in the world lack the enzyme needed to digest lactose. If this latter group drink milk, the lactose is fermented by bacteria in the gut rather than broken down by lactase, leading to flatulence and diarrhoea.

Surprisingly, right up until the mid-1960s it was generally agreed by Western nutritionists that lactose tolerance was the norm for the human species. Today, the medical establishment realises that adult lactose digestion is the result of mutations that are shared by a minority of the human population such that about 65 per cent of the global population are lactose intolerant in adulthood. But old habits die hard, and, as Andrea Wiley (2008) has documented, the tendency to think of lactose intolerance as deviant persist with such terms as 'disorder' or 'deficiency' used in the medical literature – a case of what she calls 'bio-ethnocentrism'.

There are populations, particularly in Africa and Asia, which do exhibit this typically mammalian pattern of adult lactose malabsorption (LM). There are others, however, especially people of northern European and Scandinavian descent, who exhibit lactose persistence, meaning that they can drink milk from cattle and other animals throughout their life.

To understand this distribution of genetic differences, we need to look at the role milk products played in selection pressures. One answer is that malnourishment in some environments would have helped the spread of genes that would assist in the absorption of lactose and so enable individuals to survive. This is plausible, but there are probably additional factors (Durham, 1991).

It is known that lactose, like vitamin D, also helps the absorption of calcium in the gut. Vitamin D is produced in the body by the effect of ultraviolet radiation on the skin. Vitamin D deficiency, and hence poor calcium absorption, is a serious risk for people living in high latitudes where the intensity of sunlight is low. For these people, lactose absorption had a double advantage in providing calories and essential calcium. This example neatly illustrates how a cultural change – the invention of agriculture and the domestication of animals – can bring about a genetic shift in human populations (Simoons, 2001).

If we consider populations that are lactose tolerant (LA), then a question arises as to why dairying was practised in these areas and not others. Gabrielle Bloom and Paul Sherman (2005) investigated this by compiling a massive database on LM and LA frequencies, temperature, climate and the incidence of cattle diseases. They found that LM was negatively correlated with latitude and positively with temperature; that is, LM decreased with distance away from the equator and increased with temperature. There was also a positive correlation between LM and the number of cattle diseases found in the area. Their results suggest that LA increases with latitude, since conditions away from the equator favour the keeping of cattle. The general outlines of selection for LA are clear but the distribution of the condition across the globe is complicated by additional factors such as population migrations (Figure 20.2).

Lactose tolerance as convergent evolution

The genetic basis for lactase persistence in human populations is a good example of evolutionary convergence since different mutations are found in different populations, suggesting independent selection for lactose tolerance. The lactase gene itself (often abbreviated to LCT) is located at 2q21 – meaning the

Lactase hotspots

Percentage of adult
population that can
tolerate milk

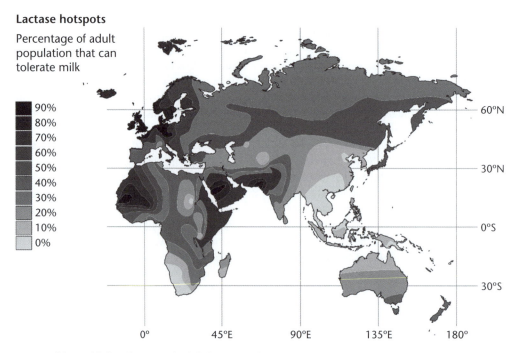

FIGURE 20.2 Old World distribution of adult lactose tolerance.
The distribution of lactase persistence is quite complex and is the product of a number of factors such as natural selection in relation to dairying practices, areas where cattle can be kept, genetic drift and gene flow caused by population migrations.

SOURCE: Compiled from various sources including: Ingram et al. (2009), Figure 1, p. 581; Curry (2013), p. 21; Bloom and Sherman (2005), Figure 2, p. 306.

long arm of chromosome 2 at position 21 – and consists of about 135,800,000 base pairs.

Table 20.1 shows the position of various mutations according to geographical region. Remarkably, all the mutations known so far are single base changes called single nucleotide polymorphisms. The enabling mutation is not in the LCT gene itself but 'upstream' in a controller region that as a result of the mutation no longer switches off the activity of the LCT gene. The European version seems to have originated in the area of modern-day Turkey around 11,000 years ago (Curry, 2013).

The use of milk products also illustrates how humans can cope with a selective pressure through both genetic and cultural adaptation. At the genetic level, as we have seen, the continued

production of lactase after childhood assists with digestion. But there are several pastoralist peoples, such as Mongolians and the Nuer and Dinka of Sudan, who produce and consume milk products

Table 20.1 Position of mutations allowing lactase synthesis persistence and hence lactose digestion in several populations.
The existence of different mutations in different regions suggests evolutionary convergence to solve a similar problem.

Area:	Europe	Ethiopia/ Sudan	Saudi Arabia	Kenya/ Tanzania
Position of SNP upstream of lactase gene (no. of base pairs)	13,910	13,907	13,915	14,010
Base substitution compared to chimpanzee. That is, ancestral–modern	C–T	C–G	T–G	G–C
SOURCE: Data taken from Ingram et al. (2009), Table 1, p. 585.				

but cannot easily digest lactose in adulthood. They have responded to the adaptive challenge at the cultural level by converting milk to products where the lactose has been removed, such as cheese where the lactose is left in the whey, or converted into other compounds as in the case of yoghurt where the lactose is converted to lactic acid. There are some Somali pastoralists who illustrate yet another way of coping with lactose. These people drink milk yet do not produce lactase as adults. Instead it seems that in these groups the gut bacteria have evolved to enable the intestinal digestion of lactose (Ingram et al., 2009).

The realisation over the last 40 years that lactose intolerance is the ancestral human condition stands in contrast to the steady growth in global milk production and consumption over this period. Between 1980 and 2007 global milk production increased about 30 per cent. Today the EU remains the largest producer of liquid milk but the growth in production has been particularly high in developing countries. Consumption of milk in the EU, the USA and Canada has remained fairly steady over the last few decades but over a 40-year period milk consumption in China has increased fifteenfold and in Thailand sevenfold – driven by a combination of rising affluence and the emulation of Western lifestyles (Wiley, 2007). A potential public health issue for the future is that growth has been strongest in precisely those areas where intolerance to lactose is high. Some of this risk is offset by the fact that milk is often converted to yoghurt (with consequent lactose reduction) but the situation will need to be monitored.

20.5.2 Adaptations to starch consumption

Diets vary around the globe often in fairly stable and predictable ways. The Hazda people of Tanzania, for example, eat a diet rich in sugar and starch from the honey and tubers they find by foraging. The Inuit of Greenland, in contrast, eat lots of fish and blubber from whales and seals and have done so for thousands of years. As well as lactase persistence, might enough time have elapsed for different populations to show other adaptations to dietary intake? The study of starch suggests some interesting possibilities.

The agricultural revolution that occurred in various parts of the world in the Neolithic period led in many areas to an increase in the consumption of carbohydrate. Archaeological evidence, for example, suggests that the domestication of rice that took place in the lower Yangtze region of China about 7,000 years ago resulted in a large increase in rice consumption. Even up to the 1960s rice accounted for on average about 40 per cent of the calories consumed by the Chinese and Japanese. Rice is about 80 per cent starch, and starch is a polysaccharide containing long chains of tens or even thousands of glucose units bound together like beads on a string. This polysaccharide is known as amylose and humans digest starch using enzymes, known as amylases, secreted in saliva and the small intestine resulting in glucose which is then delivered to cells by the bloodstream.

The gene producing salivary amylase is known as AMY1. Interestingly, the number of copies of this gene varies between 2 and 10 copies per individual, and the more copies of the gene that an individual possesses the more amylase is produced in saliva. George Perry et al. (2007) investigated whether the number of copies of this gene varied amongst peoples according to their specific and historic diets. The researchers examined seven populations grouped into low or high starch consumers. The number of copies of AMY1 was found to be significantly higher in the high starch consumers. The group also looked at AMY1 copies in chimpanzees. Chimps are of course predominantly frugivorous and ingest relatively little starch compared to humans. Also chimpanzees have not moved around the globe and so have not undergone the drastic dietary shifts experienced by some human populations. As expected, all the chimpanzees tested (N = 15) had fewer AMY1 copies than humans, with no significant variation in AMY1 number between individuals (see Figure 20.3). It looks, then, like chimpanzees, with about 2 copies of AM1 per diploid genome, represent the ancestral condition. As humans have moved around the globe so mutations causing gene duplication have occurred leading to around 6–7 copies of this gene amongst high starch consumers (such as the Japanese) and 4–5 copies amongst low starch consumers such as the Yakut people of Siberia (who eat a lot of fish, meat and dairy products).

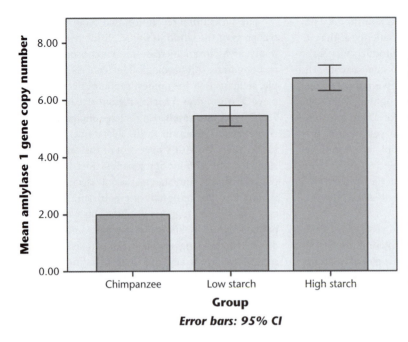

FIGURE 20.3 Duplications of the AMY1 gene in human populations and as adapted to a diet high or low in starch.

Populations with a history of consuming high starch levels tend to have more copies of the gene that produces the digestive enzyme amylase. High starch consumers are the Hadza, Japanese, and European-American populations; low starch consumers are the Datog, the Mbuti, the Biaka, and the Yakut. Error bars represent 95 per cent confidence intervals (CI) for the means of these populations.

SOURCE: Plotted from data in Perry (2007), Supplementary Table 1.

The importance of gene–culture coevolution

There are several more ancient examples where we can observe cultural change bringing about significant genetic change: hominin physique, jaw size and language. Modern *Homo sapiens* are generally less robust than some earlier species. Paleoanthropologists have argued that one reason for this might be the cultural evolution of projectiles as hunting weapons. Once an animal can be killed from a distance by throwing a spear or shaped stone, there is less need for a strong (and energetically expensive) physique. Jaw size and tooth size could have fallen in the hominin lineage following the use of fire to cook meat. Cooking breaks down the fibres of meat and reduces the need for large jaws to tear and chew meat. The evolution of language (whatever the initial cause) placed a selective pressure on the vocal tract of humans, such that it eventually became modified to produce spoken language.

So it seems increasingly likely that many physical features of the human phenotype, such as body size, dentition, loss of body hair, digestive processes and associated enzymes, manual dexterity and resistance to specific diseases, will evade a satisfactory understanding unless attention is paid to the cultural transmission of such activities as the use of fire for heat and cooking, the manufacture of clothing, agriculture and tool making. If culture can shape the evolution of physical characteristics, as many studies have convincingly demonstrated, then cognition is also a prime candidate for culturally driven selective moulding; yet, with perhaps the exception of the study of the cognitive differences between men and women, evolutionary psychology has not shown much interest in this area (see Henrich et al., 2008). For a more general review of how culture may have shaped the human genome see Laland et al. (2010).

20.6 Culture as a consequence of genotype: culture as extended phenotype

20.6.1 The extended phenotype

The notion of the **extended phenotype** was introduced by Richard Dawkins (1982) in a book of that title. The subtitle of the book, *The Long Reach of the Gene*, encapsulates the essential idea that the effect of genes can extend outside the vehicles that carry them. Dawkins realised that it was somewhat arbitrary to limit the effect of genes to protein synthesis and behaviour. Genes can bring about changes in the world outside the bodies that contain them. Some obvious examples are the mounds that termites build to control their environment, the dams built by beavers and the bowers built by bower birds (see also Section 9.5.1).

It would be foolish to suggest that the precise details of culture are described by the human genome, but it is possible with some plausibility to suggest that genes themselves give rise to development processes that predispose humans to develop certain cultural forms. This seems to be the position argued by Wilson in his pioneering book *Sociobiology: The New Synthesis* (Wilson, 1975), where he suggested that cultural forms also confer a genetic survival value. Religious systems, for example, often prescribe correct forms of behaviour in relation to food, cooperation and sex. Incest taboos may reflect an instinctive mechanism designed to avoid homozygosity for recessive alleles. In a similar vein, ethical norms may be a way of our genes persuading us to cooperate to serve their best interests (see Chapter 21).

20.6.2 Science and the arts: consilience

In *Biophilia* (1984) E. O. Wilson extended his thinking on the biological basis of human nature to examine the adaptive value of notions such as beauty. For Wilson, beautiful ideas, whether from the poet or the scientist, are those that provide elegant and cost-effective solutions to problems:

> Mathematics and beauty are devices by which human beings get through life with the limited intellectual capacity inherited by the species. (Wilson, 1984, p. 61)

It was in *Consilience* (1998), however, that Wilson expressed his most deeply held vision for the future dialogue between the arts and sciences. To conceptualise this linkage he resurrected the term **consilience**, a word first used by the British philosopher of science William Whewell and roughly meaning the use of a common and contiguous framework in which to locate ideas and knowledge. Whewell was a polymath and master of Trinity College, Cambridge, who also coined the word scientist in 1834. Ironically, Whewell's own notion of consilience only went so

far and he refused to have a copy of Darwin's *Origin of Species* (a work surely on par with Newton's *Principia* in its demonstration of the power of consilient thought) in the Library of Trinity College, Cambridge.

Although Wilson has been accused of crude reductionism (Menand, 2005), his idea is not to reduce the arts to biology but to be able to trace consistent threads linking evolutionary theory, neurophysiology and inherited mental circuitry. Then, from an understanding of brain architecture and innate dispositions we can secure a better understanding of the forms and functions of culture that humans produce. We can then work backwards to examine and perhaps even appreciate the more the particular and singular resonance of great art with the brains of humans that created it. At each level there will be a semi-autonomous set of laws and principles. This type of reductionism, for Wilson, does not 'diminish the integrity of the whole' and:

> Scholars in the humanities should lift the anathema placed on reductionism. Scientists are not conquistadors out to melt the Inca gold. (Wilson, 1998, p. 211)

The key for Wilson was to see the arts, the humanities and human biology resting on the common foundation of an evolved human nature:

> Artistic inspiration common to everyone in varying degrees rises from the artesian wells of human nature… It follows that even the greatest works of art might be understood fundamentally with knowledge of the biologically evolved epigenetic rules that guided them. (Wilson, 1998, p. 213)

From this perspective, Wilson sees the arts as an adaptive response of a highly intelligent organism to a complex and rapidly changing environment. The arts serve to prepare humans for the life they lead and the multifarious experiences they will have; by providing a vicarious account of life they serve to simulate and model reality and pass on social learning. In this view what we mean by truth and beauty in the arts is the 'precision of their adherence to human nature' (Wilson, 1998, p. 226). The oral tradition, for example, from which literature sprang, embodies generations of practical learning relevant to coping with the physical and, just as importantly for gregarious animals such as humans, the social environment. Steven Pinker made similar claims when he said that:

> The technology of fiction delivers a simulation of life that an audience can enter in the comfort of their cave, couch, or theatre seat. ... Fictional narratives supply us with a mental catalogue of the fatal conundrums we might face someday and the outcomes of strategies we could deploy in them. (Pinker, 1997, p. 543)

Since the publication of Wilson's *Consilience* there has grown up a whole movement devoted to an evolutionary understanding of aesthetics and literature. There are various positons in this movement (sometimes called biopoetics and sometimes Darwinian literary studies) but one basic thrust is to propose, as Wilson suggested, that we can see literature as the product of an evolved human nature. Culture will bear signs of this nature and also perhaps serve adaptive functions in helping humans to thrive in complex ecological and social environments. A further exploration of this exciting new field lies outside the scope of this book but it does illustrate the notion that culture may be profitably examined in terms of the concept of extended phenotype: the arts as part of a biologically generated culture that helps humans understand the world, solve recurring ecological and social problems, and act out their biologically driven imperatives of growth, survival and reproduction. The interested reader is referred to the additional reading suggested at the end of this chapter.

20.7 Case study: religion

20.7.1 The problem of religious belief

Evolutionary accounts of behaviour and culture face tough challenges when it comes to the phenomenon of religious belief and practices. We might expect, because of its obvious survival advantages, that natural selection would direct us to think and learn about the world such that our mental maps of reality would carry a reasonable degree of verisimilitude. Against this expectation we could balance the insight that at the bottom line natural selection only deals in the currency of reproductive success, not truth, and false

beliefs and biased cognitive systems will become the norm if they deliver reproductive gains (see Chapter 9). But even with this proviso, the tenets of most of the world's religions seem to test all credulity. Belief in sky gods, life after death, heaven and hell, virgin births, miracles and the periodic suspension of the laws of nature are, from a scientific perspective, simply incredible. Possibly in the past the propensity towards such strange views was fitness neutral, but many religious practices that we now see, such as celibacy, genital mutilation, self-sacrifice, and irrational restrictions on foodstuffs, seem far from fitness enhancing and have a long history.

The problem is compounded by the fact that religions are everywhere, making the study of a religion-free culture impossible. The relatively recent rise in secular values in Western European countries (such as the UK, Germany, Finland and Denmark) is not typical of the world as a whole. The European Values Study of 2008 found that for the vast majority of European countries young people in the age group 20–29 years were less religious than their older compatriots. But only in three countries, France, Germany and the Czech Republic, did belief in God fall below 40 per cent even for this age group. The USA is, on average, far more religious than Western Europe. Here over 90 per cent of the citizens claim to believe in God and about 40 per cent believe that Christ will return to Earth in the next 50 years (Appiah, 2006).

There have been numerous attempts to understand religion from an evolutionary standpoint (for example, Bulbulia, 2004; Boyer and Bergstrom, 2008; Wolpert, 2009; Wilson, 2010). A useful schema is provided by Schloss and Murray (2009) who suggest that evolutionary and biological accounts of the origins of religious belief fall into three categories: cognitivist, Darwinian and gene–culture coevolutionary (Table 20.2). We will examine these categories in turn.

20.7.2 Cognitive accounts

Cognitive accounts primarily offer proximate reasons and suggest that the capacity to accept religious beliefs is not an evolutionary adaptation by itself (our more religious ancestors did not outbreed less religious ones) but is rather a by-product of other

Table 20.2 A classification of evolutionary accounts of religious belief

Cognitive accounts (proximate mechanisms) often allied with evolutionary psychology	Darwinian accounts (ultimate causation) often allied with human behavioural ecology	Gene–culture coevolution or dual inheritance theories
Extension of attachment	Vestigial traits	Memetic pathogen
Anthropomorphic projection	Sexually selected displays and costly signalling	Memetic symbiont
Hypersensitive agency detection device	Internalisation of sanctions and promotion of cooperative behaviour	Not necessarily producing reproductive fitness benefits
Cognitive bias towards dualism	Reward of real fitness benefits to hosts	
The spandrels metaphor		
SOURCE: Adapted from Schloss and Murray (2009), p. 16.		

cognitive capacities that were selected for their survival value. In short, religion is an evolutionary accident. A similar argument might be applied to music, art or even the susceptibility to become addicted to drugs never encountered in the environment of evolutionary adaptedness (EEA) – they all 'piggy back' on neural systems selected for other purposes.

Stephen Jay Gould and Richard Lewontin (1979) used the architecture term 'spandrel' to illustrate this phenomenon. In building design a spandrel is the space left over when builders and architects combine circular domes and arches with rectangular bases. A spandrel emerges with no apparent function and was not intended as a solution to any problem but nevertheless can later be co-opted into a decorative scheme. Applying this perspective we might, for example, see religion as a product of a natural human longing for attachment to a caring other figure; or as a result of a tendency to see the world in anthropomorphic terms since anthropomorphism works perfectly well in explaining our social world.

This latter idea has been more extensively explored in the idea of a **'hypersensitive agency detection device'** (HADD). As we saw in Chapter 9 in the context of error management theory, a system that has evolved and served us well in the detection of animal and human agency at work is prone to false positives – in this case attributing perfectly natural phenomena to invisible (supernatural) agencies.

There are various sources of evidence to suggest that children have intuitive beliefs that make them receptive to a religious outlook. Margaret Evans, for example, found that children naturally tended

to favour creationist accounts of the origin of animals even when they were part of a secular family. It appeared as if creationism was the natural default setting of a young person's mind (Evans, 2001). Deborah Kelemen observed the presence of what she called 'promiscuous teleology' in the world view of children. In this mindset everything has a purpose or was made for a purpose. When, for example, children were asked to explain why some rocks were pointy children eschewed physical explanations in favour of teleological ones such as 'so children can scratch on them when they get itchy' (Kelemen, 1999). Later work showed that adults, especially when forced to make a quick judgement about an explanation, were also subject to this tendency, endorsing such teleological explanations as 'earthworms tunnel underground to aerate the soil' as correct (Kelemen and Rosset, 2009) (see also Chapter 7).

In a similar vein, Bloom (2007) explores the idea of natural cognitive biases present, he suggests, in the minds of children. He proposes that there are two fundamental and innate habits of thought that have contributed to the emergence and continuation of religious belief: agency and dualism. In proposing this he draws upon the work of the anthropologist Stewart Guthrie who in his book *Faces in the Clouds* adduced an impressive range of evidence to demonstrate how humans tend to see design in, and attribute human qualities to, natural phenomena and physical objects when purely naturalistic explanations are more reliable, or when, as he neatly put it, 'the clothes have no emperor' (Guthrie, 1993, p. 5). In respect of dualism, Bloom proposes that this emerges from the fact that children have two distinct cognitive systems, one for dealing with people and another for understanding material objects (Bloom, 2009). In essence, living things have intentions, goals, minds or even souls but objects do not. Once these thought systems are in place (and it is arguable that employing different concepts for reasoning about people and objects makes good sense) then it becomes possible to contemplate objects (such as chairs, tables and dead bodies) as existing without a soul whilst entraining the complementary idea of souls existing without a physical body. Hence there is the possibility that this intuitive dualism opens the mind to the possibility of life after death, immaterial ancestor spirits and creator gods that have no need of a visible physical presence; in short the panoply of religious belief.

One problem with this approach is that it might explain why humans are cognitively disposed to believe in spiritual beings and supernatural agencies but it does little to account for the social nature of religion: the group practices and rituals that bind religious communities together and often set up impenetrable barriers to other groups of a different religion. Indeed a thoroughgoing biological analysis of the phenomenon of religion must address the three ways in which religions affect people: what they experience, what they believe and what religiously motivated actions they perform.

20.7.3 Darwinian accounts

Under this heading there are various possibilities. A tendency towards religious beliefs may be an innate but vestigial trait: something that was adaptive in an ancestral environment but is so no longer. If this is the case then the focus should be on establishing if it is a natural human universal and what evolutionary function it may have once served that is not served any more – a demanding set of requirements.

Costly signalling theory offers the potential to explain why religious systems not only require members to express belief but also enact expensive (in terms of time, effort and sometimes resources) rituals. The argument goes that membership of a group brings about benefits from mutual cooperation, sharing and help. But such groups are vulnerable to the action of free riders – people who temporarily take benefits and fail to reciprocate by, for example, moving on to another group. To overcome this problem groups require some evidence of earnest intent, and, moreover, signals of commitment must be costly and not easy to fake – otherwise they could easily be imitated by free riders. In this view, religious obligations, such as rites of passage and time-consuming rituals, serve just such a function (Sosis and Alcorta, 2003). To assess the effect of costly signalling on the success of a group, Richard Sosis and Eric Bressler hypothesised that there would be a positive correlation between the longevity of communes and groups and the level of demands and restrictions placed on members (Sosis and Bressler, 2003). The logic of this is that the more costly the signal of membership the greater the commitment to the aims of a group, and, presumably, the more successful a group will be in surviving to serve its members since the more costly the signal the more potential benefits (that need to be shielded from outsiders) that could be expected. They collected data from 83 commune-like communities, both religious and secular, that originated in the USA in the 19th century.

From knowledge of the rules and customs of these communes it was possible to attach a quantitative measure of the restrictions and demands imposed on members in such domains as consumption (for example, avoidance of alcohol, tea, coffee, specific food types), material possessions, clothes and communication with the outside world. Figure 20.4 shows the results of this investigation. The costly signalling hypothesis is supported for religious groupings but not for secular ones – something that calls out for an explanation. One general possibility might be that religious rituals and observances enforce a stronger sense of belonging in that they reinforce a belief in supernatural entities that have an objective truth and impose rules that are transcendentally binding. Religious activities such as singing, chanting, dancing and joint prayer may also serve to psychologically bind an individual to a larger group, thereby increasing in-group solidarity and depressing selfishness. If this work is along the right lines, then it shows that costly signalling is not enough by itself to promote group success, and points to the necessity perhaps of a better theoretical integration of the belief and ritualistic components of religious observance and practice.

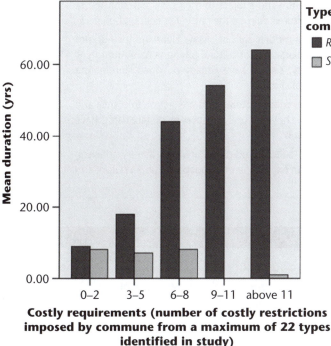

FIGURE 20.4 The duration of 30 religious communes and 53 secular communes in 19th-century America plotted according to the costly demands made on their members.

The results show a strong significant positive correlation between demands and durability for religious groups ($r^2 = 0.3697$, $p < 0.001$) but an insignificant correlation for secular ones ($r^2 = 0.0198$, $p = 0.31$). This offers some support for the costly signalling hypothesis of religious affiliation.

SOURCE: Plotted from data in Sosis and Bressler (2003); and Sosis and Alcorta (2003).

smallpox virus but harder to eradicate' (Dawkins, 1997, p. 26). But pathogenicity is not the only possible relationship between human biology and cultural evolution of religion. Religious ideas could be beneficial symbionts, in that they assist human flourishing (by encouraging practices that increase biological fitness), or simply commensalists that hitch a ride on our biology and cognitive apparatus but are fitness neutral.

The various models discussed above have many features in common. But there is the feeling that human culture is still a highly elusive and complex phenomenon. Dennett (1995) speaks of Darwinism as a 'universal acid' that passes through everything, leaving behind 'sounder versions of our most important ideas'. In the next chapter, we examine how evolutionary thinking can possibly offer a sounder version of another cultural phenomenon that has long attracted a bewildering variety of theories: our ethical and moral sensibilities and codes of conduct.

20.7.4 Gene–culture coevolution and dual inheritance theories

Religious ideas and beliefs could evolve independently as part of poorly understood cultural evolutionary processes or evolve in tandem with biological evolution in a number of possible ways. Religious beliefs could be 'memes' that are parasitical in the sense that they are transmitted like mind viruses from host to host at the expense of human biological interests. This was the view of Richard Dawkins when he suggested provocatively that 'faith is one of the world's greatest evils, comparable to the

Summary

▸ There are several models of how genes and culture could interact. Many of these models are similar and overlap in their concepts and concerns. Culture could be totally autonomous (silencing biologists); culture could be an expression of the long reach of the human genotype; culture and genes could coevolve together in a dynamic interaction; culture and genes could both be inherited and evolve independently and subject to their own selection forces (dual inheritance models); or culture could be the natural selection of memes.

- The dividing lines between these models are not always clear and they do not necessarily provide exclusive explanations of cultural phenomena. Although all these models are in their infancy, several have already provided illuminating and plausible accounts of cultural change.

- Two instructive example of gene–culture interactions are lactose tolerance and amylase gene number. Both show how cultural pressures on dietary habits have led to changes at the genetic level.

- One exciting extension of gene–culture studies is Darwinian literary criticism, which looks at how literature is an expression of an adapted human nature and in turn serves adaptive needs.

Key Words

- Conformist bias
- Consilience
- Culture
- Culturgens
- Dual inheritance theory
- Extended phenotype
- Hypersensitive agency detection device
- Meme
- Prestige bias

Further reading

Barkow, J. H., L. Cosmides and J. Tooby (1992). *The Adapted Mind*. Oxford University Press, Oxford.

An early manifesto for evolutionary psychology. See especially Chapter 1 on psychology and culture, and Chapter 18 on culture.

Blackmore, S. (1999). *The Meme Machine*. Oxford, Oxford University Press.

Bold claims are made for the power of the memes. A provocative work with an interesting analysis of the self.

Boyd, B., J. Carroll and J. Gottschall (eds.) (2010). *Evolution, Literature, and Film: A Reader*. New York, NY, Columbia University Press.

Excellent set of articles exploring how evolutionary theory helps make sense of great works of literature and film.

Carroll, J. (2004). *Literary Darwinism: Evolution, Human Nature, and Literature*. New York, NY, Routledge.

Joseph Carroll is perhaps the leading exponent exploring how Darwinism informs the study of literature.

Dunbar, R. I. M., and L. Barret (2007). *The Oxford Handbook of Evolutionary Psychology*. Oxford, Oxford University Press.

A substantial work containing many useful chapters in this whole field. Section VII covers recent ideas on cultural evolution from several authorities.

Gangestad, S. W. and J. A. Simpson (eds.) (2007). *The Evolution of Mind: Fundamental Questions and Controversies*. New York, NY, Guilford Press.

Sophisticated account of methodological and theoretical issues in the evolutionary approach to behaviour. Several chapters on culture.

Linquist, S. P. (2010). *The Evolution of Culture*. Farnham, UK, Ashgate.

Collection of articles spanning the field from 1975 onwards.

Richerson, P. J. and R. Boyd (2005). *Not by Genes Alone*. Chicago, University of Chicago Press.

Readable and interesting account of dual inheritance theory. The treatment of religion is especially interesting.

Schloss, J., and M. Murray (eds.) (2009). *The Believing Primate: Scientific, Philosophical, and Theological Reflections on the Origin of Religion*. Oxford University Press.

A series of chapters exploring scientific and philosophical dimensions of evolutionary approaches to religion.

Stewart-Williams, S. (2010). *Darwin, God and the Meaning of Life: How Evolutionary Theory Undermines Everything You Thought You Knew*. Cambridge, UK, Cambridge University Press.

Fundamental but accessible examination of the philosophical and theological implications of Darwinism.

Ethics

CHAPTER

21

In the same manner as various animals have some sense of beauty, though they admire widely different objects, so they might have a sense of right and wrong, though led by it to follow widely different lines of conduct. If, for instance, to take an extreme case, men were reared under precisely the same conditions as hive-bees, there can hardly be a doubt that our unmarried females would, like the worker-bees, think it a sacred duty to kill their brothers, and mothers would strive to kill their fertile daughters; and no one would think of interfering.

(Darwin, *The Descent of Man*, 1871, Vol 1, p. 73)

The status of moral values has been the subject of almost endless dispute. One tradition suggests they may be thoughts in the mind of God; another that they are eternal truths to be reached by rational reflection (much like the truths of mathematics); another that they are social conventions; and another that they reflect enduring features of human nature. The relationship of evolutionary biology to this question has had a troublesome history. Early efforts to draw moral guidance from the process of evolution, such as the recommendations of the social Darwinists, met with a critical response from moral philosophers. The effect of this has been that whole generations of moral philosophers have given the biological sciences a wide berth and consequently often remain poorly informed about recent advances in evolutionary thought and the neurosciences. On the other hand, scientists are beginning to obtain a good grasp of the evolution of the moral sentiments and the centres of the brain associated with emotion and motivation, but have been fearful of committing the naturalistic fallacy (see below) and so have fought shy of extrapolating their findings to ethical questions. After all, no one wants to be seen to be committing an elementary logical blunder. But in recent times this has begun to change. In this chapter, we examine the prospects for placing ethics on a Darwinian base.

21.1 Prospects for a naturalistic ethics

21.1.1 The naturalistic fallacy

Despite its legendary reputation for thwarting any attempt to link values with natural facts, the **naturalistic fallacy** is not as straightforward as might be supposed. In his *Treatise of Human Nature*, Hume wrote some passages that have been taken as proof that one cannot derive an 'ought' from an 'is' and that, accordingly, the world of facts and values must for ever remain separate. There is now a growing consensus, however, that this is far from what Hume intended (see Walter, 2006). Hume's main point was that moral sentiments are already facts (the 'is') of human nature – they exist as passions in the mind of humans. Hume's argument was that he could not see how reason alone could enable the connection between the 'is' and 'ought' to be made. Hume himself was frustrated that people misunderstood his argument. In 1752, he published a work designed to clarify his position called *An Enquiry Concerning the Principles of Morals*, in which he stated:

> The hypothesis we embrace is plain. It maintains that morality is determined by sentiment. It defines virtue

to be whatever mental action or quality give to a spectator the pleasing sentiment of approbation; and vice its contrary. (Quoted in Walter, 2006, p. 36)

So, far from wishing to place ethics beyond the reach of naturalism, Hume stands with the empiricists in opposition to the idea that morality must somehow be the process of grasping transcendental truths through the exercise of pure reason or other means. For Hume, moral truths are inherent facts about human nature.

21.1.2 A Darwinian updating of Hume

Many Darwinists (for example Curry, 2005) look to Hume for inspiration as to how ethics can be reconciled with evolution. One of the benefits of Darwinising Hume is that it helps to clarify what we mean by the term 'value' and provides an account of how values can exist in a world of facts. In Hume's view, human values are products of the passions, and passions include things such as love, hate, anger, malice, clemency and generosity. These passions or sentiments are inherent in human nature and those that promote the common good we call the 'moral passions'. In the Darwinian world, animals have adaptations that enable them to pursue their goals, such as moving towards food, moving away from predators, seeking mates, and fighting off competitors. This list of goals becomes the organism's values and the desires for these goals (and other mechanisms involved in their pursuit) are the adaptations laid down by natural selection. In the case of humans, our values are the goals of our adapted sentiments and moral values are those that serve the common good.

In this view, the question as to whether moral values are subjective (as in the tradition of emotivism) or objective (as in realism) receives an interesting answer. Moral values are subjective, in that they reside inside our heads, and objective, in that they are common to all (normally functioning) humans. They are as objective as, say, colour vision or the peacock's tail (Curry, 2005).

This perspective suggests that morality is a set of procedures to help humans, in the face of limited resources, limited sympathies, personal diversity and competitive instincts, to secure the fruits of cooperation and arrange their equitable distribution.

Morality is the name we give to those emotions and inclinations that steer us away from the temptation to cheat and reap immediate selfish rewards towards cooperative behaviour that helps us to reap the benefits of mutualism, reciprocal altruism and prisoners' dilemma-style cooperative interactions. Emotions such as sympathy, empathy and compassion enable us to experience the perspective of others and bind communities together. The crucial point is that whereas the original evolution of these tendencies took place according to the cold and ruthless logic of natural selection – they were selected for if they gave an advantage to the genes of those possessing them – we are now left enriched by these feelings that spill over into a whole range of contexts, enabling us to feel empathy with and pity for suffering even when we 'know' cognitively that any help we give is unlikely to be returned even by indirect means.

As we have discussed in previous chapters, psychological traits germane to moral theory that we are likely to possess from the study of the behaviour of humans and other animals are:

● Taking care of family and kin
● Sympathy for others
● Reciprocation
● Punishing cheats.

The essential importance of reciprocity in delivering mutual gains (of which the prisoners' dilemma is but one model; see Chapter 11) is illustrated by the importance that moral philosophers have attached to the emotions and feelings that are associated with it, such as trust, commitment, punishment, guilt and forgiveness.

This realisation that the roots of moral thinking and action must somehow lie deep in our natures explains the cogency of some of the objections often hurled against utilitarianism. Within mainstream moral philosophy, **utilitarianism** remains one of the most accessible and coherent single creeds. It is of undoubted use and its principles are widely applied in such things as cost–benefit analysis and the allocation of medical resources. Although there are various versions of utilitarianism, the underlying logic is that correct action is that which tends to promote the greatest good (pleasure or happiness) to the greatest number. It is a valuable philosophy but it is not foolproof. As an illustration, we can consider the argument William Godwin (husband of Mary Wollstonecraft) advanced in his *Political Justice* (1794). Godwin thought that morality (contra Hume) must be a product of reason. Imagine, he said, a burning building containing a famous author producing works of benefit to all mankind (we could update this argument by imagining a scientist on the verge of a cure for cancer) and a chambermaid who could be your wife or mother. You can only save one person, so who do you save? Godwin thought the answer was unquestionably the benefactor of mankind, since in the future, he or she would produce the greatest good. Moreover, it mattered not one jot if the chambermaid was related to you; as he said: 'What magic is there in the pronoun "my" to overturn the decision of impartial truth?' Most modern readers will not find the decision so easy – feeling that a wife or mother (or to update Godwin further, a husband, partner or father) surely has some pull on our obligations. Post Darwin we can answer Godwin: the magic in that 'my' is the product of hard-wired sensibilities shaped by millennia of kin selection that define our humanity and will always frustrate attempts to reduce morality to a calculus of pleasure and pain.

21.2 Game theory and moral philosophy

A promising line of research in the effort to naturalise ethics is the linkage between game theory and moral reasoning. Some interesting parallels emerge when

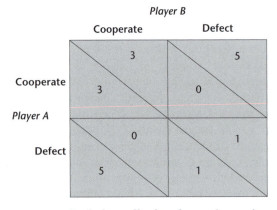

FIGURE 21.1 Typical payoff values for a prisoners' dilemma scenario.

we revisit the prisoners' dilemma in the light of some traditional ethics theories. Figure 21.1 shows a restatement of a typical prisoners' dilemma payoff matrix.

As we have observed, the problem posed by prisoners' dilemma situations is how to get individuals to cooperate for their long-term advantage. Anatol Rapoport (1965) introduced the idea of two kinds of rationality: individual and collective. Individual rationality is that which maximises the return to the individual; collective rationality is that which, if each follows, makes each better off than if each had simply used individual rationality. We have observed earlier (see Chapter 11) how it is likely that our emotional reactions in such situations serve this function of enforcing cooperation and reciprocity for the good of ourselves.

To enforce collective rationality, some have thought that what is needed is some central authority to ensure that both players cooperate for their own good. This was the position advocated by Thomas Hobbes in his famous work *Leviathan* (1651). Hobbes thought that in the natural state men were so egoistic and competitive that a restraining central authority (either a monarch or an assembly) was needed to ensure that civil society could flourish. Sadly, this does not really solve the game's dilemma, since the addition of an enforcer and a corresponding punishment regime changes the nature of the game by introducing a whole new set of payoffs (if there is a punishment for defection, then the rewards of defection have changed according to the probability of being found out). Also, the approach of Hobbes does not really help us to work out how biology solved the prisoners' dilemma, since in nature we can detect no Leviathan gently coercing organisms to cooperate. In addition, Hobbes's view of human life without government as 'solitary, poor, nasty, brutish and short' does not really accord with what we know about early human groups and hunter-gatherer societies.

It is instructive to consider the ethical system of another great philosopher, Immanuel Kant (1724–1804), in the light of game theory. Kant's approach to ethics in his *Critique of Practical Reason* (1788) had the ambition of attempting to find a universal basis for morality that is not dependent on subjective experience. He believed that morality cannot simply reside in human desires and inclinations since these vary between people. So Kant looks to the application of reason to establish universal moral laws since reason is peculiar to humans and shared by all. He then investigates what the rationally derived moral law might look like and applies the test of consistency. In his book *Groundwork of the Metaphysic of Morals* (1795) he summarised his argument as 'I am never to act otherwise than so that I could also will that my maxim should become a universal law'. Bertrand Russell responded to this by saying that it was really just a 'pompous' way of saying that we should 'do unto others as you would have them do unto you'.

Russell is perhaps a little unfair, in that Kant also thought that such maxims (or '**categorical imperatives**' as he called them) should also avoid logical self-contradiction. Hence the idea that 'I may sometimes make a promise that I will not keep' is wrong, since if everyone were to act according to this rule, then a promise would never be a promise and the rule would necessarily destroy itself. In essence, Kant's view was that morality flowed from the consideration of rational beings possessing immortal souls. We have a duty to follow the precepts that our reason reveals to us; similarly, we should distrust the emotional feelings that cloud our judgement when we face moral problems.

With game theory in mind, we can restate the categorical imperative as 'choose a strategy which, if chosen by everyone else, would lead to the best outcome from your point of view than any other'. This is consistent with Kant's insistence of treating people as ends and not means and his exhortation to only follow maxims that you can will to be a universal law without self-contradiction. We can see that 'cooperate' fulfils this criterion but 'defect' does not. If everyone defected, you would be worse off than if everyone cooperated. Kant's categorical imperative is really a version of the Biblical golden rule: 'do unto others as you would have them do unto you' (Luke, 6: 31). Kant thought that he had arrived at this through reason and that human beings as rational agents had a duty to follow its dictates. Indeed, Kant was so optimistic about the power of reason that he thought it was the answer to organising social affairs even for people who are inherently bad:

> The problem of organising a state, however hard it may seem, can be solved even for a race of devils, if only they are intelligent. (Kant, quoted in Barash, 2003, p. 132)

The problem is that Kant left open the question about what motive there is to cooperate. Individual rationality would dictate that we cheat if we can get away with it, and to cooperate we need virtues and these arise from affective not cognitive states.

Interestingly, a rule utilitarian would also come to the same conclusion that 'cooperate' is the best move to adopt. Utilitarians recommend acting so as to maximise happiness for the greatest number of people. Act utilitarians argue that each act should be considered in terms of its pros (creation of happiness units or hedons) and cons (creation of unhappiness units or dolons). For both players to cooperate yields a total of six units of benefit, higher than any other combination. Rule utilitarians argue that it is not always feasible to weigh up every action so we choose actions which, if chosen as a rule by everyone, would lead to the greatest happiness. It is clear that 'cooperate' would be the rule that utilitarians would advocate following. This contrast, and sometimes tension, between utilitarian and Kantian approaches is explored in the next section.

21.3 Trolley problems

Since Godwin's backfiring thought experiment, the use of hypothetical moral dilemmas has been a common way to explore the psychology of moral reasoning. In such studies, participants are given a scenario and asked what outcome is most desirable. One commonly explored dilemma is the 'trolley problem'.

The runaway trolley (or train) problem is a thought experiment in ethics. It was first formulated by the British philosopher Philippa Foot (1920–2010) who described it as part of a series of dilemmas where several lives can be saved but only by sacrificing one life. Since Foot's original description the original scenario and its many variants have been used extensively by psychologists, philosophers and neuroscientists; such studies have even been given the sobriquet of trolleyology.

In its basic form the following scenario is presented to a subject who is then asked to make a decision. A train is running out of control down a railway track (see Figure 21.2). On the track ahead are five people who are unable to move off the track and will be killed if the train goes ahead. You are standing near a level or switch such that if you operate it the train will be sent down a side track and only kill one person who is stuck on the line. Do you throw the switch? An important variant on this situation is the 'fat man on a footbridge' problem posed by the American philosopher Judith Thompson. In this case a train is once again hurtling out of control, down the track and destined to kill five people unless it is stopped. You are standing on a footbridge over the track next to a large overweight man. If you push the man on the tracks the trolley will stop. Do you push?

Findings tend to suggest that most people would throw the switch but not push the man (see Cushman et al., 2006). The interest lies in the reasons people give and the fact that the two situations elicit very different responses even though from a utilitarian perspective the end result is the same: one life sacrificed for five saved. It is a paradox that has exercised philosophers for several decades now. One suggestion is that people are unconsciously applying the 'doctrine of double effect' which states that you can take an action to promote the good if there are bad side effects, but you should not deliberately harm someone even for good causes. Consequentialists would tend to deny this distinction but in any case we might ask how this principle or doctrine gets into people's heads.

To investigate the roots of these two different judgements Greene and Haidt (2002) used brain imaging techniques (fMRI) to examine which bits of the brain were engaged when participants attempted to solve these two problems. The group found that in footbridge-type problems (which they called 'up close and personal') people reached a decision fairly quickly and moreover employed deontological type of reasoning (for example, it is never right to kill someone even to save others). Furthermore, this was associated with increased activity in areas of the brain associated with emotional and social information processing (for example, the medial prefrontal cortex). In the case of throwing the switch the decision took longer to reach than not pushing the man off the footbridge, and increased activity in areas of the brain associated with working memory such as the dorsolateral prefrontal cortex was observed. The conclusion that Green offers is that we make two types of moral judgements using different parts of our brains: quick emotion-based decisions which

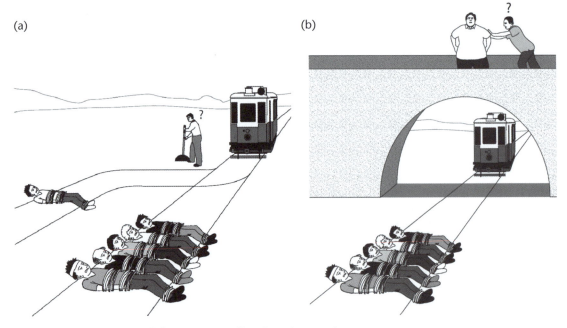

FIGURE 21.2 Two versions of the runaway trolley thought experiment.
In scenario a) the trolley will run over five people unless you throw the switch and divert it to run over one person. In scenario b) the trolley will run over five people unless you push the overweight man onto the tracks to stop it. What do you do?

are justified after the event usually through deontological styles of reasoning; and slower, more cognitive, judgements employing a utilitarian calculus.

Consequentialism and Kantianism

We should perhaps ask how evolution has come to provide us with two systems, which we might conveniently abbreviate as consequentialist (utilitarianism is the main type here) and Kantian (a form of deontology). A consequentialist system is not difficult to explain, at least in regard to actions directed towards relatives and friends. Indeed we might argue that kin-directed altruism in other animals is a form of protomorality. As we have seen in Chapter 11 animals will display kin-directed altruism and preferentially assist their own kin. But they will also sacrifice them according to the logic of inclusive fitness. Consider the taxon of insects known as the burying beetles (genus *Nicrophorus*). They get their name from the fact that they bury carcasses of small animals and use this entombed carrion to feed their larvae. Usually, both parents look after the offspring. If, however, the female has produced more young than the food supplied by the carcass can support, the parents kill some of the young, eat them and regurgitate the food to feed the remaining young. This infanticide makes good adaptive sense: better to have fewer offspring that are healthy and will survive than lots that are undernourished and not survive. The beetles are therefore applying a type of utilitarian calculus (in keeping with Williams Hamilton's ideas on altruism, discussed in Chapter 3) and presumably have no qualms about killing a few for the overall good.

The effect of inclusive fitness on the trolley dilemma has also been investigated. Bleske-Rechek et al. (2010) explored the effect of the genetic relatedness of the sacrificial victim to the observer on their

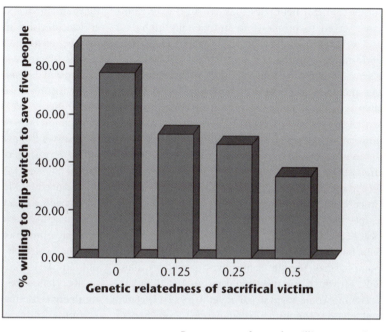

FIGURE 21.3 Percentage of people willing to sacrifice one life to save five lives in a trolley scenario in relation to the genetic relatedness between observer and victim.

SOURCE: Plotted from data in Bleske-Rechek, A., L. A. Nelson, J. P. Baker, M. W. Remiker, M. W and S. J. Brandt (2010). 'Evolution and the trolley problem: people save five over one unless the one is young, genetically related, or a romantic partner.' *Journal of Social, Evolutionary, and Cultural Psychology* **4**(3), 115. Copyright © 2010 by the American Psychological Association. Adapted with permission.

propensity to throw the switch and sacrifice one life to save five lives. The group reasoned that the higher the degree of relatedness of the victim to the observer to less likely they would be to throw the switch since here kin selection and inclusiveness fitness reasoning and feelings are in opposition to utilitarian reasoning. The study was conducted on 413 women and 239 men from an American suburb. The findings were in agreement with the inclusive fitness hypothesis (Figure 21.3).

In contrast, the Kantian style adherence to binding rules irrespective of consequences calls for a different type of explanation. If there is no calculation of loss and gain how could natural selection have shaped this response? One approach is to suggest that these deontological systems are simple and fast heuristics that averaged over time have served to promote an individual's interests and welfare in group situations (Gigerenzer, 2010). Performing a utilitarian cost–benefit analysis in each case would be too computationally demanding and so such systems are emotion based since they provide a quick-fire response. They provide 'satisficing' solutions even if they are not always accurate (see Chapter 9).

In case you are thinking that 'trolleyology' is an irrelevant academic exercise, there are real-life parallels. In the 1950s two scientists, Jonas Salk and Albert Bruce Sabin, independently produced vaccines (now named Salk and Sabin respectively) for the prevention of polio. Generally, Sabin is more effective than Salk, but, unlike Salk, in a very few cases it can cause polio. Overall though, Sabin saves more lives than Salk. So which vaccine should be used? Utilitarian reasoning would suggest Sabin; yet in 1999 a US federal advisory committee recommended a preference for Salk on the grounds that causing harm is worse than not avoiding harm. In the future as humans cede more control to machines some sort of quasi-ethical programming will need to be made. In the case of a driverless vehicle, for example, faced with an emergency situation whereby it cannot avoid a crash and different scenarios will result in different victims (for example, a mother pushing a pram, or an old man crossing the road) then what decision-making process do we program it with?

21.4 Evolutionary and affective approaches to moral development

The Triune Brain

A concern of long standing in mainstream psychology has been the development of moral thinking in the growing child and maturing adult. An influential paradigm has been the idea that just as children pass through cognitive stages so discrete stages in moral development can also be recognised and codified. But the problem with the vast majority of theories in moral psychology, such as the developmental approaches suggested by Jean Piaget (1896–1980) and later Lawrence Kohlberg (1927–87), is that they are not linked to the neurosciences; more specifically

they are unable to identify physical correlates to ethical decision-making processes. This problem may be overcome with the advent of neuroimaging studies which are able to highlight areas of the brain involved in different concepts and thought processes. At the theoretical level one novel idea is called triune ethics theory (TET), which suggests that responses to situations where we experience strong normative imperatives (for example, we feel a sense of right and wrong very often overriding other value judgements) can be aligned with the triune model of the brain as advanced by McLean (Narvaez, 2008). The TET is an intriguing idea with potential explanatory power. The stage-like development of moral thinking, for example, proposed by Piaget and Kohlberg may become explicable in terms of the differential areas of brain development through the first two decades of life. Hence 'post conventional reasoning' in the Kohlberg sense may only become possible once the prefrontal cortex is fully developed. The theory may also explain why some of our ethical intuitions seem immune to rational thought or require a laborious

Table 21.1 The triune ethics theory as envisaged by Narvaez (2008)

Area of brain	Type of ethic
Reptilian hind brain or R complex	*The ethic of security* Decisions concerned with security and physical survival. Concerned with self-interest – often the default system if others fail or are damaged. Can be inflexible and intolerant
Limbic system and related structures of mid brain	*The ethic of engagement* Gives empathy and compassion, care for others, sociability and kin-directed altruism. Can be conformist and a source of powerful emotions
Neocortex	*The ethic of imagination* Area of problem solving, rational thought and thinking (including integrating feelings into decision-making from lower centres of the brain). Coordinates and integrates inputs from the engagement ethic and the security ethic. It has the ability to (sometimes) countermand more primitive instincts and reflect upon what is virtuous behaviour. Can be slow and indecisive

post hoc justification: they emanate from parts of the brain not easily accessible to the neocortex. As a bold heuristic the theory may have some merit but work needs to be done on whether what we call ethics really does map naturally onto three categories. Also, brain imaging work may illuminate (literally) which areas of the brain really are involved in different types of ethical judgement and so support or refute TET.

The paradox of ethical theory

Cognitive and developmental approaches, with their emphasis on information processing, were not the only pathway followed by psychologists interested in ethics following the decline in behaviourism and Freudianism. Looking back, Wilson's remark in *Sociobiology* (1975) that ethics needs to be 'biologised' now seems especially prescient. Since this remark, discoveries in neuroscience and a more experimental approach to moral reasoning both suggest that we might often arrive at moral judgements without any conscious realisation of how we got there. Inspired by Wilson's call to arms and the work of neuroscientists such as Damasio and primatologists such as Frans de Wall, Jonathan Haidt (2008) has proposed a social intuitionist model (SIM) of moral development and judgement. This view suggests that moral judgement is often fast and virtually automatic and stems from parts of the brain (the limbic system) concerned with emotions. Haidt suggests that once we have passed moral judgement, or reached a decision, a common course of action is then to look for reasons to justify it:

> people engage in moral reasoning primarily to seek evidence in support of their initial intuition and also to resolve those rare but difficult cases when multiple intuitions conflict. (Haidt, 2008, p. 69)

Interestingly, a similar conclusion was reached by John Kaler (1999) who examined the usefulness of ethical theory (for example, reason-based philosophical systems designed to solve ethical problems) to the world of business. He concluded that ethical theory was in fact fairly useless, since established systems such as utilitarianism or deontology often gave unacceptable answers to ethical issues. Typical well-known examples include, in the case of

utilitarianism, sacrifice the interests of the innocent for the general good; and, in the case of deontology, follow rules and duties even when the consequences are dire. Often, argued Kaler, we somehow 'know' the outcome of the application of the theory to be wrong and so we beat a hasty retreat back along the theoretical path to make some upstream modifications to yield correct answers. Hence, as Kaler said, 'ethical theories tell us what we already know, depend on what we already know, and even end up resembling what we already know' (Kaler, 1999, p. 210). The paradox of ethical theory that Kaler identified becomes more understandable if we take a biological view of morality and appreciate that we may have made a quick moral decision based in intuitive and emotional centres of the brain. This idea is supported by experimental work on moral judgement making.

21.5 Moral foundations theory

21.5.1 Broadening the moral domain

The empirical investigation of moral psychology has made much use of questionnaires posing ethical dilemmas. Research here has been dominated by Kohlberg's idea about morality as a progressive development of concepts of justice and fairness. In her critique of Kohlberg, Carol Gilligan (1982) extended this narrow conception somewhat by pointing to the importance that people attach to such behaviour as caring for one another. More recently, behavioural economists and neuroscientists have been interested in the processes of moral decision-making as found, for example, in tackling runaway trolley problems discussed earlier. Economists have been fond of money- or credit-based games where subjects can choose to cooperate, cheat, or act fairly or unfairly in context of various levels of reward. All these approaches – Kohlberg's progressive levels, deciding what to do about a runaway trolley and acting fairly – have been primarily focused on rights, fairness and welfare and the moral quandaries that arise when interests clash.

It is tempting to think that this is after all what morality is all about: treating others with fairness and compassion, respecting their rights and autonomy, and being so treated and respected in turn. This conception, unproblematic as it sounds, is certainly in keeping with the central concerns of Enlightenment thinking as exemplified by the work of Kant, Bentham, Mill and more recently John Rawls. But recent work by Jonathan Haidt and others suggests that this conception of the moral domain is too narrow to capture what many people feel are legitimate moral questions relating to issues such as community, authority, sanctity and purity (see Greene and Haidt, 2002; Haidt and Joseph, 2004; Graham et al., 2011).

Haidt (2007b) has suggested that four principles need to guide the psychological investigation of morality:

a) The realisation that 'flashes of affect' accompany almost all human experiences. These provide a rapid basis for judgement (including moral judgement) that reason is later recruited to justify. Reason can be used to override these emotional intuitions but this is often difficult and quite rare.
b) Moral thinking supports social actions for personal benefits. As with the idea of Machiavellian intelligence, evolution teaches us that brain power and moral decision-making did not evolve to enable us to reach eternal truths, but rather to advance our own interests.
c) One of the prime functions of morality is to bind people to groups to reward prosocial actions and punish cheaters and slackers.
d) We must go beyond the two pillars view of morality (that became popular in the 1980s) that morality has two foundational concerns: fairness and justice, and protecting the weak and vulnerable. The study of what really matters to people in traditional cultures suggests that the study of morality be broadened to include such things as respect, loyalty, food taboos and disgust.

To broaden the domain of moral issues, Haidt and Joseph (2004) surveyed the literature in evolutionary psychology and anthropology looking for clusters of values and moral concerns that appeared regularly in a variety of cultures. This led to the proposal of five 'foundations' or domains which, the group argued, encapsulate the moral issues found in all cultures. Graham et al. (2007) developed these ideas which they called Moral Foundations Theory and renamed the five foundations as: harm and care,

fairness and reciprocity, in-group loyalty, authority and respect, and purity and sanctity. This model was further tested and supported by the construction and use of a Moral Foundations Questionnaire. The fourth version of this questionnaire was tested on 34,476 adults and found to be robust (Graham et al., 2011). These foundations, they argue, are akin to moral taste buds acquired through evolution and developed by experience. They appear as structures awaiting development in all (normal) newly born humans and were shaped by evolution in response to a long series of ancestral adaptive challenges. As with so many aspects of our species-typical psychology, however, the environment of modern life may be far removed from conditions to which our minds adapted causing some dispositions to lose their salience, or becoming triggered by stimuli for which they were not designed and yielding responses that have no adaptive value. Table 21.2 shows a summary of the central ideas of Moral Foundations Theory and, importantly, how our moral passions and sense of rightness and wrongness could potentially be activated by triggers in modern life for which they were not designed.

21.5.2 Applications of moral foundations theory

The idea of an innate universal moral grammar may help to explain why we respond with such intense feelings in a whole variety of situations. Stephen Pinker (2008) compares this morally charged system to a switch that can be turned on and off. Once the switch is thrown we respond according to which domain has been activated and then, if pressed to justify our actions and feelings, search for post hoc rationalisations of this 'righteous indignation'.

To experience this for yourself, consider the following set of scenarios; c) and d) are based on real events. Some are culturally specific and you may not feel very strongly that moral issues have been raised. Read each scenario and monitor your own reactions.

a) Steve and Julie enjoy each other's company and have been good friends for years. They plan a holiday together during which one night they

Table 21.2 The moral domains theory of morality according to Haidt and Joseph (2007). The theory is that five intuitive moral domains are activated by morally relevant situations. These domains were laid down by natural selection to serve adaptive needs. Now they are also activated by other triggers. See also Haidt, J. (2007b).

	FIVE MORAL DOMAINS				
	Harm and care	**Fairness and reciprocity**	**Loyalty to group**	**Respect for authority**	**Sanctity and purity**
Adaptive problem addressed	Protection of young and vulnerable, especially kin	To reap benefits of mutualism and reciprocal altruism	Advantages from group living, for example protection	Negotiate hierarchies, exploit one's position in chain of social dominance	Avoidance of infectious agents such as parasites and microbes. Intuitive microbiology
Proper domain and triggers in the EEA	Suffering of or threat to kin	Exchanges and detection of cooperators and cheaters	Maintaining group strength and cohesion	Awareness of signs of submission and dominance	Avoiding diseased people, contaminated food, waste products
Some triggers in novel environment of post-Neolithic cultures	Cruelty to innocent animals, empathy for fictional characters	Vending machines that take money but do not work	Support for nation or local team in competitive tournaments	Authority figures, class consciousness	Taboo idea, xenophobia, sanctity of religious artefacts and texts
Emotions involved in the response	Compassion, empathy	Anger, gratitude, guilt	Pride, sense of belonging	Respect, fear	Disgust
Characteristic virtues	Empathy, kindness, caring	Trustworthiness, honesty, integrity	Loyalty, patriotism 'taking one for the team'	Respectfulness, deference	Piety, temperance
Characteristic vices	Cruelty, heartlessness	Dishonesty	Cowardice, treason	Disrespect	Sacrilegiousness, heresy
EEA = Environment of evolutionary adaptedness					

decide to have consensual sexual intercourse. Steve recently had a vasectomy and Julie is taking an oral contraceptive. The sexual relationship continues for a few months and then they decide to stop. They remain good friends and see each other frequently. Steve and Julie are also brother and sister.

b) You are a vegetarian invited to a dinner party. Your host has prepared a vegetable soup just for you and a beef broth for the others. The host serves out the beef broth and then uses the same ladle to serve you your vegetable soup.

c) A British teacher working in Khartoum in Sudan allows her class to name a teddy bear Mohammed – the name of the founder of Islam.

d) In the UK in 2011 and 2012 the rising price of metal led to thieves stealing bronze plaques bearing the names of fallen soldiers from war memorials around the country.

In case a) you will probably feel the action to be morally reprehensible, even disgusting, as soon as you realise that this is incest. If prompted to explain why this should be so wrong you may resort to pointing out that it is against the law (although this does not solve the problem of why it is morally as well as legally wrong). Or you may cite religious texts where incest is condemned – essentially employing the 'divine command theory' of morality, which has its own philosophical problems. Consequentialists might be tempted to argue that incest, even if consensual, is wrong since it could produce a child with genetic abnormalities or cause psychological damage to the perpetrators. In this case, however, it is clear that children cannot result and that the couple do not seem to be adversely affected by the experience. Despite these rejoinders, you probably still have an inner voice popping up insisting that it must be wrong for some reason. In terms of Moral Foundations Theory (MFT) we might say that the sanctity/purity type of moral domain has become involved: we sense that some wrongful act has been committed violating purity concerns, we experience the force of our own moral condemnation, and then search around for a rationalisation of what is a rapid heuristic and essentially affective response.

In case b) you may have experienced something similar if you are a vegetarian or observed something similar if you are not. Many vegetarians would find the thought of even consuming a small quantity of meat products repulsive. Yet in this case the disgust emotion is not strongly linked to the rational arguments for vegetarianism. Avoiding meat may be justified intellectually on ecological, health and animal suffering grounds. But in this case the small quantity consumed has virtually zero effect on any of these dimensions (your actions will not increase animal suffering or damage ecosystems or affect your health in any measurable way). Yet the feeling is strong and real and should, of course, be respected by the host. In terms of Moral Foundations Theory this is probably another example of an issue moralised within the sanctity/purity domain.

Case c) actually happened, caused an outcry and the teacher concerned was sentenced to 15 days in prison. About 10,000 protestors took to the streets in Khartoum, many demanding the death sentence. Although it was the children who chose the name clearly the teacher's action had inflamed religious and moral sensibilities. In terms of MFT this is likely to fall under the respect for authority domain.

In case d) the actions of the thieves was vigorously condemned. The reaction was more deeply felt than for most other thefts since it touched upon many moral dimensions. In terms of MFT it may have activated the domains of fairness and reciprocity (it is wrong to take another person's property), loyalty to group (the names on the metal plaques were those who had sacrificed their lives for the wider group and the objects themselves are part of a national heritage), possibly respect for authority (since it was a public monument), and sanctity and purity

(these monuments often have a religious symbolism attached). No wonder the MP Norman Baker called the thefts 'morally repugnant' (Millward, 2011).

21.5.3 Morality: universal or culturally relative?

One of the toughest challenges for a theory of morality that sees it in terms of evolutionarily derived adaptation is to explain how a supposedly innate moral sense can be both universal (since we are a single species) and (as experience readily demonstrates) culturally variable. The answer may be that culture determines in some way what aspects of personal and social life (sex, diet, relationships, and so on) are drawn into the various moral domains such as the five foundations; and, just as importantly, culture influences how conflicting moral arguments are ranked in importance. The meticulous adherence to dietary rules by Hindus and orthodox Jews (for example, avoiding even minute quantities of certain foodstuffs such as pork), so puzzling to outsiders, become more intelligible if we see them as an evocation of purity concepts. Western business practices make a virtue out of equal opportunities and usually aim to root out nepotism in the name of fairness. But in other parts of the world, particularly Asian countries, favouring relatives over strangers is obviously the virtuous thing to do since it is in accord with notions of care for kin and group loyalty. Similarly the outrage felt by many Muslims if their prophet is insulted is intelligible in terms of the importance attached to authority and the activation of this domain by insensitive actions.

Differences between people within a culture can be interpreted in terms of the values attached to one or more of these morally relevant domains by people of different ideological groups. Haidt and Graham, for example, studying both American and UK subjects, found that liberals attached much more importance to harm and fairness considerations than did conservatives; conversely, conservatives attached more importance to in-group loyalty and purity than did liberals (Haidt and Graham, 2007). By surveying a group of 37,476 participants who completed an online Moral Foundations Questionnaire (MFQ) (www.yourmorals.org), Graham et al. (2011) argue that this liberal–conservative divide is a robust feature of many cultures. In looking at differences between cultures, the group also found that the biggest difference between Western (USA, Canada, UK and Western Europe) and Eastern (South Asia, East Asia, and SW Asia) participants came from the importance attached to the domains of community (in-group/loyalty) and purity, with Eastern respondents rating both these concerns more highly than Western ones. They only found a slight difference between East and West in the importance attached to authority, a finding that is surprising given the value attached to social hierarchy in many Eastern nations. The MFQ also revealed an interesting set of gender differences: women valued harm and care, fairness, and purity higher than men; whilst men valued authority and group loyalty marginally higher than women. These differences held even when corrected for political ideology. Gender differences were also much stronger than differences between Western and Eastern cultures.

This research opens up the exciting possibility of understanding the roots of what might seem like intransigent moral positions and irreconcilable differences. There is a need, however, for caution and further work. The sample size, though large (tens of thousands) was not perfectly representative – participants were English-speaking adults with access to the internet. The measures collected were also entirely self-reported. It is also not entirely clear how an issue becomes moralised as it were and enters one of the five domains purported to exist. Even so, the very idea of discrete and sometimes competing domains of moral concerns is surely an advance on a linear view of moral development – with its emphasis on cognitive development and progression towards a concern with a few abstract principles such as justice and fairness – and makes more of an attempt to capture the biologically rooted sources of our moral passions. The next section explores how local ecological variables, such as pathogen density, can shape moral responses.

21.5.4 Morality, pathogen prevalence and the behavioural immune system

Moral sentiments or psychological mechanisms help people survive and reproduce. They do so by encouraging people to cooperate to gain the benefits of mutualism and non-zero sum games; they ensure cheaters are punished; and they bind people appropriately to a collective good to reap individual benefits. It is also possible that they act to keep humans out of harm's way, where harm could be caused by external agents. Of the five moral domains identified by MFT, two – harm/care and fairness/reciprocity – have been called 'individuating' in that they are concerned with regulating how individuals interact with each other. The other three – in-group/loyalty, authority/respect, and purity/sanctity – have been called 'binding' in that they address behaviour that binds people to a larger social unit. Florian van Leeuwen et al. (2012) investigated if there was a cross-cultural relationship between these latter group-focused moral concerns and regional parasite stress. The logic here is that humans have evolved two types of defensive systems to cope with the constant threat from pathogens in the environment. The first is, of course, a complex immune system shared with other vertebrates; the second is what has been called a '**behavioural immune system**' (BIS). The name for this system was coined by the psychologist Mark Schaller (see Schaller and Park, 2011). He envisaged a whole suite of psychological adaptations that does three things: a) detects cues indicating the presence of pathogens, b) triggers emotional and cognitive responses, and c) motivates avoidance behaviour that reduces the threat of pathogen infection. This system is a cost-effective first line of defence since it will usually be better to avoid contamination and infection in the first place than to expend valuable biological resources fighting infection. One of the primary components of the BIS is purported to be the emotion of disgust – an emotion that is particularly effective in promoting avoidance behaviour. Disgust is an extreme emotion and directs people to avoid even the slightest contact with or ingestion of unpleasant material. Given that microbes are small we might say that disgust is a sort of intuitive microbiology. It can be activated by a whole range of stimuli such as the taste of food that has 'gone off', the odour of say faeces and vomit, and visible signs of potential contamination such as sick people or even unusual-looking strangers.

Van Leeuwen et al. hypothesised that a culture where pathogen prevalence was high would strongly endorse binding moral principles since it is these principles, such as group loyalty (and hence suspicion of strangers), respect for authority and tradition, ethnocentrism, and concerns for purity and sanctity, that are likely to reduce transmission of pathogens between individuals in a group and between individuals outside and inside a group. They collected data from 120,778 visitors to the website www.yourmorals.org who left a record of their moral outlook by filling in the MFQ and stated their country of origin. Data on pathogen prevalence in these countries was gathered from previous surveys by Murray and Schaller (2010) for historical data, and from Fincher et al. (2008) for contemporary data. Correlation measures were obtained between the strength of moral outlook along the five domains of MFT and pathogen prevalence. They found that pathogen prevalence correlated significantly with all three measures of 'binding' morality for historical pathogen prevalence but less strongly so for contemporary pathogen prevalence (see Table 21.3). This could be the result of some sort of cultural inertia linking values and environmental conditions. In both studies estimates were controlled for GDP since it is well established that low GDP is also strongly associated with pathogen stress.

If such findings as these prove robust, the question becomes how to explain why pathogen

Table 21.3 Correlations between moral concerns and pathogen prevalence.
The strong and significant correlations between historical pathogen prevalence and the three domains of binding morality lend support to the idea that these moral systems are part of a behavioural immune system.

Moral foundation domain	Historical pathogen prevalence		Contemporary pathogen prevalence	
	Correlation (r)	p	Correlation (r)	p
Harm	0.05	0.632	0.05	0.653
Fairness	0.03	0.799	0.07	0.545
In-group	0.37	0.001	0.15	0.207
Authority	0.48	< 0.001	0.21	0.062
Purity	0.42	< 0.001	0.21	0.068

SOURCE: Data taken from van Leeuwen et al. (2012), Table 2, p. 434.

prevalence predicts not only purity and sanctity concerns (as might be expected from basic hygiene considerations) but also in-group and authority concerns. Some proponents of MFT suggest that these three domains evolved in response to separate adaptive problems (see Table 21.2). But it is possible that there is overlap both in the type of adaptive problems that shaped them and any psychological mechanisms that might be in place. In support of this there is evidence that negative attitudes to out-groups among pregnant women increases with disease vulnerability (Navarrete et al., 2007).

In another study, Murray and Schaller (2012) tested the idea that a perceived threat of infection would positively influence conformist attitudes and behaviour. They allocated 217 undergraduate subjects into three groups defined as: disease threat, other threat (e.g. physical harm) and neutral. Participants were then asked to recall a time when they felt vulnerable to disease (the disease group), fearful of their physical safety (physical threat group) or simply recall the activities of the previous day (neutral group). Following this recollection process the volunteers were then asked questions to elicit their conformist values, such as their liking for people with conformist views; the value they attached to obedience; and conformity to a majority opinion. Across all measures, the value attached to conformity was higher for the disease threat group than for the other groups. It was only significant at the $p < 0.05$ level, however, for the measure of conformity to a majority view (Figure 21.4).

John Terrizzi et al. (2013) investigated this link between disease risk and conservatism more broadly through a meta-analysis of 24 studies addressing the hypothesis that the strength of the behavioural immune system (for example, fear of contamination and disgust sensitivity) is positively correlated with socially conservative values (for example, political conservatism, ethnocentrism, religious fundamentalism and right-wing authoritarianism). They found that the literature did show a moderate overall positive correlation. Clearly worries about contracting an infectious disease are not the only source of conservative values, but the work does provide an interesting perspective on the linkage between ecology, evolutionary theory and culture. The BIS promises to be an interesting phenomenon.

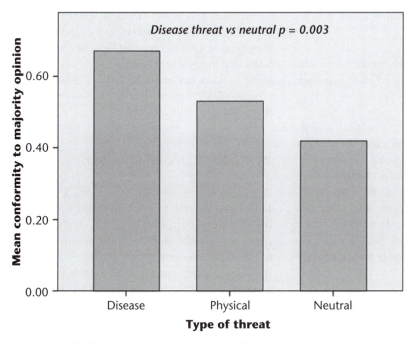

FIGURE 21.4 Conformity to a majority opinion is higher in contexts of disease threat.
Conformity was measured by tendency to express agreement with a neutral proposition for an administrative change by placing a penny in an empty jar (disagree) or a jar already full of pennies (agree). Maximum score = 1.0.
SOURCE: Data from Murray and Schaller (2012).

21.6 But is it right?

You may have noticed that we have been skirting around a central question: can this view of morality provide a sufficient basis for judging if an action is morally right?

We have been dealing with at least two layers of morality. One is the phenomenon of moral behaviour, of which a good Darwinian may be able to give a plausible account; in other words, why people erect rules and choose to live by them, and how such rules relate (or once related) to fitness gains in any given environment. Another layer is the question of whether such codes and rules are 'really' right. A large number of people, from T. H. Huxley onwards, have argued passionately that ethics transcends nature and have despaired at any attempt to draw ethical premises from evolutionary thought.

A modern exponent of this view was the Harvard biologist Stephen Jay Gould. Gould (1998, p. 21), who did much to expose the sexist and racist bias in historic attempts to capture human nature, was of the opinion that 'evolution in general (and the theory of natural selection in particular) cannot legitimately buttress any particular moral or social philosophy'.

In this whole debate, there does seem to be an entanglement of at least three different issues. The first concerns the ontological status of moral values or where the good resides; the second concerns how to justify (in the light of their ontological status) the 'rightness' of these values; and the third is how we use knowledge of this status to solve moral problems and resolve conflicts. The problem for the Darwinian approach is that it is not clear how to answer the second and third concerns from knowledge of the answer to the first. Darwinians are clear that the ontological status of the moral sense resides in those sensibilities (sense of fairness, purity, right conduct and justice) that natural selection supplied us with. But here is the crux of the matter: in justifying the rightness of these, we have to stop somewhere, otherwise we can keep asking how the justification argument (whatever it is) is also justified and so on ad infinitum (Curry, 2005). Theologians might stop at divine commands. Kantians stop at rules that can be generalised without contradiction. Relativists stop at the idea that such values are social conventions outside any transcultural justification. Utilitarians stop by equating pleasure or happiness with the good. Darwinians stop at psychology: our psychological adaptations incline us to recognise something as good or bad and these sensations of 'rightness' gained their force according to whether those dispositions once helped solved problems of social interaction and ecological survival. This is the view suggested by Larry Arnhart (2005, p. 208), when he wrote:

> Correct moral judgements are factual judgements about the species typical pattern of moral sentiments in specified circumstances.

So is morality merely a subjective experience with no objective foundation? The fact that we find ripe fruits tasty (or at least many of them) and excrement disgusting is due to fitness-related features of our evolved nervous system. Fruits are not intrinsically tasty and excrement not intrinsically disgusting since many creatures dislike fruit and hundreds of insect species seem to adore excrement. So is morality like this? Are genocide and slavery only wrong as a species-typical convention or can morality have some objective foundation?

Stephen Pinker argues that there is a case for the acceptance of a soft form of moral realism by pointing to the features of the real world that any social animal must face (Pinker, 2008). One of these features is the prevalence of non-zero sum games. The world is so structured that in many fitness-related tasks (such as finding food) two parties are often better off if they cooperate than if they act entirely selfishly or cheat (see Chapter 11). You might be good at making spears but hopeless at throwing them to catch food; your neighbour may be excellent at throwing spears at prey but lack the wherewithal to fashion them. You will both be better off by trading spears for food. Crucially, this is not just some property peculiar to a species but a reflection of the way the physical world works. Another objective anchor, according to Pinker, is the nature of reasoning. If we adopt the reasonable premise that no human individual has a vantage point privileged over all others then any rule that you would like others to follow to protect your interests (for example, do not steal my property) you must also accept as binding on your own behaviour. This idea has been expressed many times in the history of ethical thinking, including the golden rule (do unto others as you would be done by), Kant's categorical imperative and John Rawls' veil of ignorance. It appears also in less lofty idioms such as 'What's sauce for the goose is sauce for the gander'.

It is plausible, then, to think that morality, whilst rooted in our evolved natures and a product of our long experience as a social species, is also something larger and the product of natural selection acting on social animals but constrained by a distinctive set of physical and logical conditions. To some degree, then, morality is not entirely arbitrary.

21.7 Solving moral problems

So what does it mean to say that action X is wrong? In this new view, it could mean that doing X is not the way to reach agreed ends. We might wish to argue, for example, that the death penalty for adultery is morally incorrect since it is not the best way to ensure a happy marriage and a stable family and might, from a utilitarian perspective, end up making society less content and healthy. Or we might mean that X is not an accurate reflection of natural moral sentiments. Notice here that we have to use the word 'moral' in addition to natural. Killing a rival for some minor infringement might be *naturally* tempting for some people but it is hardy a morally correct course of action since it would not promote the general good.

One positive outcome of this evolution-based view of the status of morals and the nature of moral development might be the reinforcement of the realisation that someone you are in fundamental disagreement with might not be guilty of amorality (psychopaths excluded) but might be ordering the importance of the various moral dimensions to a situation differently to yourself. The person objecting to positive discrimination, to increase the proportion of ethnic minorities in specific professions, for example, might not be arguing from a position of racial prejudice but instead be stressing the imperative of fairness. Conversely, the proponent of positive discrimination might be arguing from the moral urgency of community.

One intriguing outcome of this way of thinking is the realisation that the moral sense might be subject to illusions just as our other senses are. Issues may trigger a moral response, and all the associated heat and passion, when a calmer, rational and practical response is more appropriate. As Pinker says of the moral sense: 'It is apt to reframe practical problems as moral crusades and see their solutions in punitive aggression. It imposes taboos that make certain ideas "indiscussable"' (Pinker, 2008, p. 32). This seems especially so in the context of morally irrelevant notions of purity as seen in such contexts as: the segregation policies of the USA before the 1960s and in South Africa up to 1991, where whites and blacks were required to use different swimming pools and drinking fountains; the taboo the Nazi party placed on intermarriage between Jewish and Aryan 'types'; and the legal prohibition placed on sex between consenting same-sex adults that still exists in many countries. On the positive side, practices that once pressed our moral purity buttons can become demoralised and accepted as parts of modern life. Examples here include dissection of dead bodies, blood transfusions, artificial insemination and organ transplants; all of these were at various times viewed with repugnance and are now largely accepted.

In this latter point lies the hope of not being restricted to facing the 21st century with a Stone Age morality. We could use our reason to formulate game theoretic and contractarian approaches to investigate how benefits can be maximised and spread, whilst relying on our innate moral sensibilities to drive the enterprise with some urgency and give moral force to its conclusions. We saw that with the prisoners' dilemma individually rational action led to an overall detriment. In this sense moral sentiment served the function of preventing failures in individual rationality. The reverse of this is also likely to be true: rationality can help prevent the failures of moral sentiment, for we have no reason to expect that moral passions forged in the evolutionary experience of our species will be entirely adequate for the global challenges ahead. Many aspects of modern life throw up dilemmas and polylemmas that are difficult because such anciently derived values are difficult to apply in modern situations. This may be the case with the debate over the morality of cloning and abortion, for example. Another example concerns global warming, where mutual gains (industrial development) cause future problems on a timescale that we find hard to factor into our decision-making or too abstract to comprehend with our very human-oriented value systems.

A moral response that emanates from the 'yuck factor' (as seems to be the case in many objections to genetically modified food, for example) might not be our most reliable guide to responsible action. The tricky problem, as Greene (2007) notes, is correcting the limitations of our phylogenetically contingent and emotionally dependent moral intuitions without undermining what it means to be truly human.

Summary

- ▲ The acceptance of a Darwinian approach to morality does not imply that morally responsible behaviour is impossible.

- ▲ The treatment of ethics has traditionally been dominated by transcendentalist and rationalist approaches. Darwinism seeks to understand ethical reasoning and moral behaviour empirically, relating our moral sense to natural adaptations for group living.

- ▲ Much headway in the naturalistic approach to ethics can be made by combining the insights of Darwin and Hume.

- ▲ Game theory is a useful conceptual schema for understanding moral interactions and moral dilemmas. There is evidence that we have natural dispositions to solve prisoners' dilemma-type situations in mutually beneficial ways.

- ▲ One recent advance in conceptualising moral development has been Moral Foundations Theory. This model proposes five areas of moral concern that develop in all humans, but to different degrees depending on temperament and culture. These five domains are then activated by different triggers.

➤ Interesting developments in the understanding of moral responses have come from hypothetical scenarios, or thought experiments, such as the runaway trolley problem. This has once again highlighted the paradox of ethical theory: that we often arrive at a moral judgement quickly and instinctively and then search for an intellectual theory to justify our response.

➤ On the subject of meta-ethics, Darwinists recognise no source of absolute authority and 'rightness' outside the evolved human mind.

Key Words

- **Behavioural immune system**
- **Categorical imperative**
- **Naturalistic fallacy**
- **Utilitarianism**

Further reading

Arnhart, L. (1998). *Darwinian Natural Right: The Biological Ethics of Human Nature.* New York, State University of New York Press.

Arnhart makes a brave attempt to link facts and values and show that biology can provide a sound basis for ethical thought.

Ridley, M. (1996). *The Origins of Virtue.* London, Viking.

Discusses game theory and human cooperation. Interesting and controversial application of game theory to politics and environmental issues.

Ruse, M. (2012). *The Philosophy of Human Evolution.* Cambridge, UK, Cambridge University Press.

See Chapter 6 for a discussion of morality from a leading philosopher of biology.

Schaller, M. and J. H. Park (2011). 'The behavioral immune system (and why it matters).' *Current Directions in Psychological Science* **20**(2), 99–103.

Useful overview of the impact of pathogens on moral behaviour.

Stewart-Williams, S. (2010). *Darwin, God and the Meaning of Life: How Evolutionary Theory Undermines Everything You Thought You Knew.* Cambridge, UK, Cambridge University Press.

Fundamental but accessible examination of the philosophical and theological implications of Darwinism. Part III is concerned with moral judgement.

Wilson, E. O. (1998). *Consilience: The Unity of Knowledge.* New York, Knopf.

A powerful and broad-ranging work in which Wilson airs his conviction that all knowledge must be linked together. Wilson attempts to show that ethics can be understood in evolutionary terms.

Wright, R. (1994). *The Moral Animal.* London, Little, Brown.

A readable work that explores to implication of Darwinism for understanding our moral sentiments and the problems of modern social life. Wright makes ingenious use of Darwin's own life to illustrate his arguments.

Glossary

Acclimation and acclimatisation. Terms often used interchangeably to refer to the way an organism will respond in an adaptive way to gradual environmental changes to enable it to function more effectively. Changes are often reversible.

Adaptation. A feature of an organism that has been shaped by natural selection such that it enhances the fitness of its possessor. Adaptation can also refer to the process by which the differential survival of genes moulds a particular trait so that it now appears designed for some particular survival-related purpose.

Adaptive significance. The way in which the existence of a physical or behavioural feature can be related to the function it served and may continue to serve in helping an animal to survive and reproduce.

Adaptive landscape. The idea that adaptive solutions can be understood by the metaphor of a landscape. A feature is successful if it sits near the peak of a landscape but it is difficult for features to move down an adaptive gradient to another type of solution.

Affective. In psychology, this means pertaining to emotion as opposed to cognition. Often used in the context of affective disorders that are disturbances of mood or emotion.

Affinal. Individuals related through marriage and not through genetic descent.

Allele. A particular form or variant of a single gene that exists at a given locus on the genome of an individual. There may be many forms of alleles within a population of one species. A simple example is eye colour, alleles for which exist in two forms: blue or brown. Each gene occupies a specific position on the chromosome called the **locus**. At each locus, a human possesses two alleles, one inherited from the father and one from the mother. An allele is therefore a sequence of nucleotides on the DNA molecule.

Allometry. The relationship between the size of an organism as measured by, for example, length, volume or body mass, and the size of a single feature such as brain size. The relationship can often be expressed by mathematical allometric functions or graphs showing allometric lines.

Altricial. Refers to offspring that are born in a state of helplessness, requiring constant care and attention. Humans are an altricial species and so are many types of bird.

Altruism. Self-sacrificing behaviour, whereby one individual sacrifices some component of its reproductive value for another individual. Self-sacrifice for a relative is termed kin altruism. What may appear as help or self-sacrifice may be enlightened self-interest, as in mutualism, or reciprocal altruism, where a favour is given in the expectation of some return eventually. Sacrifice at the level of the phenotype can be interpreted in terms of the 'self-interest' of the genes involved.

Amino acid. The molecular building blocks of proteins. There are 20 main amino acids in the proteins found in organisms. The particular sequence of amino acids in a protein determines its properties and is itself related to base sequences on DNA.

Amygdala. A small almond-shaped structure found in both hemispheres of the brain. The amygdalae form part of the limbic system and are involved in the processing of emotions.

Androgens. Male sex hormones such as testosterone.

Androphilia. Attraction to men or masculinity. This term and **gynephilia** refer to the object of attraction without specifying the sex of the person attracted.

Anisogamy. A situation where the gametes from sexually reproducing species are of different sizes. Males produce small, highly mobile gametes in large numbers; females produce fewer and larger eggs.

Antagonistic pleiotropy. Where genes can have multiple influences that oppose each other. For example, a set of genes might increase fertility in early life but cause disease after the reproductive period.

Archaic *Homo sapiens*. A term once given to a species of hominin that seemed intermediary between *Homo erectus* and *Homo sapiens*.

Ardipithecus ramidus. A very early member of the hominin genus (and hence an ancestor of humans). Lived in Africa about 4.4 million years ago.

Asexual reproduction. Production of offspring without sexual fertilisation of eggs (see **parthenogenesis**).

Asymmetry. A measure of the departure from symmetry of features that could in principle be symmetrical. Asymmetry is thought to be increased by poor conditions and stress.

Attachment and attachment theory. When applied to humans, this refers to the emotional bonding of one individual to another, usually a parent and child. John Bowlby developed attachment theory in a series of seminal papers published in 1958.

Australopithecines. The earliest hominids that appeared on the African plains about four million years ago.

Autism. A disorder describing people who have difficulty with language and social relationships. One recent suggestion is that autistic individuals lack a complete **theory of mind**.

Autonomic nervous system (ANS). A division of the nervous system involving important processes such as heart rate, breathing, digestion and sweating that are not normally under conscious control.

Averaged. Not to be confused with average. Usually applied to composite facial images that are combinations of multiple images, the composite being an average of the faces chosen to be compared.

Baldwin effect. A process (named after James Baldwin) whereby behavioural changes initiated by the phenotype can ultimately bring about genetic change. One example is the evolution of lactose tolerance.

Base-rate fallacy. An error of reasoning such that an individual fails to take account of the incidence (base rate) of some factor in the wider population.

Base. DNA consists of a phosphate–sugar backbone with attached nitrogenous bases. In DNA, a base can either be cytosine (C), guanine (G), adenine (A) or thymine (T). In RNA, uracil (U) is substituted for thymine. The precise sequence of bases on DNA serves as a set of instructions for building the cell.

Behaviourism. The school of psychology largely founded by Watson that suggests that observable behaviour should be the subject matter of psychology.

Bifidobacteria. Members of the genus Bifidobacterium – bacteria that have a Y shape (hence bifid) and often perform useful functions in the human intestine.

Biological determinism. A belief (often referred to pejoratively) that behaviour is caused by the biology of an individual such as physiology or genetics.

Bipolar depression. Sometimes called bipolar affective disorder. A clinical condition whereby the sufferer oscillates between periods of mania and depression.

Body mass index (BMI). A measurement of the relative weight of an individual taking into account height. It is calculated as weight/height^2.

Bounded rationality. A term that suggests people make decisions using limited and non-complete information; in contradistinction to classical rationality, which assumes that all factors are taken into account.

Broca's area. A part of the brain involved in speech production and language comprehension. The area is named after the 19th-century physician Paul Broca.

Carrier. An individual who is heterozygous for an inherited trait (often a disorder), but does not display the symptoms of the disorder.

Categorical imperative. A concept central to the whole moral philosophy of Immanuel Kant. Kant thought such imperatives, by their very nature, to be rules we have to follow, irrespective of our desires and circumstances. To arrive at such imperatives, Kant devised a number of formulations. The most commonly encountered is the idea that we should act only according to maxims that can be willed to become universal laws without self-contradiction. Applying these formulations, Kant ended up with rules such as 'it is wrong to lie and steal'. The idea that it is permissible to steal, for example, is ruled out, since the very concept of stealing presupposes the existence of private property – if stealing were permitted, then the whole notion of private property would disappear; in other words, the statement 'it is permissible to steal' runs into a self-contradiction when universalised.

Central dogma. The idea that the information flow from **DNA** through **RNA** and to the proteins of cells that make up an individual is one way and irreversible.

Central nervous system (CNS). The largest part of the human nervous system. It consists of the brain and the spinal cord. It is fundamental to the control of behaviour.

Cerebral cortex. An area of the brain resembling a folded sheet of grey tissue that covers the rest of the brain. In humans, it is associated with 'higher functions' such as speech, language and reasoning.

Cerebral hemisphere. The brain is divided into two cerebral hemispheres. The right side tends to specialise in visuo-spatial functions and emotional processing; the left side is specialised for reasoning and language.

Chromosome. Structures in the nucleus of a cell that house DNA. Chromosomes contain DNA and proteins bound to it. Chromosomes become visible to the optical microscope during meiosis and mitosis.

Cladistics. A means of producing a phylogenetic classification. In cladistics, a group shares a more recent common ancestor than members of a different group.

Codon. Three nucleotides along DNA that specify one amino acid.

Coefficient of inbreeding (F). The probability that a person with two identical copies of a gene at a specific locus on their genome acquired both copies from a single ancestor.

Coefficient of relatedness (r). The r value between two individuals is the probability that an allele chosen at random from one individual will also be present in another individual. Can also be thought of as the proportion of the total genome present in one individual present in another as a result of common ancestry.

Cognitive adaptations. Hardware and software moulded by natural selection to secure the survival and reproductive success of an organism. The term is most frequently used in the context of evolutionary psychology where it is suggested that these adaptations were responses to an ancient way of life.

Conformist bias. A bias in human reasoning and decision-making where people tend to follow the norm and the most popular solution.

Consilience. The coherent and consistent way in which systems of knowledge connect with each other.

Corpus callosum. A bundle of nerve fibres joining each cerebral hemisphere.

Crossing over. During meiosis, chromosomes in a diploid cell may exchange segments of DNA. It ensures a recombination of genetic material. The process results in highly variable gametes.

Cuckoldry. A cuckold is a married man whose wife (unknown to him) is mating with another partner. In biology, cuckoldry is the term given to a mating strategy whereby the female obtains genetic material from one male but resources from another.

Culture. An evolutionary account of culture is one of the most difficult problems facing Darwinism. As yet there is no consensus on the definition of the term. In common parlance, culture is usually taken to mean the knowledge, belief systems, art, morals and customs acquired by individuals as members of a society. More recently, evolutionists have attempted to remove behaviour from this definition, leaving behind the view that culture comprises socially transmitted information.

Culturgens. An alternative name for memes – units of cultural information that are passed on.

Cystic fibrosis. A disease that occurs in people who possess two copies of a particular recessive allele, individuals with only one copy being carriers. The condition results in the secretion of abnormally thick mucus, increased sweat electrolytes and autonomic nervous system overactivity throughout the body.

Demographic transition. A phase in the economic development of cultures such that both mortality and fertility fall, often leading to families that produce children at or below the replacement rate (two children per couple). This phase has already been reached by most advanced industrialised nations such as those of Europe and the USA.

Diploid. A diploid cell is one that possesses two sets of chromosomes; one set is obtained from the mother and one from the father. Humans are diploid organisms. Compare with **haploid**.

DNA (deoxyribonucleic acid). The molecule that contains the information needed to build cells and control inheritance.

Domain specific mental modules. The idea that the human mind contains specialised modules for dealing with specific problems, as opposed to general intelligence capabilities or domain unspecific, problem-solving mechanisms.

Dominance hierarchy. The ranking of individuals in a social group. The hierarchy is usually established by aggression and conflict but once established it can be used to settle conflict issues without fighting.

Dominant allele. An allele that is fully expressed in the phenotype. An allele A is said to be dominant if the phenotype in the heterozygotic condition Aa is the same as that in the homozygotic condition AA. In this situation, allele a is said to be recessive. Alleles can be dominant, recessive or partly dominant.

DSM system. A means of classifying mental disorders according to the *Diagnostic and Statistical Manual of Mental Disorders* published by the American Psychiatric Association. It first appeared in 1952. The current edition is referred to as DSM-5 and was published in 2013. It is widely used by health professionals, although it frequently receives criticism about the scientific validity of the categories it uses.

Dual inheritance theory. A theory that proposes that both culture and biology evolve by separate but interacting processes. The two realms can interact and affect the behaviour of humans. In these models, genes can influence the course of cultural evolution but cultural change can also bring about genetic change (see the **Baldwin effect**). In the models of Boyd and Richerson, culture enables humans to adapt to rapidly changing environments by acting as a biologically useful source of information, customs and norms.

Dualism. The belief that humans have two aspects to their lives, physical and non-physical, matter and mind, or body and soul.

Empiricism. The belief that all knowledge is or should be based on experience.

Encephalisation. The enlargement of the brain (relative to body size) over the course of evolution.

Encephalisation quotient. A relative measure of brain size that compares actual brain size with expected brain size for a species of a given body weight.

Endocrine system. A collection of glands that produce hormones that regulate a range of physiological processes such as sleep, sexual arousal, growth and metabolism.

Environment of evolutionary adaptedness (EEA). A concept highly favoured by evolutionary psychologists. That period in human evolution (over 30,000 years ago) during which the mind and body plans of humans were shaped and laid down by natural selection to solve survival problems operating then.

Environmentalism. In the science of behaviour, environmentalism is the belief that social and cultural factors are paramount in determining (human) behaviour.

Epigenetics. The study of how phenotypes respond to changes in the expression of DNA. Differences between organisms may emerge not as a result of genetic change but the modification of the way DNA is expressed by (for example) proteins binding on the surface. This epigenetic modification may be influenced by environmental variables.

Epistemology. The branch of philosophy that studies the nature, acquisition and limitations of knowledge and belief.

Ethology. A branch of biology dealing with the natural behaviour of animals.

Eugenics. A largely discredited set of beliefs that advocates selective breeding among humans to remove undesirable qualities and enhance the frequency of desirable genes.

Eusocial. A term used to describe highly socialised societies, such as found among ants and bees, where some individuals forego reproduction to assist the reproductive efforts of other members of the social group.

Evolutionarily stable strategy (ESS). A set of rules of behaviour that once adopted by members of a group is resistant to replacement by an alternative strategy.

Evolutionary epistemology. In the context of evolutionary biology this is the view that the human mind is partly genetically structured and that these structures are adaptations to past experiences and have provided successful solutions (in the sense of reproductive fitness) to cognitive problems encountered in the phylogeny of our species.

Exogamy. The practice of mating with someone outside the natal group, usually by a migration to the group of the other partner. Exogamy facilitates outbreeding. The opposite to exogamy is endogamy.

Extended phenotype. An idea popularised by Richard Dawkins suggesting that genes can shape features of the natural world outside of the organism's body. One example might be a termite mound.

Extra pair copulation. Mating by a member of one sex with another outside what appears to be the stable pair bond in a supposed monogamous relationship.

Facial averageness. By photographic and computer-assisted techniques, the 'average' appearance of a set of faces can be created. Averaged faces tend to be regarded as attractive, a finding that has spawned a number of adaptive explanations, but they may not be optimally attractive.

Fitness. The term, crucially important to evolutionary theory, continues to elude a precise and universally agreed definition. Fitness can be measured by the number of offspring that an individual leaves relative to other individuals of the same species. Direct fitness (sometimes called Darwinian fitness) can be thought of as being proportional to the number of genes contributed to the next generation by production of direct offspring. **Indirect fitness** is proportional to the number of genes appearing in the next generation by an individual helping kin that also carry those genes. Inclusive fitness is the sum of direct and indirect fitness.

Fixed action pattern. An innate or instinctive pattern of behaviour that is highly stereotyped and stimulated by some simple stimulus.

Fluctuating asymmetry (FA). FA refers to bilateral characters (that is, one feature on each side of the body) for which the population mean of asymmetry (right measure minus left measure) is zero and the variability about the mean is nearly normal. In individuals, the degree and direction of asymmetry varies (hence 'fluctuating').

Founder effect. If a new group of organisms is formed from a few in a larger population, the new group is likely to have less genetic variation and have an average genotype that may be shifted in some direction even though the shift was not the result of natural selection.

Frequency dependent (selection or strategy). A process by which the strategy that is best for an organism and is naturally selected is dependent on the strategies pursued by other competing organisms.

Function. Sometimes used as shorthand for the adaptive value of a behavioural trait. The word has experienced an ambiguous usage in the human sciences. In the early years of the 20th century, a school of functionalism grew up in psychology and sociology including William James (1842–1910) and John Dewey (1859–1952). Eventually, the evolutionarily adaptive significance of behaviours to individuals was lost within this movement. In psychology, functionalism became concerned with how individuals became adapted or adjusted in their own lives rather than how traits that had been selected over time became manifest. In sociology and anthropology, functionalists looked at how current behaviours of social practices contributed to the stability of the current order. Bronislaw Malinowski (1884–1942), in his pioneering study of kinship among the Trobriand islanders, for example, tried to show how kinship served the social order as a whole rather than the genetic fitness of individuals. The functional approach also used analogies between organs of the body and parts of a social system. The situation is further confused by a modern school of functionalism, a merger of cognitive science and artificial intelligence, which uses the word in terms analogous to mathematical functions where the mind is interpreted in terms of computer-like inputs and outputs.

Game theory. A mathematical approach to establishing what behaviour is fitness maximising by taking into account the payoffs of particular strategies in the light of how other members of the group behave.

Gamete. A sex cell. Gametes are said to be haploid in that they only contain one copy of any chromosome. A gamete can be an egg or a sperm.

Gene. A unit of hereditary information made up of specific nucleotide sequences in DNA.

Gene pool. The entire set of alleles present in a population.

Genetic drift. A change in the frequency of alleles in a population due to chance alone (as opposed to selection).

Genome. The entire set of genes carried by an organism.

Genomic imprinting. A mechanism where cells express either the maternal or paternal allele of a gene but not both.

Genotype. The term can be used in two senses: loosely, as the genetic constitution of an individual, or as the types of allele found at a locus on the genome.

Genus. In classification, the genus is the taxonomic category (taxon) above the level of a species but below that of a family. Hominids (*Homo sapiens, Homo erectus* and so on), for example, belong to the genus *Homo*.

Glycaemic index. A measure of the ability of a food stuff (carbohydrate) to break down and cause a rise in bloodstream glucose levels.

Good genes. An approach to sexual selection that suggests individuals choose mates according to the fitness potential of their genome.

Grandparental solicitude. The tendency of grandparents to invest energy and resources in the upbringing and welfare of their grandchildren.

Green beard effect. A term used by Richard Dawkins to describe the largely theoretical possibility that individuals may grow and display some phenotypic marker (the 'green beard') to indicate to others of the same species the presence of common genes. The idea is that such markers may allow individuals to direct their altruistic efforts more effectively.

Grooming. Ostensibly the cleansing of skin, fur or feathers of an animal by itself or, more significantly, by another of the same species. The function of grooming may lie deeper than that of simple hygiene and could be an indication of the formation of alliances and the resolution of conflict.

Group selection. Selection that operates between groups rather than individuals. The notion was attacked and shunned in the 1970s but may be making a comeback in studies on human evolution.

Gynephilia. Attraction to women or femininity.

Habituation. A process by which an organism learns to reduce the response to a repeated stimulus.

Haemophilia. A sex-linked genetic disorder expressed in human males and characterised by excessive bleeding following injury.

Hamilton's rule. A formula which predicts that individuals will perform altruistic acts for their relatives so long as the genetic benefit exceeds the costs (taking into account the degree of relatedness between the donor and the recipient).

Handicap. Features of an organism that seem at first sight to have a negative impact on fitness. The handicap principle was an idea advanced by Zahavi in 1975 designed to account for what appear to be maladaptive features, or handicaps, of an organism, such as the long train of a peacock or the huge antlers of deer. Zahavi suggested that these features were honest advertisements of genetic quality, since an animal, usually the male, must be strong in order to grow and bear such a burden.

Haploid. A condition where cells only contain one copy of any chromosome. Some organisms are haploid but in humans only the gametes are haploid.

Herbivore. An organism that eats only plants.

Heritability. The extent to which a difference in a character between individuals in a population is due to inherited differences in the genotype. It is often expressed as a number between zero and 1 that refers to the proportion of variance in a character in a population that is ascribable to inherited genetic differences.

Heterozygote. An individual that has two different alleles (for example Aa) at a given genetic locus.

Heterozygotic advantage. Where an organism that has Aa alleles has an advantage over AA or aa variants.

Heuristic. A fast and computationally inexpensive means of solving a problem, such as a simple rule of thumb procedure.

Homeostasis. The process whereby animals maintain a steady internal environment; usually through the use of physiological feedback mechanisms.

Homicide. The killing of one human by another

Hominid. Any member of the family Hominidae, which includes humans, chimpanzees, gorillas and all their extinct ancestors. The term is open to variable interpretation, depending on which classification system is accepted.

Hominin. A member of the subfamily Homininae. Includes extant humans and extinct precursor species following the break away from the chimpanzee lineage some seven million years ago. The term is open to variable interpretation depending on which classification system is accepted.

Homogamy. An assortative mating process whereby an individual finds members of the opposite sex that look like themselves (or resemble themselves in other culturally significant ways) attractive.

Homology. A feature present in several species and present in their common ancestor.

Homozygote. An individual having identical alleles for a given trait at a given locus (AA or aa).

Honest signal. A signal that reliably communicates the quality of an individual in terms of its fitness.

Humanistic. A term applied to a branch of psychology and psychotherapy, both of which grew out of dissatisfaction with behaviourist and psychoanalytic approaches. Humanistic psychology emphasises the holistic nature of humans, human consciousness, and the meanings, values and intentions adopted and sought by human beings. A leading exponent of this approach was Abraham Maslow (1908–70).

Huntington's disease. A debilitating disease attributable to a series of repeating DNA bases in a region of the genome that codes for the protein Huntingtin.

Hypothalamus. A structure deep within the forebrain that controls a range of autonomic functions, for example hunger, thirst, body temperature, wakefulness and sleep. Linked to the hormonal system.

Hypothesis. A conjecture set forward as a provisional explanation for a phenomenon.

Hyperinsulinemia. A disorder where levels of insulin in the bloodstream are higher than expected for the observed glucose levels. It often suggests reduced sensitivity to insulin.

Hyperglycaemia. A condition whereby excessive glucose is found in the bloodstream.

Hypertension. Elevated blood pressure.

Hypersensitive agency detection. device. The tendency of humans to attribute purposeful agency to phenomena. It is a bias in perception that may once have been advantageous.

Ideology. A set of beliefs, values and assumptions that structure understanding and inform policy decisions. An ideology usually supports the political interests of particular groups.

Imprinted genes. Genes that express their functions only when inherited from the mother or father but not both. See **genomic imprinting**.

Inbreeding. Breeding between two individuals who are close genetic relatives. The fitness of offspring produced is sometimes reduced by **inbreeding depression**.

Inbreeding depression. The phenomenon that descendants of individuals who mate with close relatives tend to be lower in fitness. It can be brought about by a lowered genetic variety or by the fact that an individual may be homozygous for recessive and deleterious alleles.

Incest. Sexual activity between two individuals who are too closely related to be allowed to marry according to local social rules and customs.

Inclusive fitness. Fitness that is measured by the number of copies of one's genes that appear in current or subsequent generations in offspring and non-offspring. Kin-directed altruism, for example, is said to increase inclusive fitness. See **fitness**.

Indirect fitness. Offspring of an individual related to a subject that carry some genes of the subject and so contribute to the subject's inclusive fitness.

Infanticide. The deliberate killing of an infant shortly after its birth.

Innate releasing mechanism. A hypothetical mechanism or model devised to help explain how an innate response is triggered by a sign stimulus.

Intensionality. A term used to express degrees of self-awareness and the awareness of the mental states of others. First-order intensionality is self-consciousness ('I know'); second order is the awareness that others may have self-awareness ('I know that you know'); third order is the knowledge that others may be aware of your thoughts ('I know that you know that I know') and so on.

Intersexual selection. A form of selection driven by the exercise of choice by one sex for specific characteristics in a mating partner of the opposite sex.

Intrasexual selection. Competition between members of the same sex (typically males) for access to the opposite sex.

Isogamy. A condition where the gametes from each partner engaged in sexual reproduction are of equal size. Isogamy is common among protists and algae. Higher plants and all animals display anisogamy.

James–Lange theory. A theory that emotional states are responses to the physical reaction of the body to external events, rather than the idea that physiological changes follow emotional states.

Jealousy. A strong human emotion probably experienced differently by males and females, and probably functionally selected to ensure paternity certainty and/or maintenance of the pair bond.

Kin. Relatives within a family.

K and r selection. K-selected species are those that produce small numbers of slow-growing and long-lived offspring that require considerable parental investment. Humans are a K-selected species. In contrast, r-selected species produce many short-lived and quickly maturing offspring that need little parental investment.

Kin discrimination. The ability of an animal to react differently to other individuals depending on their degree of genetic relatedness.

Kin selection. The suggestion that altruism can evolve because altruistic behaviour favours increases in the gene frequency of the genes responsible. A situation where altruism to relatives is favoured and spreads.

Lamarckian inheritance. Shorthand for the inheritance of acquired characteristics. A mechanism (among others) proposed by the French evolutionist Lamarck, whereby it is supposed that characters or modifications to characters acquired by the phenotype can be passed on to offspring through genetic inheritance. The mechanism is now rejected as without foundation.

Lamarckism. Doctrines associated with Jean Baptiste Lamarck. Usually taken to refer to his mechanism of inheritance.

Lek. A display site where (usually) males display and females choose a mate.

Life history theory. A theory that considers how organisms allocate resources to various life processes such as mating, growth and repair across the lifespan.

Limbic system. Part of the brain located beneath the cerebral cortex. Contains a number of structures, such as the amygdalae and the hippocampus, involved in learning, memory and emotion.

Lineage. A sequence showing how species are descended from one another.

Locus. The particular site where a particular gene is found on DNA.

Machiavellian intelligence. The idea that one of the prime factors leading to the growth of intelligence in primates was the need for an individual to manipulate its social world through a mixture of cunning, deceit and political alliances.

Maternal–foetal conflict. The theory that the foetus and the mother that carries it may have different genetic interests and so engage in a biological tug-of-war over the level of resources passed to the foetus as it develops.

Matrilocal. A social system where women tend to remain in their local group after marriage, while men tend to leave their family group to join that of their wives.

Meiosis. A type of cell division whereby diploid cells produce haploid gametes. The double set of chromosomes in a diploid cell is thereby reduced to a single set in the resulting gametes. Crossing over and recombination occur during meiosis.

Meme. An activity or unit of information that can be passed on by imitation.

Memetics. The scientific study of memes.

Menarche. The first menstrual period or bleeding experienced by a girl. In the USA, the average age of menarche is now about 12 years. It is a key feature of puberty.

Mendelian genetics. A mode of inheritance first described by Gregor Mendel. In **diploid** species showing Mendelian inheritance, genes are passed to offspring in the same form as they were inherited from the previous generation. At each locus, an individual has two genes (**haploid**), one inherited from its mother and one from its father.

Menopause. The cessation of monthly ovulation experienced by

women, usually in their late forties. Women are infertile after the menopause. The adaptive significance of the menopause is probably related to the risks of childbirth and the need to care for existing children.

MHC (Major histocompatibility complex). A set of genes coding for antigens responsible for the rejection of genetically different tissue. The antigens are known as histocompatibility (that is, tissue compatibility) antigens. There are at least 30 histocompatibility gene loci. The genes of the MHC are subject to simple Mendelian inheritance and are co-dominantly expressed; that is, alleles from both parents are equally expressed. Each cell, therefore, in any offspring has maternal and paternal MHC molecules on its surface. The human MHC is found on the short arm of chromosome 6.

Mitochondrial DNA. A short section of DNA found only in the mitochondria of cells and in humans only inherited from the mother.

Mitochondrial Eve. The name given to the woman who is the most recent matrilineal common ancestor to all humans. The mitochondrial DNA in all humans can be traced back to her.

Mitosis. Part of the process where one cell divides into another identical one with the same number of chromosomes as the original. The growth of an organism is through mitosis. Sexual reproduction requires **meiosis**.

Modularity. The belief that the human brain consists of a number of discrete problem-solving units, or areas of specialised function (such as theory of mind and face recognition) shaped by natural selection to solve specific problems. See **domain specific modules**.

Monism. The belief that the universe is composed of one basic substance. In the context of human behaviour (materialistic), monism would assert that behaviour has physical causes.

Monogamy. The mating of a single male with a single female. In annual monogamy, the bond is dissolved each year and a fresh partner found. In perennial monogamy, the bond lasts for the reproductive life of the organisms.

Morgan's canon. The assertion that explanations for the behaviour of animals should be kept as simple as possible. Higher mental activities should not be attributed if lower ones will suffice.

Multiregional model. A theory that proposes that *Homo sapiens* arose in various parts of the world from pre-existing ancient stock that had already reached there (such as *Homo erectus*).

Mutation. In modern genetics, a mutation is a heritable change in the base sequences in the DNA of a genome. Most mutations are deleterious.

Mutation–selection balance. The dynamic equilibrium established when new mutations arise as fast as old ones are removed by selection from the gene pool. The net result will be a steady state level of background mutations present in genotypes.

Mutualism. A symbiotic relationship between individuals of two species such that both partners benefit.

Nativism. The idea that children are born with a basic understanding of some features of the world (such as number, space and time).

Naturalism. The belief that all phenomena can be explained by scientific laws and principles without recourse to the supernatural or entities outside the remit of science.

Naturalistic fallacy. A fallacy of reasoning that supposes it is possible to derive 'ought'-type statements from factual statements. It was introduced by the British philosopher

G. E. Moore and then (confusingly) applied by others to the 'is/ought' problem identified by Hume.

Neocortex. The outer layers (about 2–4 mm in thickness) covering the entire cerebral cortex. A late addition to the brain in evolutionary terms. Involved with higher brain functions such as sensory perception, conscious thought and language. A highly folded structure, it comprises about 75 per cent of the brain volume. Contains about 20 per cent of all the neurons of the brain.

Neonatal adiposity. The high percentage of body fat in newly born humans and other animals.

Neoteny. The retention of typically juvenile features into adulthood.

Neuron. A cell found in the nervous system (brain and spinal cord).

Oestrus. A period of heightened interest in copulation experienced by female animals.

Oligosaccharides. Long sugar molecules containing many repeating monosaccharide units.

Ontogeny (coming into being). The development and growth of an organism from a fertilised cell, through the foetus to an adult.

Oogenesis. The formation of female sex cells (ova). See also **spermatogenesis**.

Operant conditioning. A type of learning whereby an action (operant) is carried out more frequently if it is rewarded.

Operational sex ratio. The ratio of sexually receptive males to females in a particular area or over a particular time.

Optimality. The idea that the behaviour of animals will be that which is ideally suited to bring maximum gain for minimum cost or, more precisely, where gain minus cost is maximised. The assumption that all behaviour must be optimal can be misleading.

Orbitofrontal cortex. The portion of the outer layer of the brain that lies just above the eye sockets. Involved in directing socially appropriate responses to emotional events.

Order. A unit of classification above the family and below the class. Humans belong to the order of primates.

Out of Africa hypothesis. The idea that the species *Homo sapiens* arose once in Africa and that members of this species migrated out of Africa about 150,000 years ago to eventually populate the globe.

Ovarian cycle. A cycle of events controlled by hormones in the ovary of mammals that leads to **ovulation**.

Ovulation. The release of a female sex cell or ovum (plural ova) from an ovarian follicle.

Paradigm. A cluster of ideas and theories that are consistent and form part of a distinct way of understanding the world. Evolutionary psychology can be regarded as a paradigm.

Parasite. An organism in a symbiotic relationship with another (host) such that the parasite gains in fitness at the expense of the host.

Parental investment. Actions that increase the survival chances of one set of offspring but at the expense of the parent procuring more offspring.

Parent–offspring conflict. Disputes between parents and their children over the allocation of resources. Such conflicts are the focus of attention by evolutionary psychologists, in that they may have a basis in the different reproductive interests of the parties concerned. Typically, the best reproductive interests of young parents may be served by allocating resources to future children – a decision that may not be optimum from the point of view of an existing child.

Parthenogenesis. Asexual reproduction. Production of offspring by virgin birth.

Paternity certainty. An expression of the confidence that a male has that he is really the genetic father of any offspring.

Pathogen. A disease-causing organism.

Patrilocal. A social system whereby women leave their family group upon marriage to live with the natal group of their husband (contrasts with **matrilocal**).

Phenotype. The characteristics of an organism as they have been shaped by both the **genotype** and environmental influences.

Phenotypic plasticity. A debated term. The average value for a phenotypic character between two individuals with identical genomes or two populations with similar gene frequencies may be different because of different environmental influences. Phenotypic plasticity is often used to refer to irreversible change and phenotypic flexibility to refer to reversible change. One key question is whether phenotypic plasticity can serve adaptive purposes.

Phobia. An irrational and persistent fear of specific objects or situations. The symptoms of the disorder include an unreasonable desire to avoid the object or situation. When the phobia becomes acute, the sufferer is often diagnosed as suffering from an anxiety disorder. Examples of phobias include arachnophobia (fear of spiders), xenophobia (fear of strangers) and agoraphobia (fear of open spaces outside home).

Phylogenetic inertia. The process by which features are inherited from an ancestor but do not necessarily now (in contemporary environments) serve any adaptive purpose.

Phylogeny. The branching history of a species showing its relationship to ancestral species. A phylogenetic tree can be used to infer relationships between existing species and their evolutionary history.

Phylum. A taxonomic category above class but below kingdom. Humans belong to the order of primates, the class of mammals and the phylum Chordata.

Placenta. The organ that provides nutrients and oxygen to an embryo.

Pleiotropy. The ability of a single gene to have a number of different effects on the phenotype.

Polyandry. A type of mating system such that a single female mates with more than one male in a given breeding season.

Polygamy. The mating of one member of one sex with more than one member of the other sex. The two varieties are polyandry and polygyny.

Polygynandry. A mating system in which males and females mate with each other in a promiscuous way.

Polygyny. A mating system whereby a single male mates with more than one female in a given breeding season.

Polymorphism. A condition where a population may have more than one allele (at significant frequencies) at a particular locus. The different forms of the alleles may all be adaptive in their own right.

Population. A group of individuals, usually geographically localised and interacting, that all belong to the same species.

Positivism. The belief that science represents the positive and more advanced state of knowledge. Also the supposition that only objects that can be experienced directly form part of the proper process of scientific enquiry. Positivism seeks to repudiate metaphysics.

Precocial. A term applied to species whose young are born relatively mature and mobile from the moment of birth. The opposite of precocial is **altricial**.

Pregnancy sickness. Nausea during pregnancy.

Preparedness theory. A theory that humans are born with biases and propensities to quickly learn specific responses. In psychology, it is often used to explain the basis for some **phobias**.

Prestige bias. A bias in human reasoning and decision-making whereby individuals tend to copy the solutions of prestigious members of their group.

Prisoners' dilemma. A model of the problem faced by individuals in knowing how to act to serve their own best interests in the context of uncertain knowledge of how others, who may be rivals, will also act. The analogy is drawn with two prisoners who if they cooperate with each other will receive a lighter sentence.

Prosimians. A diverse group of small-bodied primates. Most species live in Africa. Best-known examples include bushbabies and lemurs. Thought to be more 'primitive' than the simians.

Proximate cause. In behavioural terms, the immediate mechanism or stimuli that initiates or triggers a pattern of behaviour.

Pseudopathology. A category of illness that may result from the operation of a response or strategy that was once adaptive but is no longer so in current environments.

Psychodynamic. A view of the human mind that studies behaviour in terms of the action of motivational forces, desires and drives (hence 'dynamic'). Its application to therapy often draws upon the work of Sigmund Freud, who saw early childhood experiences and repressed emotions as the key to adult behaviour.

Psychopathy. A term closely related to sociopathy. A psychopath is someone who lacks empathy, is impulsive, self-centred and manipulative. The condition does not feature in DSM-IV but is closely related to antisocial personality disorder. The relationship and differences between psychopathy, sociopathy and antisocial personality disorder are the subjects of an ongoing debate.

Recessive allele. See **dominant allele**.

Reciprocal altruism. Altruism that is favoured because it is returned by the recipient at a later date.

Reciprocity. The donation of assistance in the expectation that favours will be returned at some future date.

Recombination. An event whereby **chromosomes** cross over and exchange genetic material during **meiosis**. Recombination tends to break up **genes** that are linked together.

Reductionism. The attempt to explain a wide range of phenomena by employing a smaller range of concepts and principles that are more basic.

Relativism. The view that there is no privileged knowledge or belief system that can claim supremacy over another. A relativist would argue that the validity of belief systems cannot be decided by universal criteria.

Reptilian brain. A section of the brain labelled by McLean to describe the oldest (in evolutionary terms) part of the brain that deals with basic survival and reproductive functions.

Reverse engineering. A way of thinking about the consequences of evolution. It starts with a contemporary understanding of the function of adaptive behavioural or physiological attributes and tries to infer what problems our ancestors faced to give rise to these adaptive solutions.

RNA (ribonucleic acid). Molecules that act as intermediaries as the hereditary code in **DNA** is converted into proteins. Messenger RNA

(mRNA) is the molecule that carries information from DNA in the nucleus to the sites in the cytoplasm where protein synthesis takes place.

Saturated fats. Fats where there are no carbon–carbon double bonds in the long fatty chain – in essence the carbon atoms are 'saturated' with hydrogen.

Schizophrenia. A psychological disorder characterised by hallucinations, delusional behaviour and disordered thoughts. Thought to be present in some form in about 1 per cent of any population.

Selection. The differential survival of organisms (or genes) in a population as a result of some selective force. Selectionist thinking is the approach that looks for how features of organisms can be interpreted as the result of years of selection acting upon ancestral populations. **Directional** selection tends to favour an extreme measure of the natural variability in a population and the average measure will gradually move in this direction. **Disruptive** selection tends to favour more than one phenotype. **Stabilising** selection tends to favour the means values currently found and ensures that variation is reduced. See also **Sexual selection**.

Sex chromosomes. Chromosomes that determine if an individual is male or female. In humans, a female possesses two X chromosomes and a male one X and one Y chromosome.

Sex ratio. The ratio of males to females at any one time. At birth for humans the ratio is about 1.06.

Sexual crypsis. Sometimes known as cryptic oestrus or concealed ovulation. The suggestion that in human females the time of **oestrus** is concealed.

Sexual dimorphism. Differences in morphology, physiology or behaviour between the sexes in a single species.

Sexual imprinting. A process where a young organism acquires

a template for what is desirable in the opposite sex by exposure to a member of that sex in early youth (typically a parent) that thereafter serves as a standard.

Sexual selection. Selection that takes place as a result of mating behaviour. Intrasexual selection occurs as a result of competition between members of the same sex. Intersexual selection occurs as a result of choices made by one sex for features of another.

Sexually antagonistic selection. A genetically based feature that enhances the fitness of one sex but decreases for the other sex.

Siblicide. The killing of a sibling (brother or sister) by a brother or sister.

Smoke detector principle. A term popularised by Nesse to argue that it may be natural to demonstrate higher levels of anxiety than necessary, since the cost of reacting to a false alarm is much less than not reacting to a rarer true alarm.

Speciation. The creation of a new species through the splitting of an existing species into two or more new species.

Species. Using the biological species concept, a species is a set of organisms that possess similar inherited characteristics and crucially have the potential to interbreed to produce fertile offspring.

Sperm competition. Competition between sperm from two or more males that is present in the reproductive tract of the female.

Spermatogenesis. The formation of male sex cells (sperm).

Strategy. A pattern of behaviour or rules guiding behaviour shaped by natural selection to increase the fitness of an animal. A strategy is not taken to be a set of conscious decisions in non-human animals. A strategy may also be flexible, in that different biotic and abiotic factors may trigger different forms of optimising behaviour.

Stem cells. Undifferentiated cells that have the ability to reproduce and differentiate in other more specialised cells or more stem cells.

Supernormal stimuli. Stimuli that are of greater magnitude than would normally be encountered yet still produce an effect (often enhanced) on the recipient.

Symbiosis. A relationship between organisms of different species that live in close association and interact. See **mutualism** and **parasite**.

Symmetry. A state where the physical features of an organism on one side of its body are matched in size and shape by those on the other. Symmetry may be an indication of physiological health since stress increases asymmetry. Animals may, therefore, use symmetry as an honest signal of fitness. See **fluctuating asymmetry**.

Taxon (plural taxa). A named group in classification. It may be a species, a genus, a family, an order or other category.

Taxonomy. The theory and practice of classifying organisms.

Teleology. The belief that nature has purposes, that events are shaped by intended outcomes.

Teleonomy. The idea that natural selection and resulting adaptations may give rise to effects that look as if they occur for a purpose but avoiding the suggestion that the natural world is shaped by goals.

Testis (plural testes). Male sex organ that produces sperm and associated hormones.

Theory of mind. The suggestion that an important component of the mental and emotional life of humans and some other primates is the ability to be self-aware and to appreciate that others also have awareness. The theory of mind also implies that an individual is capable of distinguishing between the real intentional or emotional states of others and those that may be feigned. Having a theory of

mind is an essential component of **Machiavellian intelligence**.

Thrifty genotype. An idea originally proposed to account for the rise of diabetes in Western populations. The suggestion is that some populations have genotypes that efficiently convert calories to fat to prepare humans for periodic bouts of famine. These genes, it is suggested, cause diabetes and obesity in populations where food is abundant.

Thrifty phenotype. An idea proposed in the light of difficulties faced by the **thrifty genotype** hypothesis. The idea here is that the phenotype is prepared for periods of poor (or indeed abundant) nutrition not through a set of genes but by the calibration of physiological systems during development (e.g. *in utero*).

Tit for tat. A strategy that can be played by individuals stuck in a **prisoners' dilemma**. The strategy is to never be the first to defect but to follow defection or cheating by the other player with a retaliation. The tit for tat strategy, on average, works well against a variety of other strategies.

Tragedy of the commons. A term introduced by Garrett Hardin (1968) to describe a system where the logical strategy of individual self-aggrandisement leads to a failure of resources when this strategy is pursued by all.

Trans fatty acids. Synthesised fatty acids that are distinguished from natural fatty acids by a kink in the carbon chain.

Ultimate causation. The explanation for the behaviour of an organism that reveals its adaptive value.

Unipolar depression. Sometimes called major depression or major depressive disorder. It may be a single episode of a severely depressed mood that lasts over two weeks, or recurrent periods that may occur over many years. If the patient experiences episodes of mania or elevated mood, then a diagnosis of bipolar disorder is usually made.

Utilitarianism. A branch of moral philosophy that suggests that the moral value of an action is related to its contribution to the overall utility. When faced with an ethical dilemma, utilitarians would identify the best course of action as that which delivers the maximum happiness (or pleasure) to the greatest number.

Vestigial structures. Features that are relics of a previous stage in the evolutionary history of the species. Such features are essentially relics and may no longer serve any purpose.

Virulence. A measure of the damage caused to the host by an infectious disease. It is also used to indicate the infectivity of a **pathogen** – how likely it is to cause the disease in the first place.

Waist-to-hip ratio. The circumference of the waist divided by that of the hips.

Wernicke's area. A part of the brain involved in language, named after the 19th-century Polish physician Karl Wernicke. Like **Broca's area**, it lies in the left hemisphere of the brain. It is connected to Broca's area by the arculate fasciculus.

Westermarck effect. Named after the Finnish anthropologist Edvard Westermarck, a mechanism whereby males and females co-socialised from an early age tend not to develop sexual feelings for each other.

Zoonosis. The process of the transfer of disease from an animal to a human.

Zygote. A fertilised cell formed by the fusion of two gametes. Monozygotic refers to (identical) twins that result from the separation of a single fertilised egg or zygote and are therefore identical in terms of their DNA. Dizygotic refers to twins that result from two independent fertilisations of two eggs by two sperm and are therefore non-identical.

Further reading

Many of the terms above are uncontentious, but some have disputed meanings. The reader is referred to Keller, E. F. and E. A. Lloyd (eds.) (1994) *Keywords in Evolutionary Biology*. Cambridge, MA, Harvard University Press. It considers 37 key terms in some depth.

References

Abbot, P., J. Abe, J. Alcock et al. (2011). 'Inclusive fitness theory and eusociality.' *Nature* **471**(7339): E1–E4.

Aggleton, J. P., R. W. Kentridge and N. J. Neave (1993). 'Evidence for longevity differences between left handed and right handed men: an archival study of cricketers.' *Journal of Epidemiology and Community Health* **47**: 206–9.

Aiello, L. C. and P. Wheeler (1995). 'The expensive-tissue hypothesis: the brain and the digestive system in human primate evolution.' *Current Anthropology* **36**: 199–221.

Aiello, L. C., N. Bates and T. Joffe (2001). In defence of the expensive tissue hypothesis. In D. Falk and K. R. Gibson (eds.), *Evolutionary Anatomy of the Primate Cerebral Cortex*. Cambridge, Cambridge University Press.

Alba, D. M. (2010). 'Cognitive inferences in fossil apes (Primates, Hominoidea): does encephalization reflect intelligence?' *J Anthropol Sci* **88**: 11–48.

Alexander, R. and K. Noonan (1979). Concealment of ovulation, parental care and human social evolution. In N. I. A. Chagnon and W. Irons (eds.), *Evolutionary Biology and Human Social Behavior: An Anthropological Perspective*. North Scituate, MA, Duxbury.

Altmann, J. (2001). *Baboon Mothers and Infants*. Chicago, University of Chicago Press.

Altmann, J. and S. C. Alberts (2003). Variability in reproductive success viewed from a life-history perspective in baboons. *American Journal of Human Biology* **15**(3): 401–9.

Almond, D. and L. Edlund (2007). 'TriversWillard at birth and one year: evidence from US natality data 1983–2001.' *Proceedings of the Royal Society B: Biological Sciences* **274**(1624): 2491–6.

Alvard, M. S. and D. A. Nolin (2002). 'Rousseau's Whale Hunt?' *Current Anthropology* **43**(4): 533–59.

Alvarez, G., F. C. Ceballos and C. Quinteiro (2009). 'The role of inbreeding in the extinction of a European royal dynasty.' *PLoS One* **4**(4): e5174.

Alvarez, L. and K. Jaffe (2004). 'Narcissism guides mate selection: humans mate assortatively, as revealed by facial resemblance, following an algorithm of "self seeking like".' *Evolutionary Psychology* **2**: 177–94.

Anderson, K. (2006). 'How well does paternity confidence match actual paternity?' *Current Anthropology* **47**(3): 513–20.

Anderson, J. W., P. Baird, R. H. Davis, S. Ferreri, M. Knudtson, A. Koraym, and C. L. Williams (2009). 'Health benefits of dietary fiber.' *Nutrition Reviews* **67**(4): 188–205.

Anderson, M. J. and A. F. Dixson (2002). 'Sperm competition: motility and the midpiece in primates.' *Nature* **416**(6880): 496.

Angel, J. L. (1975). Paleoecology, paleodemography and health. In S. Polgar (ed.), *Population, Ecology and Social Evolution*. The Hague, Moulton Publishers, pp. 167–90.

Annett, M. (1964). 'A model of the inheritance of handedness and cerebral dominance.' *Nature* **204**: 59–60.

Antfolk, J., M. Karlsson, A. Bäckström and P. Santtila (2012). 'Disgust elicited by third-party incest: the roles of biological relatedness, co-residence, and family relationship.' *Evolution and Human Behavior* **33**(3): 217–23.

Antfolk, J., B. Salo, K. Alanko, E. Bergen, J. Corander, N. K. Sandnabba and P. Santtila (2015). 'Women's and men's sexual preferences and activities with respect to the partner's age: evidence for female choice.' *Evolution and Human Behavior* **36**(1): 73–9.

Apicella, C. L. and F. W. Marlowe (2004). 'Perceived mate fidelity and paternal investment resemblance predicts men's investment in children.' *Evolution and Human Behavior* **25**(6): 371–9.

Appiah, K. A. (2006). *Cosmopolitanism: Ethics in a World of Strangers (Issues of Our Time)*. London, UK. WW Norton & Company.

Archer, J. (1992). *Ethology and Human Development*. Hemel Hempstead, Harvester Wheatsheaf.

Archer, J. (2013). 'Can evolutionary principles explain patterns of family violence?' *Psychological Bulletin* **139**(2): 403.

Arechiga, J., C. Prado, M. Canto and H. Carmenati (2001). 'Women in transition – menopause and body composition in different populations.' *Collective Anthropology* 25: 443–8.

Armstrong, E. (1983). 'Relative brain size and metabolism in mammals.' *Science* 220(4603): 1302–4.

Arner, G. B. L. (1908). *Consanguineous Marriages in the American Population.* Columbia University, Longmans, Green & Co., Agents.

Arnhart, L. (2005). Incest taboo as Darwinian natural right. In A. P. Wolf and W. H. Durham (eds.), *Inbreeding, Incest and the Incest Taboo.* Stanford, CA, Stanford University Press.

Ash, J. and G. Gallup (2007). Brain size, intelligence and paleoclimatic variation. In G. Geher and G. Millier (eds.), *Mating Intelligence.* New York, Psychology Press.

Ash, J. and G. G. Gallup Jr (2007). 'Paleoclimatic variation and brain expansion during human evolution.' *Human Nature* 18(2): 109–24.

Audette, R. (1995). *NeanderThin: Eat Like a Caveman to Achieve a Lean, Strong, Healthy Body.* Dallas, Paleolithic Press.

Austad, S. N. (1993). 'Retarded senescence in an insular population of Virginia opossums.' *Journal of Experimental Zoology* 229: 695–708.

Austad, S. N. and K. E. Fisher (1991). 'Mammalian aging, metabolism and ecology: Evidence from bats and marsupials.' *Journal of Gerontology* 46: B47–53.

Avila, M., G. Thaker and H. Adami (2001). 'Genetic epidemiology and schizophrenia: a study of reproductive fitness.' *Schizophrenia Research* 47(2): 233–41.

Axelrod, R. (1984). *The Evolution of Cooperation.* New York, Basic Books.

Axelrod, R. and W. D. Hamilton (1981). 'The Evolution of Cooperation.' *Science* 211: 1390–6.

Badcock, C. (1991). *Evolution and Individual Behaviour: An Introduction to Human Sociobiology.* Oxford, Blackwell.

Badcock, C. (2013). *Evolutionary Psychology: A Clinical Introduction.* Cambridge, Blackwell.

Bagatell, C. J. and W. J. Bremner (1990). 'Sperm counts and reproductive hormones in male marathoners and lean controls.' *Fertility and Sterility* 53: 688–92.

Bailey, D. H. and D. C. Geary (2009). 'Hominid brain evolution.' *Human Nature* 20(1): 67–79.

Bailey, J. M. (1998). Can behaviour genetics contribute to evolutionary behavioural science? In C. Crawford and D. L. Krebs (eds.), *Handbook of Evolutionary Psychology.* Mahwah, NJ, Lawrence Erlbaum Associates, Inc.

Bailey, J. M. and R. C. Pillard (1991). 'A genetic study of male sexual orientation.' *Archives of General Psychiatry* 48(12): 1089.

Baillargeon, R. (1987). 'Object Permanence in 3 1/2- and 4 1/2-Month-Old Infants.' *Developmental Psychology* 23: 655–64.

Baillargeon, R. and J. DeVos (1991). 'Object Permanence in Young Infants: Further Evidence.' *Development* 62: 1227–46.

Baillargeon, R., E. Spelke and S. Wasserman (1985). 'Object permanence in five-month-old infants.' *Cognition* 20: 191–208.

Baker, R. R. and M. A. Bellis (1989). 'Number of sperm in human ejaculates varies in accordance with sperm competition theory.' *Animal Behaviour* 37: 867–9.

Baker, R. R. and M. A. Bellis (1995). *Human Sperm Competition.* London, Chapman and Hall.

Ball, H. L. and K. Klingaman (2008). Breastfeeding and mother–infant sleep proximity. In Wenda R. Trevathan, E. O. Smith and James J. McKenna (eds.), *Evolutionary Medicine and Health: New Perspectives.* New York, Oxford University Press, pp. 226–41.

Barash, D. (1982). *Sociobiology and Behaviour.* New York, Elsevier.

Barash, D. (2003). *The Survival Game.* New York, NY, Times Books.

Barker, D. (2007). 'The origins of the developmental origins theory.' *Journal of Internal Medicine* 261: 412–7.

Barkow, J. (1989). *Darwin, Sex, and Status.* Toronto, University of Toronto Press.

Barkow, J. H., L. Cosmides and J. Tooby (1992). *The Adapted Mind.* Oxford, Oxford University Press.

Barratt, C., V. Kay and S. K. Oxenham (2009). 'The human spermatozoon – a stripped down but refined machine.' *Journal of Biology* 8(7): 63.

Barrett, D. (2010). *Supernormal Stimuli: How Primal Urges Overran Their Evolutionary Purpose.* WW Norton & Company.

Barrett, L. and R. I. M. Dunbar (1994). 'Not now dear, I'm busy.' *New Scientist* 142: 30–4.

Barton, S. C., M. A. Surani and M. L. Norris (1984). 'Role of paternal and maternal genomes in mouse development.' *Nature* 311(5984): 373–6.

Bateson, M., D. Nettle and G. Roberts (2006). 'Cues of being watched enhance cooperation in a real-world setting.' *Biology Letters* 2(3): 412–4.

Bateson, P. (1980). 'Optimal outbreeding and the development of sexual preferences in Japanese quail.' *Zeitschrift für Tierpsychologie* 53: 321–49.

Bateson, P. (1982). 'Preferences for cousins in Japanese Quail.' *Nature* 295: 236–7.

Beachy, P. A., S. S. Karhadkar and D. M. Berman (2004). 'Tissue repair and stem cell renewal in carcinogenesis.' *Nature* 432(7015): 324–31.

Beaulieu, D. A. and D. Bugental (2008). 'Contingent parental investment: An evolutionary framework for understanding early interaction between mothers and children.' *Evolution and Human Behavior* 29(4): 249–55.

Beer, J. M. and J. M. Horn (2000). 'The influence of rearing order on personality development within two adoption cohorts.' *Journal of Personality* 68: 769–819.

Bell, A. P. and M. S. Weinberg (1978). *Homosexualities: A Study of Diversity Among Men and Women*. New York, Simon & Schuster.

Belsky, J., L. Steinberg and P. Draper (1991). 'Childhood experience, interpersonal development, and reproductive strategy: an evolutionary theory of socialization.' *Child Development* 62(4): 647–70.

Belt, T. (1874). *The Naturalist in Nicaragua*. London, Bumpus.

Bennett, C. M., E. Boye and E. J. Neufeld (2008). 'Female monozygotic twins discordant for hemophilia A due to nonrandom X chromosome inactivation.' *American Journal of Hematology* 83(10): 778–80.

Benoit, D. and K. C. Parker (1994). 'Stability and transmission of attachment across three generations.' *Childhood Development* 65: 1444–56.

Benshoof, L. and R. Thornhill (1979). 'The evolution of monogamy and concealed ovulation in humans.' *Journal of Social and Biological Structures* 2: 95–106.

Bereczkei, T., P. Gyuris, P. Koves and L. Bernath (2002). 'Homogamy, genetic similarity, and imprinting; parental influence on mate choice preferences.' *Personality and Individual Differences* 33(5): 677–90.

Bereczkei, T., P. Gyuris and G. E. Weisfeld (2004). 'Sexual imprinting in human mate choice.' *Proceedings of the Royal Society of London, Series B: Biological Sciences* 271(1544): 1129–34.

Berger, L. R., J. Hawks, D. J. de Ruiter, S. E. Churchill, P. Schmid, L. K. Delezene, T. L. Kivell, H. M. Garvin, S. A. Williams and J. M. DeSilva (2015). 'Homo naledi, a new species of the genus Homo from the Dinaledi Chamber, South Africa.' *eLife* 4: e09560.

Berra, T. M., G. Alvarez and F. C. Ceballos (2010). 'Was the Darwin/Wedgwood dynasty adversely affected by consanguinity?' *BioScience* 60(5): 376–83.

Betzig, L. (1998). Not whether to count babies, but which. In C. Crawford and D. L. Krebs (eds.), *Handbook of Evolutionary Psychology*. Mahwah, NJ, Lawrence Erlbaum.

Bingham, S. A., N. E. Day, R. Luben, P. Ferrari, N. Slimani, T. Norat, F. Clavel-Chapelon, E. Kesse, A. Nieters and H. Boeing (2003). 'Dietary fibre in food and protection against colorectal cancer in the European Prospective Investigation into Cancer and Nutrition (EPIC): an observational study.' *The Lancet* 361(9368): 1496–1501.

Bishop, G. D., A. L. Alva, L. Cantu and T. K. Rittiman (1991). 'Responses to persons with AIDS: fear of contagion or stigma?' *Journal of Applied Social Psychology* 21(23): 1877–88.

Biswas, G., P. Gupta and D. Das (2010). 'Wisdom teeth – a major problem in young generation, study on the basis of types and associated complications.' *Journal of College of Medical Sciences – Nepal* 6(3): 24–8.

Bittles, A. H. (1995). The influence of consanguineous marriage on reproductive behaviour in India and Pakistan. In C. G. N. Mascie-Taylor and A. J. Boyce (eds.), *Mating Patterns*. Cambridge, Cambridge University Press.

Bittles, A. H. (2004). Genetic aspects of inbreeding and incest. In A. P. Wolf and W. H. Durham (eds.), *Inbreeding, Incest and the Incest Taboo*. Stanford, CA, Stanford University Press.

Bittles, A. H. and U. Makov (1988). Inbreeding in human populations: assessment of the costs. In C. G. N. Mascie-Taylor and A. J. Boyce (eds.), *Mating Patterns*. Cambridge, Cambridge University Press.

Bittles, A. H. and J. V. Neel (1994). 'The costs of human inbreeding and their implications for variation at the DNA level.' *Nature Genetics* 8: 117–21.

Björkstén, B. (1999). 'Allergy priming early in life.' *The Lancet* 353(9148): 167–8.

Björkstén, B. (2009). 'The hygiene hypothesis: do we still believe in it?' *Nestle Nutr Ser Pediatr Program* 64: 11–8.

Blackmore, S. (1999). *The Meme Machine*. Oxford, Oxford University Press.

Blanchard, R. (1997). 'Birth order and sibling sex ratio in homosexual versus heterosexual males and females.' *Annual Review of Sex Research* 8: 27.

Blanchard, R. (2004). 'Quantitative and theoretical analyses of the relation between older brothers and homosexuality in men.' *Journal of Theoretical Biology* 230(2): 173–87.

Blanchard, R. and P. Klassen (1997). 'HY antigen and homosexuality in men.' *Journal of Theoretical Biology* 185(3): 373–8.

Blatteis, C. M. (2003). 'Fever: pathological or physiological, injurious or beneficial?' *Journal of Thermal Biology* **28**(1): 1–13.

Bleske-Rechek, A., L. A. Nelson, J. P. Baker, M. W. Remiker and S. J. Brandt (2010). 'Evolution and the trolley problem: People save five over one unless the one is young, genetically related, or a romantic partner.' *Journal of Social, Evolutionary, and Cultural Psychology* **4**(3): 115.

Bloom, G. and P. W. Sherman (2005). 'Dairying barriers affect the distribution of lactose malabsorption.' *Evolution and Human Behavior* **26**(4): 301–13.

Bloom, P. (2007). 'Religion is natural.' *Developmental Science* **10**(1): 147–51.

Bloom, P. (2009). 'Religious belief as an evolutionary accident.' In J. Schloss and M. J. Murray (eds.), *The Believing Primate: Scientific, Philosophical, and Theological Reflections on the Origin of Religion*. Oxford, Oxford University Press, pp. 118–27.

Blurton Jones, N. and R. M. Sibly (1978). Testing adaptiveness of culturally determined behaviour: do Bushman women maximize their reproductive success by spacing births widely and foraging seldom? In N. Blurton Jones and V. Reynolds (eds.), *Human Behaviour and Adaptation*. London, Taylor & Francis.

Blurton Jones, N., K. Hawkes and J. F. O'Connell (1999). Some current ideas about the evolution of the human life history. In P. C. Lee (ed.), *Comparative Primate Socioecology*. Cambridge, Cambridge University Press, pp. 140–66.

Boaz, N. T. and A. J. Almquist (1997). *Biological Anthropology*. Englewood Cliffs, NJ, Prentice Hall.

Bobrow, D. and J. M. Bailey (2001). 'Is male homosexuality maintained via kin selection?' *Evolution and Human Behavior* **22**(5): 361–8.

Bogin, B. (1999). *Patterns of Human Growth*. Cambridge, Cambridge University Press.

Bogin, B. (2010). Evolution of human growth. In M. P. Muehlenbein (ed.), *Human Evolutionary Biology*, Cambridge, Cambridge University Press, pp. 379–95.

Bolhuis, J. J., G. R. Brown, R. C. Richardson and K. N. Laland (2011). 'Darwin in mind: new opportunities for evolutionary psychology.' *PLoS Biology* **9**(7): e1001109.

Bollinger, R. R., A. S. Barbas, E. L. Bush, S. S. Lin and W. Parker (2007). 'Biofilms in the large bowel suggest an apparent function of the human vermiform appendix.' *Journal of Theoretical Biology* **249**(4): 826–31.

Borgi, M., I. Cogliati-Dezza, V. Brelsford, K. Meints and F. Cirulli (2014). 'Baby schema in human and animal faces induces cuteness perception and gaze allocation in children.' *Frontiers in Psychology* **5**: 411.

Bornstein, M. H. and D. L. Putnick (2007). 'Chronological age, cognitions, and practices in European American mothers: a multivariate study of parenting.' *Developmental Psychology* **43**(4): 850.

Bourke, A. (2011). 'The validity and value of inclusive fitness theory.' *Proceedings of the Royal Society, Series B* **278**: 3313–20.

Bovine HapMap Consortium (2009). 'Genome-wide survey of SNP variation uncovers the genetic structure of cattle breeds.' *Science* **324**(5926): 528–32.

Bowlby, J. (1969). *Attachment Theory, Separation, Anxiety and Mourning*. New York, NY, Basic Books.

Bowlby, J. (1980). *Attachment and Loss, vol 3. Loss: Sadness and Depression*. New York, NY, Basic Books.

Boyd, R., and P. J. Richerson (1988). *Culture and the Evolutionary Process*. Chicago, University of Chicago Press.

Boyd, R. and P. J. Richerson (1995). 'Why does culture increase human adaptability?' *Ethology and Sociobiology* **16**: 125–43.

Boyer, P. (1994). *The Naturalness of Religious Ideas: A Cognitive Theory of Religion*. Berkeley and Los Angeles, University of California Press.

Boyer, P. and B. Bergstrom (2008). 'Evolutionary perspectives on religion.' *Annual Review of Anthropology* **37**: 111–30.

Bramble, D. M. and D. E. Lieberman (2004). 'Endurance running and the evolution of Homo.' *Nature* **432**(7015): 345–52.

Brewer, M. B. (1979). 'In-group bias in the minimal intergroup situation: a cognitive-motivational analysis.' *Psychological Bulletin* **86**(2): 307.

Brewis, A. and M. Meyer (2005). 'Demographic evidence that human ovulation is undetectable (at least in pair bonds).' *Current Anthropology* **46**: 465–71.

British Association of Aesthetic Surgeons (2014). *Britain Sucks*. http://baaps.org.uk/about-us/press-releases/1833-britain-sucks, accessed January 2015.

Brooks, R. (2011). '"Asia's missing women" as a problem in applied evolutionary psychology?' *Evolutionary Psychology: An International Journal of Evolutionary Approaches to Psychology and Behavior* **10**(5): 910–25.

Brooks, R., I. M. Scott, A. A. Maklakov, M. M. Kasumovic, A. P. Clark and I. S. Penton-Voak (2011). 'National income inequality predicts women's preferences for masculinized faces better than health does.' *Proceedings of the Royal Society of London B: Biological Sciences* **278**(1707): 810–12.

Brosch, T., D. Sander and K. R. Scherer (2007). 'That baby caught my eye… attention capture by infant faces.' *Emotion* **7**(3): 685–9.

Brown, G. R., K. N. Laland and M. B. Mulder (2009). 'Bateman's principles and human sex roles.' *Trends in Ecology & Evolution* **24**(6): 297–304.

Brown, P., T. Sutikna, M. J. Morwood, R. P. Soejon, Jatmiko, E. Wayhu Saptomo and R. A. Due (2004). 'A new small-bodied hominin from the Late Pleistocene of Flores, Indonesia.' *Nature* **431**: 1055–61.

Browne, K. (2002). *Biology at Work: Rethinking Sexual Equality*. Rutgers University Press.

Browne, K. R. (2006). 'Sex, power, and dominance: the evolutionary psychology of sexual harassment.' *Managerial and Decision Economics* **27**(2–3): 145–58.

Brownell, K. D. and J. L. Pomeranz (2014). 'The trans-fat ban – food regulation and long-term health.' *New England Journal of Medicine* **370**(19): 1773–5.

Bryant, G. A. and M. G. Haselton (2009). 'Vocal cues of ovulation in human females.' *Biology Letters* **5**: 12–15.

Bugental, D. B. and K. Happaney (2004). 'Predicting infant maltreatment in low-income families: the interactive effects of maternal attributions and child status at birth.' *Developmental Psychology* **40**(2): 234.

Bugental, D. B., D. A. Beaulieu and A. Silbert-Geiger (2010). 'Increases in parental investment and child health as a result of an early intervention.' *Journal of Experimental Child Psychology* **106**(1): 30–40.

Bulbulia, J. (2004). 'The cognitive and evolutionary psychology of religion.' *Biology and Philosophy* **19**(5): 655–86.

Bulik, C. M., P. F. Sullivan, A. Pickering, A. Dawn and M. McCullan (1999). 'Fertility and reproduction in women with anorexia nervosa: a controlled study.' *Journal of Clinical Psychiatry* **60**: 130–5.

Bundey, S. and H. Alam (1993). 'A five-year prospective study of the health of children in different ethnic groups, with particular reference to the effect of inbreeding.' *European Journal of Human Genetics* **1**: 206–19.

Burch, R. L. and G. G. Gallup (2000). 'Perceptions of paternal resemblance predict family violence.' *Evolution and Human Behavior* **21**(6): 429–37.

Burkhardt, R. W. (1983). The development of an evolutionary ethology. In D. S. Bendall (ed.), *Evolution from Molecules to Men*. Cambridge, Cambridge University Press.

Burley, N. (1979). 'The evolution of concealed ovulation.' *American Naturalist* **114**: 835–58.

Burnham, T. C. and B. Hare (2007). 'Engineering human cooperation.' *Human Nature* **18**(2): 88–108.

Burton, R. (1973). 'Folk theory and the incest taboo.' *Ethos* **1**: 504–16.

Buss, D. M. (1989a). 'Conflict between the sexes: strategic interference and the evocation of anger and upset.' *Journal of Personality and Social Psychology* **56**(5): 735.

Buss, D. M. (1989b). 'Sex differences in human mate preferences: evolutionary hypotheses tested in 37 cultures.' *Behavioural and Brain Sciences* **12**: 1–49.

Buss, D. M. (2009). 'How can evolutionary psychology successfully explain personality and individual differences?' *Perspectives on Psychological Science* **4**(4): 359–66.

Buss, D. M. (2014). *Evolutionary Psychology: The New Science of the Mind*. Harlow, Pearson.

Buss, D. M., R. J. Larsen, D. Westen and J. Semmelroth (1992). 'Sex differences in jealousy: evolution, physiology and psychology.' *Psychological Science* **3**(4): 251–5.

Bussière, L. F., M. C. Tinsley and A. T. Laugen (2013). 'Female preferences for facial masculinity are probably not adaptations for securing good immunocompetence genes.' *Behavioral Ecology* **24**(3): 593–4.

Bygren L. O., G. Kaati and S. Edvinsson (2001). 'Longevity determined by ancestors' overnutrition during their slow growth period.' *Acta Biotheoretica* **49**: 53–9.

Byrne, R. W. and A. Whiten (1988). *Machiavellian Intelligence: Social Expertise and the Evolution of Intellect in Monkeys, Apes and Humans*. Oxford, Clarendon Press.

Calvin, W. H. (1982). *The Throwing Madonna: Essays on the Brain*. New York, McGraw–Hill.

Cameron, N. and B. Bogin (2012). *Human Growth and Development*, Access online via Elsevier.

Campbell, A. (2009). Gender and crime. In A. Walsh and K. M. Beaver (eds.), *Biosocial Criminology: New Directions in Theory and Research*. New York, Routledge.

Campbell, D. (1974). Evolutionary epistemology. In P. A. Schilpp (ed.), *The Philosophy of Karl Popper*. LaSalle, IL, Open Court.

Camperio-Ciani, A., F. Corna and C. Capiluppi (2004). 'Evidence for maternally inherited factors favouring male homosexuality and promoting female fecundity.' *Proceedings of the Royal Society of London. Series B: Biological Sciences* **271**(1554): 2217–21.

Cann, R. L., M. Stoneking and A. C. Wilson (1987). 'Mitochondrial DNA and human evolution.' *Nature* **325**: 31–6.

Canudas-Romo, V. (2008). 'The modal age at death and the shifting mortality hypothesis.' *Demographic Research* **19**: 1179–1204.

Cárdenas, R. A., L. J. Harris and M. W. Becker (2013). 'Sex differences in visual attention toward infant faces.' *Evolution and Human Behavior* **34**(4): 280–7.

Cardno, A. G., F. V. Rijsdijk, P. C. Sham, R. M. Murray and P. McGuffin (2002). 'A twin study of genetic relationships between psychotic symptoms.' *American Journal of Psychiatry* **159**(4): 539–45.

Carrera-Bastos, P., M. Fontes-Villalba, J. H. O'Keefe, S. Lindeberg and L. Cordain (2011). 'The Western diet and lifestyle and diseases of civilization.' *Res Rep Clin Cardiol* **2**: 15–35.

Carroll, J. (2004). *Literary Darwinism: Evolution, Human Nature, and Literature*. New York, Routledge.

Carruthers, P. (2006). *The Architecture of the Mind*. Oxford, Oxford University Press.

Cars, O., S. Mölstad and A. Melander (2001). 'Variation in antibiotic use in the European Union.' *The Lancet* **357**(9271): 1851–53.

Carter, A. and A. Nguyen (2011). 'Antagonistic pleiotropy as a widespread mechanism for the maintenance of polymorphic disease alleles.' *BMC Medical Genetics* **12**(1): 160.

Carter, G. G. and G. S. Wilkinson (2013). 'Food sharing in vampire bats: reciprocal help predicts donations more than relatedness or harassment.' *Proceedings of the Royal Society B: Biological Sciences* **280**(1753).

Cashdan, E. (1998). 'Adaptiveness of food learning and food aversions in children.' *Social Science Information* **37**(4): 613–32.

Cashdan, E., F. W. Marlowe, A. Crittenden, C. Porter and B. M. Wood (2012). 'Sex differences in spatial cognition among Hadza foragers.' *Evolution and Human Behavior* **33**(4): 274–84.

Casscells, W., A. Schoenberger and T. Grayboys (1978). 'Interpretation by physicians of clinical laboratory results.' *New England Journal of Medicine* **299**: 999–1000.

Cavalli-Sforza, L. L. and M. W. Feldman (1981). *Cultural Transmission and Evolution: A Quantitative Approach* (No. 16). Princeton, NJ, Princeton University Press.

Cavalli-Sforza, L. L., P. Menozzi and A. Piazza (1993). 'Demic expansions and human evolution.' *Science* **259**(5095): 639–46.

Chagnon, N. I. A. and W. Irons (1979). *Evolutionary Biology and Human Social Behavior: An Anthropological Perspective*. North Scituate, MA, Duxbury.

Champagne, F. A. (2008). 'Epigenetic mechanisms and the transgenerational effects of maternal care.' *Frontiers in Neuroendocrinology* **29**: 386–97.

Chapman, D. and K. Scott (2001). 'The impact of intergenerational risk factors on adverse developmental outcomes.' *Developmental Review* **21**: 305–25.

Chevalier-Skolnikoff, S. (1973). Facial expression of emotion in nonhuman primates. In P. Ekman (ed.), *Darwin and Facial Expression*. New York and London, Academic Press.

Chisholm, J. S. (1996). 'The evolutionary ecology of attachment organization.' *Human Nature* **7**: 1–38.

Chisholm, J. S., J. A. Quinlivan, R. W. Petersen and D. A. Coall (2005). 'Early stress predicts age at menarche and first birth, adult attachment, and expected lifespan.' *Human Nature* **16**(3): 233–65.

Chomsky, N. (1959). 'A Review of B. F. Skinner's Verbal Behavior.' *Language* **35**(1): 26–58.

Chung, W. and M. D. Gupta (2007). 'The decline of son preference in South Korea: The roles of development and public policy.' *Population and Development Review* **33**(4): 757–83.

CIA (2010). *The World Factbook 2010*. Washington, DC, Central Intelligence Agency.

CIA (2013). *The World Factbook 2013–14*. Washington, DC, Central Intelligence Agency.

Civitello, L. (2011). *Cuisine and Culture: A History of Food and People*. Hoboken, John Wiley & Sons.

Clatworthy, A. E., E. Pierson and D. T. Hung (2007). 'Targeting virulence: a new paradigm for antimicrobial therapy.' *Nature Chemical Biology* **3**(9): 541–8.

Clutton-Brock, T. (2009). 'Cooperation between non-kin in animal societies.' *Nature* **462**(7269): 51–7.

Clutton-Brock, T. H. and A. C. Vincent (1991). 'Sexual selection and the potential reproductive rates of males and females.' *Nature* **351**(6321): 58–60.

Collard (2002) "Grades and transitions in human evolution." *Proceedings of the British Academy* **106**, 61–100.

Conrad, P. and S. Markens (2001). 'Constructing the "gay gene" in the news: optimism and skepticism in the US and British press.' *Health* **5**(3): 373–400.

Cordain, L. (2007). 'Implications of plio-pleistocene hominin diets for modern humans.' In P. Ungar (ed.), *Early Hominin Diets: The Known, The Unknown, and the Unknowable*. Oxford, Oxford University Press, pp. 363–83.

Cordain, L., S. Eaton, J. Brand Miller, N. Mann and K. Hill (2002). 'Original communications – the paradoxical nature of hunter-gatherer diets: meat-based, yet non-atherogenic.' *European Journal of Clinical Nutrition* 56(1): S42.

Cordain, L., S. B. Eaton, A. Sebastian, N. Mann, S. Lindeberg, B. A. Watkins, J. H. O'Keefe and J. Brand-Miller (2005). 'Origins and evolution of the Western diet: health implications for the 21st century.' *The American Journal of Clinical Nutrition* 81(2): 341–54.

Cordain, L., J. B. Miller, S. B. Eaton, N. Mann, S. H. Holt and J. D. Speth (2000). 'Plant–animal subsistence ratios and macronutrient energy estimations in worldwide hunter-gatherer diets.' *The American Journal of Clinical Nutrition* 71(3): 682–92.

Cornwell, R. E., C. Palmer, P. M. Guinther and H. P. Davis (2005). 'Introductory psychology texts as a view of sociobiology/evolutionary psychology's role in psychology.' *Human Nature* 3: 355–74.

Cosmides, L. and J. Tooby (1992). Cognitive adaptations for social exchange. In J. H. Barkow, L. Cosmides and J. Tooby (eds.), *The Adapted Mind*. Oxford, Oxford University Press.

Cosmides, L. and J. Tooby (1996). 'Are humans good intuitive statisticians after all? Rethinking some conclusions from the literature on judgement under uncertainty.' *Cognition* 58: 1–73.

Cosmides, L., H. C. Barrett and J. Tooby (2010). 'Adaptive specializations, social exchange, and the evolution of human intelligence.' *Proceedings of the National Academy of Sciences* 107(Supplement 2): 9007–14.

Crawford, C. (1998a). Environments and adaptations: then and now. In C. Crawford and D. L. Krebs (eds.), *Handbook of Evolutionary Psychology*. Mahwah, NJ, Lawrence Erlbaum.

Crawford, C. (1998b). The theory of evolution in the study of human behavior. In C. Crawford and D. L. Krebs (eds.), *Handbook of Evolutionary Psychology*. Mahwah, NJ, Lawrence Erlbaum.

Crivelli, C., P. Carrera and J. Fernadez-Dols (2015). 'Are smiles signs of happiness? Spontaneous expressions of judo winners.' *Evolution and Human Behavior* 36: 52–8.

Cronin, H. (1991). *The Ant and the Peacock*. Cambridge, Cambridge University Press.

Cross-Disorder Group of the Psychiatric Genomics Consortium (2013). 'Genetic relationship between five psychiatric disorders estimated from genome-wide SNPs.' *Nature Genetics* 45(9): 984–94.

Crow, J. F. (2003). 'There's something curious about paternal-age effects.' *Science* 301(5633): 606–7.

Crow, J. F. (2006). 'Age and sex effects on human mutation rates: an old problem with new complexities.' *Journal of Radiation Research* 47(Suppl B): B75–82.

Cunnane, S. C. and M. A. Crawford (2003). 'Survival of the fattest: fat babies were the key to evolution of the large human brain.' *Comparative Biochemistry and Physiology Part A: Molecular & Integrative Physiology* 136(1): 17–26.

Cunningham, M. R. (1986). 'Measuring the physical in physical attractiveness: quasi-experiments on the sociobiology of female facial beauty.' *Journal of Personality and Social Psychology* 50(5): 925.

Cunningham, M. R., A. R. Roberts, A. P. Barbee, P. B. Druen and C.-H. Wu (1995). '"Their ideas of beauty are, on the whole, the same as ours": consistency and variability in the cross–cultural perception of female physical attractiveness.' *Journal of Personality and Social Psychology* 68(2): 261.

Curley, J. P., F. Moshoodh and F. A. Champagne (2011). 'Epigenetics and the origins of paternal effects.' *Hormones and Behaviour* 59: 306–14.

Curry, A. (2013). 'The milk revolution.' *Nature* 500(7460): 20–2.

Curry, O. (2005). *Morality as Natural History*. PhD. University of London.

Cutler, S. J., A. R. Fooks and W. H. van der Poel (2010). 'Public health threat of new, re-emerging, and neglected zoonoses in the industrialized world.' *Emerging Infectious Diseases* 16(1): 1.

Daly, M. (1997). Introduction. In G. Bock and G. Cardew (eds.), *Characterizing Human Psychological Adaptations*. New York, NY, John Wiley.

Daly, M. and M. Wilson (1988a). *Homicide*. New York, Aldine De Gruyter.

Daly, M. and M. Wilson (1988b). 'Evolutionary Social Psychology and Family Homicide.' *Science* 242: 519–24.

Daly, M. and M. Wilson (1997). Evolutionary social psychology and family homicide. In S. Baron-Cohen (ed.), *The Maladapted Mind: Classic Readings in Evolutionary Psychopathology*. East Sussex, UK, Psychology Press, p. 115.

Daly, M. and M. Wilson (1998). *The Truth About Cinderella*. London, Orion.

Daly, M. and M. I. Wilson (1994). 'Some differential attributes of lethal assaults on small children by stepfathers versus genetic fathers.' *Ethology and Sociobiology* 15(2): 207–17.

Daly, M. and M. I. Wilson (1999). 'Human evolutionary psychology and animal behaviour.' *Animal Behaviour* 57: 509–519.

Daly, M., M. I. Wilson and S. J. Weghorst (1982). 'Male sexual jealousy.' *Ethology and Sociobiology* **3**: 11–27.

Damasio, A. R. (2000). 'A neural basis for socio-pathy.' *Archives of General Psychiatry* **57**(2): 128–9.

Danielsbacka, M., A. O. Tanskanen, M. Jokela and A. Rotkirch (2011). 'Grandparental child care in Europe: evidence for preferential investment in more certain kin. *Evolutionary Psychology* **9**(1). doi: 147470491100900102.

Darwin, C. (1858). Letter to Charles Lyell 18th June 1858. In F. Burkhardt and S. Smith (eds.), *The Correspondence of Charles Darwin*. Cambridge, Cambridge University Press, 1991.

Darwin, C. (1859a). Letter to Alfred Russell Wallace. In F. Burkhardt and S. Smith (eds.), *The Correspondence of Charles Darwin*. Cambridge, Cambridge University Press, 1991.

Darwin, C. (1859b). *On the Origin of Species by Means of Natural Selection*. London, John Murray.

Darwin, C. (1871/1981). *The Descent of Man and Selection in Relation to Sex*. Princeton, NJ, Princeton University Press.

Darwin, C. (1872). *The Expression of the Emotions in Man and Animals*. London, UK, John Murray.

Darwin, C. (1874). *The Descent of Man and Selection in Relation to Sex*, 2nd edition. London, John Murray.

Darwin, C. (1899). *The Descent of Man and Selection in Relation to Sex*. London, John Murray.

Darwin, G. H. (1875). 'Marriages between first cousins in England and their effects.' *Journal of the Statistical Society of London* **38**(2): 153–184.

Das, G., S. Crocombe, M. McGrath, J. L. Berry and M. Mughal (2006). 'Hypovitaminosis D among healthy adolescent girls attending an inner city school' *Archives of Disease in Childhood* **91**(7): 569–72.

Davies, J. H. and N. J. Shaw (2010). 'Preventable but no strategy: vitamin D deficiency in the UK.' *Archives of Diseases in Childhood* **96**, 614–5.

Davis, J. N. (1997). 'Birth order, sibship size, and status in modern Canada.' *Human Nature* **8**: 205–30.

Dawkins, R. (1976). *The Selfish Gene*. Oxford, Oxford University Press.

Dawkins, R. (1982). *The Extended Phenotype*. Oxford, W.H. Freeman.

Dawkins, R. (1986). *The Blind Watchmaker*. London, Longman.

Dawkins, R. (1989). *The Selfish Gene*. Oxford, Oxford University Press.

Dawkins, R. (1997). 'Is science a religion?' *The Humanist* **57**(1): 26–29.

Dawkins, R. (2006). *The God Delusion*. London, Bantam Press.

Dawkins, R. (2012). 'The descent of Edward Wilson.' *Prospect* (June): 66–9.

D'Costa, V. M., K. M. McGrann, D. W. Hughes and G. D. Wright (2006). 'Sampling the antibiotic resistome.' *Science* **311**(5759): 374–7.

De Backer, C., J. Braeckman and L. Farinpour (2008). Mating intelligence in personal ads. In G. Geher and G. Miller (eds.), *Mating Intelligence: Sex, Relationships, and the Mind's Reproductive System*. New York, Taylor & Francis, pp. 77–101.

de Rooij, S. R., H. Wouters, J. E. Yonker, R. C. Painter and T. J. Roseboom (2010). 'Prenatal under-nutrition and cognitive function in late adulthood.' *Proceedings of the National Academy of Sciences* **107**(39): 16881–6.

de Waal, F. B. M. (1997). 'The chimpanzee's service economy: food for grooming.' *Evolution and Human Behavior* **18**: 375–86.

Deacon, T. (1997). *The Symbolic Species*. London, Penguin.

DeBruine, L. M., B. C. Jones, J. R. Crawford, L. L. Welling and A. C. Little (2010a). 'The health of a nation predicts their mate preferences: cross-cultural variation in women's preferences for masculinized male faces.' *Proceedings of the Royal Society of London B: Biological Sciences* **277**(1692): 2405–10.

De Bruine, L. M., B. C. Jones, D. A. Frederick, M. G. Hasleton, I. S. Penton-Voak and D. I. Perrett (2010b). 'Evidence for menstrual cycle shifts in women's preferences for masculinity: a response to Harris (in press) "Menstrual Cycle and Facial Preferences Reconsidered".' *Evolutionary Psychology* **8**(4): 768–75.

DeBruine, L. M., B. C. Jones, A. C. Little, L. G. Boothroyd, D. I. Perrett, I. S. Penton-Voak, P. A. Cooper, L. Penke, D. R. Feinberg and B. P. Tiddeman (2006). 'Correlated preferences for facial masculinity and ideal or actual partner's masculinity.' *Proceedings of the Royal Society of London B: Biological Sciences* **273**(1592): 1355–60.

DeKay, W. T. (1995). 'Grandparental investment and the uncertainty of kinship.' Seventh Annual Meeting of the Human Behaviour and Evolution Society, Santa Barbara, CA.

Dennett, D. C. (1995). *Darwin's Dangerous Idea*. New York, Simon and Schuster.

Desmond, A. and J. Moore (1991). *Darwin*. London, Michael Joseph.

DeSouza, M. J. and D. A. Metzger (1991). 'Reproductive dysfunction in amenorrheic athletes and anorexic patients.' *Medicine and Science in Sports and Exercise* 56: 20–7.

Diamond, J. (1987). 'The worst mistake in the history of the human race.' *Discover* 8(5): 64–6.

Diamond, J. (1991). *The Rise and Fall of the Third Chimpanzee*. London, Vintage.

Diamond, M. (2010). Sexual orientation and gender identity. In I. B. Weiner and W. E. Craighead (eds.), *The Corsini Encyclopedia of Psychology, Volume 4*. New York, John Wiley and Sons.

Dias, B. G. and K. J. Ressler (2014). 'Parental olfactory experience influences behavior and neural structure in subsequent generations.' *Nature Neuroscience* 17(1): 89–96.

Dixson, A. F. (1987). 'Baculum length and copulatory behavior in primates.' *American Journal of Primatology* 13(1): 51–60.

Dixson, B. J., A. F. Dixson, B. Li and M. J. Anderson (2007). 'Studies of human physique and sexual attractiveness: sexual preferences of men and women in China.' *American Journal of Human Biology* 19(1): 88–95.

do Amaral, L. Q. (1996). 'Loss of body hair, bipedality and thermoregulation. Comments on recent papers in the Journal of Human Evolution.' *Journal of Human Evolution* 30(4): 357–66.

Domb, L. G. and M. Pagel (2001). 'Sexual swellings advertise female quality in wild baboons.' *Nature* 410: 204–6.

Downie, J. (2004). 'Evolution in health and disease: the role of evolutionary biology in the medical curriculum.' *BioScience Education* (4).

Doyle, J. F. (2009). 'A woman's walk: attractiveness in motion.' *Journal of Social, Evolutionary, and Cultural Psychology* 3(2): 81.

Doyle, J. F. and F. Pazhoohi (2012). 'Natural and augmented breasts: is what is not natural most attractive?' *Human Ethology Bulletin* 27(4): 4–14.

Driscoll, C. A., D. W. Macdonald and S. J. O'Brien (2009). 'From wild animals to domestic pets, an evolutionary view of domestication.' *Proceedings of the National Academy of Sciences* 106(Supplement 1): 9971–78.

Dunbar, R. (1980). 'Determinants and evolutionary consequences of dominance among female Gelada baboons.' *Behavioural Ecology and Sociobiology* 7: 253–65.

Dunbar, R. I. and S. Shultz (2007). 'Evolution in the social brain.' *Science* 317(5843): 1344–47.

Dunbar, R. I. M. (1993). 'Coeveolution of neocortical size, group size and language in humans.' *Behavioural and Brain Sciences* 16: 681–735.

Dunbar, R. I. M. (1996a). Determinants of group size in primates: a general model. In W. G. Runciman, J. Maynard Smith and R. I. M. Dunbar (eds.), *Evolution of Social Behaviour in Primates and Man*. Oxford, Oxford University Press.

Dunbar, R. I. M. (1996b). *Grooming, Gossip and the Evolution of Language*. London, Faber and Faber.

Dunn, M. J., S. Brinton and L. Clark (2010). 'Universal sex differences in online advertisers' age preferences: comparing data from 14 cultures and 2 religious groups.' *Evolution and Human Behavior* 31(6): 383–93.

Dunson, D. B., B. Colombo and D. D. Baird (2002). 'Changes with age in the level and duration of fertility in the menstrual cycle.' *Human reproduction* 17(5): 1399–1403.

Dunsworth, H., A. G. Warrener, T. Deacon, P. T. Ellison and H. Pontzer (2012). 'Metabolic hypothesis for human altriciality.' *Proceedings of the National Academy of Sciences of America*.

Durant, J. R. (1986). 'The making of ethology: the association for the study of animal behaviour, 1936–1986.' *Animal Behaviour* 34: 1601–16.

Durham, W. H. (1991). *Coevolution: Genes, Culture, and Human Diversity*. Stanford University Press.

Dyer, O. (2005). 'MP is criticised for saying that marriage of first cousins is a health problem.' *British Medical Journal*, 331(7528): 1292.

Easton, J. A., L. D. Schipper and T. K. Shackelford (2007). 'Morbid jealousy from an evolutionary psychological perspective.' *Evolution and Human Behavior* 28(6): 399–402.

Eaton, S. B., M. Konner and N. Paleolithic (1985). 'A consideration of its nature and current implications.' *New England Journal of Medicine* 312(5): 283–89.

Eaton, S. B., M. Konner and M. Shostak (1988). 'Stone agers in the fast lane: chronic degenerative disease in evolutionary perspective.' *American Journal of Medicine* 84: 739–49.

Edlund, L., H. Li, J. Yi and J. Zhang (2013). Sex ratios and crime: Evidence from China. *Review of Economics and Statistics*, 95(5), 1520–1534.

Eibl-Eibesfeldt, I. (1970). *Ethology: The Biology of Behaviour*. New York, Holt, Rinehart and Winston.

Eibl-Eibesfeldt, I. (1989). *Human Ethology*. New York, Aldine de Gruyter.

Einon, D. (1998). 'How many children can one man have?' *Evolution and Human Behavior* **19**: 413–26.

Ekman, P. (1973). Cross cultural studies of facial expressions. In P. Ekman (ed.), *Darwin and Facial Expression*. New York and London, Academic Press.

Elia, M. (1992). Organ and tissue contribution to metabolic rate. In J. M. Kinney and H. N. Tucker (eds.), *Energy Metabolism: Tissue Determinants and Cellular Corollaries*. New York, Raven Press, pp. 19–60.

Ellis, B. J. and J. Garber (2000). 'Psychological antecedents of variation in girls' pubertal timing: maternal depression, stepfather presence, and marital and family stress.' *Child Development* **71**: 485–501.

Ellis, B. J., M. Del Giudice, T. J. Dishion, A. J. Figueredo, P. Gray, V. Griskevicius, P. H. Hawley, W. J. Jacobs, J. James, A. A. Volk and D. S. Wilson (2012). 'The evolutionary basis of risky adolescent behavior: implications for science, policy, and practice.' *Developmental Psychology* **48**(3): 598–623.

Ellis, L. and A. Walsh (2000). *Criminology: A Global Perspective*. Boston, Allyn & Bacon

Ellison, P. (2001). *On Fertile Ground: A Natural History of Human Reproduction*. Cambridge, MA, Harvard University Press.

Ernest-Jones, M., D. Nettle and M. Bateson (2011). 'Effects of eye images on everyday cooperative behavior: a field experiment.' *Evolution and Human Behavior* **32**(3): 172–78.

Euler, H. A. and B. Weitzel (1996). 'Discriminating grandparental solicitude as reproductive strategy.' *Human Nature* **7**: 39–59.

Evans, E. M. (2001). 'Cognitive and contextual factors in the emergence of diverse belief systems: creation versus evolution.' *Cognitive Psychology* **42**(3): 217–66.

Ewald, P. W. (1991). 'Transmission modes and the evolution of virulence.' *Human Nature* **2**(1): 1–30.

Ewald, P. W. (1994). *Evolution of Infectious Disease*. New York, Oxford University Press.

Eysenk, M. (2004). *Psychology: An International Perspective*. Hove, East Sussex, Psychology Press.

Falk, D. (1983). 'Cerebral cortices of East African early hominids.' *Science* **221**: 1072–74.

Farris, C., T. A. Treat, R. J. Viken and R. M. McFall (2008). 'Sexual coercion and the misperception of sexual intent.' *Clinical Psychology Review* **28**(1): 48–66.

Faulkner, J., M. Schaller, J. H. Park and L. A. Duncan (2004). 'Evolved disease-avoidance mechanisms and contemporary xenophobic attitudes.' *Group Processes & Intergroup Relations* **7**(4): 333–53.

Faurie, C., A. Alvergne, S. Bonenfant, M. Goldberg, S. Hercberg, M. Zins and M. Raymond (2006). 'Handedness and reproductive success in two large cohorts of French adults.' *Evolution and Human Behavior* **27**(6): 457–72.

Faurie, C. and M. Raymond (2004). 'Handedness frequency over more than ten thousand years.' *Proceedings of the Royal Society, London* **271**: 43–5.

Faurie, C. and M. Raymond (2005). 'Handedness, homicide and negative frequency-dependent selection.' *Proceedings of the Royal Society B* **272**: 25–8.

Faurie, C., W. Schiefenhövel, S. le Bomin, S. Billiard and M. Raymond (2005). 'Variation in the frequency of left-handedness in traditional societies.' *Current Anthropology* **46**(1): 142–7.

Fehr, E. and S. Gaechter (2002). 'Altruistic punishment in humans.' *Nature* **10**: 137–40.

Feig, D. S., B. Zinman, X. Wang and J. E. Hux (2008). 'Risk of development of diabetes mellitus after diagnosis of gestational diabetes.' *Canadian Medical Association Journal* **179**(3): 229–34.

Feinberg, D., B. Jones, M. Law Smith, F. R. Moore, L. DeBruine, R. Cornwell, S. Hillier and D. Perrett (2006). 'Menstrual cycle, trait estrogen level, and masculinity preferences in the human voice.' *Hormones and Behavior* **49**(2): 215–22.

Feldman, M. W. and K. N. Laland (1996). 'Gene–culture coevolutionary theory.' *Trends in Evolution and Ecology* **11**: 453–7.

Felson, R. B. (1997). 'Anger, aggression, and violence in love triangles.' *Violence and Victims* **12**(4): 345–62.

Fernández–Dols, J.-M. and C. Crivelli (2013). 'Emotion and expression: naturalistic studies.' *Emotion Review* **5**(1): 24–9.

Fessler, D. M. T. and C. D. Navarrete (2004). 'Third-party attitudes towards sibling incest. Evidence for Westermarck's hypotheses.' *Evolution and Human Behavior* **25**: 277–94.

Fessler, D. T. (2002). 'Reproductive immunosuppression and diet.' *Current Anthropology* **43**(1): 19–61.

Fifer, F. (1987). 'The adoption of bipedalism by the hominids: a new hypothesis.' *Human Evolution* **2**(2): 135–47.

Finch, C. E. and R. M. Sapolsky (1999). 'The evolution of Alzheimer disease, the reproductive schedule, and apoE isoforms.' *Neurobiology and Aging* **20**: 407–28.

Fincher, C. L., R. Thornhill, D. R. Murray and M. Schaller (2008). 'Pathogen prevalence predicts human cross-cultural variability in individualism/

collectivism.' *Proceedings of the Royal Society B: Biological Sciences* **275**(1640): 1279–85.

Finlay, B. L., R. B. Darlington and N. Nicastro (2001). 'Developmental structure in brain evolution.' *Behavioral and Brain Sciences* **24**(02): 263–78.

Firman, R. C. and L. W. Simmons (2010). 'Sperm midpiece length predicts sperm swimming velocity in house mice.' *Biology Letters* **6**(4): 513–16.

Fischbein, S. (1980). 'IQ and social class.' *Intelligence* **4**(1): 51–63.

Fisher, H. E. (1992). *Anatomy of Love*. New York, W.W. Norton.

Fisher, H. E. (2012). Serial monogamy and clandestine adultery: evolution and consequences of the dual human reproductive strategy. In S. Craig Roberts (ed.), *Applied Evolutionary Psychology*. Oxford, Oxford University Press.

Fisher, R. A. (1930). *The Genetical Theory of Natural Selection*. Oxford, Clarendon Press.

Fitzgerald, C. J. and M. B. Whitaker (2010). 'Examining the acceptance of and resistance to evolutionary psychology.' *Evolutionary Psychology* **8**(2), 284–96.

Flaxman, S. M. and P. W. Sherman (2000). 'Morning sickness: a mechanism for protecting mother and embryo.' *Quarterly Review of Biology*: 113–48.

Fodor, J. (1983). *The Modularity of Mind*. Cambridge, MA, MIT Press.

Foley, R. (1987). *Another Unique Species*. Harlow, Longman.

Foley, R. A. (1989). The evolution of hominid social behaviour. In V. Standen and R. A. Foley (eds.), *Comparative Socioecolgy*. Oxford, Blackwell Scientific.

Folstad, I. and A. J. Karter (1992). 'Parasites, bright males, and the immunocompetence handicap.' *American Naturalist*: 603–22.

Fonseca-Azevedo, K. and S. Herculano-Houzel (2012). 'Metabolic constraint imposes tradeoff between body size and number of brain neurons in human evolution.' *Proceedings of the National Academy of Sciences* **109**(45): 18571–6.

Ford, L., V. Graham, A. Wall and J. Berg (2006). 'Vitamin D concentrations in an inner-city multicultural outpatient population.' *Annals of Clinical Biochemistry* **43**: 469–73.

Foster, K. R., T. Wenseleers and F. L. W. Ratnieks (2001). 'Spite: Hamilton's unproven theory.' *Ann. Zool. Fennici* **38**: 229–38.

Fox, M., L. A. Knapp, P. W. Andrews and C. L. Fincher (2013). 'Hygiene and the world distribution of Alzheimer's disease: epidemiological evidence for a relationship between microbial environment and age-adjusted disease burden.' *Evolution, Medicine, and Public Health* **2013**(1): 173–86.

Fraga, M. F., E. Ballestar, M. F. Paz, S. Ropero, F. Setien, M. L. Ballestar, D. Heine-Suñer, J. C. Cigudosa, M. Urioste and J. Benitez (2005). 'Epigenetic differences arise during the lifetime of monozygotic twins.' *Proceedings of the National Academy of Sciences of the United States of America* **102**(30): 10604–9.

Frank, R. (1988). *Passions within Reason: The Strategic Role of the Emotions*. New York, W.W. Norton.

Frassetto, L., R. Morris Jr, D. Sellmeyer, K. Todd and A. Sebastian (2001). 'Diet, evolution and aging.' *European Journal of Nutrition* **40**(5): 200–13.

Frayser, S. (1985). *Varieties of sexual experience: an anthropological perspective of human sexuality*. New Haven, CT, HRAF Press.

Frazer, J. G. (1910). *Totemism and Exogamy: A Treatise on Certain Early Forms of Superstition and Society*. Macmillan and Co., Limited.

Frederick, D. A. and M. G. Haselton (2007). 'Why is muscularity sexy? Tests of the fitness indicator hypothesis.' *Personality and Social Psychology Bulletin* **33**(8): 1167–83.

Freud, S. (1913/1950). *Totem and Taboo*. New York, W.W. Norton.

Friday, A. E. (1992). Human evolution: the evidence from DNA sequencing. In S. Jones, R. Martin and D. Pilbeam (eds.), *The Cambridge Encyclopedia of Human Evolution*. Cambridge, Cambridge University Press.

Fridlund, A. J. (1994). *Human Facial Expression: An Evolutionary View*. Academic Press.

Friedman, D. (2005). *Economics and Evolutionary Psychology*. Emerald Group Publishing Limited.

Frisancho, A. R. (1993). *Human Adaptation and Accommodation*. Ann Arbor, University of Michigan Press.

Frisancho, A. R. (2010). The study of human adaptation. In M. P. Muehlenbein (ed.), *Human Evolutionary Biology*. Cambridge, Cambridge University Press.

Frisch, R. E. and R. Revelle (1970). 'Height and weight at menarche and a hypothesis of critical body weights and adolescent events.' *Science* **169**(3943): 397–9.

Frisch, R. E. and R. Revelle (1971). 'Height and weight at menarche and a hypothesis of menarche.' *Archives of Disease in Childhood* **46**(249): 695–701.

Gadgil, M. and W. H. Bossert (1970). 'Life historical consequences of natural selection.' *American Naturalist* **104**: 1–24.

Gagnon, A., K. R. Smith, M. Tremblay, H. Vézina, P. P. Paré and B. Desjardins (2009). 'Is there a trade-off between fertility and longevity? A comparative study of women from three large historical databases accounting for mortality selection.' *American Journal of Human Biology* **21**(4): 533–40.

Gangestad, S. W. and R. Thornhill (1994). 'Facial attractiveness, developmental stability and fluctuating asymmetry.' *Ethology and Sociobiology* **15**: 73–85.

Gangestad, S. W. and R. Thornhill (1998). 'Menstrual cycle variation in women's preferences for the scent of symmetrical men.' *Proceedings of the Royal Society of London B: Biological Sciences* **265**(1399): 927–33.

Gangestad, S. W., R. Thornhill and C. Garver (2002). 'Changes in women's sexual interests and their partner's mate-retention tactics across the menstrual cycle: evidence for shifting conflicts of interest.' *Proceedings of the Royal Society B* **269**: 975–82.

Gangestad, S. W., R. Thornhill and C. Garver-Apgar (2005). Adaptations to ovulation. In D. Buss (ed.), *The Handbook of Evolutionary Psychology*. Hoboken, NJ, John Wiley and Sons Inc.

Gates, G. J. and F. Newport (2012). 'Special report: 3.4% of US adults identify as LGBT.' Washington: Gallup.

Gaulin, S. C. and D. Sailer (1985). 'Are females the ecological sex?' *American Anthropology* **87**: 111–9.

Gaulin, S. J. C. and H. A. Hoffman (1988). Evolution and development of sex differences in spatial ability. In L. Betzig, M. B. Mulder and P. Turke (eds.), *Human Reproductive Behaviour: A Darwinian Perspective*. Cambridge, Cambridge University Press.

Geary, D. C. (1998). *Male, Female. The Evolution of Human Sex Differences*. Washington, DC, American Psychological Association.

Geary, D. C. and K. J. Huffman (2002). 'Brain and cognitive evolution: forms of modularity and functions of mind.' *Psychological Bulletin* **128**(5): 667–98.

George, S. M. (2006). 'Millions of missing girls: from fetal sexing to high technology sex selection in India.' *Prenatal Diagnosis* **26**(7): 604–9.

George, S. M. and R. S. Dahiya (1998). 'Female foeticide in rural Haryana.' *Economic and Political Weekly*: 2191–8.

Geschwind, N. (1984). 'Cerebral dominance in biological perspective.' *Neuropsychologia* **22**: 675–83.

Gigerenzer, G. (1994). 'Why the distinction between single-event probabilities and frequencies is important for psychology (and vice versa).' *Subjective Probability*: 129–61.

Gigerenzer, G. (2000). *Adaptive Thinking*. New York, Oxford University Press.

Gigerenzer, G. (2001). The adaptive toolbox. In G. Gigerenzer and R. Selten (eds.), *Bounded Rationality. The Adaptive Toolbox*. Cambridge, MA, MIT Press.

Gigerenzer, G. (2010). 'Moral satisficing: rethinking moral behavior as bounded rationality.' *Topics in Cognitive Science* **2**(3): 528–54.

Gigerenzer, G. and U. Hoffrage (1995). 'How to improve Bayesian reasoning without instruction: frequency formats.' *Psychological Review* **102**(4): 684.

Gigerenzer, G., P. M. Todd and the ABC Research Group (eds.) (1999). *Simple Heuristics That Make Us Smart*. New York, Oxford University Press.

Gilbert, S. F. and Z. Zevit (2001). 'Congenital human baculum deficiency: the generative bone of Genesis 2: 21–3.' *American Journal of Medical Genetics* **101**(3): 284–5.

Gilligan, C. (1982). *In a Different Voice*. Cambridge, MA, Harvard University Press.

Gluckman, P. and M. Hanson (2005). *The Fetal Matrix: Evolution, Development And Disease*. Cambridge, Cambridge University Press.

Gluckman, P. D. and M. A. Hanson (2006). 'Evolution, development and timing of puberty.' *Trends in Endocrinology & Metabolism* **17**(1): 7–12.

Gluckman, P. D., M. A. Hanson and A. S. Beedle (2007). 'Feature article early life events and their consequences for later disease: a life history and evolutionary perspective.' *American Journal of Human Biology* **19**: 1–19.

Gluckman, P. D., M. A. Hanson, T. Buklijas, F. M. Low and A. S. Beedle (2009). 'Epigenetic mechanisms that underpin metabolic and cardiovascular diseases.' *Nature Reviews Endocrinology* **5**(7): 401–8.

Gluckman, P. D., F. M. Low, T. Buklijas, M. A. Hanson and A. S. Beedle (2011). 'How evolutionary principles improve the understanding of human health and disease.' *Evolutionary Applications* **4**(2): 249–63.

Goldstein, L., C. D. Bushnell, R. J. Adams, L. J. Appel, L. T. Braun, S. Chaturvedi, M. Creager, A. Culebras, R. Eckel and R. Hard (2011). 'On behalf of the American Heart Association Stroke Council, Council on Cardiovascular Nursing, Council on Epidemiology and Prevention, Council for High Blood Pressure Research, and Council on Peripherdal Vascular Disease, and Interdisciplinary Council on Quality of Care and Outcomes Research.' *Stroke* **42**(2): 517–84.

Gollisch, T. and M. Meister (2010). 'Eye smarter than scientists believed: neural computations in circuits of the retina.' *Neuron* **65**(2): 150–64.

Gonzaga, G., M. G. Haselton, M. S. Davies, J. Smurda and J. C. Poore (2008). 'Love, desire, and the suppression of thoughts of romantic alternatives.' *Evolution and Human Behavior* **29**: 119–126.

Goodman, A., I. Koupil and D. W. Lawson (2012). 'Low fertility increases descendant socioeconomic position but reduces long-term fitness in a modern post-industrial society.' *Proceedings of the Royal Society B: Biological Sciences* **279**(1746): 4342–51.

Goossens, H., M. Ferech, R. Vander Stichele and M. Elseviers (2005). 'Outpatient antibiotic use in Europe and association with resistance: a cross-national database study.' *The Lancet* **365**(9459): 579–87.

Gopnik, M., J. Dalalakis, S. E. Fukuda, S. Fukuda and E Kehayia (1996). 'Genetic language impairment: unruly grammars.' *Proceedings of the British Academy* **88**: 223–50.

Gottlieb, G. (1971). *Development of Species Identification in Birds*. Chicago, University of Chicago Press.

Gottschall, J. and D. S. Wilson (2005) *The Literary Animal: Evolution and the Nature of Narrative*. Northwestern University Press.

Gould, R. G. (2000). 'How many children could Moulay Ismail have had?' *Evolution and Human Behavior* **21**(4): 295.

Gould, S. J. (1982). 'A biographical homage to Mickey Mouse.' *The Panda's Thumb*: 95–107.

Gould, S. J. (1998, 29 May). 'Let's leave Darwin out of it.' *New York Times*.

Gould, S. J. and R. C. Lewontin (1979) 'The spandrels of San Marco and the Panglossian paradigm: a critique of the adaptionist programme.' *Proceedings of the Royal Society of London* **205**: 581–98.

Graham, J., B. A. Nosek, J. Haidt, R. Iyer, S. Koleva and P. H. Ditto (2011). 'Mapping the moral domain.' *Journal of Personality and Social Psychology* **101**(2): 366.

Graham-Kevan, N. and J. Archer (2011). 'Violence during pregnancy: investigating infanticidal motives.' *Journal of Family Violence* **26**(6): 453–58.

Grainger, S. and J. Beise (2004). 'Menopause and post-generative longevity: testing the "stopping early" and "grandmother" hypotheses.' *Max Planck Institute for Demographic Research Working Paper* **2004**: 3.

Grammer, K. (1992). 'Variations on a theme: age dependent mate selection in humans.' *Behaviour and Brain Sciences* **15**: 100–2.

Grammer, K. and R. Thornhill (1994). 'Human facial attractiveness and sexual selection: the roles of averageness and symmetry.' *Journal of Comparative Psychology* **108**: 233–42.

Gray, P. (2003). Marriage, parenting, and testosterone variation among Kenyan Swahili men. *American Journal of Physical Anthropology*, **122**(3), 279–286.

Gray, P. B., S. M. Kahlenberg, E. S. Barrett and S. F. Lipson (2002). 'Marriage and fatherhood are associated with lower testosterone in males.' *Evolution and Human Behavior* **23**: 193–201.

Gray, R. D., M. Heaney and S. Fairhall (2003). Evolutionary psychology and the challenge of adaptive explanation. In J. F. K. Sterelny (ed.), *From Mating to Mentality: Evaluating Evolutionary Psychology* London, Psychology Press, pp. 247–68.

Gray, S. J. (1996). 'Ecology of weaning among nomadic Turkana pastoralists of Kenya: maternal thinking, maternal behavior, and human adaptive strategies.' *Human Biology*: 437–65.

Greene, J. D. (2007). The secret joke of Kant's soul. In W. Sinnott-Armstrong (ed.), *Moral Psychology, Vol. 3: The Neuroscience of Morality: Emotion, Disease, and Development*. Cambridge, MA, MIT Press, pp. 35–80.

Greene, J. and J. Haidt (2002). 'How (and where) does moral judgment work?' *Trends in Cognitive Sciences* **6**(12): 517–23.

Greenless, I. A. and W. C. McGrew (1994). 'Sex and age differences in preferences and tactics of mate attraction: analysis of published advertisements.' *Ethology and Sociobiology* **15**: 59–72.

De Gregoria et al. (2005). Lesch–Nyhan disease in a female with a clinically normal monozygotic twin. *Molecular genetics and metabolism*, **85**(1), 70–77.

Griffin, A. S. and S. A. West (2003). 'Kin discrimination and the benefit of helping in cooperatively breeding vertebrates.' *Science* **302**(5645): 634–6.

Griffiths, P. E. (2001). From adaptive heuristic to phylogenetic perspective: some lessons from the evolutionary psychology of emotion. In H. R. Holcomb III (ed.), *Conceptual Challenges in Evolutionary Psychology*. Springer.

Griggs, R. C., W. Kingston, R. F. Jozefowicz, B. E. Herr, G. Forbes and D. Halliday (1989). 'Effect of testosterone on muscle mass and muscle protein synthesis.' *Journal of Applied Physiology* **66**(1): 498–503.

Gross, C. G. (1993). 'Huxley versus Owen: the hippocampus minor and evolution.' *Trends in Neurosciences* **16**(12): 493–8.

Grotto, D. and E. Zied (2010). 'The standard American diet and its relationship to the health status of Americans.' *Nutrition in Clinical Practice* **25**(6): 603–12.

Grouios, G., H. Tsorbatzoudis, K. Alexandris and V. Barkoukis (2000). 'Do left-handed competitors have an innate superiority in sports?' *Perceptual and Motor Skills* **90**(3c): 1273–82.

Gruber, H. E. (1974). *Darwin on Man: A Psychological Study of Scientific Creativity, Together With Darwin's Early and Unpublished Notebooks Transcribed and Annotated by Paul H. Barrett*. London, Wildwood House.

Gueguen, N. (2009a). 'Menstrual cycle phases and female receptivity to a courtship solicitation: an evaluation in a nightclub.' *Evolution and Human Behavior* **30**(5): 351–55.

Gueguen, N. (2009b). 'The receptivity of women to courtship solicitation across the menstrual cycle: a field experiment.' *Biological Psychology* **80**: 321–4.

Gurven, M. and H. Kaplan (2007). 'Longevity among hunter-gatherers: a cross-cultural examination.' *Population and Development Review* **33**(2): 321–65.

Guthrie, S. (1993). *Faces in the Clouds*. Oxford, Oxford University Press.

Guthrie, S. (2001). Why gods? A cognitive theory. In J. Andresen (ed.), *Religion in Mind: Cognitive Perspectives on Religious Belief, Ritual, and Experience*. Cambridge, Cambridge University Press, p. 94.

Hage, P. and J. Marck (2003). 'Matrilineality and the Melanesian origin of Polynesian Y chromosomes.' *Current Anthropology* **44**(s5): 121–27.

Hagen, E. (1999). 'The functions of postpartum depression.' *Evolution and Human Behavior* **20**: 325–59.

Hagen, E. H. and P. Hammerstein (2006). Game theory and human evolution: a critique of some recent interpretations of experimental games.' *Theoretical Population Biology* **69**(3): 339–48.

Hagen, E. H., H. C. Barrett and M. E. Price (2006). 'Do human parents face a quantity–quality trade-off?: evidence from a Shuar community.' *American Journal of Physical Anthropology* **130**(3): 405–18.

Haider-Markel, D. P. and M. R. Joslyn (2008). 'Beliefs about the origins of homosexuality and support for gay rights an empirical test of attribution theory.' *Public Opinion Quarterly* **72**(2): 291–310.

Haidt, J. (2007a). 'Moral psychology and the misunderstanding of religion.' *Edge: The Third Culture*. www.edge.org.

Haidt, J. (2007b). 'The new synthesis in moral psychology.' *Science* **316**(5827): 998–1002.

Haidt, J. (2008). 'Morality.' *Perspectives on Psychological Science* **3**(1): 65–72.

Haidt, J. and J. Graham (2007). 'When morality opposes justice: conservatives have moral intuitions that liberals may not recognize.' *Social Justice Research* **20**(1): 98–116.

Haidt, J. and C. Joseph (2004). 'Intuitive ethics: how innately prepared intuitions generate culturally variable virtues.' *Daedalus* **133**(4): 55–66.

Haidt J. and C. Joseph (2007). The moral mind: how 5 sets of innate intuitions guide the development of many culture-specific virtues, and perhaps even modules. In P. Carruthers, S. Laurence and S. Stich (eds.), *The Innate Mind*. Vol. 3. New York, Oxford University Press, pp. 367–91.

Haig, D. (1993). 'Genetic Conflicts in Human Pregnancy.' *The Quarterly Review of Biology* **68**(4): 495–532.

Haig, D. (2007). 'Weismann rules! OK? Epigenetics and the Lamarckian temptation.' *Biology & Philosophy* **22**(3): 415–28.

Halberstadt, J. and G. Rhodes (2000). 'The attractiveness of nonface averages: implications for an evolutionary explanation of the attractiveness of average faces.' *Psychological Science* **11**(4): 285–89.

Haldane, J. B. S. (1932). *The Causes of Evolution*. London, Longman, Green.

Haldane, J. B. S. (1955). Population genetics. In M. L. Johnson, M. Abercrombie and G. E. Fogg (eds.), *New Biology*. London, Penguin Books.

Haley, K. J. and D. M. Fessler (2005). 'Nobody's watching?: subtle cues affect generosity in an anonymous economic game. *Evolution and Human Behavior* **26**(3): 245–56.

Hamer, D. H., S. Hu, V. L. Magnuson, N. Hu and A. M. Pattatucci (1993). 'A linkage between DNA markers in the X chromosome and male sexual orientation.' *Science* **261**: 321–7.

Hamilton, W. D. (1964). 'The genetical evolution of social behaviour, 1.' *Journal of Theoretical Biology* **7**: 1–16.

Hamilton, W. D. (1970). 'Selfish and spiteful behaviour in an evolutionary model.' *Nature* **228**: 1218–20.

Harcourt, A. (2012). *Human Biogeography*. Berkeley and Los Angeles, University of California.

Harcourt, A. H. (1991). 'Sperm competition and the evolution of non-fertilizing sperm in mammals.' *Evolution* **45**(2): 314–28.

Harcourt, A. H., P. H. Harvey, S. G. Larson and R. V. Short (1981). 'Testis weight, body weight and breeding system in primates.' *Nature* **293**: 55–7.

Hardin, G. (1968). 'The tragedy of the commons.' *Science* **162**: 1243–8.

Harper R. M. J. (1975). *Evolutionary Origins of Disease.* Barnstaple, G. Mosdell.

Harris, C. R. (2003). 'A review of sex differences in sexual jealousy, including self-report data, psychophysiological responses, interpersonal violence, and morbid jealousy.' *Personality and Social Psychology Review* **7**(2): 102–28.

Harris, C. R. (2011). 'Menstrual cycle and facial preferences reconsidered.' *Sex Roles* **64**(9–10): 669–81.

Harris, G. T., N. Z. Hilton, M. E. Rice and A. W. Eke (2007). 'Children killed by genetic parents versus stepparents.' *Evolution and Human Behavior* **28**(2): 85–95.

Harris, H. (1974). 'Lionel Sharples Penrose (1898–1972).' *Journal of Medical Genetics* **11**(1): 1.

Hartung, J. (1985). 'Review of Shepher's Incest. A biosocial view.' *American Journal of Physical Anthropology* **67**(2): 169–71.

Harvey, P. H. and J. W. Bradbury (1991). Sexual selection. In J. R. Krebs and N. B. Davies (eds.), *Behavioural Ecology*. Oxford, Blackwell Scientific.

Haselton, M. G. (2003). 'The sexual overperception bias: evidence of a systematic bias in men from a survey of naturally occurring events.' *Journal of Research in Personality* **37**(1): 34–47.

Haselton, M. G. and D. M. Buss (2000). 'Error management theory: a new perspective on biases in cross-sex mind reading.' *Journal of Personality and Social Psychology* **78**(1): 81.

Haselton, M. G. and S. W. Gangestad (2006). 'Conditional expression of women's desires and men's mate guarding across the ovulatory cycle.' *Hormones and Behavior* **49**(4): 509–18.

Haselton, M. G. and K. Gildersleeve (2011). 'Can men detect ovulation?' *Current Directions in Psychological Science* **20**(2): 87–92.

Haselton, M. G. and T. Ketelaar (2006). 'Irrational emotions or emotional wisdom? The evolutionary psychology of affect and social behavior.' *Affect in Social Thinking and Behavior* **8**: 21.

Haselton, M. G. and D. Nettle (2006). 'The paranoid optimist: an integrative evolutionary model of cognitive biases.' *Personality and Social Psychology Review* **10**(1): 47–66.

Haselton, M. G., D. M. Buss, V. Oubaid and A. Angleitner (2005). 'Sex, lies and strategic interference: the psychology of deception between the sexes.' *Personality and Social Psychology Bulletin* **31**: 3–23.

Haselton, M. G., M. Mortezaie, E. G. Pilsworth, A. Bleske–Rechek and D. A. Frederick (2007). 'Ovulatory shifts in human female ornamentation: near ovulation women dress to impress.' *Hormones and Behaviour* **51**: 40–5.

Hau, M. (2007). 'Regulation of male traits by testosterone: implications for the evolution of vertebrate life histories.' *BioEssays* **29**(2): 133–44.

Haukka, J., J. Suvisaari and J. Lönnqvist (2003). 'Fertility of patients with schizophrenia, their siblings, and the general population: a cohort study from 1950 to 1959 in Finland.' *American Journal of Psychiatry* **160**(3): 460–3.

Havlíček, J. and S. Craig Roberts (2009). 'MHC-correlated mate choice in humans: a review.' *Psychoneuroendocrinology* **34**: 497–512.

Havlíček, J., S. C. Roberts and J. Flegr (2005). 'Women's preference for dominant male odour: effects of menstrual cycle and relationship status.' *Biology Letters* **1**(3): 256–9.

Havlíček, J., R. Dvořáková, L. Bartoš and J. Flegr (2006). 'Non-advertized does not mean concealed: body odour changes across the human menstrual cycle.' *Ethology* **112**(1): 81–90.

Hawkes, K., J. F. O'Connell and N. G. B. Jones (2001). 'Hunting and nuclear families.' *Current Anthropology* **42**(5): 681–709.

Hayward, A. D. and V. Lummaa (2013). 'Testing the evolutionary basis of the predictive adaptive response hypothesis in a preindustrial human population.' *Evolution, Medicine, and Public Health* **2013**(1): 106–17.

Healey, M. D. and B. J. Ellis (2007). 'Birth order, conscientiousness, and openness to experience: tests of the family-niche model of personality using a within-family methodology.' *Evolution and Human Behavior* **28**(1): 55–9.

Heider, F. and M. Simmel (1944). 'An experimental study of apparent behavior.' *The American Journal of Psychology* **57**(2): 243–59.

Heijmans, B. T., E. W. Tobi, A. D. Stein, H. Putter, G. J. Blauw, E. S. Susser, P. E. Slagboom and L. Lumey (2008). 'Persistent epigenetic differences associated with prenatal exposure to famine in humans.' *Proceedings of the National Academy of Sciences* **105**(44): 17046–9.

Helle, S., V. Lummaa and J. Jokela (2005). 'Are reproductive and somatic senescence coupled in humans? Late, but not early, reproduction correlated with longevity in historical Sami women.' *Proceedings of the Royal Society B: Biological Sciences* **272**(1558): 29–37.

Henrich, J. and R. Boyd (1998). 'The evolution of conformist transmission and the emergence of between-group differences.' *Evolution and Human Behavior* **19**(4): 215–41.

Henrich, J., R. Boyd and P. J. Richerson (2008). 'Five misunderstandings about cultural evolution.' *Human Nature* **19**(2): 119–37.

Henrich, J., R. Boyd and P. J. Richerson (2012). 'The puzzle of monogamous marriage.' *Philosophical Transactions of the Royal Society B: Biological Sciences* **367**(1589): 657–69.

Henrich, J., P. Young, R. Boyd, K. McCabe,W. Albers, A. Ockenfels and G. Gigerenzer (2001). What is the role of culture in bounded rationality? In G. Gigerenzer and R. Selten (eds.), *Bounded Rationality. The Adaptive Toolbox*. Cambridge, MA, MIT Press.

Herculano–Houzel, S. (2009). 'The human brain in numbers: a linearly scaled-up primate brain.' *Frontiers in Human Neuroscience* **3**.

Hertwig, R. and G. Gigerenzer (1999). 'The "conjunction fallacy revisited": how intelligent inferences look like reasoning errors.' *Journal of Behavioural Decision Making* **12**: 275–305.

Hesketh, T. and Z. W. Xing (2006). 'Abnormal sex ratios in human populations: causes and consequences.' *Proceedings of the National Academy of Sciences* **103**(36): 13271–5.

Hesketh, T., L. Lu and Z. W. Xing (2011). 'The consequences of son preference and sex-selective abortion in China and other Asian countries.' *Canadian Medical Association Journal* **183**(12): 1374–7.

Hill, K. (1982). 'Hunting and human evolution.' *Journal of Human Evolution* **11**: 521–44.

Hill, K. (1993). 'Life history theory and evolutionary anthropology.' *Evolutionary Anthropology* **2**: 78–88.

Hill, K. and A. M. Hurtado (1991). 'The evolution of premature reproductive senescence and menopause in human females.' *Human Nature* **2**(4): 313–50.

Hill, K. and A. M. Hurtado (1997). The evolution of premature reproductive senescence and menopause in human females: an evaluation of the grandmother hypothesis. In L. Betzig (ed.), *Human Nature: a Critical Reader*. New York, Oxford University Press.

Hill, K. and H. Kaplan (1988). Tradeoffs in male and female reproductive strategies among the Ache. In L. Betzig, M. B. Mulder and P. Turke (eds.), *Human Reproductive Behaviour*. Cambridge, Cambridge University Press, pp. 277–89.

Hilton, N. Z., G. T. Harris and M. E. Rice (2015). 'The step-father effect in child abuse: comparing discriminative parental solicitude and antisociality.' *Psychology of Violence* **5**(1): 8.

Hinde, R. A. (1982). *Ethology*. Oxford, Oxford University Press.

Hinde, R. A. and L. A. Barden (1985). 'The evolution of the teddy bear.' *Animal Behaviour* **33**(4): 1371–3.

Hintzpeter, B., C. Scheidt-Nave, M. J. Müller, L. Schenk and G. B. Mensink (2008). 'Higher prevalence of vitamin D deficiency is associated with immigrant background among children and adolescents in Germany.' *The Journal of Nutrition* **138**(8): 1482–90.

Hirschi, T. and M. Gottfredson (1983). 'Age and the explanation of crime.' *American Journal of Sociology*: 552–84.

Hoffman, E., K. A. McCabe and V. L. Smith (1998). 'Behavioral foundations of reciprocity: experimental economics and evolutionary psychology.' *Economic Inquiry* **36**(3): 335–52.

Holloway, R. (1983). 'Human paleontological evidence relevant to language behaviour.' *Human Neurobiology* **2**: 105–14.

Holloway, R. L. (1975). 'The role of human social behavior in the evolution of the brain' (James Arthur lecture on the evolution of the human brain, no. 43, 1973).

Holloway, R. L. (1996). Evolution of the human brain. In A. Lock and C. R. Peters (eds.), *Handbook of Human Symbolic Evolution*. New York, Oxford University Press, pp. 74–116.

Holt, S., J. Miller and P. Petocz (1997). 'An insulin index of foods: the insulin demand generated by 1000–kJ portions of common foods.' *The American Journal of Clinical Nutrition* **66**(5): 1264–76.

Holzleitner, I. J., D. W. Hunter, B. P. Tiddeman, A. Seck, D. E. Re and D. I. Perrett (2014). 'Men's facial masculinity: when (body) size matters.' *Perception* **43**: 1191–1202.

Hopkins, W. D., K. A. Bard, A. B. Jones and S. L. Bales. (1993). 'Chimpanzee hand preference in throwing and infant cradling: implications for the origin of human handedness.' *Current Anthropology* **34**(5): 786–90.

Hrdy, S. B. (1979). 'Infanticide among animals: a review, classification and examination of the implications for the reproductive strategies of females.' *Ethology and Sociobiology* **1**: 13–40.

Hu, S., A. M. Pattatucci, L. L. Chavis Patterson, D. W. Fulker, S. S. Cherny, L. Kruglyak and D. H. Hamer (1995). 'Linkage between sexual orientation and chromosome Xq28 in males but not in females.' *Nature Genetics* **11**(3): 248–56.

Huang, Z., W. C. Willet and G. A. Colditz (1999). 'Waist circumference, waist:hip ratio, and risk of breast cancer in the Nurses' Health Study.' *American Journal of Epidemiology* **150**: 1316–24.

Hume, D. (1739/1985). *A Treatise of Human Nature*. London, Penguin Classics.

Humphreys, L. G. (1939). 'Acquisition and extinction of verbal expectations in a situation analogous to conditioning.' *Journal of Experimental Psychology* **25**: 294–301.

Hurtado, A. M. and K. R. Hill (1992). 'Paternal effect on offspring survivorship among Ache and Hiwi hunter-gatherers: implications for modeling pair-bond stability.' *Father–child Relations: Cultural and Biosocial Contexts*: 31–55.

Hutchinson, G. E. (1959). 'A speculative consideration of certain possible forms of sexual selection in man.' *American Naturalist*: 81–91.

Iemmola, F. and A. Camperio Ciani (2009). 'New evidence of genetic factors influencing sexual orientation in men: female fecundity increase in the maternal line.' *Archives of Sexual Behavior* **38**(3): 393–9.

Ingman, M., H. Kaessmann, S. Pääbo and U. Gyllensten (2000). 'Mitochondrial genome variation and the origin of modern humans.' *Nature* **408**: 708–13.

Ingram, C. J. E., C. A. Mulcare, Y. Itan, M. G. Thomas and D. M. Swallow (2009). 'Lactose digestion and the evolutionary genetics of lactase persistence.' *Human Genetics* **124**(6): 579–91.

Iredale, W., M. Van Vugt and R. I. M. Dunbar (2008). 'Showing off in humans: male generosity as a mating signal.' *Evolutionary Psychology* **6**(3): 386–92.

Irons, W. (1983). Human female reproductive strategies. In S. K. Wasser (ed.), *Social Behavior of Female Vertebrates*. New York, Academic Press, pp. 169–213.

Isbell, L. A. and T. P. Young (1996). 'The evolution of bipedalism in hominids and reduced group size in chimpanzees: alternative responses to decreasing resource availability.' *Journal of Human Evolution* **30**(5): 389–97.

Isler, K. and C. van Schaik (2012a). 'Allomaternal care, life history and brain size evolution in mammals.' *Journal of Human Evolution* **63**: 52–63.

Isler, K. and C. van Schaik (2012b). 'How our ancestors broke through the gray ceiling.' *Current Anthropology* **53**(S6): S453–65.

Jablonka, E. and G. Raz (2009). 'Transgenerational epigenetic inheritance: prevalence, mechanisms and implications for the study of heredity and evolution.' *Quaternary Review of Biology* **84**(2): 131–76.

Jablonski, N. G. (2004). 'The evolution of human skin and skin color.' *Annual Review of Anthropology*, 585–623.

Jablonski, N. G. and G. Chaplin (2000). 'The evolution of human skin coloration.' *Journal of Human Evolution* **39**(1): 57–106.

Jaffe, K. (2001). 'On the relative importance of haplo-diploidy, assortative mating and social synergy on the evolutionary emergence of social behavior.' *Acta Biotheoretica* **49**(1): 29–42.

James, W. (1884). 'II.—What is an emotion?' *Mind* **34**: 188–205.

James, W. H. (1987). 'The human sex ratio. Part 1: a review of the literature.' *Human Biology*: 721–52.

Jamison, K. R. (1994). *Touched With Fire: Manic Depressive Illness and the Artistic Temperament*. New York, Free Press Paperbacks.

Jang, K. L., W. J. Livesley and P. A. Vemon (1996). 'Heritability of the big five personality dimensions and their facets: a twin study.' *Journal of Personality* **64**(3): 577–92.

Jefferson, T., J. H. Herbst and R. R. McCrae (1998). 'Associations between birth order and personality traits: evidence from self-reports and observer ratings.' *Journal of Research in Personality* **32**(4): 498–509.

Jerison, H. J. (1973). *Evolution of the Brain and Intelligence*. New York, Academic Press.

Jobling, M., E. Hollox, M. Hurles, T. Kivisild and C. Tyler-Smith (2014). *Human Evolutionary Genetics*. Garland Science.

Joffres, M., E. Falaschetti, C. Gillespie, C. Robitaille, F. Loustalot, N. Poulter, F. A. McAlister, H. Johansen, O. Baclic and N. Campbell (2013). 'Hypertension prevalence, awareness, treatment and control in national surveys from England, the USA and Canada, and correlation with stroke and ischaemic heart disease mortality: a cross-sectional study.' *BMJ Open* **3**(8): e003423.

Johnson, A. K., A. Barnacz, T. Yokkaichi, J. Rubio, C. Racioppi, T. K. Shackelford, M. L. Fisher and J. P. Keenan (2005). 'Me, myself, and lie: the role of self-awareness in deception.' *Personality and Individual Differences* **38**(8): 1847–53.

Johnson, D. D., D. T. Blumstein, J. H. Fowler and M. G. Haselton (2013). 'The evolution of error: error management, cognitive constraints, and adaptive decision–making biases.' *Trends in Ecology & Evolution* **28**(8): 474–81.

Johnson-Laird, P. N. and K. Oatley (1992). 'Basic emotions, rationality, and folk theory.' *Cognition & Emotion* **6**(3–4): 201–23.

Johnston, V. S., R. Hagel, M. Franklin, B. Fink and K. Grammer (2001). 'Male facial attractiveness. Evidence for hormone-mediated adaptive design.' *Evolution and Human Behavior* **22**: 251–69.

Jones, B. C., L. M. DeBruine, D. I. Perrett, A. C. Little, D. R. Feinberg and M. J. L. Smith (2008). 'Effects of menstrual cycle phase on face preferences.' *Archives of Sexual Behavior* **37**(1): 78–84.

Jørgensen, A., J. Philip, W. Raskind, M. Matsushita, B. Christensen, V. Dreyer and A. Motulsky (1992). 'Different patterns of X inactivation in MZ twins discordant for red–green color-vision deficiency.' *American Journal of Human Genetics* **51**(2): 291.

Juliano, A. and S. J. Schwab (2000). 'Sweep of sexual harassment cases.' *The Cornell Lit. Rev.* **86**: 548.

Kaati, G., L. O. Bygren and S. Edvinsson (2002). 'Cardiovascular and diabetes mortality determined by nutrition during parents' and grandparents' slow growth period.' *European Journal of Human Genetics* **10**(11): 682–8.

Kaati, G., L. O. Bygren, M. Pembrey, M. Sjöström (2007). 'Transgenerational response to nutrition, early life circumstances and longevity.' *European Journal of Human Genetics* **15**: 784–90.

Kahneman, D. (2011). *Thinking, Fast and Slow*. New York, NY, Farrah, Straus & Gidoux.

Kahneman, D., P. Slovic and A. Tversky (eds.) (1982). *Judgement under Uncertainty*. Cambridge, Cambridge University Press.

Kaler, J. (1999). 'What's the good of ethical theory?' *Business Ethics: A European Review* **8**(4): 206–13.

Kanazawa, S. (2001). 'Why father absence might precipitate early menarche: the role of polygyny.' *Evolution and Human Behavior* **22**(5): 329–34.

Kanazawa, S. (2009). Evolutionary psychology and crime. In A. Walsh and K. M. Beaver (eds.), *Biosocial Criminology*. New York, Routledge.

Kanazawa, S. and M. C. Still (2000). 'Why men commit crimes (and why they desist).' *Sociological Theory* **18**(3): 434–47.

Kaňková, Š., J. Šulc, K. Nouzová, K. Fajfrlík, D. Frynta and J. Flegr (2007). 'Women infected with parasite Toxoplasma have more sons.' *Naturwissenschaften* **94**(2): 122–7.

Kaplan, H. S. and S. W. Gangestad (2004). Life history theory and evolutionary psychology. In D. Buss (ed.), *The Handbook of Evolutionary Psychology*. Hoboken, NJ, John Wiley and Sons.

Kaplan, H. S. and J. B. Lancaster (1999). Skills-based competitive labour markets, the demographic transition, and the interaction of fertility and parental human capital in the determination of child outcomes. In L. Cronk, W. Irons and N. Chagnon (eds.), *Human Behavior and Adaptation: An Anthropological Perspective*. New York, Aldine de Gruyter.

Kaplan, H. S., K. Hill, J. B. Lancaster and A. M. Hurtado (2000). 'A theory of human life history evolution: diet, intelligence, and longevity.' *Evolutionary Anthropology* **9**: 156–85.

Kaplan, H. S., J. B. Lancaster, J. A. Bock and S. E. Johnson (1995). 'Does observed fertility maximize fitness among New Mexican men? A test of an optimality model and a new theory of parental investment in the embodied capital of offspring.' *Human Nature* **6**: 325–60.

Kaptijn, R., F. Thomese, A. C. Liefbroer and M. Silverstein (2013). 'Testing evolutionary theories of discriminative grandparental investment.' *Journal of Biosocial Science* **45**(03): 289–310.

Karremans, J. C., W. E. Frankenhuis and S. Arons (2010). 'Blind men prefer a low waist-to-hip ratio.' *Evolution and Human Behavior* **31**(3): 182–6.

Katzmarzyk, P. T. and W. R. Leonard (1998). 'Climatic influences on human body size and proportions: ecological adaptations and secular trends.' *American Journal of Physical Anthropology* **106**(4): 483–503.

Keating, C. F., D. W. Randall, T. Kendrick and K. A. Gutshall (2003). 'Do babyfaced adults receive more help? The (cross-cultural) case of the lost resume.' *Journal of Nonverbal Behavior* **27**(2): 89–109.

Keenan, J. P., G. G. Gallup Jr, N. Goulet and M. Kulkarni (1997). 'Attributions of deception in human mating strategies.' *Journal of Social Behavior & Personality* **12**(1): 45–52.

Kelemen, D. (1999). 'Why are rocks pointy? Children's preference for teleological explanations of the natural world.' *Developmental Psychology* **35**(6): 1440.

Kelemen, D. and E. Rosset (2009). 'The human function compunction: teleological explanation in adults.' *Cognition* **111**(1): 138–43.

Keller, M. C. and G. Miller (2006). 'Resolving the paradox of common, harmful, heritable mental disorders: which evolutionary genetic models work best?' *Behavioral and Brain Sciences* **29**(04): 385–404.

Keller, M. C., R. M. Nesse and S. Hofferth (2001). 'The Trivers–Willard hypothesis of parental investment: no effect in the contemporary United States.' *Evolution and Human Behavior* **22**(5): 343–60.

Kendrick, K. M., M. R. Hinton, K. Atkins, M. A. Haupt and J. D. Skinner (1998). 'Mothers determine sexual preferences.' *Nature* 395(6699): 229–30.

Kenny, A. (1986). *Rationalism, Empiricism and Idealism.* Oxford, Oxford University Press.

Kenrick, D. T. and R. C. Keefe (1992). 'Age preferences in mates reflect sex differences in human reproductive strategies.' *Behavioral and Brain Sciences* 15(01): 75–91.

Ketelaar, T. and W. T. Au (2003). 'The effects of guilty feelings on the behaviour of uncooperative individuals in repeated social bargaining games: an affect-as-information interpretation of the role of emotion in social interaction.' *Cognition and Emotion* 17: 429–53.

Ketelaar, T. and A. S. Goodie (1998). 'The satisficing role of emotions in decision making.' *Psykhe: Revista de la Escuela de Psicologia* 7: 63–77.

Ketterson, E. D. and V. Nolan Jr. (1999). 'Adaptation, exaptation, and constraint: a hormonal perspective.' *The American Naturalist* 154(S1): S4–25.

Keverne, E. B. and J. P. Curley (2008). Epigenetics, brain evolution and behaviour. *Frontiers in neuroendocrinology,* 29(3), 398–412.

Khan, R. and D. J. Cooke (2008). 'Risk factors for severe inter-sibling violence: a preliminary study of a youth forensic sample.' *Journal of Interpersonal Violence* 23(11): 1513–30.

Kijas, J. W., J. A. Lenstra, B. Hayes, S. Boitard, L. R. P. Neto, M. San Cristobal, B. Servin, R. McCulloch, V. Whan and K. Gietzen (2012). 'Genome-wide analysis of the world's sheep breeds reveals high levels of historic mixture and strong recent selection.' *PLoS Biology* 10(2): e1001258.

Killian, J. K., J. C. Byrd, J. V. Jirtle, B. L. Munday, M. K. Stoskopf, R. G. MacDonald and R. L. Jirtle (2000). 'Imprinting evolution in mammals.' *Molecular Cell* 5(4): 707–16.

Kimura, D. (1999). *Sex and Cognition.* Cambridge, MA, MIT Press.

King, M. and E. McDonald (1992). 'Homosexuals who are twins. A study of 46 probands.' *The British Journal of Psychiatry* 160(3): 407–9.

Kinsey, A. C., W. B. Pomeroy, C. E. Martin and P. Gebhard (1953). *Sexual Behaviour in the Human Female.* Philadelphia, PA, W.B. Saunders.

Kiple, K. F. (2000). *The Cambridge World History of Food.* Cambridge, Cambridge University Press.

Kipling, R. (1967). *Just So Stories.* London, Macmillan.

Kirby, J. (2003). 'A new group-selection model for the evolution of homosexuality.' *Biology and Philosophy* 18(5): 683–94.

Kittler, R., M. Kayser and M. Stoneking (2003). 'Molecular evolution of *Pediculus humanus* and the origin of clothing.' *Current Biology* 13(16): 1414–17.

Klinnert, M. D., J. Campos, J. Sorce, R. N. Emde and M. Svejda (1982). The development of social referencing in infancy. In R. Plutchik and H. Kellerman (eds.), *Emotion: Theory, Research and Experience, Vol. 2: Emotion in Early Development.* New York, Academic Press.

Kluger, M. J., W. Kozak, C. A. Conn, L. R. Leon and D. Soszynski (1998). 'Role of fever in disease.' *Annals of the New York Academy of Sciences* 856(1): 224–33.

Knowler, W. C., D. J. Pettitt, M. F. Saad and P. H. Bennett (1990). 'Diabetes mellitus in the Pima Indians: incidence, risk factors and pathogenesis.' *Diabetes/Metabolism Reviews* 6(1): 1–27.

Knudson, A. G., L. Wayne and W. Y. Hallett (1967). 'On the selective advantage of cystic fibrosis heterozygotes.' *American Journal of Human Genetics* 19(3 Pt 2): 388.

Kondrashov, A. S. (2003). 'Direct estimates of human per nucleotide mutation rates at 20 loci causing Mendelian diseases.' *Human Mutation* 21(1): 12–27.

Kong, A., M. L. Frigge, G. Masson, S. Besenbacher, P. Sulem, G. Magnusson, S. A. Gudjonsson, A. Sigurdsson, A. Jonasdottir and A. Jonasdottir, et al. (2012). 'Rate of de novo mutations and the importance of father's age to disease risk.' *Nature* 488(7412): 471–45.

Konner, M. and S. B. Eaton (2010). 'Paleolithic nutrition twenty-five years later.' *Nutrition in Clinical Practice* 25(6): 594–602.

Kopp, W. (2006). 'The atherogenic potential of dietary carbohydrate.' *Preventive Medicine* 42(5): 336–42.

Kościński, K. (2012). 'Mere visual experience impacts preference for body shape: evidence from male competitive swimmers.' *Evolution and Human Behavior* 33(2): 137–46.

Kramer, M. (2000). 'Balanced protein/energy supplementation in pregnancy.' *Cochrane Database Syst Rev* 2(2): CD000032.

Krasnow, M. M., D. Truxaw, S. J. Gaulin, J. New, H. Ozono, S. Uono, T. Ueno and K. Minemoto (2011). 'Cognitive adaptations for gathering-related navigation in humans.' *Evolution and Human Behavior* 32(1): 1–12.

Krebs, J. R. and N. B. Davies (1991). *Behavioural Ecology.* Oxford, Blackwell Scientific.

Kröger, R. H. and O. Biehlmaier (2009). 'Space-saving advantage of an inverted retina.' *Vision Research* 49(18): 2318–21.

Kruger, D. J. and R. M. Nesse (2006). 'An evolutionary life-history framework for understanding sex differences in human mortality rates.' *Human Nature* **17**(1): 74–97.

Kuhn, T. S. (1962). *The Structure of Scientific Revolutions*. Chicago, University of Chicago Press.

Kuper, A. (2009). 'Commentary: a Darwin family concern.' *International Journal of Epidemiology* **38**(6): 1439–42.

Kuper, A. (2010). *Incest and Influence: The Private Life of Bourgeois England*. Harvard University Press.

Kuzawa, C. W. (1998). 'Adipose tissue in human infancy and childhood: an evolutionary perspective.' *American Journal of Physical Anthropology* **107** (s 27): 177–209.

Kuzawa, C. W. (2005). 'Fetal origins of developmental plasticity: are fetal cues reliable predictors of future nutritional environments?' *American Journal of Human Biology* **17**(1): 5–21.

Kuzawa, C. W. and E. A. Quinn (2009). 'Developmental origins of adult function and health: evolutionary hypotheses.' *Annual Review of Anthropology* **38**: 131–47.

Lack, D. (1943). *The Life of the Robin*. Witherby, London.

Laham, S. M., K. Gonsalkorale and W. von Hippel (2005). 'Darwinian grandparenting: preferential investment in more certain kin.' *Personality and Social Psychology Bulletin* **31**(1): 63–72.

Laitman, J. T. (1984). 'The anatomy of human speech.' *Natural History* (August): 20–7.

Laland, K. N. and G. R. Brown (2002). *Sense and Nonsense: Evolutionary Perspectives on Human Behaviour*. Oxford, Oxford University Press.

Laland, K. N., J. Odling–Smee and S. Myles (2010). 'How culture shaped the human genome: bringing genetics and the human sciences together.' *Nature Reviews Genetics* **11**(2): 137–48.

Laland, K. N., P. J. Richerson and R. Boyd (1996). Developing a theory of animal social learning. In C. M. Heyes and B. G. Galef (eds.), *Social Learning in Animals: The Roots of Culture*. New York, Academic Press.

Lancaster, J. B. (1997). An evolutionary history of human reproductive strategies and the status of women in relation to population growth and social stratification. In P. A. Gowaty (ed.), *Evolution and Feminism*. New York, Chapman Hall, pp. 466–88.

Lane, N. (2015). *The Vital Question: Energy, Evolution, and the Origins of Complex Life*. London, UK, WW Norton & Company.

Langlois, J. H. and L. A. Roggmam (1990). 'Attractive faces are only average.' *Psychological Science* **1**: 115–21.

Langlois, J. H., L. E. Kalakanis, A. J. Rubenstein, A. D. Larson, M. J. Hallam and M. T. Snoot (2000). 'Maxims or myths of beauty: a meta-analytic and theoretical review.' *Psychological Bulletin* **126**: 390–423.

Långström, N., Q. Rahman, E. Carlström and P. Lichtenstein (2010). 'Genetic and environmental effects on same-sex sexual behavior: a population study of twins in Sweden.' *Archives of Sexual Behavior* **39**(1): 75–80.

Lanska, J. D., M. J. Lanska, A. J. Hartz and A. A. Rimm (1985). 'Factors influencing anatomical location of fat tissue in 52,953 women.' *International Journal of Obesity* **9**: 29–38.

Larson, G., K. Dobney, U. Albarella, M. Fang, E. Matisoo–Smith, J. Robins, S. Lowden, H. Finlayson, T. Brand and E. Willerslev (2005). 'Worldwide phylogeography of wild boar reveals multiple centers of pig domestication.' *Science* **307**(5715): 1618–21.

Larson, G., R. Liu, X. Zhao, J. Yuan, D. Fuller, L. Barton, K. Dobney, Q. Fan, Z. Gu and X.-H. Liu (2010). 'Patterns of East Asian pig domestication, migration, and turnover revealed by modern and ancient DNA.' *Proceedings of the National Academy of Sciences* **107**(17): 7686–91.

Lassek, W. D. and S. J. Gaulin (2009). 'Costs and benefits of fat-free muscle mass in men: relationship to mating success, dietary requirements, and native immunity.' *Evolution and Human Behavior* **30**(5): 322–8.

Laurin, M., M. L. Everett and W. Parker (2011). 'The cecal appendix: one more immune component with a function disturbed by post-industrial culture.' *The Anatomical Record* **294**(4): 567–79.

Law, R. (1979). 'Ecological determinants in the evolution of life histories.' *Population Dynamics*. Oxford: Blackwell Scientific Publications, pp. 81–103.

Lawson, D. W. and R. Mace (2009). 'Trade-offs in modern parenting: a longitudinal study of sibling competition for parental care.' *Evolution and Human Behavior* **30**(3): 170–83.

Leakey, R. (1994). *The Origin of Humankind*. London, Weidenfeld and Nicolson.

Leavitt, G. (1990). 'Sociobiological explanations of incest avoidance: a critical review of evidential claims.' *American Anthropologist* **92**: 971–93.

Lee, R. B. (1979). *The !Kung San: Men, Women, and Work in a Foraging Society*. Cambridge, Cambridge University Press.

Lee-Thorp, J. A. (2008). 'On Isotopes and Old Bones.' *Archaeometry* **50**(6): 925–50.

Lehrman, D. S. (1953). 'A critique of Konrad Lorenz's theory of instinctive behaviour.' *Quarterly Review of Biology* **28**: 337–63.

Lench, H. C., S. W. Bench, K. E. Darbor and M. Moore (2015). 'A functionalist manifesto: goal-related emotions from an evolutionary perspective.' *Emotion Review* **7**(1): 90–8.

Leonard, W. R. (2007). 'Lifestyle, diet, and disease: comparative perspectives on the determinants of chronic health risks.' *Evolution in Health and Disease*: 265–76.

Leslie, A. M. (1982). 'The perception of causality in infants.' *Perception* **11**: 173–86.

Leslie, A. M. (1984). 'Spatiotemporal continuity and the perception of causality in infants.' *Perception* **13**: 287–305.

Lévi-Strauss, C. (1956). *The Family. Man, Culture and Society*. H. L. Shapiro. London, Oxford University Press.

Lewin, R. (2005). *Human Evolution: An Illustrated Introduction*. Oxford, Blackwell.

Lewis, K. (2013). 'Platforms for antibiotic discovery.' *Nat Rev Drug Discov* **12**(5): 371–87.

Lichtenstein, A. H., L. J. Appel, M. Brands, M. Carnethon, S. Daniels, H. A. Franch, B. Franklin, P. Kris-Etherton, W. S. Harris and B. Howard (2006). 'Diet and lifestyle recommendations revision 2006. A scientific statement from the American Heart Association nutrition committee.' *Circulation* **114**(1): 82–96.

Lie, H. C., G. Rhodes and L. W. Simmons (2008). 'Genetic diversity revealed in human faces.' *Evolution* **62**(10): 2473–86.

Lieberman, D. and T. Lobel (2012). 'Kinship on the kibbutz: coresidence duration predicts altruism, personal sexual aversions and moral attitudes among communally reared peers.' *Evolution and Human Behavior* **33**(1): 26–34.

Lieberman, D., J. Tooby, and L.Cosmides (2007). 'The architecture of human kin detection.' *Nature* **445**: 727–31.

Lieberman, D., J. Tooby and L. Cosmides (2003). 'Does morality have a biological basis? An empirical test of the factors governing moral sentiments relating to incest.' *Proceedings of the Royal Society of London. Series B* **270**: 819–26.

Lindeberg S. (2010). *Food and Western Disease: Health and Nutrition from an Evolutionary Perspective*. Chichester, UK, Wiley-Blackwell.

Lindeberg, S., M. Eliasson, B. Lindahl and B. Ahrén (1999). 'Low serum insulin in traditional Pacific Islanders – the Kitava study.' *Metabolism* **48**(10): 1216–9.

Lindeberg, S., E. Berntorp, P. Nilsson-Ehle, A. Terént and B. Vessby (1997). 'Age relations of cardiovascular risk factors in a traditional Melanesian society: the Kitava study.' *The American Journal of Clinical Nutrition* **66**(4): 845–52.

Lishman, W. (1996). *Father Goose*. Canada, Little, Brown.

Little, A., I. Penton-Voak, D. Burt and D. Perrett (2003). 'Investigating an imprinting-like phenomenon in humans: partners and opposite-sex parents have similar hair and eye colour.' *Evolution and Human Behavior* **24**(1): 43–51.

Little, A. C., B. C. Jones and R. P. Burriss (2007). 'Preferences for masculinity in male bodies change across the menstrual cycle.' *Hormones and Behavior* **51**(5): 633–39.

Little, A. C., I. S. Penton-Voak, D. M. Burt and D. I. Perrett (2000). Evolution and individual differences in the perception of attractiveness: how cyclic hormonal changes and self-perceived attractiveness influence female preferences for male faces. In G. Rhodes and L. Zebrowitz (eds.), *Facial Attractiveness*. Westport, Ablex Publishing.

Little, A. C., B. C. Jones, I. S. Penton-Voak, D. M. Burt and D. I. Perrett (2002). 'Partnership status and the temporal context of relationships influence human female preferences for sexual dimorphism in male face shape.' *Proceedings of the Royal Society of London B: Biological Sciences* **269**(1496): 1095–100.

Lobmaier, J. S., R. Sprengelmeyer, B. Wiffen and D. I. Perrett (2010). 'Female and male responses to cuteness, age and emotion in infant faces.' *Evolution and Human Behavior* **31**(1): 16–21.

Lorenz, K. (1943). 'Die angeborenen formen möglicher erfahrung.' *Zeitschrift für Tierpsychologie* **5**(2): 235–409.

Lorenz, K. (1953). *King Solomon's Ring*. London, The Reprint Society.

Low, B. (1989). 'Cross-cultural patterns in the training of children: an evolutionary perspective.' *Journal of Comparative Psychology* **103**(4): 311–9.

Lozano, R., M. Naghavi, K. Foreman, S. Lim, K. Shibuya, V. Aboyans, J. Abraham, T. Adair, R. Aggarwal and S. Y. Ahn (2013). 'Global and regional mortality from 235 causes of death for 20 age groups in 1990 and 2010: a systematic analysis for the Global Burden of Disease Study 2010.' *The Lancet* **380**(9859): 2095–128.

Lubelchek, R. J. and R. A. Weinstein (2008). Antibiotic resistance and nosocomial infections. In K. H. Mayer and H. F. Pizer (eds.), *The Social Ecology of Infectious Diseases*. Massachusetts, Academic Press, pp. 241–74.

Lukaszewski, A. W. and J. R. Roney (2011). 'The origins of extraversion: joint effects of facultative calibration and genetic polymorphism.' *Personality and Social Psychology Bulletin* 37(3): 409–21.

Lumsden, C. J. and E. O. Wilson (1981). *Genes, Mind and Culture*. Cambridge, MA, Harvard University Press.

Luo, Z. C., K. Albertsson-Wikland and J. Karlberg (1998). 'Target height as predicted by parental heights in a population-based study.' *Pediatric Research* 44(4): 563–71.

Maestripieri, D. and S. Pelka (2002). 'Sex differences in interest in infants across the lifespan.' *Human Nature* 13(3): 327–44.

Magnus, P., H. Gjessing, A. Skrondal and R. Skjaerven (2001). 'Paternal contribution to birth weight.' *Journal of Epidemiology and Community Health* 55(12): 873–7.

Magurran, A. E. (2005). *Evolutionary Ecology: The Trinidadian Guppy*. Oxford, Oxford University Press.

Mahoney, S. A. (1980). 'Cost of locomotion and heat balance during rest and running from 0 to 55 degrees C in a patas monkey.' *Journal of Applied Physiology* 49(5): 789–800.

Mäki, P., J. Veijola, P. B. Jones, G. K. Murray, H. Koponen, P. Tienari, J. Miettunen, P. Tanskanen, K.-E. Wahlberg and J. Koskinen (2005). 'Predictors of schizophrenia – a review.' *British Medical Bulletin* 73(1): 1–15.

Malaspina, D., S. Harlap, S. Fennig, D. Heiman, D. Nahon, D. Feldman and E. S. Susser (2001). 'Advancing paternal age and the risk of schizophrenia.' *Archives of General Psychiatry* 58(4): 361–7.

Malaspina, D., C. Corcoran, C. Fahim, A. Berman, J. Harkavy-Friedman, S. Yale, D. Goetz, R. Goetz, S. Harlap and J. Gorman (2002). 'Paternal age and sporadic schizophrenia: evidence for de novo mutations.' *American Journal of Medical Genetics* 114(3): 299–303.

Mann, G., B. M. Lippe, M. E. Geffner, T. J. Merimee, B. Hewlett, L. Cavalli-Sforza and J. Zapf (1987). 'The riddle of pygmy stature.' *New England Journal of Medicine* 317(11): 709–10.

Manning, J. T., K. Koukourakis and D. A. Brodie (1997). 'Fluctuating asymmetry, metabolic rate and sexual selection in human males.' *Evolution and Human Behavior* 18: 15–21.

Manning, J. T., D. Scutt, G. H. Whitehouse, S. J. Leinster and J. M. Walton (1996). 'Asymmetry and the menstrual cycle in women.' *Ethology and Sociobiology* 17: 129–43.

Marks, I. M. and R. M. Nesse (1994). 'Fear and fitness: an evolutionary analysis of anxiety disorders.' *Ethology and Sociobiology* 15: 247–61.

Marlowe, F. W. (2004). 'Marital residence among foragers.' *Current Anthropology* 45: 277–84.

Marlowe, F. W. (2005). 'Hunter-gatherers and human evolution.' *Evolutionary Anthropology: Issues, News, and Reviews* 14(2): 54–67.

Marlowe, F. W. and A. Wetsman (2001). 'Preferred waist-to-hip ratio and ecology.' *Personality and Individual Differences* 30: 481–9.

Marlowe, F. W., C. L. Apicella and D. Reed (2005). 'Men's preferences for women's profile waist-to-hip ratio in two societies.' *Evolution and Human Behavior* 26(6): 458–69.

Marr, D. (1982). *Vision*. New York, W. H. Freeman.

Martin, R. D. (1981). 'Relative brain size and basal metabolic rate in terrestrial vertebrates.' *Nature* 293: 57–60.

Martin, R. D. (2007). 'The evolution of human reproduction: a primatological perspective.' *American Journal of Physical Anthropology* 134(S45): 59–84.

Mascaro, J. S., P. D. Hackett and J. K. Rilling (2013). 'Testicular volume is inversely correlated with nurturing-related brain activity in human fathers.' *Proceedings of the National Academy of Sciences* 110(39): 15746–51.

Masters, R. D. and M. Gruter (1992). *The Sense of Justice: Biological Foundations of Law*. Sage.

Mathews, T. and B. E. Hamilton (2005). 'Trend analysis of the sex ratio at birth in the United States.' *National Vital Statistics Reports* 53(20): 1–17.

Maynard Smith, J. (1974). 'The theory of games and the evolution of animal conflicts.' *Journal of Theoretical Biology* 47: 209–21.

Maynard Smith, J. (ed.) (1982). *Evolution Now: A Century after Darwin*. London, Macmillan.

Maynard Smith, J. (1989). *Evolutionary Genetics*. Oxford, Oxford University Press.

Maynard Smith, J. and E. Szathmary (1995). *The Major Transitions in Evolution*. Oxford, Oxford University Press.

Mazur, A. and J. Michalek (1998). 'Marriage, divorce, and male testosterone.' *Social Forces* 77(1): 315–30.

McGowan, P. O., A. Sasaki, A. C. D'Alessio, S. Dymov, B. Labonté, M. Szyf ... and M. J. Meaney

(2009). 'Epigenetic regulation of the glucocorticoid receptor in human brain associates with childhood abuse.' *Nature Neuroscience* **12**(3): 342–8.

McHenry (1991). Sexual dimorphism in Australopithecus afarensis. *Journal of Human Evolution,* **20**(1), 21–32.

McKnight, J. (1997). *Straight Science?: Homosexuality, Evolution and Adaptation.* Routledge.

McKusick, V. A. (2000). 'Ellis–van Creveld syndrome and the Amish.' *Nature Genetics* **24**: 203–4.

McKusick, V. A., J. A. Egeland, R. Eldridge and D. E. Krusen (1964). 'Dwarfism in the Amish. The Ellis–van Creveld syndrome.' *Bull. Johns Hopkins Hosp* **115**: 306–36.

McLellan, B. and S. J. McKelvie (1993). 'Effects of age and gender on perceived facial attractiveness.' *Canadian Journal of Behavioural Science/Revue canadienne des sciences du comportement* **25**(1): 135.

Mealey, L., R. Bridgestock and G. Townsend (1999). 'Symmetry and perceived facial attractiveness.' *Journal of Personality and Social Psychology* **76**: 151–8.

Mealey, L., C. Daood and M. Krage (1996). 'Enhanced memory for faces of cheaters.' *Ethology and Sociobiology* **17**: 119–28.

Meiri, S. and T. Dayan (2003). 'On the validity of Bergmann's rule.' *Journal of Biogeography* **30**(3): 331–51.

Menand, L. (2005). Dangers within and without. In R. G. Feal (ed.), *Profession 2005.* New York, Modern Language Association, pp. 10–17.

Mendell, D. and J. Bigness (1998, 6 September). 'For clerks, smiles can go too far.' *Chicago Tribune.*

Menozzi, P., A. Piazza and L. Cavalli-Sforza (1978). 'Synthetic maps of human gene frequencies in Europeans.' *Science* **201**(4358): 786–92.

Merlo, L. M., J. W. Pepper, B. J. Reid and C. C. Maley (2006). 'Cancer as an evolutionary and ecological process.' *Nature Reviews Cancer* **6**(12): 924–35.

Mesoudi, A., A. Whiten and K. N. Laland (2004). 'Perspective: is human cultural evolution Darwinian? Evidence reviewed from the perspective of *The Origin of Species*.' *Evolution* **58**(1): 1–11.

Migliano, A. B., L. Vinicius and M. M. Lahr (2007). 'Life history trade-offs explain the evolution of human pygmies.' *Proceedings of the National Academy of Sciences* **104**(51): 20216–9.

Mildvan, A. S. and B. L. Strehler (1960). A critique of theories of mortality. In B. L. Strehler, J. D. Ebert, H. B. Glass and N. W. Shock (eds.), *The Biology of Aging.* Washington, DC, American Institute of Biological Sciences, pp. 216–35.

Miller, E. M. (2000). 'Homosexuality, birth order, and evolution: toward an equilibrium reproductive economics of homosexuality.' *Archives of Sexual Behavior* **29**(1): 1–34.

Miller, G. (2000). *The Mating Mind.* London, Heinemann/Doubleday.

Miller, G., J. M. Tybur and B. D. Jordan (2007). 'Ovulatory cycle effects on tip earnings by lap dancers: economic evidence for human estrus?' *Evolution and Human Behavior* **28**(6): 375–81.

Millward, D. (2011, 31 October). 'Rapid rise in thefts from war memorials.' *The Telegraph Online* http://www.telegraph.co.uk/news/uknews/8858346/Rapid-rise-in-thefts-from-war-memorials.html, accessed 10 February 2016.

Milton, K. (1988). Foraging behaviour and the evolution of primate intelligence. In R. W. Byrne and A. Whiten (eds.), *Machiavellian Intelligence.* Oxford, Oxford University Press.

Misra, A. and N. Vikram (2003). 'Clinical and pathophysiological consequences of abdominal adiposity and abdominal adipose tissue deposits.' *Nutrition* **19**: 456–7.

Mock, D. W. and M. Fujioka (1990). 'Monogamy and long-term pair bonding in vertebrates.' *Trends in Ecology & Evolution* **5**(2): 39–43.

Mock, D. W. and G. A. Parker (1997). *The Evolution of Sibling Rivalry.* Oxford, Oxford University Press.

Moller, A. P. (1987). 'Behavioural aspects of sperm competition in swallows (Hirundo rustica).' *Behaviour* **100**: 92–104.

Morris, D. (1967). *The Illustrated Naked Ape: A Zoologist's Study of the Human Animal.* Jonathan Cape.

Morris, P. H., V. Reddy and R. Bunting (1995). 'The survival of the cutest: who's responsible for the evolution of the teddy bear?' *Animal Behaviour* **50**(6): 1697–1700.

Morris, P. H., J. White, E. R. Morrison and K. Fisher (2013). 'High heels as supernormal stimuli: how wearing high heels affects judgements of female attractiveness.' *Evolution and Human Behavior* **34**(3): 176–81.

Morrow, A. L., G. M. Ruiz–Palacios, M. Altaye, X. Jiang, M. L. Guerrero, J. K. Meinzen-Derr, T. Farkas, P. Chaturvedi, L. K. Pickering and D. S. Newburg (2004). 'Human milk oligosaccharides are associated with protection against diarrhea in breast-fed infants.' *The Journal of Pediatrics* **145**(3): 297–303.

Mount, L. E. and L. Mount (1979). *Adaptation to Thermal Environment: Man and His Productive Animals.* Edward Arnold, London.

Muehlenbein, M. P. (2010). *Human Evolutionary Biology*. Cambridge, Cambridge University Press.

Muehlenbein, M. P. and R. G. Bribiescas (2005). 'Testosterone-mediated immune functions and male life histories.' *American Journal of Human Biology* **17**(5): 527–58.

Mulder, M. B. and K. L. Rauch (2009). 'Sexual conflict in humans: variations and solutions.' *Evolutionary Anthropology: Issues, News, and Reviews* **18**(5): 201–14.

Murdock, G. P. and C. Provost (1973). 'Factors in the division of labor by sex: a cross-cultural analysis.' *Ethnology*: 203–25.

Murdock, G. P. and D. R. White (1969). 'Standard cross-cultural sample.' *Ethnology* 9: 329–69.

Murray, D. R. and M. Schaller (2010). 'Historical prevalence of infectious diseases within 230 geopolitical regions: a tool for investigating origins of culture.' *Journal of Cross-Cultural Psychology* **41**(1): 99–108.

Murray, D. R. and M. Schaller (2012). 'Threat(s) and conformity deconstructed: perceived threat of infectious disease and its implications for conformist attitudes and behavior.' *European Journal of Social Psychology* **42**(2): 180–8.

Mustanski, B. S., M. L. Chivers and J. M. Bailey (2003). 'A critical review of recent biological research on human sexual orientation.' *Annual Review of Sex Research* 13: 89–140.

Mustanski, B. S., M. G. DuPree, C. M. Nievergelt, S. Bocklandt, N. J. Schork and D. H. Hamer (2005). 'A genomewide scan of male sexual orientation.' *Human Genetics* **116**(4): 272–8.

Myers, D. G. (1993). *Social Psychology*. New York, McGraw-Hill Inc.

Johnson-Laird, P. N. and K. Oatley (1992). 'Basic emotions, rationality, and folk theory.' *Cognition & Emotion* 6(3-4): 201–23.

Nachman, M. W. and S. L. Crowell (2000). 'Estimate of the mutation rate per nucleotide in humans.' *Genetics* **156**(1): 297–304.

Narvaez, D. (2008). 'Triune ethics: the neurobiological roots of our multiple moralities.' *New Ideas in Psychology* **26**(1): 95–119.

Nascimento, J. M., L. Z. Shi, S. Meyers, P. Gagneux, N. M. Loskutoff, E. L. Botvinick and M. W. Berns (2008). 'The use of optical tweezers to study sperm competition and motility in primates.' *Journal of the Royal Society Interface* 5(20): 297–302.

National Heart, Lung and Blood Institute (1998). *Clinical Guidelines on the Identification, Evaluation and Treatment of Overweight and Obesity in Adults: The Evidence Report*. Bethesda, MD, National Institute of Health.

Navarrete, A., C. P. van Schaik and K. Isler (2011). 'Energetics and the evolution of human brain size.' *Nature* **480**(7375): 91–3.

Navarrete, C. D. and D. M. Fessler (2006). 'Disease avoidance and ethnocentrism: the effects of disease vulnerability and disgust sensitivity on intergroup attitudes.' *Evolution and Human Behavior* **27**(4): 270–82.

Neave, N. and K. Shields (2008). 'The effects of facial hair manipulation on female perceptions of attractiveness, masculinity, and dominance in male faces.' *Personality and Individual Differences* **45**(5): 373–7.

Neel, J. V. (1962). 'Diabetes Mellitus: a thrifty genotype rendered detrimental by "progress".' *American Journal of Human Genetics* **14**(4): 353–62.

Nesse, R. and G. C. Williams (1995). *Why We Get Sick: The New Theory of Darwinian Medicine*. Random House, New York.

Nesse, R. M. (2001). *Evolution and the Capacity for Commitment*. New York, Russell Sage.

Nesse, R. M. (2005a). 'Maladaptation and natural selection.' *The Quarterly Review of Biology* **80**(1): 62–70.

Nesse, R. M. (2005b). 'Natural selection and the regulation of defenses: a signal detection analysis of the smoke detector principle.' *Evolution and Human Behavior* **26**(1): 88–105.

Nesse, R. M. and S. C. Stearns (2008). 'The great opportunity: evolutionary applications to medicine and public health.' *Evolutionary Applications* **1**(1): 28–48.

Nettle, D. (2004a). Adaptive illusions: optimism, control and human rationality. In D. Evans and P. Cruse (eds.), *Emotion, Evolution and Rationality*. Oxford, Oxford University Press, pp. 193–208.

Nettle, D. (2004b). 'Evolutionary origins of depression: a review and reformulation.' *Journal of Affective Disorders* **81**: 91–102.

Nettle, D., T. E. Dickins, D. A. Coall and P. de Mornay Davies (2013a). 'Patterns of physical and psychological development in future teenage mothers.' *Evolution, Medicine, and Public Health* (1): 187–96.

Nettle, D., M. A. Gibson, D. W. Lawson and R. Sear (2013b). 'Human behavioral ecology: current research and future prospects.' *Behavioral Ecology* **24**(5): 1031–40.

Neubauer S. and J. Hublin (2012). 'The evolution of human brain development.' *Evolutionary Biology* **39**(4): 568–86.

Neuhoff, J. G. (2001). 'An adaptive bias in the perception of looming auditory motion.' *Ecological Psychology* **13**(2): 87–110.

Neuhoff, J. G., R. Planisek and E. Seifritz (2009). 'Adaptive sex differences in auditory motion perception: looming sounds are special.' *Journal of Experimental Psychology: Human Perception and Performance* **35**(1): 225.

Nicholson, N. (2000). *Executive Instinct: Managing the Human Animal in the Information Age*. Crown Business, New York.

Niemitz, C. (2010). 'The evolution of the upright posture and gait – a review and a new synthesis.' *Naturwissenschaften* **97**(3): 241–63.

Novella, S. (2008). 'Suboptimal optics: vision problems as scars of evolutionary history.' *Evolution: Education and Outreach* **1**(4): 493–7.

Nowak, M. and K. Sigmund (1998). 'Evolution of indirect reciprocity by image scoring.' *Nature* **393**: 573–6.

Nowak, M. A., C. E. Tarmita and E. O. Wilson (2010). 'The evolution of eusociality.' *Nature* **466**: 1057–62.

Nowell, P. C. (1976). 'The clonal evolution of tumor cell populations.' *Science* **194**(4260): 23–28.

Oakley, K. P. (1959). *Man the Toolmaker*. Chicago, University of Chicago Press.

Oberzaucher, E., S. Katina, S. F. Schmehl, I. J. Holzleitner, I. Mehu-Blantar and K. Grammer (2012). 'The myth of hidden ovulation: shape and texture changes in the face during the menstrual cycle.' *Journal of Evolutionary Psychology* **10**(4): 163–75.

Office for National Statistics (2012). *Integrated Household Survey*. http://www.ons.gov.uk/ons/dcp171778_280451.pdf, accessed 28 February 2013.

Office for National Statistics (2013). *Divorces in England and Wales*. http://www.ons.gov.uk/ons/rel/vsob1/divorces-in-england-and-wales/2011/stb-divorces-2011.html, accessed 8 February 2013.

Ohman, A. and S. Mineka (2001). 'Fears, phobias, and preparedness: toward an evolved module of fear and fear learning.' *Psychological Review* **108**(3): 483–522.

Oliver, W. J., E. L. Cohen and J. V. Neel (1975). 'Blood pressure, sodium intake, and sodium related hormones in the Yanomamo Indians, a "no–salt" culture.' *Circulation* **52**(1): 146–51.

Omenn, G. S. (2010). 'Evolution and public health.' *Proceedings of the National Academy of Sciences* **107**(suppl 1): 1702–9.

Oppenheimer, S. (2012). 'Out-of-Africa, the peopling of continents and islands: tracing uniparental gene trees across the map.' *Philosophical Transactions of the Royal Society B: Biological Sciences* **367**(1590): 770–84.

Ottenheimer, M. (1996). *Forbidden Relatives*. Urbana IL, University of Illinois Press.

Pagel, M. and W. Bodmer (2003). 'A naked ape would have fewer parasites.' *Proceedings of the Royal Society of London. Series B: Biological Sciences* **270** (Suppl 1): S117–9.

Painter, R., C. Osmond, P. Gluckman, M. Hanson, D. Phillips and T. Roseboom (2008). 'Transgenerational effects of prenatal exposure to the Dutch famine on neonatal adiposity and health in later life.' *BJOG: An International Journal of Obstetrics & Gynaecology* **115**(10): 1243–9.

Paley, W. (1836). *The Works of William Paley*. Philadelphia, J. J. Woodward.

Parent, A.-S., G. Teilmann, A. Juul, N. E. Skakkebaek, J. Toppari and J.-P. Bourguignon (2003). 'The timing of normal puberty and the age limits of sexual precocity: variations around the world, secular trends, and changes after migration.' *Endocrine Reviews* **24**(5): 668–93.

Park, J. H. (2007). 'Persistent misunderstandings of inclusive fitness and kin selection: their ubiquitous appearance in social psychology textbooks.' *Evolutionary Psychology* **5**(4): 860–73.

Parker, G. A., R. R. Baker and V. G. F. Smith (1972). 'The origin and evolution of gamete dimorphism and the male–female phenomenon.' *Journal of Theoretical Biology* **36**: 529–53.

Parkinson, B. (2005). 'Do facial movements express emotions or communicate motives?' *Personality and Social Psychology Review* **9**(4): 278–311.

Pashos, A. (2000). 'Does paternal uncertainty explain discriminative grandparental solicitude?' *Evolution and Human Behavior* **21**: 97–111.

Pasquali, R., A. Gambineri, B. Anconetani, V. Vicennati, D. Colitta, E. Caramelli, F. Casimirri and M. Morselli–Labali (1999). 'The natural history of the metabolic syndrome in young women with the polycystic ovary syndrome and the effect on long-term oestrogen–progestagen treatment.' *Clinical Endocrinology* **50**: 517–27.

Paul, R. A. (1991). 'Psychoanalytic theory and incest avoidance rules.' *Behavioral and Brain Sciences* **14**(02): 276–7.

Pavé, R., M. M. Kowalewski, S. M. Peker and G. E. Zunino (2010). 'Preliminary study of mother–offspring conflict in black and gold howler monkeys (Alouatta caraya).' *Primates* **51**(3): 221–6.

Pearce, D., A. Markandya and E. B. Barbier (1989). *Blueprint for a Green Economy.* London, Earthscan.

Pembrey, M., L. O. Bygren, G. Kaati, et al. (2006). 'Sex–specific, male line transgenerational responses in humans.' *European Journal of Human Genetics* **14**: 159–66.

Penn, D. J., K. Damjanovich and W. K. Potts (2002). 'MHC heterozygosity confers a selective advantage against multiple-strain infections.' *Proceedings of the National Academy of Sciences* **99**(17): 11260–4.

Pennington, R. (1992). 'Did food increase fertility? An evaluation of !Kung and Herero history.' *Human Biology* **64**: 497–521.

Pennisi, E. (2001). 'Tracking the sexes by their genes.' *Science* **291**(5509): 1733.

Penton-Voak, I. and J. Chen (2004). 'High salivary testosterone is linked to masculine male facial appearance in humans.' *Evolution and Human Behavior* **25**(4): 229–42.

Penton-Voak, I. S. and D. I. Perrett (2000a). 'Consistency and individual differences in facial attractiveness judgements: an evolutionary perspective.' *Social Research*: 219–44.

Penton-Voak, I. S. and D. I. Perrett (2000b). 'Female preference for male faces changes cyclically: further evidence.' *Evolution and Human Behavior* **21**(1): 39–49.

Perrett, D. I., K. A. May and S. Yoshikawa (1994). 'Facial shape and judgements of female attractiveness.' *Nature* **368**(17 March): 239–42.

Perrett, D. I., D. M. Burt, I. S. Penton-Voak, K. J. Lee and D. A. Rowland (1999). 'Symmetry and human facial attractiveness.' *Evolution and Human Behavior* **20**: 295–307.

Perrett, D. I., K. J. Lee et al. (1998). 'Effects of sexual dimorphism on facial attractiveness.' *Nature* **394**: 884–7.

Perry, G. H. and N. J. Dominy (2009). 'Evolution of the human pygmy phenotype.' *Trends in Ecology & Evolution* **24**(4): 218–25.

Perry, G. H., N. J. Dominy, K. G. Claw et al. (2007). 'Diet and the evolution of human amylase gene copy number variation.' *Nature Genetics* **39**: 1256–60.

Perusse, D. (1993). 'Cultural and reproductive success in industrial societies: testing the relationship at the proximate and ultimate levels.' *Behavioural and Brain Sciences* **16**: 267–323.

Pew Research Center (2003). *Religious Beliefs Underpin Opposition to Homosexuality.* www.people-press. org/2003/11/18/religious-beliefs-underpin-opposition-to-homosexuality, accessed 5 March 2013.

Pflüger, L. S., E. Oberzaucher, S. Katina, I. J. Holzleitner and K. Grammer (2012). 'Cues to fertility: perceived attractiveness and facial shape predict reproductive success.' *Evolution and Human Behavior* **33**(6): 708–14.

Pietrzak, R. H., J. D. Laird, D. A. Stevens and N. S. Thompson (2002). 'Sex differences in human jealousy: a coordinated study of forced-choice, continuous rating-scale, and physiological responses on the same subjects.' *Evolution and Human Behavior* **23**(2): 83–95.

Pigliucci, M. (2007). 'Do we need an extended evolutionary synthesis?' *Evolution and Human Behavior* **61**: 2743–9.

Pinker, S. (1994). *The Language Instinct.* London, Penguin.

Pinker, S. (1997). *How the Mind Works.* New York, Norton.

Pinker, S. (2008, 13 January). 'The moral instinct.' *The New York Times*, p. 32 (magazine section).

Pinker, S. and P. Bloom (1990). 'Natural language and natural selection.' *Behavioural and Brain Sciences* **13**: 707–84.

Platek, S. M., R. L. Burch, I. S. Panyavin, B. H. Wasserman and G. G. Gallup Jr (2003). 'Reactions to children's faces: resemblance affects males more than females.' *Evolution and Human Behavior* **23**: 159–66.

Platek, S. M., D. M. Raines, G. G. Gallup Jr, F. B. Mohamed, J. W. Thomson, T. E. Myers, I. S. Panyavin, S. L. Levin, J. A. Davis, C. M. Fonteyn and D. R. Arigo (2004). 'Reactions to children's faces: males are more affected by resemblance than females are, and so are their brains.' *Evolution and Human Behavior* **25**(6): 394–406.

Plomin, R. (1990). *Behavioral Genetics. A Primer.* New York, Freeman and Co.

Plomin, R., A. Caspi, L. Pervin and O. John (1999). 'Behavioral genetics and personality.' *Handbook of Personality: Theory and Research* **2**: 251–76.

Pollick, F. E., J. W. Kay, K. Heim and R. Stringer (2005). Gender recognition from point-light walkers.' *Journal of Experimental Psychology: Human Perception and Performance* **31**(6): 1247.

Poolman, E. M. and A. P. Galvani (2007). 'Evaluating candidate agents of selective pressure for cystic fibrosis.' *Journal of the Royal Society Interface* **4**(12): 91–8.

Popper, K. R. (1963). *Conjectures and Refutations. The Growth of Scientific Knowledge (Essays and Lectures).* London, Routledge & Kegan Paul.

Portmann, A. (1969). *A Zoologist Looks At Mankind* (trans. Schaefer, J., 1990). New York, Columbia University Press.

Postma, E., L. Martini and P. Martini (2010). 'Inbred women in a small and isolated Swiss village have fewer children.' *Journal of Evolutionary Biology* **23**(7): 1468–74.

Poulsen, P., M. Esteller, A. Vaag and M. F. Fraga (2007). 'The epigenetic basis of twin discordance in age-related diseases.' *Pediatric Research* **61**: 38R–42R.

Pound, N., I. S. Penton-Voak and A. K. Surridge (2009). 'Testosterone responses to competition in men are related to facial masculinity.' *Proceedings of the Royal Society of London B: Biological Sciences* **276**(1654): 153–9.

Powell, K. L., G. Roberts and D. Nettle (2012). 'Eye images increase charitable donations: evidence from an opportunistic field experiment in a supermarket.' *Ethology* **118**(11): 1096–1101.

Prechtl, H. (1986). 'New perspectives in early human development.' *European Journal of Obstetrics & Gynecology and Reproductive Biology* **21**(5): 347–55.

Preuschoft, S. (1992). 'Laughter and smile in Barbary macaques.' *Ethology* **91**: 220–36.

Preuschoft, S. (2000). 'Primate faces and facial expressions.' *Social Research* **67**: 245–71.

Previc, F. H. (2009). *The Dopaminergic Mind in Human Evolution and History*. Cambridge, Cambridge University Press.

Price, J., L. Sloman, R. Gardner, P. Gilbert and P. Rohde (1997). The social competition hypothesis of depression. In S. Baron-Cohen (ed.), *The Maladapted Mind*. Hove, Psychology Press.

Price, M. (2011). Cooperation as a classic problem in behavioural biology. In V. Swami (ed.), *Evolutionary Psychology*. Chichester, Blackwell, pp. 73–107.

Profet, M. (1988). 'The evolution of pregnancy sickness as protection to the embryo against Pleistocene teratogens.' *Evolutionary Theory* **8**(3): 177–90.

Profet, M. (1992). Pregnancy sickness as adaptation: a deterrent to maternal ingestion of teratogens. In L. Cosmides, J. H. Barkow, J. Tooby (eds.), *The Adapted Mind: Evolutionary Psychology and the Generation of Culture*. New York, Oxford University Press.

Profet, M. (1993). 'Menstruation as a defense against pathogens transported by sperm.' *Quarterly Review of Biology* **68**: 335–86.

Prokop, P., M. J. Rantala, M. Usak and I. Senay (2012). 'Is a woman's preference for chest hair in men influenced by parasite threat?' *Archives of Sexual Behavior*: 1–9.

Provost, M. P., V. L. Quinsey and N. F. Troje (2008). 'Differences in gait across the menstrual cycle and their attractiveness to men.' *Archives of Sexual Behavior* **37**(4): 598–604.

Provost, M. P., N. F. Troje and V. L. Quinsey (2008). 'Short-term mating strategies and attraction to masculinity in point-light walkers.' *Evolution and Human Behavior* **29**(1): 65–9.

Purcell, S. M., N. R. Wray, J. L. Stone, P. M. Visscher, M. C. O'Donovan, P. F. Sullivan, P. Sklar, D. M. Ruderfer, A. McQuillin and D. W. Morris (2009). 'Common polygenic variation contributes to risk of schizophrenia and bipolar disorder.' *Nature* **460**(7256): 748–52.

Pusey, A. (2004). Inbreeding avoidance in primates. In A. P. Wolf and W. H. Durham (eds.), *Inbreeding, Incest and the Incest Taboo*. Stanford, CA, Stanford University Press.

Puts, D. A. (2010). 'Beauty and the beast: mechanisms of sexual selection in humans.' *Evolution and Human Behavior* **31**(3): 157–75.

Puts, D. A., S. J. Gaulin and K. Verdolini (2006). 'Dominance and the evolution of sexual dimorphism in human voice pitch.' *Evolution and Human Behavior* **27**(4): 283–96.

Quillian, L. and D. Pager (2001). 'Black neighbors, higher crime? The role of racial stereotypes in evaluations of neighborhood crime 1.' *American Journal of Sociology* **107**(3): 717–67.

Quinlan, R. J. (2007). 'Human parental effort and environmental risk.' *Proceedings of the Royal Society B: Biological Sciences* **274**(1606): 121–5.

Rahman, Q. and M. S. Hull (2005). 'An empirical test of the kin selection hypothesis for male homosexuality.' *Archives of Sexual Behavior* **34**(4): 461–7.

Raihani, N. J. and R. Bshary (2012). 'A positive effect of flowers rather than eye images in a large-scale, cross-cultural dictator game.' *Proceedings of the Royal Society B: Biological Sciences* **279**(1742): 3556–64.

Ramachandran, V. S. (1997). 'Why do gentlemen prefer blondes?' *Medical Hypotheses* **48**(1): 19–20.

Rantala, M. (2007). 'Evolution of nakedness in Homo sapiens.' *Journal of Zoology* **273**(1): 1–7.

Rantala, M. J. (1999). 'Human nakedness: adaptation against ectoparasites?' *International Journal for Parasitology* **29**(12): 1987–9.

Rantala, M. J. and U. M. Marcinkowska (2011). 'The role of sexual imprinting and the Westermarck effect in mate choice in humans.' *Behavioral Ecology and Sociobiology* **65**(5): 859–73.

Rapoport, A. (1965). *Prisoner's Dilemma: A Study In Conflict and Cooperation.* University of Michigan Press.

Ravussin, E., M. E. Valencia, J. Esparza, P. H. Bennett and L. O. Schulz (1994). 'Effects of a traditional lifestyle on obesity in Pima Indians.' *Diabetes Care* **17**(9): 1067–74.

Reaven, G. M. (1995). 'Pathophysiology of insulin resistance in human disease.' *Physiological Reviews* **75**(3): 473–86.

Reed, W., M. Clark, P. Parker, S. Raouf, N. Arguedas, D. Monk, E. Snajdr, V. Nolan Jr and E. Ketterson (2006). 'Physiological effects on demography: a long-term experimental study of testosterone's effects on fitness.' *The American Naturalist* **167**(5): 667–83.

Regan, P. C. (1996). 'Rhythms of desire: the association between menstrual cycle phases and female sexual desire.' *Canadian Journal of Human Sexuality* **5**: 145–56.

Reich, G. (2003). 'Depression and couples relationship.' *Psychotherapeut* **48**(1): 2–14.

Reiches, M. W., P. T. Ellison, S. F. Lipson, K. C. Sharrock, E. Gardiner and L. G. Duncan (2009). 'Pooled energy budget and human life history.' *American Journal of Human Biology* **21**(4): 421–9.

Reisenzein, R., M. Studtmann and G. Horstmann (2013). 'Coherence between emotion and facial expression: evidence from laboratory experiments.' *Emotion Review* **5**(1): 16–23.

Reynolds, J. D. and P. H. Harvey (1994). Sexual selection and the evolution of sex differences. In R. V. Short and E. Balban (eds.), *The Differences Between the Sexes.* Cambridge, Cambridge University Press.

Rhodes, G. (2006). 'The evolutionary psychology of facial beauty.' *Annu. Rev. Psychol.* **57**: 199–226.

Rice, G., C. Anderson, N. Risch and G. Ebers (1999). 'Male homosexuality: absence of linkage to microsatellite markers at Xq28.' *Science* **284**(5414): 665–7

Richards, C., S. Watkins, E. Hoffman, N. Schneider, I. Milsark, K. Katz, J. Cook, L. Kunkel and J. Cortada (1990). 'Skewed X inactivation in a female MZ twin results in Duchenne muscular dystrophy.' *American Journal of Human Genetics* **46**(4): 672.

Richards, R. J. (1993). 'Birth, death, and resurrection of evolutionary ethics.' *Evolutionary ethics*: 113–131.

Richerson, P. J. and R. Boyd (2005). *Not by Genes Alone.* Chicago, University of Chicago Press.

Ridley, M. (1993). *The Red Queen.* London, Viking.

Ridley, M. (1996). *The Origins of Virtue.* London, Viking (Penguin Group).

Roberts, D. F. (1953). 'Body weight, race and climate.' *American Journal of Physical Anthropology* **11**(4): 533–58.

Roberts, D. F. and D. P. Kahlon (1976). 'Environmental correlations of skin colour.' *Annals of Human Biology* **3**: 11–22.

Roberts, M. L., K. L. Buchanan and M. Evans (2004). 'Testing the immunocompetence handicap hypothesis: a review of the evidence.' *Animal Behaviour* **68**(2): 227–39.

Roberts, S. C., L. M. Gosling, V. Carter and M. Petrie (2008). 'MHC-correlated odour preferences in humans and the use of oral contraceptives.' *Proceedings of the Royal Society B: Biological Sciences* **275**(1652): 2715–22.

Roberts, S. C., J. Havlíček, J. Flegr, M. Hruskova, A. C. Little, B. C. Jones, D. I. Perrett and M. Petrie (2004). 'Female facial attractiveness increases during the fertile phase of the menstrual cycle.' *Proceedings of the Royal Society of London B: Biological Sciences* **271**(Suppl 5): S270–72.

Roberts, S. C., A. C. Little, L. M. Gosling, B. C. Jones, D. I. Perrett, V. Carter and M. Petrie (2005a). 'MHC-assortative facial preferences in humans.' *Biology Letters* **1**(4): 400–3.

Roberts, S. C., A. C. Little, L. M. Gosling, D. I. Perrett, V. Carter, B. C. Jones, I. Penton-Voak and M. Petrie (2005b). 'MHC-heterozygosity and human facial attractiveness.' *Evolution and Human Behavior* **26**(3): 213–26.

Robins, A. H. (1991). *Biological Perspectives on Human Skin Pigmentation.* Cambridge, Cambridge University Press.

Robson, S. L. and B. Wood (2008). 'Hominin life history: reconstruction and evolution.' *Journal of Anatomy* **212**(4): 394–425.

Rodin, J. and N. Radke-Sharpe (1991). 'Changes in appetitive variables as a function of pregnancy.' *Chemical Senses. Volume 4: Appetite and Nutrition*: 325.

Rodman, P. S. and H. M. McHenry (1980). 'Bioenergetics of hominid bipedalism.' *Journal of Physical Anthropology* **52**: 103–6.

Rogers, A. (2004). 'Genetic Variation at the MC1R Locus and the time since loss of human body hair 1.' *Current Anthropology* **45**(1): 105–8.

Rogers, A. R. (1989). 'Does biology constrain culture?' *American Anthropologist* **90**: 819–31.

Roseboom, T. J., R. C. Painter, A. F. van Abeelen, M. V. Veenendaal and S. R. de Rooij (2011). 'Hungry in the womb: what are the consequences? Lessons from the Dutch famine.' *Maturitas* **70**(2): 141–5.

Rosenberg, K. and W. Trevathan (2002). 'Birth, obstetrics and human evolution.' *BJOG: An International Journal of Obstetrics & Gynaecology* **109**(11): 1199–1206.

Rosenfeld, C. S. and R. M. Roberts (2004). 'Maternal diet and other factors affecting offspring sex ratio: a review.' *Biology of Reproduction* **71**(4): 1063–70.

Roth, G. and U. Dicke (2005). 'Evolution of the brain and intelligence.' *Trends in Cognitive Science* **9**(5): 250–7.

Roy, R. P. (2003). 'A Darwinian view of obstructed labor.' *Obstetrics and Gynecology* **101**: 397–401.

Rozin, P. and A. E. Fallon (1987). 'A perspective on disgust.' *Psychological Review* **94**(1): 23.

Rubin, P. H. (2002). *Darwinian Politics: The Evolutionary Origin of Freedom*. Rutgers University Press.

Rudski, J. (2001). 'Competition, superstition and the illusion of control.' *Current Psychology* **20**(1): 68–84.

Rudski, J. (2004). 'The illusion of control, superstitious belief, and optimism.' *Current Psychology* **22**(4): 306–315.

Rudski, J. M. and A. Edwards (2007). 'Malinowski goes to college: factors influencing students' use of ritual and superstition.' *The Journal of General Psychology* **134**(4): 389–403.

Ruff, C. (2002). 'Variation in human body size and shape.' *Annual Review of Anthropology* **31**: 211–32.

Ruse, M. (1988). *Homosexuality: A Philosophical Inquiry*. Oxford, UK, Blackwell.

Ruse, M. (1993). "The New Evolutionary Ethics" in Nitecki, D. and Nitecki, M. eds. Evolutionary Ethics. Albany, NY, SUNY Press.

Ruston, D. et al. (2004). *The National Diet & Nutrition Survey: Adults Aged 19 to 64 Years: Volume 4: Nutritional Status (Anthropometry and Blood Analytes), Blood Pressure and Physical Activity*. TSO.

Ruvolo, M., D. Pan, S. Zehr, T. Goldberg, T. R. Disotell and M. Von Dornum (1994). 'Gene trees and hominid phylogeny.' *Proceedings of the National Academy of Science* **91**: 8900–4.

Ruxton, G. D. and D. M. Wilkinson (2011). 'Avoidance of overheating and selection for both hair loss and bipedality in hominins.' *Proceedings of the National Academy of Sciences* **108**(52): 20965–9.

Ryan, C. and C. Jethá (2010). *Sex at Dawn: The Prehistoric Origins of Modern Sexuality*. New York, NY, Harper.

Saal, F. E., C. B. Johnson and N. Weber (1989). 'Friendly or sexy?: it may depend on whom you ask.' *Psychology of Women Quarterly* **13**(3): 263–76.

Sabeti, P. C., P. Varilly, B. Fry, J. Lohmueller, E. Hostetter, C. Cotsapas, X. Xie, E. H. Byrne, S. A. McCarroll and R. Gaudet (2007). 'Genome-wide detection and characterization of positive selection in human populations.' *Nature* **449**(7164): 913–8.

Sacher, G. A. (1959). Relation of lifespan to brain weight and body weight in mammals. In G. E. W. Wolstenholme and M. O'Connor (eds.), *CIBA Foundation Colloquia on Ageing* **5**: 115–33.

Sagarin, B. J., A. L. Martin, S. A. Coutinho, J. E. Edlund, L. Patel, J. J. Skowronski and B. Zengel (2012). 'Sex differences in jealousy: a meta-analytic examination.' *Evolution and Human Behavior* **33**(6): 595–614.

Saino, N., A. M. Bolzern and A. P. Moller (1997). 'Immunocompetence, ornamentation, and viability of male barn swallows (Hirundo rustica).' *Proceedings of the National Academy of Science, USA* **94**: 549–52.

Salmon, C. A. (1999). 'On the impact of sex and birth order on contact with kin.' *Human Nature* **10**: 183–97.

Sankararaman, S., S. Mallick, M. Dannemann, K. Prüfer, J. Kelso, S. Pääbo, N. Patterson and D. Reich (2014). 'The genomic landscape of Neanderthal ancestry in present-day humans.' *Nature* **507**(7492): 354–7.

Santos, P. S. C., J. A. Schinemann, J. Gabardo and M. D. Bicalho (2005). 'New evidence that the MHC influences odor perception in humans: a study with 58 Southern Brazilian students.''' *Hormones and Behaviour* **47**(4): 384–8.

Santrock, J. W. (2008). Motor, sensory, and perceptual development. In M. Ryan (ed.), *A Topical Approach to Life-Span Development*. Boston, MA, McGraw–Hill Higher Education, pp. 172–205.

Santtila, P., A.-L. Högbacka, P. Jern, A. Johansson, M. Varjonen, K. Witting, B. Von Der Pahlen and N. K. Sandnabba (2009). 'Testing Miller's theory of alleles preventing androgenization as an evolutionary explanation for the genetic predisposition for male homosexuality.' *Evolution and Human Behavior* **30**(1): 58–65.

Saroglou, V. and L. Fiasse (2003). 'Birth order, personality, and religion: a study among young adults from a three-sibling family.' *Personality and Individual Differences* **35**: 19–29.

Sawai, H., H. L. Kim, K. Kuno, S. Suzuki, H. Gotoh, M. Takada, N. Takahata, Y. Satta and F. Akishinomiya (2010). 'The origin and genetic variation of domestic chickens with special reference to junglefowls Gallus g. gallus and G. varius.' *PloS One* **5**(5): e10639.

Scelza, B. A. (2013). 'Jealousy in a small-scale, natural fertility population: the roles of paternity, investment and love in jealous response.' *Evolution and Human Behavior* **35**(2): 103–8.

Schaller, M. and J. H. Park (2011). 'The behavioral immune system (and why it matters).' *Current Directions in Psychological Science* 20(2): 99–103.

Scheib, J. E., S. W. Gangestad and R. Thornhill (1999). 'Facial attractiveness, symmetry and cues to good genes.' *Proceedings of the Royal Society of London, Series B* 266: 1913–7.

Schlomer, G. L. and J. Belsky (2012). 'Maternal age, investment, and parent–child conflict: a mediational test of the terminal investment hypothesis.' *Journal of Family Psychology* 26(3): 443.

Schloss, J. and M. Murray (2009). *The Believing Primate: Scientific, Philosophical, and Theological Reflections on the Origin of Religion.* Oxford, Oxford University Press.

Schmidt J (1920). 'Racial investigations IV. The genetic behavior of a secondary sexual character.' *Comptes rendus des travaux du Laboratoire Carlsberg* 14(227).

Schmidt, K. L. and J. F. Cohn (2001). 'Human facial expressions as adaptations: evolutionary questions in facial expression research.' *Yearbook of Physical Anthropology* 44: 3–24.

Schneider, M. A. and L. Hendrix (2000). 'Olfactory sexual inhibition and the Westermarck effect.' *Human Nature* 11(1): 65–91.

Schultz, A. (1969). *The Life of Primates.* London, Weidenfeld & Nicolson.

Schwarz, S. and M. Hassebrauck (2012). 'Sex and age differences in mate-selection preferences.' *Human Nature* 23(4): 447–66.

Scott, I. M., A. P. Clark, L. G. Boothroyd and I. S. Penton-Voak (2013). 'Do men's faces really signal heritable immunocompetence?' *Behavioral Ecology* 24(3): 579–89.

Scott-Phillips, T. C., T. E. Dickins and S. A. West (2011). 'Evolutionary theory and the ultimate–proximate distinction in the human behavioral sciences.' *Perspectives on Psychological Science* 6(1): 38–47.

Sear, R. and R. Mace (2008). 'Who keeps children alive? A review of the effects of kin on child survival.' *Evolution and Human Behavior* 29(1): 1–18.

Sear, R., R. Mace and I. A. McGregor (2000). 'Maternal grandmothers improve nutritional status and survival of children in rural Gambia.' *Proceedings of the Royal Society, London, B Series.* 267: 1641–7.

Seligman, M. E. P. (1971). 'Phobias and preparedness.' *Behaviour Therapy* 2: 307–20.

Sharot, T. (2011). 'The optimism bias.' *Current Biology* 21(23): R941–5.

Shepard, R. N. and J. Metzler (1971). 'Mental rotation of three-dimensional objects.' *Science* 171: 701–3.

Shepher, J. (1971). 'Mate selection among second generation kibbutz adolescents and adults: incest avoidance and negative imprinting.' *Archives of Sexual Behavior* 1(4): 293–307.

Sherman, P. W. and J. Billing (1999). 'Darwinian gastronomy: why we use spices.' *BioScience* 49: 453–63.

Sherman, P. W. and S. M. Flaxman (2002). 'Nausea and vomiting of pregnancy in an evolutionary perspective.' *American Journal of Obstetrics and Gynecology* 186(5): S190–7.

Sherman, P. W. and G. A. Hash (2001). 'Why vegetable recipes are not very spicy.' *Evolution and Human Behavior* 22(3): 147–64.

Sherman, P. W. and H. K. Reeve (1997). Forward and backward: alternative approaches to studying human behaviour. In L. Betzig (ed.), *Human Nature.* Oxford, Oxford University Press.

Sherwood, C. C., F. Subiaul and T. W. Zawidzki (2008). 'A natural history of the human mind: tracing evolutionary changes in brain and cognition.' *Journal of Anatomy* 212(4): 426–54.

Shively, M. G. and J. P. De Cecco (1977). 'Components of sexual identity.' *Journal of Homosexuality* 3(1): 41–8.

Shokeir, M. (1975). 'Investigation on Huntington's disease in the Canadian Prairies.' *Clinical Genetics* 7(4): 349–53.

Shor, E. and D. Simchai (2009). 'Incest avoidance, the incest taboo, and social cohesion: revisiting Westermarck and the case of the Israeli Kibbutzim 1.' *American Journal of Sociology* 114(6): 1803–42.

Short, R. V. (1979). 'Sexual selection and its component parts, somatic and genital selection, as illustrated by man and the great apes.' *Advances in the Study of Behavior* 9, 131–58.

Short, R. V. (1994). Why sex? In R. V. Short and E. Balaban (eds.), *The Differences between the Sexes.* Cambridge, Cambridge University Press.

Short, R. V. and E. Balban (eds.) (1994). *The Differences Between the Sexes.* Cambridge, Cambridge University Press.

Shultz, S. and M. Maslin (2013). 'Early Human Speciation, Brain Expansion and Dispersal Influenced by African Climate Pulses.' *PloS One* 8(10): e76750.

Siegel, S. and L. G. Allan (1996). 'The widespread influence of the Rescorla–Wagner model.' *Psychonomic Bulletin & Review* 3(3): 314–21.

Sillen-Tullberg, B. and A. Moller (1993). 'The relationship between concealed ovulation and mating systems in anthropoid primates: a phylogenetic analysis.' *The American Naturalist* **141**: 1–25.

Silventoinen, K. (2003). 'Determinants of variation in adult body height.' *Journal of Biosocial Science* **35**(02): 263–85.

Silverman, I. and M. Eals (1992). Sex differences in spatial abilities: evolutionary theory and data. In J. H. Barkow, L. Cosmides and J. Tooby (eds.), *The Adapted Mind*. Oxford, Oxford University Press.

Silverman, I., J. Choi and M. Peters (2007). 'The hunter-gatherer theory of sex differences in spatial abilities: data from 40 countries.' *Archives of Sexual Behavior* **36**(2): 261–8.

Silverman, I., J. Choi, A. Mackewn and M. Fisher (2000). 'Evolved mechanisms underlying wayfinding: further studies on the hunter-gatherer theory of spatial sex differences.' *Evolution and Human Behavior* **21**(3): 201–15.

Simon, H. (1956). 'Rational choice and the structure of environments.' *Psychology Review* **63**: 129–38.

Simon, H. (1990). 'Invariants of human behaviour.' *Annual Review of Psychology* **41**: 1–19.

Simoons, F. J. (2001). 'Persistence of lactase activity among northern Europeans: a weighing of the evidence for the calcium absorption hypothesis.' *Ecology of Food and Nutrition* **40**: 397–469.

Singh, D. (1993). 'Adaptive significance of female attractiveness.' *Journal of Personality and Social Psychology* **65**: 293–307.

Singh, D. (1995). 'Female judgement of male attractiveness and desirability for relationships: role of waist-to-hip ratios and financial status.' *Journal of Personality and Social Psychology* **69**(6): 1089–1101.

Singh, D. and P. M. Bronstad (2001). 'Female body odour is a potential cue to ovulation.' *Proceedings of the Royal Society of London B: Biological Sciences* **268**(1469): 797–801.

Singh, D. and S. Luis (1995). 'Ethnic and gender consensus for the effect of waist-to-hip ratio on judgement of women's attractiveness.' *Human Nature* **6**(1): 51–65.

Sipos, A., F. Rasmussen, G. Harrison, P. Tynelius, G. Lewis, D. A. Leon and D. Gunnell (2004). 'Paternal age and schizophrenia: a population based cohort study.' *British Medical Journal* **329**(7474): 1070.

Skinner, B. F. (1948). '"Superstition" in the pigeon.' *Journal of Experimental Psychology* **38**(2): 168.

Skinner, B. F. (1974). *Walden Two*. Indianapolis, Hackett Publishing.

Slob, A. K., C. M. Bax, H. W. C., D. L. Rowland and J. J. van der Werflen Bosch (1996). 'Sexual arousability and the menstrual cycle.' *Pychoneuroendocrinology* **21**: 545–58.

Sloboda, D. M., R. Hart, D. A. Doherty, C. E. Pennell and M. Hickey (2007). 'Age at menarche: influences of prenatal and postnatal growth.' *Journal of Clinical Endocrinology & Metabolism* **92**(1): 46–50.

Sloboda, D. M., G. J. Howie, A. Pleasants, P. D. Gluckman and M. H. Vickers (2009). 'Pre- and postnatal nutritional histories influence reproductive maturation and ovarian function in the rat."' *PloS One* **4**(8): e6744.

Smith, E. A. (2010). 'Communication and collective action: language and the evolution of human cooperation.' *Evolution and Human Behavior* **31**(4): 231–45.

Smith, E. A. and R. L. Bliege Bird (2000). 'Turtle hunting and tombstone opening: public generosity as costly signalling.' *Evolution and Human Behavior* **21**: 245–61.

Smith, H. F., W. Parker, S. H. Kotzé and M. Laurin (2013). 'Multiple independent appearances of the cecal appendix in mammalian evolution and an investigation of related ecological and anatomical factors.' *Comptes Rendus Palevol*, **12**(6): 339–54.

Smith, R. (1997). *The Fontana History of the Human Sciences*. London, Fontana.

Smith, R. J. and J. M. Cheverud (2002). 'Scaling of sexual dimorphism in body mass: a phylogenetic analysis of Rensch's rule in primates.' *International Journal of Primatology* **23**(5): 1095–1135.

Smith, T. M., Z. Machanda, A. B. Bernard, R. M. Donovan, A. M. Papakyrikos, M. N. Muller and R. Wrangham (2013). 'First molar eruption, weaning, and life history in living wild chimpanzees.' *Proceedings of the National Academy of Sciences* **110**(8): 2787–91.

Smoller, J. W. and M. T. Tsuang (1998). 'Panic and phobic anxiety: defining phenotypes for genetic studies.' *American Journal of Psychiatry* **155**(9): 1152–62.

Sniegowski, P. D., P. J. Gerrish, T. Johnson and A. Shaver (2000). 'The evolution of mutation rates: separating causes from consequences.' *Bioessays* **22**(12): 1057–66.

Snyder, H. N. (2012). *Arrest in the United States, 1990–2010*. US Department of Justice, Office of Justice Programs, Bureau of Justice Statistics.

Sokal, R. R., N. L. Oden and C. Wilson (1991). 'Genetic evidence for the spread of agriculture in Europe by demic diffusion.' *Nature* **351**: 143–5.

Sørensen, S. A., K. Fenger and J. H. Olsen (1999). 'Significantly lower incidence of cancer among patients with Huntington disease.' *Cancer* **86**(7): 1342–6.

Soronen, P., M. Laiti, S. Törn, P. Härkönen, L. Patri-kainen, Y. Li, A. Pulkka, R. Kurkela, A. Herrala and H. Kaija (2004). 'Sex steroid hormone metabolism and prostate cancer.' *The Journal of Steroid Biochemistry and Molecular Biology* **92**(4): 281.

Sosis, R. and C. Alcorta (2003). 'Signaling, solidarity, and the sacred: the evolution of religious behavior.' *Evolutionary Anthropology: Issues, News, and Reviews* **12**(6): 264–74.

Sosis, R. and E. R. Bressler (2003). 'Cooperation and commune longevity: a test of the costly signaling theory of religion.' *Cross-Cultural Research* **37**(2): 211–39.

Southam, L., N. Soranzo, S. B. Montgomery, T. M. Frayling, M. I. McCarthy, I. Barroso and E. Zeggini (2009). 'Is the thrifty genotype hypothesis supported by evidence based on confirmed type 2 diabetes- and obesity-susceptibility variants?' *Diabetologia* **52**(9): 1846–51.

Spain, D. H. (1991). 'Muddled theory and misinterpreted data: comments on yet another attempt to identify a so-called Westermarck effect and, in the process, to refute Freud.' *Behavioral and Brain Sciences* **14**(02): 278–9.

Sparks, A. and P. Barclay (2013). 'Eye images increase generosity, but not for long: the limited effect of a false cue.' *Evolution and Human Behavior* **34**(5): 317–22.

Sprengelmeyer, R., D. Perrett, E. Fagan, R. Cornwell, J. Lobmaier, A. Sprengelmeyer, H. Aasheim, I. Black, L. Cameron and S. Crow (2009). 'The cutest little baby face: a hormonal link to sensitivity to cuteness in infant faces.' *Psychological Science* **20**(2): 149–54.

Stanford, C. B. (1996). 'The hunting ecology of wild chimpanzees: implications for the evolutionary ecology of Pliocene hominids.' *American Anthropologist* **98**(1): 96–113.

Stearns, S. C. and D. Ebert (2001). 'Evolution in health and disease: work in progress.' *Quarterly Review of Biology*: 417–32.

Stein, T. S. (1998). 'Social constructionism and essentialism.' *Journal of Gay & Lesbian Psychotherapy* **2**(4): 29–49.

Sternglanz, S. H., Gray, J. L. and Murakami, M. (1977). 'Adult preferences for infantile facial features: an ethological approach.' *Animal Behaviour* **25**: 108–15.

Stini, W. A. (1981). 'Body composition and nutrient reserves in evolutionary perspective.' *World Review of Nutrition and Dietetics* **37**: 55.

Strachan, T. and A. P. Read (1996). *Human Molecular Genetics*. Oxford, Bios Scientific.

Strassman, B. I. (1996a). 'Menstrual huts visits by Dogon women: a hormonal test distinguishes deceit from honest signalling.' *Behavioural Ecology* **7**(3): 304–15.

Strassmann, B. I. (1996b). 'Energy economy in the evolution of menstruation.' *Evolutionary Anthropology: Issues, News, and Reviews* **5**(5): 157–64.

Strassmann, B. I. and R. I. Dunbar (1999). 'Human evolution and disease: putting the Stone Age in perspective.' *Evolution in Health and Disease*: 91–101.

Streeter, S. A. and D. H. McBurney (2003). 'Waist-hip ratio and attractiveness: new evidence and a critique of "a critical test".' *Evolution and Human Behavior* **24**(2): 88–99.

Stringer, C. B. and P. Andrews (1988). 'Genetic and fossil evidence for the origin of modern humans.' *Science* **239**: 1263–8.

Studd, M. V. and U. E. Gattiker (1991). 'The evolutionary psychology of sexual harassment in organizations.' *Ethology and Sociobiology* **12**(4): 249–90.

Stumpf, R. and C. Boesch (2005). 'Does promiscuous mating preclude female choice? Female sexual strategies in chimpanzees (*Pan troglodytes verus*) of the Taï National Park, Côte d'Ivoire.' *Behavioral Ecology and Sociobiology* **57**(5): 511–24.

Sugiyama, L. S. (2004). 'Is beauty in the context-sensitive adaptations of the beholder?: Shiwiar use of waist-to-hip ratio in assessments of female mate value.' *Evolution and Human Behavior* **25**(1): 51–62.

Sulloway, F. J. (1996). *Born to Rebel: Birth Order, Family Dynamics and Creative Lies*. New York, Pantheon.

Surbey, M. K. (1990). Family composition, stress, and the timing of human menarche. In T. E. Ziegler and F. B. Bercovitvch (eds.), *Socioendocrinology of Primate Reproduction*. New York, Wiley-Liss.

Swami, V. and A. Furnham (2007). *The Psychology of Physical Attraction*. London, Psychology Press.

Swami, V. and A. Furnham (2008). *The Psychology of Physical Attraction*. Routledge/Taylor & Francis Group.

Symons, D. (1979). *The Evolution of Human Sexuality*. Oxford, Oxford University Press.

Symons, D. (1992). On the use and misuse of Darwinism. In J. H. Barkow, L. Cosmides and J. Tooby (eds.), *The Adapted Mind*. Oxford, Oxford University Press.

Tang-Martinez, Z. (2010). 'Bateman's principles: original experiment and modern data for and against.' *Encyclopedia of Animal Behavior* **1**: 166–76.

Taylor, S. E. and J. D. Brown (1988). 'Illusion and well-being: a social psychological perspective on mental health.' *Psychological Bulletin* **103**(2): 193.

Taylor, L. H., S. M. Latham and E. J. Mark (2001). 'Risk factors for human disease emergence.' *Philosophical Transactions of the Royal Society of London B: Biological Sciences* **356**(1411): 983–9.

Temrin, H., J. Nordlund, M. Rying and B. S. Tullberg (2011). 'Is the higher rate of parental child homicide in stepfamilies an effect of non-genetic relatedness?' *Current Zoology* **57**(3).

ten Cate, C., M. N. Verzijden and E. Etman (2006). 'Sexual imprinting can induce sexual preferences for exaggerated parental traits.' *Current Biology* **16**(11): 1128–32.

Terrizzi, J. A., Shook, N. J., & McDaniel, M. A. (2013). The behavioral immune system and social conservatism: A meta-analysis. *Evolution and Human Behavior, 34*(2), 99–108.

Thorburn, A. W., J. Brand and A. Truswell (1987). 'Slowly digested and absorbed carbohydrate in traditional bushfoods: a protective factor against diabetes?' *The American Journal of Clinical Nutrition* **45**(1): 98–106.

Thornhill, R., J. F. Chapman and S. W. Gangestad (2013). 'Women's preferences for men's scents associated with testosterone and cortisol levels: patterns across the ovulatory cycle.' *Evolution and Human Behavior* **34**(3): 216–21.

Thornhill, R. and S. W. Gangestad (1993). 'Human facial beauty: averageness, symmetry and parasite resistance.' *Human Nature* **4**: 237–69.

Thornhill, R. and S. W. Gangestad (1994). 'Human fluctuating asymmetry and sexual behaviour.' *Psychological Science* **5**: 297–302.

Thornhill, R., S. W. Gangestad, R. Miller, G. Scheyd, J. McCullough and M. Franklin (2003). 'MHC, symmetry and body scent attractiveness in men and women.' *Behavioural Ecology* **14**: 668–78.

Thorpe, S. K. and R. H. Crompton (2006). 'Orangutan positional behavior and the nature of arboreal locomotion in Hominoidea.' *American Journal of Physical Anthropology* **131**(3): 384–401.

Thorpe, W. H. (1961). *Bird-Song: The Biology of Vocal Communication and Expression in Birds*. Cambridge University Press.

Tiddeman, B., M. Burt and D. Perrett (2001). 'Prototyping and transforming facial textures for perception research.' *Computer Graphics and Applications, IEEE* **21**(5): 42–50.

Tienari, P. (1991). Interaction between genetic vulnerability and family environment: the Finnish adoptive family study of schizophrenia.' *Acta Psychiatrica Scandinavia* **84**: 460–65.

Tiggemann, M. and S. Hodgson (2008). 'The hairlessness norm extended: reasons for and predictors of women's body hair removal at different body sites.' *Sex Roles* **59**(11–12): 889–97.

Tiggemann, M. and C. Lewis (2004). 'Attitudes toward women's body hair: relationship with disgust sensitivity.' *Psychology of Women Quarterly* **28**(4): 381–7.

Tinbergen, N. (1952). 'The Curious Behaviour of the Stickleback.' *Scientific American* **187**(Dec): 22–6.

Tinbergen, N. (1963). 'On the aims and methods of ethology.' *Zeitschrift fur Tierpsychologie* **20**: 410–33.

Todd, P. and G. Miller (1999). From pride and prejudice to persuasion. In G. Gigerenzer, P. Todd and T. A. R. Group (eds.), *Simple Heuristics that Make Us Smart*. Oxford, Oxford University Press.

Tomasello, M. and J. Call (1997). *Primate Cognition*. Oxford, Oxford University Press.

Tomasetti, C. and B. Vogelstein (2015). 'Variation in cancer risk among tissues can be explained by the number of stem cell divisions.' *Science* **347**: 78–81.

Tooby, J. and L. Cosmides (1990a). 'On the universality of human nature and the uniqueness of the individual: the role of genetics and adaptation.' *Journal of Personality* **58**(1): 17–67.

Tooby, J. and L. Cosmides (1990b). 'The past explains the present: adaptations and the structure of ancestral environments.' *Ethology and Sociobiology* **11**: 375–424.

Tooby, J. and L. Cosmides (1992). The psychological foundations of culture. In J. H. Barkow, L. Cosmides and J. Tooby (eds.), *The Adapted Mind*. Oxford, Oxford University Press.

Tooby, J. and L. Cosmides (1996). Friendship and the banker's paradox. In W. G. Runciman, J. Maynard Smith and R. I. M. Dunbar (eds.), *Evolution of Social Behaviour Patterns in Primates and Man*. Oxford, Oxford University Press, pp. 119–43.

Tooby, J. and L. Cosmides (2005). Conceptual foundations of evolutionary psychology. In D. Buss (ed.), *The Handbook of Evolutionary Psychology*. Hoboken, John Wiley and Sons.

Tooke, W. and L. Camire (1991). 'Patterns of deception in intersexual and intrasexual mating strategies.' *Ethology and Sociobiology* **12**(5): 345–64.

Tovee, M. J. and P. L. Cornelissen (1999). 'The mystery of human beauty.' *Nature* **339**: 215–6.

Tovee, M. J. and P. L. Cornelissen (2001). 'Female and male perceptions of female attractiveness in front-view and profile.' *British Journal of Psychology* **92**: 391–402.

Tovee, M. J., V. Swami, A. Furnham and R. Mangalparsad (2006). 'Changing perceptions of attractiveness as observers are exposed to a different culture.' *Evolution and Human Behavior* **27**(6): 443–57.

Tracy, J. L. and D. Matsumoto (2008). 'The spontaneous expression of pride and shame: evidence for biologically innate nonverbal displays.' *Proceedings of the National Academy of Sciences* **105**(33): 11655–60.

Trevathan, W., E. O. Smith and J. J. McKenna (eds.) (2008). *Evolutionary Medicine and Health: New Perspectives*. Oxford University Press, New York.

Trivers, R. (1985). *Social Evolution*. California, Benjamin-Cummings.

Trivers, R. L. (1971). 'The evolution of reciprocal altruism.' *Quarterly Review of Biology* **46**: 35–57.

Trivers, R. L. (1972). Parental investment and sexual selection. In B. Campbell (ed.), *Sexual Selection and the Descent of Man*. Chicago, Aldine.

Trivers, R. L. (1974). 'Parent–offspring conflict.' *American Zoologist* **14**: 249–64.

Trivers, R. L. and D. E. Willard (1973). 'Natural selection of parental ability to vary the sex ratio of offspring.' *Science* **179**(4068): 90–2.

Tullberg, B. S. and V. Lummaa (2001). 'Induced abortion ratio in Sweden falls with age, but rises again before menopause.' *Evolution and Human Behavior* **22**: 1–10.

Turke, P. W. (1989). 'Evolution and the demand for children.' *Popul. Dev. Rev.* **15**: 61–90.

Turkheimer, E., A. Haley, M. Waldron, B. D'Onofrio and I. I. Gottesman (2003). 'Socioeconomic status modifies heritability of IQ in young children.' *Psychological Science* **14**(6): 623–8.

Tutin, C. (1979). 'Responses of chimpanzees to copulation, with special reference to interference by immature individuals.' *Animal Behaviour* **27**: 845854.

Tversky, A. and D. Kahneman (1983). 'Extensional versus intuitive reasoning: the conjunction fallacy in probability judgement.' *Psychological Review* **90**: 293–315.

Uher, R. (2009). 'The role of genetic variation in the causation of mental illness: an evolution-informed framework.' *Molecular Psychiatry* **14**(12): 1072–82.

Underhill, P. A., G. Passarino, A. A. Lin, P. Shen, M. Mirazon Lahr, R. A. Foley, P. J. Oefner and L. L. Cavalli-Sforza (2001). 'The phylogeography of Y chromosome binary haplotypes and the origins of modern human populations.' *Annals of Human Genetics* **65**(1): 43–62.

US Department of Health and Human Services (2005). *Dietary Guidelines for Americans, 2005*.

Valles, S. A. (2010). 'The mystery of the mystery of common genetic diseases.' *Biology & Philosophy* **25**(2): 183–201.

Van Dongen, S. and S. W. Gangestad (2011). 'Human fluctuating asymmetry in relation to health and quality: a meta-analysis.' *Evolution and Human Behavior* **32**(6): 380–98.

Van Hooff, J. A. R. A. M. (1971). *Aspects of the Social Behavior and Communication in Human and Higher Non-human Primates*. Rotterdam, Bronderoffset.

Van Hooff, J. A. R. A. M. (1972). A comparative approach to the phylogeny of laughter and smiling. In R. A. Hinde (ed.), *Non-Verbal Communication*. Cambridge, Cambridge University Press.

Van Hooff, M. H., F. J. Voorhorst, M. B. Kaptein, R. A. Hirasing, C. Koppenaal and J. Schoemaker (2000). 'Insulin, androgen and gonadotrophin concentration, body mass index, and waist-to-hip ratio in the first years after menarche in girls with regular menstrual cycle, or oligomenorrhea.' *Journal of Clinical Endocrinology and Metabolism* **85**: 1394–1400.

van Leeuwen, F., J. H. Park, B. L. Koenig and J. Graham (2012). 'Regional variation in pathogen prevalence predicts endorsement of group-focused moral concerns.' *Evolution and Human Behavior* **33**(5): 429–37.

van Oers, K. and D. L. Sinn (2011). Toward a basis for the phenotypic gambit: advances in the evolutionary genetics of animal personality. In *From Genes to Animal Behavior*. Toyko, Springer, pp. 165–83.

Vasey, P. L., D. S. Pocock and D. P. VanderLaan (2007). 'Kin selection and male androphilia in Samoan fa'afafine.' *Evolution and Human Behavior* **28**(3): 159–67.

Verzijden, M. N. and C. ten Cate (2007). 'Early learning influences species assortative mating preferences in Lake Victoria cichlid fish.' *Biology Letters* **3**(2): 134–6.

Vick, S.-J., B. Waller, L. Parr, M. Smith-Pasqualini and K. Bard (2006). *ChimpFACS: The Chimpanzee Facial Action Coding System*. http://www.chimpfacs.com/video_clips/Manual2006.pdf, accessed February 2016.

Visscher, P. M., W. G. Hill and N. R. Wray (2008). 'Heritability in the genomics era – concepts and misconceptions.' *Nature Reviews Genetics* **9**(4): 255–66.

Voland, E. (1990). 'Differential reproductive success within the Krummhorn population.' *Behavioural Ecology and Sociobiology* **26**: 65–72.

Voland, E. and C. Engel (1989). Women's reproduction and longevity in a premodern population. In E. Rasa, C. Vogel and E. Voland (eds.), *The Sociobiology of Sexual and Reproductive Strategies*. London, Chapman and Hall.

Voland, E. and K. Grammer (2003). *Evolutionary Aesthetics*. Berlin, Springer.

von Cramon-Taubadel, N. (2011). 'Global human mandibular variation reflects differences in agricultural and hunter-gatherer subsistence strategies.' *Proceedings of the National Academy of Sciences* **108**(49): 19546–51.

Walker, C. L. and S. M. Ho (2012). 'Developmental reprogramming of cancer susceptibility.' *Nature Reviews Cancer* **12**(7): 479–86.

Walker, D., P. Harper, R. Newcombe and K. Davies (1983). 'Huntington's chorea in South Wales: mutation, fertility, and genetic fitness.' *Journal of Medical Genetics* **20**(1): 12–17.

Walker, R., O. Burger, J. Wagner, and C. R. Von Rueden (2006a). 'Evolution of brain size and juvenile periods in primates.' *Journal of Human Evolution* **51**: 480–9.

Walker, R., M. Gurven, K. Hill, A. Migliano, N. Chagnon, R. De Souza, G. Djurovic, R. Hames, A. M. Hurtado and H. Kaplan (2006b). 'Growth rates and life histories in twenty-two small-scale societies.' *American Journal of Human Biology* **18**(3): 295–311.

Wallace, A. R. (1905). *My Life: A Record of Events and Opinions*. London, Chapman & Hall.

Waller, B. and R. I. M. Dunbar (2005). 'Differential behavioural effects of silent bared teeth display and relaxed open mouth display in chimpanzees (Pan Troglodytes)." *Ethology* **111**: 129–42.

Walsh, A. and K. M. Beaver (2009). *Biosocial Criminology*. New York, Routledge.

Walter, A. (2006). 'The anti–naturalistic fallacy: evolutionary moral psychology and the insistence of brute facts.' *Evolutionary Psychology* **4**: 33–48.

Walter, A. and S. Buyske (2003). 'The Westermarck effect and early childhood co-socialization: sex differences in inbreeding-avoidance.' *British Journal of Developmental Psychology* **21**(3): 353–65.

Walther, B. A. and P. W. Ewald (2004). 'Pathogen survival in the external environment and the evolution of virulence.' *Biological Reviews* **79**(4): 849–69.

Wang, Y. C., K. McPherson, T. Marsh, S. L. Gortmaker and M. Brown (2011). 'Health and economic burden of the projected obesity trends in the USA and the UK.' *The Lancet* **378**(9793): 815–25.

Warner, H., D. E. Martin and M. E. Keeling (1974). 'Electroejaculation of the great apes.' *Annals of Biomedical Engineering* **2**: 419–32.

Warshawsky, B., I. Gutmanis, B. Henry, J. Dow, J. Reffle, G. Pollett, R. Ahmed, J. Aldom, D. Alves and A. Chagla (2002). 'Outbreak of Escherichia coli 0157: H7 related to animal contact at a petting zoo.' *The Canadian Journal of Infectious Diseases* **13**(3): 175.

Watson, J., R. Payne, A. Chamberlain, R. Jones and W. Sellers (2008). 'The energetic costs of load-carrying and the evolution of bipedalism.' *Journal of Human Evolution* **54**(5): 675–83.

Watson, J. B. (1930). *Behaviourism*. New York, Norton.

Watson, P. J. and P. W. Andrews (2002). 'Towards a revised evolutionary adaptationist analysis of depression: the social navigation hypothesis.' *Journal of Affective Disorders* **72**: 1–14.

Waynforth, D. (2011). Mate choice and sexual selection. In V. Swami (ed.), *Evolutionary Psychology. A Critical Introduction*. Chichester, Blackwell.

Waynforth, D. and R. I. Dunbar (1995). 'Conditional mate choice strategies in humans: evidence from "Lonely Hearts" advertisements.' *Behaviour* **132**(9): 755–79.

Webb, R., K. Abel, A. Pickles, L. Appleby, S. King-Hele and P. Mortensen (2006). 'Mortality risk among offspring of psychiatric inpatients: a population-based follow-up to early adulthood.' *American Journal of Psychiatry* **163**(12): 2170–7.

Webster, G. D. (2007). 'Evolutionary theory in cognitive neuroscience: a 20-year quantitative review of publication trends.' *Evolutionary Psychology* **5**(3): 520–30.

Wedekind, C. and S. Furi (1997). 'Body odour preferences in men and women: do they aim for specific MHC combinations or simply heterozygosity?' *Proceedings of the Royal Society of London Series B* **264**: 1471–9.

Wedekind, C., T. Seebeck, F. Bettens and A. J. Paepke (1995). 'MHC-dependent mate preferences in humans.' *Proceedings of the Royal Society of London B: Biological Sciences* **260**(1359): 245–9.

Weeden, J. and J. Sabini (2005). 'Physical attractiveness and health in Western societies: a review.' *Psychological Bulletin* **131**(5): 635.

Weigel, R. M. and M. Weigel (1989). 'Nausea and vomiting of early pregnancy and pregnancy outcome. A meta-analytical review.' *BJOG: An International Journal of Obstetrics & Gynaecology* **96**(11): 1312–8.

Weiss, R. (2009). 'Apes, lice and prehistory.' *Journal of Biology* **8**(2): 20.

Wells, J. C., J. M. DeSilva and J. T. Stock (2012). 'The obstetric dilemma: an ancient game of Russian roulette, or a variable dilemma sensitive to ecology?' *American Journal of Physical Anthropology* **149**(S55): 40–71.

Werbowetski-Ogilvie, T. E., L. C. Morrison, A. Fiebig-Comyn and M. Bhatia (2012). 'In vivo generation of neural tumors from neoplastic pluripotent stem cells models early human pediatric brain tumor formation.' *Stem Cells* **30**(3): 392–404.

West, S. A., C. El Mouden and A. Gardner (2011). 'Sixteen common misconceptions about the evolution of cooperation in humans.' *Evolution and Human Behavior* **32**(4): 231–62.

West–Eberhard, M. J. (1975). 'The evolution of social behaviour by kin selection.' *Quarterly Review of Biology* **50**: 1–33.

Westermarck, E. A. (1891). *The History of Human Marriage.* New York, Macmillan.

Wharton, C. H. (2001). *Metabolic Man: Ten Thousand Years from Eden.* Winmark Pub.

Wheeler, P. E. (1984). 'The evolution of bipedality and loss of functional body hair in hominids.' *Journal of Human Evolution* **13**(1): 91–8.

Wheeler, P. E. (1991). 'The thermoregulatory advantages of hominid bipedalism in open equatorial environments: the contribution of increased convective heat loss and cutaneous evaporative cooling.' *Journal of Human Evolution* **21**(2): 107–15.

Whitam, F. L., M. Diamond and J. Martin (1993). 'Homosexual orientation in twins: a report on 61 pairs and three triplet sets.' *Archives of Sexual Behavior* **22**(3): 187–206.

White, C. R. and R. Seymour (2003). 'Mammalian basal metabolic rate is proportional to body mass, 2/3.' *Proceedings of the National Academy of Sciences* **100**(7): 4046–9.

Whiten, A., R. A. Hinde, K. N. Laland and C. B. Stringer (2011). 'Culture evolves.' *Philosophical Transactions of the Royal Society of London B: Biological Sciences* **366**(1567): 938–48.

Wierson, M., P. J. Long and R. L. Forehand (1993). 'Toward a new understanding of early menarche: the role of environmental stress in pubertal timing.' *Adolescence* **23**: 913–24.

Wijendran, V. and K. Hayes (2004). 'Dietary n–6 and n–3 fatty acid balance and cardiovascular health.' *Annu. Rev. Nutr* **24**: 597–615.

Wilcox, A. J., D. D. Baird, D. B. Dunson, D. R. McConnaughey, J. S. Kesner and C. R. Weinberg (2004). 'On the frequency of intercourse around ovulation: evidence for biological influences.' *Human Reproduction* **19**(7): 1539–43.

Wiley, A. S. (2007). 'The globalization of cow's milk production and consumption: biocultural perspectives.' *Ecology of Food and Nutrition* **46**(3–4): 281–312.

Wiley, A. S. (2008). Cow's milk consumption and health: an evolutionary perspective. In W. Trevathan, E. O. Smith and J. J. McKenna (eds.), *Evolutionary Medicine and Health: New Perspectives.* Oxford, UK, Oxford University Press, pp. 116–33.

Wilkinson, G. (1984). 'Reciprocal food sharing in vampire bats.' *Nature* **308**: 181–4.

Wilkinson, G. S. (1990). 'Food sharing in vampire bats.' *Scientific American* **262**: 76–82.

Williams, G. (1957). 'Pleitropy, natural selection and the evolution of senescence.' *Evolution* **11**: 398–411.

Williams, G. C. (1966). *Adaptation and Natural Selection.* Princeton, Princeton University Press.

Williams, G. C. and R. M. Nesse (1991). 'The dawn of Darwinian medicine.' *Quarterly Review of Biology*: 1–22.

Williams, M. A., S. H. Ambrose, S. van der Kaars, C. Ruehlemann, U. Chattopadhyaya, J. Pal and P. R. Chauhan (2009). 'Environmental impact of the 73ka Toba super-eruption in South Asia.' *Palaeogeography, Palaeoclimatology, Palaeoecology* **284**(3): 295–314.

Williamson, S. H., M. J. Hubisz, A. G. Clark, B. A. Payseur, C. D. Bustamante and R. Nielsen (2007). 'Localizing recent adaptive evolution in the human genome.' *PLoS Genetics* **3**(6): e90.

Wilson, D. S. (2010). *Darwin's Cathedral: Evolution, Religion, and the Nature of Society.* Chicago, University of Chicago Press.

Wilson, E. O. (1975). *Sociobiology: The New Synthesis.* Cambridge, MA, Harvard University Press.

Wilson, E. O. (1998). *Consilience: The Unity of Knowledge.* London, Little, Brown and Co.

Wilson, E. O. (1984). *Biophilia.* Cambridge, MA, Harvard University Press.

Wilson, E. O. (2012). *The Social Conquest of Earth.* New York, WW Norton & Company.

Winchester, B., E. Young, S. Geddes, S. Genet, J. Hurst, H. Middelton-Price, N. Williams, M. Webb, A. Habel and S. Malcolm (1992). 'Female twin with hunter disease due to nonrandom inactivation of the X-chromosome: a consequence of twinning.' *American Journal of Medical Genetics* **44**(6): 834–8.

Wirtz, P. (1997). 'Sperm selection by females.' *Trends in Ecology and Evolution* **12**(5): 172–3.

Wolf, A. (2004). Introduction. In A. P. Wolf and W. H. Durham (eds.), *Inbreeding, Incest and the Incest Taboo*. Stanford, CA, Stanford University Press.

Wolf, A. P. (1970). 'Childhood association and sexual attraction: a further test of the Westermarck hypothesis.' *American Anthropologist* **72**(June): 503–15.

Wolf, A. P. (1993). 'Westermarck redivivus.' *Annual Review of Anthropology* **22**: 157–75.

Wolpert, L. (2009). 'The relationship between science and religion.' In C. W. Du Toit (ed.), *The Evolutionary Roots of Religion: Cultivate, Mutate or Eliminate?* South African Science and Religion Forum. Volume 13. Pretoria, Research Institute for Theology and Religion, Unisa, pp. 35–51.

Wolpoff, M. H., X. Wu and A. G. Thorne (1984). Modern Homo sapiens origins: a general theory of hominid evolution involving the fossil evidence from East Asia. In F. Smith and F. Spencer (eds.), *The Origins of Modern Humans: A World Survey of the Fossil Evidence*. New York, Alan R. Liss.

Wong, A. H., I. I. Gottesman and A. Petronis (2005). 'Phenotypic differences in genetically identical organisms: the epigenetic perspective.' *Human Molecular Genetics* **14**(suppl 1): R11–8.

Wood, J. W. (1989). 'Fecundity and natural fertility in humans.' *Oxford Reviews Of Reproductive Biology* **11**: 61–109.

Wood, J. W. (1990). 'Fertility in anthropological populations.' *Annual Review of Anthropology* **19**: 211–42.

Woodhouse, M. and R. Antia (2008). Emergence of new infectious diseases. In S. C. Stearns and J. Koella (eds.), *Evolution in Health and Disease*. Oxford, Oxford University Press.

Workman, L. and W. Reader (2004). *Evolutionary Psychology*. Cambridge, Cambridge University Press.

Wright, R. (1994). *The Moral Animal: Evolutionary Psychology and Everyday Life*. London, Little, Brown and Co.

Wynn, T. (1988). Tools and the evolution of human intelligence. In R.W. Byrne and A. Whiten (eds.), *Machiavellian Intelligence*. Oxford, Oxford University Press.

Wynne-Edwards, V. C. (1962). *Animal Dispersion in Relation to Social Behaviour*. Edinburgh, Oliver and Boyd.

Xiang, H., J. Gao, B. Yu, H. Zhou, D. Cai, Y. Zhang, X. Chen, X. Wang, M. Hofreiter and X. Zhao (2014). 'Early Holocene chicken domestication in northern China.' *Proceedings of the National Academy of Sciences* **111**(49): 17564–9.

Yamazaki, K., G. K. Beauchamp, D. Kupniewski, J. Bard, L. Thomas and E. A. Boyse (1988). 'Familial imprinting determines H-2 selective mating preferences.' *Science* **240**: 1331–2.

Yehuda, R., N. P. Daskalakis, L. M. Bierer, H. N. Bader, T. Klengel, F. Holsboer and E. B. Binder (2015). 'Holocaust exposure induced intergenerational effects on FKBP5 methylation.' *Biological Psychiatry*. doi:10.1016/j.biopsych.2015.08.005.

Young, R. W. (2003). 'Evolution of the human hand: the role of throwing and clubbing.' *Journal of Anatomy* **202**(1): 165–74.

Zaadstra, B. M., J. C. Seidell, P. A. van HNoord, E. R. te Velde, J. D. F. Habbema, B. Vrieswijk and J. Karbaat (1993). 'Fat and female fecundity: prospective study of effect of body fat distribution on conception rates.' *British Medical Journal* **306**: 484–7.

Zahavi, A. (1975). 'Mate selection – a selection for handicap.' *Journal of Theoretical Biology* **53**: 205–14.

Zhao, Q., C. L. Tan and W. Pan (2008). 'Weaning age, infant care, and behavioral development in Trachypithecus leucocephalus.' *International Journal of Primatology* **29**(3): 583–91.

Zietsch, B. P., K. I. Morley, S. N. Shekar, K. J. Verweij, M. C. Keller, S. Macgregor, M. J. Wright, J. M. Bailey and N. G. Martin (2008). 'Genetic factors predisposing to homosexuality may increase mating success in heterosexuals.' *Evolution and Human Behavior* **29**(6): 424–33.

Zihlman, A. (1989). Woman the gatherer: the role of women in early hominid evolution. In S. Morgen (ed.), *Gender and Anthropology: Critical Reviews for Research and Teaching*. Washington, DC, APA, pp. 21–40.

Zihlman, A. L. (1982). *The Human Evolution Colouring Book*. Harper Resource.

Zivkovic, A. M., J. B. German, C. B. Lebrilla and D. A. Mills (2011). 'Human milk glycobiome and its impact on the infant gastrointestinal microbiota.' *Proceedings of the National Academy of Sciences* **108**(Supplement 1): 4653–8.

Index